Management of Information Technology

Second Edition

Management of Information Technology

Second Edition

Carroll W. Frenzel

A DIVISION OF COURSE TECHNOLOGY
ONE MAIN STREET, CAMBRIDGE, MA 02142

an International Thomson Publishing company I(T)P

Albany • Bonn • Boston • Cincinnati • London • Madrid • Melbourne • Mexico City
New York • Paris • San Francisco • Singapore • Tokyo • Toronto • Washington

This book is dedicated to John and Lisa, with love.

Managing Editor	DeVilla Williams
Product Manager	Lisa Strite
Production Editor	Barbara Worth
Production Services	Books By Design, Inc.
Interior Design and Composition	Gex, Inc.
Cover Design	Diana Coe, Ko Design

For more information contact:

Course Technology
One Main Street
Cambridge, MA 02142

International Thomson Publishing Europe
Berkshire House 168-173
High Holborn
London WCIV 7AA
England

Thomas Nelson Australia
102 Dodds Street
South Melbourne, 3205
Victoria, Australia

Nelson Canada
1120 Birchmount Road
Scarborough, Ontario
Canada M1K 5G4

International Thomson Editores
Campos Eliseos 385, Piso 7
Col. Polanco
11560 Mexico D.F. Mexico

International Thomson Publishing GmbH
Kônigswinterer Strasse 418
53227 Bonn
Germany

International Thomson Publishing Asia
211 Henderson Road
#05-10 Henderson Building
Singapore 0315

International Thomson Publishing Japan
Hirakawacho Kyowa Building, 3F
2-2-1 Hirakawacho
Chiyoda-ku, Tokyo 102
Japan

ISBN 0-7895-0413-8

Printed in the United States of America

10 9 8 7 6 5

Contents

PART SIX *PREPARING FOR IT ADVANCES*

Preface

Management of Information Technology, Second Edition, is written for students and practitioners who desire to understand the framework of sound management principles and its application to the rapidly changing arena of information technology (IT)—computer and telecommunication systems. Since the first edition appeared in late 1991, IT has developed and expanded exponentially. Accompanying this expansion are proportionately growing management challenges that cause some to despair of obtaining reasonable solutions or finding people willing and qualified to address them. This book's purpose is to teach principles and applications with which students and managers can grow and prosper in this turbulent environment.

Since 1991, more than 100 million personal computers have been sold; client-server architecture has obsoleted many centralized mainframes; and organizations have downsized, rightsized, outsourced, and reengineered, constantly struggling to maintain or gain advantage from advancing technology. Capitalizing on digital signal processing technology, telecommunications firms are presenting an onslaught of new capabilities as they develop products and services capable of maintaining everyone on-line all the time regardless of geographic location. Firms are frantically struggling to position themselves for the future via alliances, partnerships, joint ventures, and mergers. The importance of the information technology revolution seems impossible to overestimate. To informed and thoughtful observers, it is enormously profound.

In the United States, the penetration of personal computers or workstations in businesses is nearly complete and is gaining momentum in households. Consumers purchased $8 billion worth of PCs in 1994, just slightly less than they spent on TVs. The number of households with computers increased 50 percent from 1990 to 1994 and is expected to increase another one-third to 48 million by 1998. By 1998, nearly 50 percent of U.S. households will own PCs; more than half will be networked. About one million Americans each month connect to networks, principally the Internet. The age of the wired society is rapidly approaching.

In the first edition of *Management of Information Technology*, I asserted that students and prospective managers must regard information technology as the integration of computer and telecommunications systems. Then, the coalescence of these technologies was a strong trend; today it is a reality. The distinction between computing and communication systems has disappeared for information system users and is becoming increasingly blurred for technical professionals. For managers, concepts such as strategizing, planning, and controlling information systems apply without distinction.

Management of Information Technology, Second Edition, focuses on the management issues surrounding information technology and lays the foundations of management knowledge required by successful managers. The text teaches sound, proven IT management basics, and offers processes and procedures for applying management knowledge. It describes what managers must do and how they can best do it. Successful managers must know management principles: they must know when and how to apply them. This text teaches both.

Because technology's rapid pace has had a confounding influence on management, this edition emphasizes management fundamentals, addresses environmental change, and expands on telecommunications basics. Important new topics such as CD-ROM, RAID storage, repeaters, bridges, routers, downsizing, outsourcing, reengineering, and others are discussed. The text focuses on important management concepts applied to new architectures such as client/server implementation. As the information industry develops and introduces new technology, it is undergoing rapid change and becoming more tightly linked to business generally. The text highlights the importance of these developments for managers.

The text focuses on these complex topics for advanced undergraduates or graduate students and current practitioners. To take full advantage of this material, students are expected to have a background in information processing systems through course work in computer science, telecommunications, management information systems, or engineering. However, this subject is important to students from fields as diverse as economics, law, and military science.

SCOPE OF THIS TEXT

This text considers information technology from the perspective of several management levels—from first-line managers to chief executive officers. Managers at several levels, or aspiring managers, can use the text's frameworks and principles to attain benefits from the rapidly advancing technology.

Information technology includes the mainframes owned and operated by the central information systems organization, the information processing capabilities dispersed throughout the firm, and the voice, video, and data transmission networks linking these facilities internally and externally. Because of technology dispersion, the firm's IT managers are acquiring considerable staff responsibility. And, as IT line and staff responsibilities grow at many firms, the senior IT manager's job is expanding in scope and responsibility. These trends and others demand that IT managers' skills closely resemble those of general managers. General management skills are important, but a keen awareness and knowledge of technical issues and technology trends is vital. IT executives have many weighty responsibilities and need a broad array of finely tuned management skills to be successful. This text prepares individuals for these responsibilities.

CONTENT OVERVIEW

Part One: Foundations of IT Management

Part One lays the foundations of information technology in organizations and discusses its management issues. Readers gain perspectives on IT management through examples and management frameworks. Part One also explains information technology's strategic importance and teaches the essential ingredients for developing strategies and translating them into strategic plans. Strategic planning processes are logically extended to tactical and operational planning models. This part models the planning processes relating tools, information, and organizations interacting over time.

Part Two: Technology and Industry Trends

Part Two explores information delivery systems, technology trends, and industry dynamics. Advances in microcomputers, mainframes, and supercomputers are correlated with trends in semiconductor logic and memory. This part discusses the importance of advanced programming, telecommunications, and workstation systems. Students gain perspectives on technology's pace and direction that translate into management opportunities. This part focuses extensively on the value of telecommunications for modern businesses and reveals the importance of information industry dynamics to business managers.

Part Three: Managing Application Portfolio Resources

The firm's application portfolio is a major asset and a profound source of difficulty. Large development backlogs, extended completion dates, and increasing expenditures plague many managers. Large, growing, and complex databases that support the firm's application programs further complicate the manager's job. This part develops procedures and techniques that help managers handle this complex and exasperating problem. It also focuses on application development and acquisition in a client/server, distributed computing environment. Finally, it teaches managers how to achieve success in this challenging, critical business activity.

Part Four: Tactical and Operational Considerations

The fourth part develops disciplined processes for managing tactical and operational IT activities in centralized and distributed environments. Part Four develops the basis for managing customer expectations and IT's response to user requirements. Problem avoidance, problem management, and change management systems are developed in detail followed by contingency management and disaster recovery planning. This part teaches computer operations management and the disciplines for managing networks. Management reporting systems unite these operational disciplines by responding to established customer expectations. Students will also learn successful tactical and operational information and telecommunications systems management.

Part Five: Controlling Information Resources

Rapid infusion of information technology into organizations demands careful controls for success. Part Five relates IT investment and return measurements to financial control, customer relations, and client expectations. Students learn the need for internal controls and audits in application systems and acquire methods to satisfy these needs. IT managers must secure information assets, develop reasonable business controls, and assure senior executives that security and control procedures are effective. This part teaches IT managers how to protect and control critical firm assets successfully.

Part Six: Preparing for IT Advances

Part Six explains how high-performance organizations restructure business processes and introduce new technology while managing talented people with sensitivity and skill. As technology offers attractive new options and organizations restructure to capture its advantages, first-line and senior managers must understand and manage individual transitions skillfully. This part teaches the elements of effective people management, how effective managers achieve high morale, and examines ethical and legal considerations in managing employees. Part Six summarizes the management principles and practices developed throughout the text into a cohesive management system to serve the firm's managers. The final chapter unites the text's themes by teaching how successful CIOs develop policy, introduce technology, facilitate change, and use management systems effectively.

A NOTE TO THE READER

This text's subject matter is broad and comprehensive. Students and aspiring managers who devote sufficient time to the vignettes, text material, questions, assignments, and references will be rewarded with significant, valuable knowledge. However, this book does not pretend to answer all the questions about this complex subject and is not intended to portray managers as responding to stimuli with conditioned reflexes. People and technology are incredibly complex. Even after extensive study and prolonged experience, a wise manager maintains a high level of humility and flexibility. Management can be a highly satisfying profession; I hope this text will bring success and increased satisfaction to students and managers.

ACKNOWLEDGMENTS

I derived much of this material from my experiences as an IBM IT manager and from many enriching interchanges with faculty and students at the University of Colorado. IBM gave me a great deal of formal training during my tenure, and I am grateful for the educational opportunities. My management knowledge and experience was enriched by associations with many wonderful people at IBM, the University of Colorado, and other organizations. I was taught by superiors, colleagues, students, clients, and subordinates alike. With this text, I reflect their input and acknowledge their contributions to my education.

Management of Information Technology

Second Edition

Part
One

Foundations of IT Management

Rapid information and telecommunications systems advances have created enormous opportunities for skillful managers. The management of information technology (IT) demands general management expertise focusing on strategy development, long-range planning, and business controls. Successful information technology managers exhibit outstanding resource and people management skills. This part includes chapters on Management in the Information Age, Information Technology's Strategic Importance, Developing the Firm's IT Strategy, and Information Technology Planning.

Managing information technology is an exceptionally difficult task. The ability to develop IT strategic directions and plan effective implementations is a vital first step toward successful IT management.

1 Management in the Information Age

The typical large business 20 years hence will have fewer than half the levels of management of its counterpart today, and no more than a third of the managers. In its structure, and in its management problems and concerns, it will bear little resemblance to the typical manufacturing company, circa 1950, which our textbooks still consider the norm. Instead it is far more likely to resemble organizations that neither the practicing manager nor the management scholar pays much attention to today: the hospital, the university, the symphony orchestra. For like them, the typical business will be knowledge-based, an organization composed largely of specialists who direct or discipline their own performance through organized feedback from colleagues, customers, and headquarters. For this reason, it will be what I call an information-based organization.[1]

INTRODUCTION

New technology designed to process and transport data and information has been developing at an exceptional rate for more than four decades. This "information revolution" has significantly affected employees, managers, and their organizations. New technology has created countless opportunities and challenges for millions of individuals and organizations. In particular, the challenges facing managers responsible for introducing and using this technology have been especially high. In our information-based society, management must learn to maximize the advantages offered by information technology, while avoiding the many pitfalls associated with rapid technological change.

Information technology (IT) is radically altering the balance of power between institutions, governments, and people by broadly disseminating important information. Power bases dependent on information are being built, transformed, and destroyed as critical information virtually flows around the globe without restriction. Information technology has altered the way many people do their jobs, and has changed the nature of work in industrialized nations: The practice of management has been greatly affected, and aspiring managers must be fluent in new management trends and techniques in order to succeed.

The technological revolution that we are experiencing has emerged over the past 40 years. Many individuals entering the workplace today have directly experienced this technological phenomenon, and perhaps they take it for granted. Employees completing their careers have seen the whole spectrum of events unfold during their lifetimes. For many people, the growth of information technology has been an unmitigated blessing. "New" technology was a major part of their formal education. It forms the basis of their employment, and serves as a platform on which their future depends. For others, information technology has been a complicating factor, creating apprehension or outright fear. To a greater or lesser degree, information technology has brought change to nearly everyone immersed in today's marketplace.

Information technology (IT) has become increasingly vital for creating and delivering the products and services in industrialized nations. For the individuals managing information technology within a larger enterprise, the rapid pace of innovation has meant an unprecedented growth of job opportunities fueled by an ever-increasing need for skilled managers. Senior executives who can manage these complex tasks remain in great demand. This demand has been driven by the growth of computer applications such as decision support systems and expert systems, and by computer-aided design, computer-aided manufacturing, electronic imaging, and multimedia applications. Demand for skilled executives has also been created by the rapid advances in telecommunications which have spawned electronic data interchange (EDI), local and wide-area networking, and interorganizational systems. The outlook for skilled IT managers remains challenging and bright.

James Sutter, eleven-year veteran vice president and general manager of information systems (IS) at Rockwell International Corporation (defense and space), is an example of a highly successful IT manager. He manages an IS budget of about $400 million in a company that focuses on productivity improvements. He consolidated IT outposts to a single data center reducing computer operations people by 50 percent but still spends 15 percent of software development resources on client/server applications. His goal is to fund new projects out of productivity improvements thus coping with difficult times in the defense business. Rockwell is one of the Premier 100 companies as measured by *Computerworld's* productivity index.[2]

This introductory chapter sets the scene for the management task in the field of information technology. It analyzes the foundations of information technology and relates them to the challenges of management. Individuals within modern organizations have visions or expectations of what the technology can do for them, but they may have less appreciation for the impact the technology will have on them. The significance of advanced technology must be acknowledged and understood, however, and must be well managed for the firm to use it to its full advantage. Issues facing senior IT executives are presented in this chapter, and frameworks for dealing effectively with them are found throughout the text. Top IT managers need to have strong foundations in general management principles. They must understand how to apply these principles to the complex tasks of introducing, implementing, and managing technological change.

Although the opportunities available to highly skilled managers are bright indeed, the challenges of the profession remain equally high. IT trends suggest that these challenges will increase dramatically in the coming years, and that the successful IT manager will learn to keep pace starting now.

Many challenges arise from the volatility of the technology and from the relationship of the evolving technology to the structure of the firm. Other sources of challenge are the expectations held by members of the firm at many levels. IT managers must anticipate these challenges, and create thoughtful plans to manage related issues to the firm's advantage. Well-prepared managers will seek to understand the issues, and will prepare to control their consequences.

For IT managers to succeed within their organizations, they must do more than cope with issues as they arise; they must take a strong leadership position in formulating and shaping them. Their insight and vision of future technology is a valuable resource that executives use to develop strategies that permit the firm to gain competitive advantage. IT managers need to supply executives in the firm with their technological *and* their business vision so that the firm can anticipate future structural changes and prepare for them well in advance. With the CEO's financial and strategic company goals in mind, IT managers must champion their technological vision from the general manager's perspective while inspiring the executive team with their realistic, practical, and innovative view of the future.

IT management must be at its best when dealing with expectations. Senior executives have expectations that include using information technology for competitive advantage and attaining bottom-line results for the firm. Given that corporations spend anywhere from one to five percent of revenue on information technology, these expectations are entirely reasonable.[3] CEOs have every reason to require the IT branch of the organization to conduct its affairs in a businesslike manner and to conform to business practices common to their corporations. CEOs reasonably require their senior IT managers to perform their jobs with executive level skills.

CEOs are influenced by many sources of information originating both inside and outside the firm. These information sources include trade associations, government agencies, and informal communication with their peers, in addition to routine communication with the firm's officers and directors. Information is a basis for expectations, and senior IT managers must respond to these expectations in a disciplined and professional manner.

IT managers must have a good understanding of the corporate culture. Corporate culture or corporate philosophy consists of the basic beliefs or ideas that guide members of the organization in their behavior within the organization. As Marvin Bower puts it, these behavior patterns describe "how we do things around here."[4] IT managers must also have a clear and realistic view of technology trends and an appreciation for the degree of technological maturity of the firm. This knowledge is essential for providing the CEO and the executive staff with information upon which reasonable expectations can be built.

Expectations held by the firm's senior executives constitute a yardstick against which IT managers will ultimately be measured. It doesn't matter where the expectations originated or whether or not they were realistic. But in most organizations, completely fulfilled expectations lead to a satisfactory performance appraisal. Skillful IT managers, those with a general management view of the business in which they are operating, are likely to position themselves and their organizations to maximize this eventuality. Superior IT managers understand the importance of expectations, and are proactive in their efforts to manage them effectively.

Less-skilled managers set and cope with expectations less effectively. They frequently find themselves and their organizations overcommitted, or operating reactively. Unskilled managers and the organizations they lead create expectations they are unable to fulfill. Through lack of discipline or excess enthusiasm, such managers sow the seeds of their own demise. Executives rely upon the IT manager to set and manage expectations skillfully. They understand that the success of the IT team is integral to their personal success, and to the overall success of their organizations. In today's climate of rapid technological change, substandard IT management is rarely tolerated, and savvy IT managers will strive to position themselves as an indispensable executive resource for achieving expectations.

Skills alone, however, are not enough. Tools and processes, along with a management system in which these processes can operate effectively within the corporate culture, are necessary for success. These processes engage various members of the firm in activities ranging from long-term strategic considerations at one extreme, to very short-term considerations at the other. In short, successful IT management can flourish only in a supportive environment. All players engaged in this activity bear some responsibility for the success of the processes.

ORGANIZATIONS AND INFORMATION

The IT Organization Within the Firm

The IT organization performs a vital function for the firm and usually enjoys commensurate status. IT holds a position similar to that of marketing, accounting, engineering, or manufacturing. The IT organization is structured with departments that have specific responsibilities reporting through department managers to the senior IT manager. In some firms, the senior manager reports within the finance organization, but there are various alternate reporting relationships resulting from the evolution of the Chief Information Officer, or CIO. IT organizational alignment will be discussed later in this text.[5]

The IT organization operates as a business within a business, supporting many other functional units in a variety of ways. Table 1.1 illustrates some of these important functional units, and the IT applications that typically support them.

TABLE 1.1 Functions and Applications Supporting Them

Functions	Applications Supported
Product development	Design automation, parts catalog
Manufacturing	Materials logistics, factory automation
Distribution	Warehouse automation, shipping and receiving
Sales	Order entry, sales analysis, commission accounting
Servic	Call dispatching, parts logistics, failure analysis
Finance and accounting	Ledger, planning, accounts payable
Administration	Office systems, personnel records

The typical large firm uses information technology extensively to accomplish its business functions. The examples in Table 1.1 are rather common and easily recognized, but represent only a fraction of the total portfolio of computerized applications. In many firms, applications are developed within functional areas and become part of the application portfolio. A medium-size corporation may hold several thousand computer programs in its portfolio.[6]

Most IT functions are both line and staff, and most senior IT managers have line and staff responsibilities. Line functions are those that have direct responsibility for accomplishing the objectives of the enterprise. Staff functions are elements of the organization that help the line effectively accomplish the primary objectives of the enterprise.[7]

Usually the IT function contains some core activities representing its line responsibilities and some activities to discharge various staff responsibilities. For example, IT functions usually have departments to perform applications development, operate centralized computer facilities, develop and maintain IT plans and strategies, and offer technical support. Some IT functions maintain departments to handle staff responsibilities. For example, there may be a group to support client computer operations or applications development, or a department to support telecommunications in client functions, or persons who develop internal standards for client functions.[8] The corporate marketplace and the firm's culture are key factors that govern the scope and construct of the IT functions from among these variations. For example, in high-tech firms, the technical support role may be expanded, while in others this activity may be delegated to the development group or to computer operations. Structures vary with corporate and client needs, and they evolve as the firm reengineers to improve operations. Subsequent portions of this text will develop these concepts further.

Structures and Information

Organizational structure has been studied actively for many years, and theorists agree that some sort of structure is necessary for the effective operation of human enterprise. The pyramid structure, with the nonmanagerial employees forming the base and the chief executive at the top, is one familiar structure, as shown in Figure 1.1. Among many other topics, discussions of organizational structure include the concepts of levels of management and span of control.

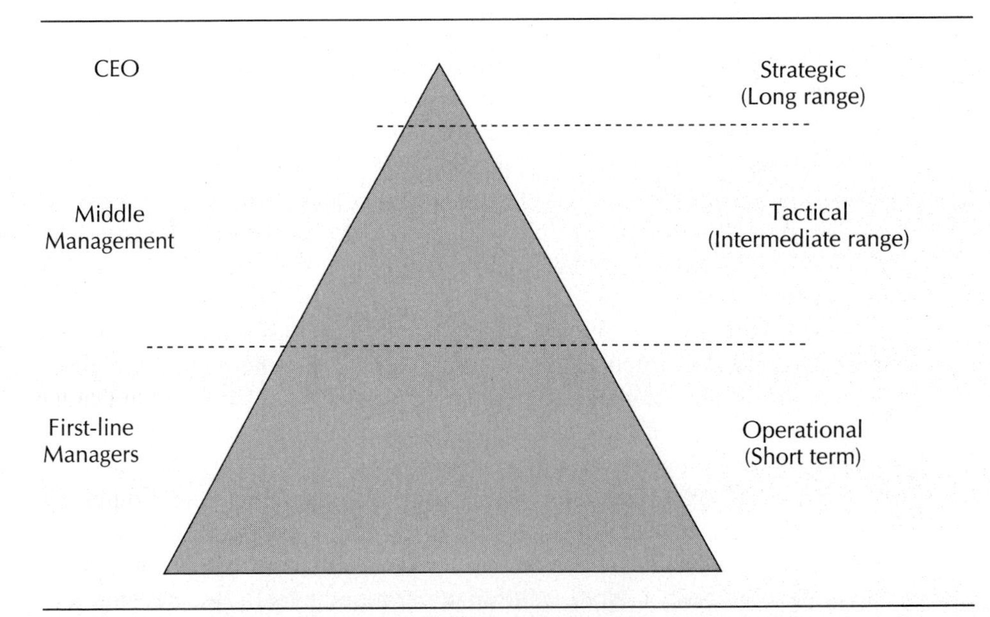

FIGURE 1.1 The Traditional Managerial Hierarchy

The concept of levels of management implies differences in responsibility and degrees of authority: Workers in positions at the base of the pyramid have the least responsibility and authority, while senior executives at the top enjoy the most. The responsibility differences can be distinguished in part by the characteristics of the data and information utilized by the encumbents and the manner in which they are used. In particular, three facets help differentiate the levels of management and their relation to and use of information:

1. The sources of information applicable to the position
2. The degree of judgment involved in the application of the information
3. The time span in which the information is considered pertinent for making decisions.

There are other characteristics, as well, such as level of aggregation, currency, and frequency of use.[9]

These differentiating characteristics are useful for understanding the manner and degree that information technology is applied within the firm's structure. Keep in mind, however, that the introduction of information technology within a structure changes that structure and, as Drucker points out in the opening passage, these alterations may be large and very important, indeed.

Information employed by individuals at the base of the pyramid is most likely to originate within the firm. The information will have value over a relatively short time measured in hours or days, and will require relatively little judgment in its operational application. Systems to do this type of work are generally called transaction processing systems or structured decision support systems.

In contrast, information of value to the CEO is more likely to originate outside the firm. This information will have value or meaning for a much longer period—one to five years or more—and will require a high degree of experience and judgment in its application. Executive activity involves information and knowledge. To the limited extent they exist, the systems that assist in this work are called executive information systems.

At lower levels in the firm, data and information with short-term value is applied operationally. But senior managers use data and information that has long-term value tactically or strategically. To utilize this commodity fully requires increasing amounts of judgment as one's responsibilities broaden. Middle managers utilize information originating inside and outside the firm which requires modest amounts of judgment for application. The period applicable to this information is typically one to two years.

We will return later to further explore organizational structure, but for now keep in mind the transformations that are taking place. For a number of reasons, including the use of information technology, organizational structures are changing. Peter Drucker, Tom Peters, and others have pointed out the tendency for levels of management to decrease and for the organization to become "flatter." For example, Nucor, the billion-dollar sales steel company, has only three layers between the millworker and the chairman of the board. Wal-Mart, with sales of $25 billion, also has three layers in its hierarchy. It operates a satellite network with more than 1100 terminals and more than 1800 of its 5000 suppliers are connected to Wal-Mart through electronic data interchange (EDI). Wal-Mart is investing in IT at the rate of $100 million per year.[10]

We know from the flatter structure that these very successful companies operate with a style of management entirely different from conventionally organized companies. As the structure flattens, managers do more professional work and have fewer supervisory responsibilities but employees have more opportunity for leadership. Common goals are more important in flatter organizational structures. In all organizations, however, the executive functions differ markedly from those performed at other levels. The next few paragraphs discuss some of the characteristics of these positions and describe how they vary within the organization.

This text adopts Drucker's definition of information: "Information is data endowed with relevance and purpose. Converting data into information thus requires knowledge. Knowledge, by definition, is specialized."[11] Knowledge is information that has been distilled via study or research and augmented by judgment and experience. Actually, these distinctions become blurred because one person's information may become someone else's data, as communication occurs within the firm.

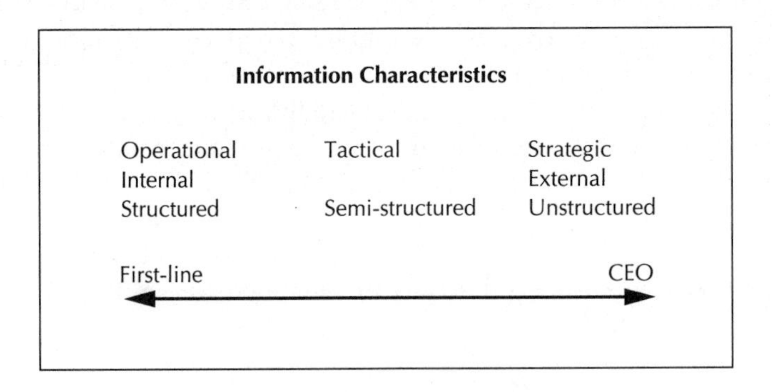

FIGURE 1.2 Information and Levels of Management

Figure 1.2 illustrates the relationships among the management levels in an organization and the timeframes within which managers or executives normally operate. In this traditional view of organizational structure, the first-line managers are responsible for the firm's activities on a day-to-day basis, while higher levels of management maintain a longer term perspective. First-line managers plan tomorrow's activities and the senior executives plan for the next several years or the next decade. These perspectives require vastly different types of information. The required information delivery system also varies considerably among the managerial levels.

The terms strategic and tactical are used in this text to describe the kind of activity and the time periods involved at the CEO level and the middle-management level, respectively. There is also a relationship among these terms, the organizational levels, and the quantity of judgment and experience that may be required. For example, the application of data to the operational activities of loading a supertanker may require little or no judgment. But the tactical use of data or information by middle managers to schedule the tanker fleet requires more sophistication and experience. The CEO, who must decide whether to expand or contract the tanker fleet, requires an exquisite, strategic blend of knowledge, experience, intuition, and judgment. Management control or middle-management activities combine some attributes from each end of the management spectrum. This structure or taxonomy was first described by Anthony.[12]

Another useful framework for understanding these concepts was developed by Gorry and Scott-Morton.[13] Their model relates levels of management to structured or unstructured decision making. Structured decisions are those based on a relatively firm understanding of the underlying parameters. Determining how much inventory is on hand is structured because all the factors in the determination can be quantified. Deciding how many products to schedule in the production line tomorrow is an example of a semi-structured decision because customer demand, work in process, and other necessary factors are predictable but not usually completely known. Deciding how much money to spend on research is unstructured because the factors on which to base this decision are not crisply known and are difficult to predict. The results of the research activity are an uncertain, but important, factor. Table 1.2 illustrates the Gorry and Scott-Morton model.

TABLE 1.2 The Gorry and Scott-Morton Model

	Operational Control	Tactical Control	Strategic Planning
Structured	Accounts receivable	Budget analysis	Tanker fleet mix
	Order entry Inventory control	Short-term forecasting	Factory location
Semi-structured	Production scheduling	Variance analysis	Mergers and acquisitions
	Cash management	Budget preparation	New product planning
Unstructured	PERT/COST systems	Sales and production	R & D planning

As one changes focus from the first-line management position toward the CEO position, the sources of information change from mostly internal to mostly external. First-line managers have little need to seek information from external sources in most cases. Their major sources of information are the databases supporting the transaction processing systems at their command. The CEO, on the other hand, utilizes and relies on external sources of information to a far greater extent. For the most part, this external information cannot be found in the databases internal to the firm.

First-line managers and nonmanagerial employees are concerned mostly with operational considerations and data. Their view of the firm is operational and near-term. CEOs must take the long-term or strategic view. They are not dealing with operational issues in most of their activities. This description does not imply a smooth continuum in the variables as one moves from the first-line position to the CEO post. Additionally, the situation is not the same for all firms and may differ for various functions within the same firm.

For example, in some firms the CEO has developed very capable subordinates who assume nearly complete responsibility for the operational activities of the firm. In these firms, the CEO develops and manages external, long-range relationships with customers, suppliers, governments, and others. In other firms, particularly those in distress, the CEO and other senior executives concentrate their energy and talents on the immediate, mostly internal actions that must be accomplished successfully for survival. Likewise, a smoothly operating manufacturing function will receive little executive attention, but a marketing organization, facing stiff competition, will undoubtedly attract an abundance of executive attention.

As Drucker noted in the opening passage, future organizations will experience major structural changes as they transform themselves into knowledge-based organizations through the pervasive employment of information technology. Technology will permit CEOs to interact directly and efficiently with more members of the management team. Their effective span of control will increase. The span of control of the other managers in the firm will also increase. And, in networked organizations, communication among employees in all positions will increase substantially. Consequently, the firm will require fewer levels of management and fewer managers.

Information technology is having a profound impact on business enterprises and their internal structures and operations. Peters describes IT's stunning organizational implications: the "informating" of employees (providing increased information and broadened responsibilities), the creation of electronic highway systems, "extended family" project management, smaller is better (better communication makes smaller units plausible), "network size" (uses temporary collection of skilled people from around the globe to accomplish specified tasks), and knowledge-based organizations linked electronically to each other and to learning centers.[14] Noel Tichy, Professor, School of Business Administration, University of Michigan, uses the words "corporate transformation" to describe the organizational restructuring taking place. This restructuring "is like trying to change the fan belt while the motor is running," he claims.[15] The challenges and the opportunities for managers at all levels and in all functions will be enormous. These important considerations will be explored at appropriate points throughout this text.

INFORMATION TECHNOLOGY MANAGEMENT

Information technology is the term that describes the disciplines encompassing computer systems, telecommunication networks, and multimedia (combined audio, text, and video information) applications. It came into common use in the late 1980s supplanting earlier terms such as electronic data processing (EDP), management information systems (MIS), and information systems (IS), although the latter term is still in use. Other terms such as information management (IM) and information resource management (IRM) have also been used to describe computer information systems organizations, particularly in government agencies.

The Evolution of Information Technology

Electronic information processing has been used in business processes for four decades and has evolved through several identifiable phases; its evolution continues.[16] In the late 1950s and throughout the 1960s, routine business data handling was being automated through punched card equipment, electronic accounting machines (EAM), and by physically large but relatively low-power electronic computers. Because these early capabilities lent themselves to the automation of accounting and financial activities, the early DP or EDP (electronic data processing) departments were usually placed under the accounting or finance functions. Typically, early EDP departments were isolated from the remainder of the organization and appeared to be more interested in using technology for its own sake than in solving business problems with it. Some viewed EDP apprehensively because they feared job loss through automation.

During this period, investments in DP were based on traditional budgeting processes and, in many cases, were handled as overhead. When attempts were made to recover costs through charge-outs, the frustrations of those served by DP were heightened as they found themselves paying for an organization only marginally serving their needs. At the same time, firms found EDP to be essential to their operations. Typically, EDP managers slighted the managerial dimensions of their jobs and failed to collaborate with other functions as EDP brought change to the organization.

A decade later, terminals were being connected to mainframes and database management systems were introduced as large business data files were accumulated. The information systems (IS) function became a centralized entity and information infrastructures began to emerge during the 1970s. Massive management reports of financial and production data proliferated as IS struggled to assist managers. This was the era of management information systems (MIS) although there remained much confusion over what management information really was. Decision support systems were fashioned to operate on the massive data stores and end-user computing developed through terminals and later, via personal computers. In many organizations, MIS managers still slighted their managerial role and failed to develop partnerships with their clients. Determining MIS's effectiveness became a concern for most organizations.

During the 1980s, telecommunications and networking flourished as distributed data processing, office systems, and personal computers were introduced. Firms discovered infrastructure fragmentation with the growth of incompatible systems and the accompanying loss of discipline as operations became disbursed. The potential for enormous gain from information systems was recognized and firms searched for competitive advantage through information systems development and change. During this time, the subject began to be called information technology (IT) as telecommunications merged with computers into one discipline under the leadership of the emerging position, the chief information officer (CIO).

Strategic planning, competitive advantage, organizational learning, and the role and contribution of IT were the central issues facing most organizations as they struggled with the advancing technology, and restructured to take advantage of it. Communications between IT and other executives and functions remained an issue as IT-induced changes swept the firm. Getting functional managers involved in using IT to reshape business processes became important to senior executives.

Now, in the 1990s, as technology advances at a torrid pace, firms are using it to streamline structures and to link with suppliers and customers. Business process reengineering, downsizing, outsourcing, and restructuring take on new meaning as firms emphasize quick response and flexible infrastructures to improve effectiveness. Business integration and restructuring define the era of the 1990s. Executives have embraced networked organizations as they decentralize operations while maintaining centralized coordination through IT systems. IT is receiving top-level attention as it plays an increasingly important role in corporate policy.

Internally, IT is struggling with systems integration issues—open systems are not necessarily as open as advertised—and internal standards must be rigorously established.[17] Outsourcing considerations and application programs make-vs.-buy decisions greatly complicate life for CIOs and their peers. IT is creating tremendous opportunities and daunting challenges which today's corporate executives must routinely and effectively manage.

THE CHALLENGES OF IT MANAGEMENT

For varied new and emerging reasons, information technology managers have difficult tasks to perform. Although many IT managers have established brilliant performance records, others have fared less well. Mediocre performance and a lack of success have dominated the careers of many IT managers who have fallen short in their effort to manage rapid technological change effectively.

Many information technology managers find themselves and their organizations in untenable positions. They are implementing new technologies that have high potential value to the organization but that carry commensurately high risk. They are supporting clients who are requesting increased services while senior executives, observing productivity rising slowly or not at all, are questioning the increasing expense levels. The unmanaged demand for services is manifested in expanding backlogs of work. Expectations in many parts of the firm are rising, and many senior IT managers are unable to meet concomitant challenges effectively.

Many IT managers have been trained in technology and lack the general management skills demanded by their organizational role. They and their organizations are under constant scrutiny. Their jobs demand knowledge of people management and organizational considerations rather than programming or hardware expertise. Their jobs also require them to demonstrate productivity improvements in return for the resources they consume. In addition, mergers, acquisitions, and reorganizations of all kinds are threatening the stability of their position.

The root cause of the difficulties of information technology management are ill-prepared managers in charge of complex, rapidly changing technology. Executives expect to restructure and streamline their organizations around information technology, but the technology itself causes structural and personal dislocations and generates high expectations within the organization and among its managers. To address this situation, IT managers must acquire general management skills that will allow them to contend successfully with these complex issues. Their skills must include managing expectations and coping with the personal and structural changes. In short, the IT manager of the future must be the organization's technological leader and a superb generalist as well. They must capitalize on their technological expertise while solidifying their platform of managerial experience and skill.

In addition, it is imperative that IT managers have extensive management systems and processes to assist them in discharging their responsibilities. Understanding these management systems and processes and their application and use within the firm are necessary conditions for success. This text devotes considerable attention to the development of these systems and processes.

Technological Complexity

Tiny, semiconductor computer chips are commercially available for several hundred dollars each. These chips carry millions of electronic circuits and cost $200–300 million or more to develop. The people of the world are linked electronically through strands of fiber optic cable capable of carrying 100,000 simultaneous telephone conversations. Hundreds of earth satellites relay enormous amounts and kinds of data while thousands more are being planned. High-tech storage devices maintain huge databases at very modest costs, serving as sources and sinks of data for telecommunications and computer systems around the globe. Not only is the complexity of this technology increasing over time, but the rate of change is accelerating as well. The development of modern computer systems is the most complex human activity ever undertaken, and their application is also monumentally complex.

The application of modern computer and communication systems is highly beneficial for firms of all kinds. However, the complexities inherent in technology adoption and utilization raise the management challenges to new heights. The opportunities for individuals expertly prepared to manage these new challenges are equally unprecedented.

Pervasiveness

Computer and telecommunications technology is deeply infused and widely disbursed in today's organizations. Business users of information technology are installing intelligent workstations at a rapid pace; most professional employees have their own individual workstation or terminal.[18] Workstations for office and professional workers in our organizations are mostly interconnected via local area networks that access servers connected to other nets or to large computer datastores. The servers function as message-switching systems connecting to sophisticated telecommunication networks. These internets or information highways link the firm's sites nationally and internationally, electronically uniting these firms with their customers and suppliers around the globe. The penetration of information technology into the fabric of human activity will continue unabated into the foreseeable future.

Applications and Data

Rapidly growing, advanced computer and telecommunication systems are facilitating large, sophisticated, and very valuable applications programs for today's firm. The firm's databases are growing in size and importance. The acquisition and maintenance of this vast program and data resource demands very careful attention from many members of the firm's senior management team. Application programs and databases in the U.S. valued at $1 trillion or more are expensed during development, and are not recorded on the corporate balance sheet. But these and other intellectual assets are the foundation of today's information-based organizations.[19]

Computer Operations

Skillfully managed computer and network operations are vital for the efficient performance of the firm. Today, many firms process hundreds of revenue producing transactions per second that must be executed promptly and flawlessly. In effect, the firm's "information system" is in series with the firm's operations. Loss of service for even a few seconds has serious consequences to the firm's financial health and reputation. Management performance demanded by this kind of operation is extremely challenging.

Controls and Environmental Factors

Knowledge-based organizations must rely upon careful, constant control. Weak, ineffective control, or loss of control in highly valuable and highly automated operations, rapidly propagate errors. External and internal threats to sophisticated human and machine business operations must be neutralized through careful attention to business controls of all types. The control issue grows in importance as networked systems and internetworking increase.

The social and political environment surrounding the technological evolution is also critically important. Internationalization of business and growing international competition, partly enabled by technological advances, is altering the way firms conduct their affairs. International governmental activity is shaping business enterprise worldwide, and the pace of change in the business sector is increasing with time. This moving backdrop against which business strategy and planning take place greatly increases the management task for the firm's executives, and for the IT executive as well.

Strategic Considerations

Information technology has great strategic value and significance for most organizations today. Information technologists and their organizations are expected to provide the tools with which firms can capture strategic competitive advantage. Consequently, many information technology organizations and their managers are on their firm's critical path to success. The IT organization may be limiting its parent organization's long-term performance. Obviously, this poses a very special challenge for the firm and its executives.

People and Organizations

Information technology alters the nature of work in industrialized societies. The technology impacts organizations, managers, and workers at every level. Not all the consequences are perceived favorably; many are considered threatening or intimidating to managers and workers alike. These personal and organizational dislocations further heighten the management challenges accompanying the technology.

To cope successfully with these challenges, effective managers rely upon their increased awareness and keen appreciation of the phenomena, and fine tune their people-management skills. High-performance organizations have found ways to maximize their human resources. They recognize that people are the key to capturing the benefits of technological advances in today's information age. Effective executives understand this extremely well.

TECHNOLOGY ASSIMILATION

Organizations go through predictable stages of growth as they adopt and implement technology. The stage hypothesis was first developed in 1974 by Nolan and Gibson; the number of identifiable stages was later expanded from four to six.[20] These writings provide an important basis for understanding the introduction and assimilation of new technology. Table 1.3 identifies the six stages.

TABLE 1.3 Six Stages of Growth

1. Initiation
2. Contagion
3. Control
4. Integration
5. Data Administration
6. Maturity

The six stages of growth are briefly summarized in the paragraphs that follow:

Stage 1. *Initiation.* In this stage, the technology is initially introduced into the organization, and some users begin to find applications. The use grows slowly as people become familiar with the technology and its applications.

Stage 2. *Contagion.* As more individuals and departments become acquainted with new technology, demand increases and use of the technology proliferates. Enthusiasm for the new technology builds rapidly during this stage.

Stage 3. *Control.* During the control stage, the issue of cost versus benefit intensifies and management becomes increasingly concerned about the economics of the technology.

Stage 4. *Integration.* As systems proliferate within the organization and databases continue to grow, the notion of systems integration becomes dominant. Management becomes interested in leveraging integrated systems and their databases.

Stage 5. *Data Administration.* During the data administration stage, management is concerned with the value of data resources. Functions are created to manage and control the databases and to ensure that they are utilized effectively.

Stage 6. *Maturity.* In this stage, if it ever occurs, the technology and the management process are integrated into an efficiently functioning entity.

The stages-of-growth concept is useful because it operates at several different levels within an industry or within an organization, and because the concept provides predictability. Not all firms within an industry are in the same stage of growth, and not all functions within a corporation are at the same stage. Executives can observe where their organizations are in the stage theory and can make reasonable assertions concerning the future behavior of their organizations. Thoughtful managers can prepare themselves and their organizations for the future according to their observations of the present.

The stages-of-growth concept is also useful in information technology planning. It can provide guidance in planning future activities, and it forms the basis for one framework of IT planning. As new technology becomes available and is introduced into the organization, the stage theory comes into play again. Managers can observe the technology adoption process. They can implement management practices required to capitalize on the technology's potential while minimizing the effects of the inevitable pitfalls. This text explores these notions in later chapters in connection with IT planning, the introduction of personal workstations, and other topics.

The stage hypothesis is an important management concept because it provides insight into the technology adoption process. Skillful IT managers are cognizant of these processes and they utilize this knowledge to improve their performance and that of the organization.

INFORMATION TECHNOLOGY ISSUES

The task of managing the information technology function remains difficult for a variety of important reasons. The issues and concerns relate to very basic concepts involving the inner workings of the firm and the utilization of information technology within the organization. For the most part, these are managerial issues rather than technical concerns. Management skills are much more valuable in coping with the difficulties than are technical skills.

Various organizations and individuals have conducted key issue surveys among senior IT executives, their superiors, consultants, and educators in the field. These analyses categorize the concerns facing IT managers, their peers, and their superiors, and rank them according to importance. For example, university researchers have conducted key issue surveys of corporate members of the Society of Information Management approximately every three years. These surveys identify current issues while revealing trend information. CSC Index, Inc., Datamation, Digital Equipment Company, and several consulting firms have developed lists of issues and performed analyses of the trends and considerations behind the issues.

In recent years, specific issues have varied in importance, as can be expected; however, some key concerns have shown remarkable consistency. For example, strategic planning for information technology has been one of the top five issues since the mid 1980s. During this period, aligning corporate and IT goals has also been a prominent issue. Recently, reengineering business processes has risen in importance as firms struggle to improve efficiency by reorganizing and integrating technology more deeply into their activities.

A survey of 202 managers taken in November 1993 by *Datamation* revealed that more than half the respondents believed that using IT to improve productivity and quality, creating competitive advantage through information technology, reengineering IT to better reflect company strategy, and reengineering business processes were critical or very important issues for them.[21] Other important issues were reducing IT costs and budgets, and selling IT to top management. This portends more change, realignment and restructuring, decentralization, and managing within tight financial constraints.

In a survey of 603 senior IS executives in 1995 by CSC Consulting Group, Cambridge, Mass., the top five issues were 1) aligning IS and corporate goals, 2) instituting cross-functional systems, 3) organizing and using data, 4) implementing business reengineering, and 5) improving the IS human resource.[22] Reengineering business processes has been in the top five in recent years but did not make the top ten when this research was begun in 1988. According to these researchers, the main goals of reengineering are to increase productivity, improve customer satisfaction, and improve quality. Additional goals of reengineering are to reduce costs, to reduce cycle time, and to increase revenue. Reengineering is consistently at the heart of business effectiveness.

At their annual conference in September 1994, the Society for Information Management reported that IS organizations' "main focuses now are on business strategies, setting enterprisewide standards, managing partnerships with users and vendors, and building and managing state-of-the-art information infrastructures." And, according to the reporter, "if they aren't focused on these areas, they should be."[23]

Other issues seem to be declining in rank although they remain important. Some of these are technology integration, end-user computing, and the IS role and contribution. Studies support the idea that key issues overseas tend to be similar to those in the U.S.

What is the fundamental meaning in all of this? The following section helps glean some basic understanding of this information generally useful to IT managers.

The Maturation of IT Management

One emerging theme is that the issues and concerns of IT managers are moving from the specific to the general, from technical to managerial, as IT moves from the technology orientation of the 1960s and 1970s to the business orientation of the late 1980s and 1990s. Telecommunications, competitive advantage, organizational learning, role and contribution of IS, and acting as change agents dominated IT managers' thinking in the late 1980s. In the mid 1990s, organizations are restructuring, reengineering, and realigning to bring IT closer to internal clients and the firm closer to external customers and suppliers. Now, as firms restructure, improving business effectiveness dominates manager's thinking. And to reduce costs and improve internal efficiency, corporations and IT organizations are downsizing and outsourcing.[24]

In the 1990s, firms and their IT organizations are struggling with plan alignment issues, reengineering, downsizing, establishing standards, and resolving policy issues. In part, the rise of the CIO reflects these new roles for information technology. As Information Technology becomes a mature management discipline in tomorrow's organizations, governance—establishing the rules of conduct in the firm—will become a more dominate concern for IT executives.

Today, the typical IT organization is struggling in its relationship with senior executives and other functions as it contends with distributed architectures, user-developed systems, and rapid changes in the business environment. The firm is affected by the increasing pace of change both here and abroad, and by the growing complexities wrought by the technology. IT executives are expected to contribute to the firm's financial results and are struggling with these responsibilities. Meanwhile, they are experiencing some of the same old problems: backlogs of work, mergers and reorganizations, training and retraining employees, and managing with constrained financial resources. The IT executive needs a keen awareness of business issues and refined general management skills to contend successfully with this difficult environment.

Managing Mature IT Organizations

As IT becomes a more mature discipline in tomorrow's organizations, IT managers must develop more sophisticated, more mature models of behavior. They must be knowledgeable about technology and its trends, but this knowledge alone is not enough. Broad-based business experience is critical to IT managers but, by itself, is insufficient for success. Tomorrow's successful IT manager needs solid business skills *and* a sound understanding of technology, its trend, and its implications.[25] In addition, IT managers need models of behavior and frameworks of business management.

In an important work, Paul Strassmann presents a model of Information Management Superiority premised on the idea that IT management only has value within the context of business management. He goes on to say that "the benefits of investments in information technology can be assessed only as seen from the standpoint of a business plan."[26] His model depicts information management superiority being sustained by five reinforcing and interacting ideas.

Information Management Superiority depends on:

A. *Governance.* Governance, or information politics, is used not only to exercise authority but is an art for achieving corporate consensus. It guides how individuals and groups cooperate to achieve business objectives.

B. *Business Plan Alignment.* IT business plans must be congruent with the organization's business plans or the worth of IT plans will always be suspect.

C. *Process Improvement.* All IT and business activities should be regularly scrutinized to identify areas where improvements, however small, can be made.

D. *Resource Optimization.* Managers must always be questioning whether money, space, time, or people can be used more effectively to further corporate goals.

E. *Operating Excellence.* All the operational details of the business must be performed in a superior fashion. Quality in the business process must be an overriding consideration throughout the business.

According to Strassmann, information management superiority is achieved by constant management interaction among these five activities. For example, a manufacturing process improvement may require alterations to business plans. Suppose, for instance, that the inspection of arriving chemicals will be delegated to an independent testing lab and, therefore, the receiving inspection program is no longer needed. The business plans for manufacturing and for IT will be altered as receiving inspection systems are no longer required. Later, this may lead to a policy change regarding all incoming supplies, thus changing the governance. The importance of this model will become clearer as we explore subsequent topics later in this text.

Strassmann's model helps explain the developing maturity of IT management. Thirty years ago, EDP managers were concerned with improving their operations, getting more work through CPUs, and keeping systems running around the clock. Management tasks were secondary for many. Twenty years ago, optimizing hardware systems reached a fine art and IT concentrated on improving IT processes. More attention was given to applications development. During the 1980s, IT executives concentrated on end-user systems, teaching organizations about technology, and developing strategies aligned with the business strategy.

Today, as IT matures, executives are concentrating on the rules governing technology disbursement and are establishing policies regarding distributed computing, portable computing (notebooks and laptop computers), and wireless networks. Policy making and the establishment of authority for technology acquisition and deployment are now more critical than owning and operating wide-area networks. Increasingly, IT executives are delegating routine development, maintenance, and system operation to information servicing firms in order to concentrate on the larger, more important policy issues.

SKILLS THAT FAVOR SUCCESS

With a knowledge of the important problems facing IT executives, and with a perspective or model for understanding them, managers are positioned to describe the factors necessary for success. What actions must IT executives carry out successfully? What management systems and processes are vital for their personal success, and for the success of the IT organization?

Critical Success Factors

The answers to some of these questions are found in the notion of Critical Success Factors (CSF). The idea of critical success factors was developed by

Rockart in the late 1970s to help executives define their information needs.[27] The concepts are useful in information technology planning and will be discussed further in succeeding chapters. Critical success factors identify those few areas where things must go right; they are the executive's necessary conditions for success. They apply to IT executives, to their subordinate managers, and to senior executives in the firm.

Rockart identified four sources or areas where executives should search for critical factors: the industry in which the firm operates, the company itself, the environment, and time-dependent organizational areas. This last source accounts for the possibility that some organizational activity may be outside the bounds of normal operations and require intense executive attention for a short period. In addition, Rockart identified two types of CSFs: the monitoring type and the building type. The monitoring type keeps track of the ongoing operation. The building type initiates activity designed to change the functions of the organization in some way.

Critical success factors apply to the IT manager. What are the conditions necessary for the IT manager's success? What tasks must be carried out very well in order for the manager to succeed? The critical issues studied previously in conjunction with Strassmann's model provide the information needed to answer these questions. The factors necessary for success can be grouped into several classes, as shown in Table 1.4.

TABLE 1.4 Critical Areas for IT Managers

1. Business management issues
2. Strategic and competitive issues
3. Planning and implementation concerns
4. Operational items

The firm's IT management team must take action in these critical areas. The IT manager must have goals and objectives to solve problems, if there are problems, and to prevent the development of issues. Outstanding managers may use a roadmap of critical success factors (CSF) to assess their posture on these vital topics. The following CSF list serves as a roadmap that successful IT managers might follow:

1. Business Management Issues
   a. Obtain agreement with the firm's executives on how information technology will be managed within the firm.
   b. Operate the IT function within the parent organization's cultural norms.
   c. Attract and retain highly skilled people.

d. Practice good people management skills.

e. Improve IT productivity.

2. Strategic and Competitive Issues (long range)

a. Develop IT strategies aligned with the firm's strategic goals and objectives.

b. Provide leadership in technology applications to attain advantage for the firm.

c. Educate the management team about the opportunities and challenges surrounding technology introduction.

d. Ensure realism in long-term expectations.

3. Planning and Implementation Concerns (intermediate range)

a. Develop plans supporting the firm's goals and objectives.

b. Provide effective communication channels so that plans and variances are widely understood.

c. Establish partnerships with client IT organizations during planning and implementation.

d. Maintain realism within the organization regarding intermediate term expectations.

4. Operational Items (short range)

a. Provide customer service with high reliability and availability.

b. Deliver service of all kinds on schedule and within planned costs.

c. Respond to unusual customer demands and to emergencies.

d. Maintain management processes that align operational expectations with IT capabilities.

Not all of these items will be on every IT manager's list of critical success factors. In many cases, some of the critical areas will be operating smoothly and routine attention will maintain high-quality operation. In other instances, managers must add temporal organizational factors or company-specific factors to the list. In all cases, superior managers remain attentive to those factors necessary for their success.

Information technology managers who can accomplish the tasks outlined above have a good chance of becoming highly successful. They have developed general management skills which will prepare them for increased future responsibility. Throughout the remainder of this text, management tools, techniques, and processes will be developed to enable IT managers to accomplish these critical tasks successfully. Achieving success in these key areas is a high-priority goal for IT managers.

A MODEL FOR THE STUDY OF IT MANAGEMENT

The study of information technology management concentrates on business results, attaining efficiency and effectiveness, and achieving competitiveness with the external environment. The intended result is improved profitability for the firm. The structure of this text is related to business results and is portrayed in Figure 1.3.

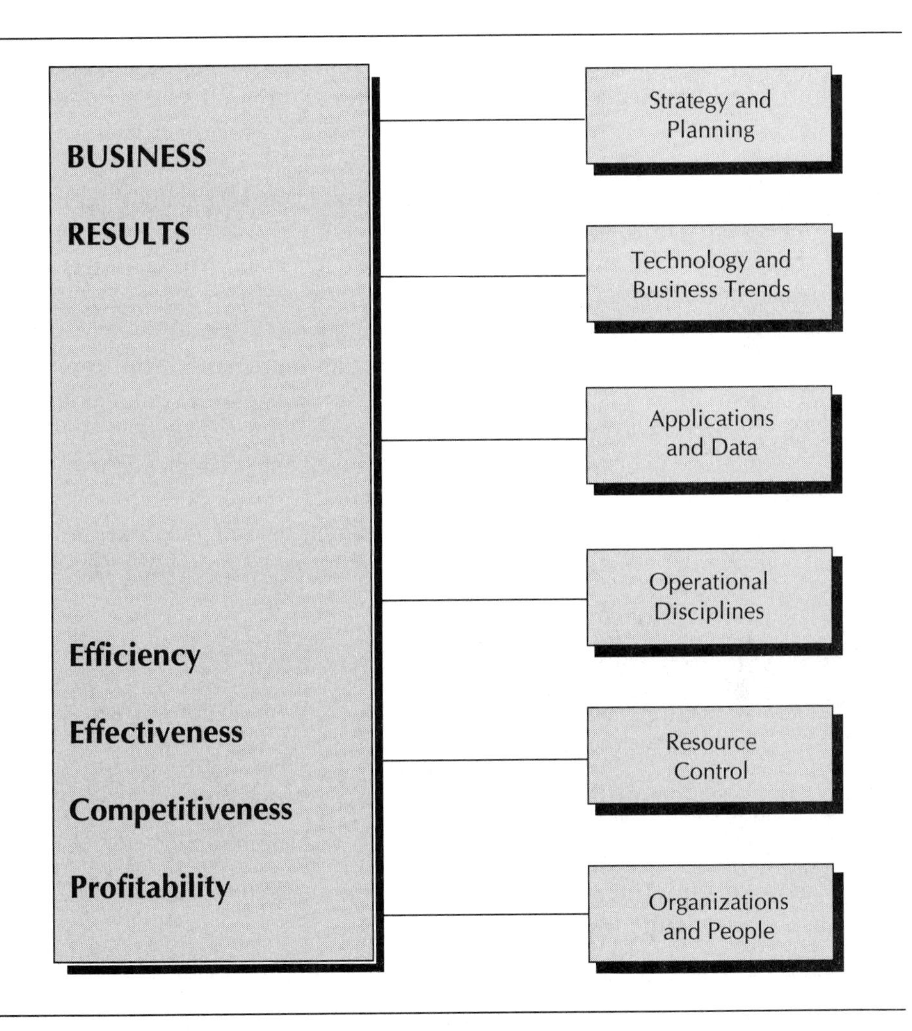

FIGURE 1.3 A Model for the Study of IT Management

Each of the topics on the right side of Figure 1.3 contributes to the success of the firm in an essential manner. Each represents a tangible or intangible asset to be deployed effectively for the firm's benefit. In every area, the IT executive needs reciprocal cooperation with managers throughout the organization in order to maximize these assets. Each part in this text addresses one of these vital assets.

SUMMARY

Information technology is a powerful force for change in our industrialized world. The change is fundamental in nature, profoundly affecting people, organizations, industries, and nations. Managers in all spheres of endeavor must adjust their behavior in order to achieve success in these turbulent times.

This chapter concentrated on the basic ideas surrounding the management of organizations, and introduced topics and issues important for the managers of information technology. The subjects introduced in this chapter, expectations, stages of growth, important current issues, and critical success factors, are interwoven through the remaining text material.

Information technology is an important ingredient in the strategies of nearly all firms today. Consequently, IT managers play a vital role in achieving long-term success for the firm. *Management of Information Technology*, 2nd edition, illuminates the general management principles that are necessary for success in this endeavor.

The rest of this text focuses on attaining business success through the application of management principles to information processing technology.[28]

Review Questions

1. According to Drucker's thesis, how will information technology alter the future structure and operation of the firm?
2. What causes the challenges facing IT managers to remain high?
3. Why are expectations so important to IT managers?
4. Relate levels of management to sources of information and to types of decision making.
5. What stages has electronic information processing gone through in the last 40 years?
6. What were the main characteristics of information systems during the 1970s?

7. What are the root causes of the difficulties of IT management?
8. What kind of preparation improves the IT manager's chances of success?
9. Identify the challenges of IT management outlined in the chapter.
10. What are the six stages of growth identified by Nolan, and during which stage does senior management take a serious interest in the technology and why?
11. Why is it unlikely that a firm will achieve the stage of maturity?
12. Why is an understanding of stages of growth useful to management?
13. What important points are revealed by the study of IT issues?
14. What changes are occurring as the IT discipline grows toward maturity?
15. What are the five ideas of Strassmann's model?
16. What is the difference between necessary conditions for success and sufficient conditions for success?
17. Why doesn't the text present a set of sufficient conditions for success?
18. Why does the IT manager's list of critical success factors contain statements regarding expectations?
19. What are the causes of the dramatic structural changes predicted by Drucker and others?

Discussion Questions

1. How does Drucker's hypothesis bear on the concepts displayed in Figures 1.1 and 1.2?
2. Information technology has high potential value for most firms. Why does this cause the IT manager's job to be so demanding?
3. Discuss the role that expectations play in IT management as discussed in this chapter. Do you think expectations are as important to managers in other functions in the firm? If so, why?
4. Describe the role that IT organizations and their managers play in firms today. Discuss the meaning of "line" and "staff."

5. Discuss the Gorry and Scott-Morton model and relate it to Figure 1.1 and to Drucker's hypothesis.
6. Discuss the evolutionary stages of information technology by describing the characteristics of each. In particular, differentiate the 1980s from the 1990s.
7. Discuss the challenges of IT management. Which challenge do you think is most difficult and why?
8. Discuss the significance of the stages-of-growth concept. Can you think of an example of this concept in operation?
9. What personal characteristics would you expect to find in the individuals responsible for the initiation of a technological thrust?
10. Why is it possible that not all parts of a corporation will be at the same stage at the same time? What opportunities does this present management?
11. There may be some individuals in an organization who are extremely reluctant to embrace a new technology. What special problems does this raise? What tools are available to management to solve this problem?
12. What factors cause one to believe that IT management is moving toward maturity? What factors might cause one to reach a different conclusion?
13. Discuss the relationship between Strassmann's model and the list of IT critical success factors.
14. Discuss the relationship between the idea of corporate culture and that of the critical success factor 1b.
15. Why do you think it is important that IT managers prepare themselves for increased future responsibility?

Assignments

1. Read the referenced article by Peter Drucker and prepare a written summary of his thesis, concentrating on the meaning for future organizations and managers. Specifically, how do you think this will affect the managers of IT organizations?
2. Using library reference material, find an IT manager's success story and prepare a synopsis for class presentation. Identify the chief factors contributing to the manager's success.
3. Which of Nolan's stages of growth best describes your school's level of information processing maturity? Prepare your answer by discussing this subject with a knowledgeable individual in your school's administration.
4. Read the Nolan and Gibson article, "Managing the Four Stages of EDP Growth," referenced in footnote 20, and prepare a one-page summary of its main points.

ENDNOTES

[1] Peter F. Drucker, "The Coming of the New Organization," *Harvard Business Review*, January-February, 1988, 45.

[2] Premier 100, Supplement to *Computerworld*, September 19, 1994, 25.

[3] "The Premier 100," Supplement to *Computerworld*, September 13, 1993. This article presents expense-to-revenue ratios for information systems. For the 100 companies reported on, the ratio ranges from a low of 0.4 percent to a high of 16.3 percent.

[4] Marvin Bower, *The Will to Manage* (New York: McGraw-Hill Book Company, 1966), 22. Also Terrence E. Deal and Allen A. Kennedy, *Corporate Culture* (Reading, MA: Addison-Wesley Publishing Co., 1982).

[5] The term Chief Information Officer was coined in 1981 by William Synott, senior director of the Yankee Group, an industry consulting group based in Boston. See Chapter 18 for trends in IT organizations and Chapter 19 for discussions of the concept and role of the Chief Information Officer.

[6] The total number of applications is increasing very rapidly as firms purchase applications from vendors and as clients develop programs at their individual workstations. In most firms the investments and benefits of the complete portfolio cannot be estimated at this time.

[7] Harold Koontz and Cyril O'Donnell, *Principles of Management* (New York: McGraw-Hill, 1964), 262.

[8] IT people are professionals and, therefore, those they support are properly termed clients and colleagues, not users as they are often labeled.

[9] Gordon Davis and Margarethe Olson, *Management Information Systems* (New York: McGraw-Hill, 1985). In particular, refer to Chapter 7, Concepts of Information, and Chapter 11, Organizational Structure and Management Concepts, for a detailed presentation of these topics.

[10] Tom Peters, "Tomorrow's Companies," *The Economist*, March 4, 1989, 19. See also *Liberation Management* (New York: Alfred A. Knopf, 1992) in which Peters states "The way we organize to get things done in the public and private sectors, and in general, is undergoing the most profound shift since (at least) the Industrial Revolution."

[11] Peter Drucker, "Coming of the New Organization," 46.

[12] Robert N. Anthony, *Planning and Control Systems: A Framework for Analysis* (Cambridge, MA: Harvard University Press, 1965). For a clear perspective on how this model relates to management information systems, or MIS, see David Kronke, *Management Information Systems* (Santa Cruz, CA: Mitchell Publishing, Inc., 1992), 101–102.

[13] G. A. Gorry and Michael Scott-Morton, "A Framework for Management Information Systems," *Sloan Management Review*, Spring, 1989, 49. This article updates the classic article with the same title by these authors in *Sloan Management Review*, Fall 1971, 55.

[14] Tom Peters, *Liberation Management* (New York: Alfred A. Knopf, 1992), 121–2. Informated first-line employees are those who are empowered to deal directly with customers and have all the firm's information at their disposal to do so, according to Peters.

[15] *Executive Summary*, The 1990 Society for Information Management Institutional Member Conference, March 14–16, 1990, 1.

[16] For more detail on this evolution see Peter G. W. Keen, *Every Manager's Guide to Information Technology* (Cambridge, MA: Harvard Business School Press, 1991), 7.

[17] Open systems means an electronic environment that permits hardware, software and network components to co-exist and evolve toward increased capacity.

[18] "The Premier 100," *Computerworld*, September 13, 1993. About 75 percent of employees in these firms use PCs or terminals.

[19] For a fascinating discussion of intellectual capital and its importance in relation to federal government activities see Walter B. Wriston, *The Twilight of Sovereignty* (New York: Charles Scribner's Sons, 1992), 11–12.

[20] Richard L. Nolan and Cyrus F. Gibson, "Managing the Four Stages of EDP Growth," *Harvard Business Review*, January-February 1974, 76. Also Richard L. Nolan, "Managing the Crisis in Data Processing," *Harvard Business Review*, March-April, 1979, 115.

[21] Jeff Moad, "IS Rises To The Competitiveness Challenge," *Datamation*, January 7, 1994, 16.

[22] Julia King, "Reengineering Focus Slips," *Computerworld*, March 13, 1995, 6. See also, "Top Concerns of Information Managers," *InformationWeek*, January 30, 1995, 70.

[23] Julia King, "You Can't Do It All," *Computerworld*, October 10, 1994, 102.

[24] Downsizing—reducing numbers of people, middle managers in many cases, and outsourcing—moving work to outside specialty firms are current trends.

[25] Current debates about whether CIOs should be technologist or business generalists miss the point: CIOs must combine the skills of both.

[26] Paul Strassmann, *The Politics of Information Management* (New Canaan, CT: The Information Economics Press, 1995), 10. Policy is to management what law is to governance according to Strassmann.

[27] John F. Rockart, "Chief Executives Define Their Own Data Needs," *Harvard Business Review*, March-April, 1979, 81.

[28] The concepts in this text are applicable to institutions, agencies, government entities, and not-for-profit organizations. The use of terms such as bottom-line results and business results is intended to encompass the results of all forms of organizations, not just profit-making firms.

2 *Information Technology's Strategic Importance*

A Business Vignette

Microsoft Invades the Information Economy†

In the two decades since Bill Gates and a boyhood pal founded Microsoft, sales have grown to $4.65 billion and the market capitalizes its shares at more than $38 billion. Gates's net worth is about $9 billion. And it seems that the end to his success is nowhere in sight.

Gates wants to move Microsoft software beyond the desktop and the den and into the guts of the Information Economy. Within the decade, he wants Microsoft code to become the genes of new corporate computer infrastructures that will replace today's crazy quilt of incompatible PC and workstation networks, minicomputers, mainframes, and supercomputers. Simply put, Microsoft must transform itself from a maker of packaged goods into something more like a utility company, or even the old IBM. But it must stick to its strategy of controlling the key chokepoints of a much heralded, nascent industry known as the information highway.

Microsoft is moving to make sure that its code will touch every kind of commerce conducted on this network, from the usual consumer services to managing factories, inventories, and corporate databases, trading stocks, and verifying credit cards. "This new electronic world of the information highway will generate a higher volume of transactions than anything has to date, and we're proposing that Windows be at the center, servicing all those transactions," says Gates.

The key to understanding just about everything in the information economy is a technological axiom called Moore's law, named for Gordon Moore, the chairman of Intel. Moore's law states that, measured against its price, the performance of semiconductor technology doubles every 18 months or so. Microsoft bases its business model on Moore's law. It asked, for example, what if computers were free? The answer: individuals would own them and software standards would be critical. In response, Microsoft built new and more versatile programs and expanded its business from operating systems to applications and now to multimedia.

Today, the exponential increase in bandwidth is bringing Microsoft to another threshold. According to Gates the new question is: What if digital communications were free? The answer is that the way we learn, buy, socialize, do business, and entertain ourselves will be very different, again making software

† Brent Schlender, "What Bill Gates Really Wants," *Fortune*, January 16, 1995, 343–63.

standards important. Microsoft hopes to identify those chokepoints in nascent businesses that could give it new opportunities. Microsoft must evolve from the packaged-goods model to the utility model in order to control the strategic components of the information highway against the giants of the telecommunications industry.

Microsoft has to develop technologies that enable the company to seize the strategic leverage points in the information economy, then allow those technologies to transform the very nature of the company.

But Microsoft has a bigger agenda. Why not go beyond the usual on-line service strategy of setting up an electronic forum, and also try to be a middleman, providing transaction-clearing services to make on-line commerce a natural and reliable way of doing business? Talk about repeat revenue! Microsoft could collect "rent" from electronic vendors and services who set up shop on its network, and bill individual customers for the time they spend browsing and doing business. Better yet, it could charge businesses, banks, and credit card companies fees for helping to handle each commercial transaction.

The attempted acquisition of Intuit, maker of the most popular line of personal finance software for PCs, was indicative of Microsoft's coordinated strategy to gain a toehold in electronic commerce. Many in the industry call personal finance software the next "killer app"—a product so universally useful that everyone has to have it, like word processors. Microsoft has already cut deals with banks and credit card companies and plans to put it all on-line with the launch of its own network in competition with the likes of America On-line and CompuServe. It's an opening move to make the coming information highway into both a giant digital shopping mall and an even larger arena for conducting business-to-business transactions.

Gates is betting that his technology is the best platform for building the information highway—better than the telephone business because it already boasts multimedia capability, and better than the cable TV business because it's already a medium for doing real work.

But Microsoft's real goal is to develop a much hardier version of Windows NT—code named Cairo—that can marshal the power of thousands of linked microcomputers to provide the performance, security, and reliability required by corporate customers for their "mission critical" data-processing activities. Cairo would obviate the need for a central corporate computer by distributing data over the whole network of microcomputers.

What most distinguishes Cairo from its competitors is that it is being designed with the information highway in mind. From the beginning, Cairo could accommodate not only PCs but all manner of nodes plugged into the high-bandwidth, video-spewing networks that will compose the information highway—everything from TV set-top boxes to supercomputers. That means Cairo networks must be able to efficiently sort, search, and transmit a staggering volume of all kinds of data and transactions: video on demand, videotelephony, interactive games, on-line commerce, still images, or traditional text. Microsoft is spending more than $600 million a year on research and development. Cairo is the linchpin of the entire plan.

In the next four years, Gates expects Microsoft to double in size to nearly $10 billion in sales, and he sees no reason Microsoft can't grow into a $20 billion company in the future. Value Line projects Microsoft's 1995 revenue to be $5.7 billion with profits of $1.33 billion.[1]

INTRODUCTION

Information technology is more profoundly impacting the operation of organizations than any other technology ever has. Advances in space travel, nuclear energy, improved medical technology and pharmacology, advances in chemical fertilizers, and breakthroughs in plant and animal genetics have all been highly important to the world and its people but none has affected organizations in the fundamental way that information technology has.

Power shifts from secretive governments to informed people when CNN broadcasts real-time bombings in Chechnya just as Moscow proclaims bombings are not part of its plan. Power shifts from governments to people when intellectual property moves across national boundaries without intervention by customs agents. And power (money) shifts from the U.S. Post Office to Federal Express, United Parcel Service, and to numerous telecommunications service providers of E-mail and Fax as the Post Office deliberately ignores advancing technology. Executives in organizations throughout the world who disregard information technology do so at their peril and at considerable risk to their organizations and employees.

Senior executives in most firms today expect to use information technology to improve business processes, streamline operations and organizations, and link their organizations more tightly to customers, suppliers, and business partners. They want to decentralize operational decision making while retaining centralized control over critical functions. They know information technology can help them do this. Although their expectations include additional automation and cost reduction in the more routine processes in the organization, they clearly have larger expectations for themselves and for their organization as a whole. Their vision includes significant additional support for the activities they conduct personally and, more importantly, they are increasingly questioning whether their information technology strategy is providing sufficient leverage to their firms or to their employees.

Walter Wriston, former chairman of Citicorp, described the goal of an information strategy neatly when he said, "The essence of an information strategy is to turn the burden of burgeoning business data into a bounty of business opportunity. The business organization has to be rebuilt around the goal of managing information productively. The object of the game is to get information to the person or company that needs and can use it in a timely way."[2] His vision of rebuilding the organization, turning information overload into opportunities, and empowering companies or individuals with timely, vital information is central to the information technology paradigm of the 1990s.

STRATEGIC ISSUES FOR SENIOR EXECUTIVES

Increasingly, senior managers in most firms are using information systems for strategic purposes. There are several reasons why this is happening. These forces are listed in Table 2.1. For the last several decades, executives have authorized expenditures for the development of systems that automate the relatively routine transaction-processing activities of the firm. They have invested in systems to help employees make better decisions (DSS) and to help managers operate the business better (MIS). They have also authorized investments in distributed systems designed to bring information closer to the person who needs it. And they also note that significant resources are being deployed on more sophisticated applications such as expert systems, wireless networks, and many others. In some cases, senior executives are openly questioning whether these investments have done more than keep pace with their industry because they believe IT investments should give them a competitive edge in their industry.

TABLE 2.1 Forces Driving Today's Strategic Vision

1. Obtain competitive advantage for the organization
2. Interorganizational linking via networking technologies
3. Desire for decentralized autonomy and central coordination
4. Obtain flexible and responsive infrastructures
5. Capitalize on fleeting but critical business information

As competition drives firms to grow and expand globally via alliances, joint ventures, and mergers, CEOs insist on tight coupling and coordination between operational units regardless of location. They know that network technologies can achieve the desired interorganizational linking. They want the hub or corporate central organization to monitor operations in near real-time and perform coordinating activities yet permit decentralized operational decision making.

Time is of the essence in today's competitive world and change is a way of life, so the infrastructure, particularly the information infrastructure, must be adaptable and responsive to change. Today's term for this is agile operations. Information and opportunities are frequently short lived; their value deteriorates rapidly in most cases. Thus, getting important information to the people who need it is critical. Executives expect that, when properly applied, information technology can help achieve this goal for their firm. In short, senior executives expect IT to contribute significantly to business results.

Senior executives believe that innovative information technology holds great promise for improved corporate or organizational leadership. They desire to employ technology innovatively and to strive for a leadership position for the firms they head. Executives believe that past and future investments in information technology should result in leadership via innovative applications of the technology. This executive viewpoint has great importance for IT managers. It must direct their thinking if they are to achieve success.

Over the years, but particularly in the recent past, CEOs in many firms have observed the productive use of information technology in virtually every aspect of their operations. The funds deployed for this technology have been invested across the board, significantly affecting nearly every facet of their organizations. Office automation, E-mail, client/server implementations, and EDI links to sales, service, supplier, and customer locations are almost universal corporate appendages. The very pervasiveness of the technology demands strategic thinking about its future use.

Finally, the belief that information systems can have a very significant impact on a firm's strategic direction and its long-range position in the industry is not speculative but is validated by many examples from current experience. Senior executives have specific, concrete precedents on which to base their desires and aspirations. For example, they are familiar with the success enjoyed by firms in the airline industry which own important and vital reservation systems. They are aware of significant systems in the brokerage industry and of highly valuable order entry systems. These precedents validate their desire to see their firms enjoy comparable successes.

For these reasons, senior executives, IT managers, controllers, and others are focusing their attention on the long-range, strategic implications of the technology and are concentrating their energies on attaining competitive advantage from their IT investments. Therefore, utilizing information technology for competitive advantage is a high-priority issue for business managers in most corporations today.

Relation to Other Critical Issues

Strategic concerns have occupied the attention of senior IT executives for many years, but these issues are not fully appreciated if considered in isolation. Strategic issues are frequently accompanied by or occur in conjunction with other important questions such as: "Is there an effective planning process in the firm and in the IT organization?" "Does the firm have a well-defined information architecture?" and "Do the data resources support the functions and goals of the firm in an effective manner?"[3] Missions, goals, and organizational alignment among and between the functional units within the firm all have a bearing on the strategic concern. In particular, the general, internal effectiveness of the IT organization and its effectiveness in relating to other functions are very important considerations.

The task of improving in any one of these issues generally involves making improvements in all of them. The mutuality of these topics demands that management make progress across a broad front. It is misleading for organizations to believe that competition can be overtaken or beaten by the development and implementation of one new strategic system. A significant leap forward using information technology appears very unlikely if the organization has major weakness in planning, alignment, or other critical areas.

STRATEGIC INFORMATION SYSTEMS DEFINED

Columbia University professor Charles Wiseman makes the following observation concerning strategic information systems:

> Strategic information systems are information systems in which the primary function of the system is either to process predefined transactions and produce fixed-format reports on schedule or to provide query and analysis capabilities. The primary use of SIS is to support or shape the competitive strategy of the enterprise, its plan for gaining or maintaining competitive advantage or reducing the advantage of its rivals.[4]

This definition mostly addresses the business activities of a firm but, for completeness, SIS must also include technical systems that generate competitive advantage for the firm. For instance, support to product development and manufacturing with sophisticated, unique electronic design or automated production systems must be included within the scope of SIS. Strategic systems shape the competitive posture and strategy of the firms that own them whether they are administrative or technical in nature.

Major Types of Systems

There are many types of systems defined in the IT literature that are prominent in the discussions and the thinking of information systems professionals. These include Transaction Processing Systems (TPS), Decision Support Systems (DSS), Office Automation Systems (OAS), Management Information Systems (MIS), and End-User Computing Systems (EUC). Transaction processing systems handle routine information items, usually manipulating the data in some useful way as it enters or leaves the firm's database. Many transaction processing systems are on-line, which means that many users interact with the database simultaneously, performing updates or retrievals from workstations at their desks.

Decision support systems are analytic models used to increase managerial or professional decision making by bringing important data to view. But, according to Keen, "the term DSS is now so vague as to be indistinguishable from executive information system, end-user computing, expert systems and even personal computing.[5] Management information systems provide a focused view of information flow as it develops during the course of business activities. The information is useful in operating the business. Report generation is a critical end-result of management information systems.

Office automation systems provide electronic mail, word processing, electronic filing, scheduling, calendaring capability, and other support to office workers. End-user computing, a form of distributed data processing, places computational capability into the hands of users to develop or execute programs.

Some of these systems are operational in nature; they support the firm in the short term. Others aid managers in making intermediate or long-term decisions. All of these systems have been designed to improve the manner in which employees and managers carry out the firm's functions.

The Character of SIS

To the extent that any system gains or maintains competitive advantage for its firm, it forms part of the spectrum of strategic information systems. The attribute of competitive advantage is what distinguishes a strategic system from all others. In practice, every system the firm uses should provide some advantage or else its operation should be discontinued.

Forces Governing Competition

To understand the possible roles that information systems can play in shaping or altering the competitive posture of a firm, managers must visualize business competition in its broadest terms. A succinct and lucid view of the forces shaping competition was first presented by Michael Porter in 1979.[6] Porter believes that industry consists of firms jockeying for preferred positions while being impacted by the bargaining power of suppliers, the bargaining power of customers, the threat of new entrants, and the threat of substitute products or services. In order for the firm to prosper and grow, it must contend strategically with these forces governing the competitive industrial climate.

Porter suggests that companies need to address strategic actions based on the factors identified in his model. These actions consist of diminishing customer or supplier power, lowering the possibility of substitute products entering the marketplace, discouraging new entrants, or gaining a competitive edge within the existing industry.

Porter's model, shown in Figure 2.1, is very helpful in thinking about competition in its broadest terms and suggests areas in which a firm may need to examine its competitive posture. The model is a framework for judging the firm's position versus its competitors and for analyzing various strategies the firm may elect to employ.

Strategic Thrusts

Wiseman presents a detailed addition to the framework of strategy development. His discussion gives insight into the theory of strategic thrusts, and is complete with numerous examples of strategic information systems.[7] The ingredients of Wiseman's strategic moves are contained in five basic thrusts. I have added "time" as a sixth strategic thrust because reductions in time or increases in responsiveness are critical factors in business competition today. Time must be considered a strategic thrust.[8] Its inclusion as a basic strategic thrust is mandated by the increasing importance of telecommunications in strategic systems today and by the growing number of firms that consider themselves to be engaged in time-based competition. The six basic thrusts discussed in this chapter are presented in Table 2.2.

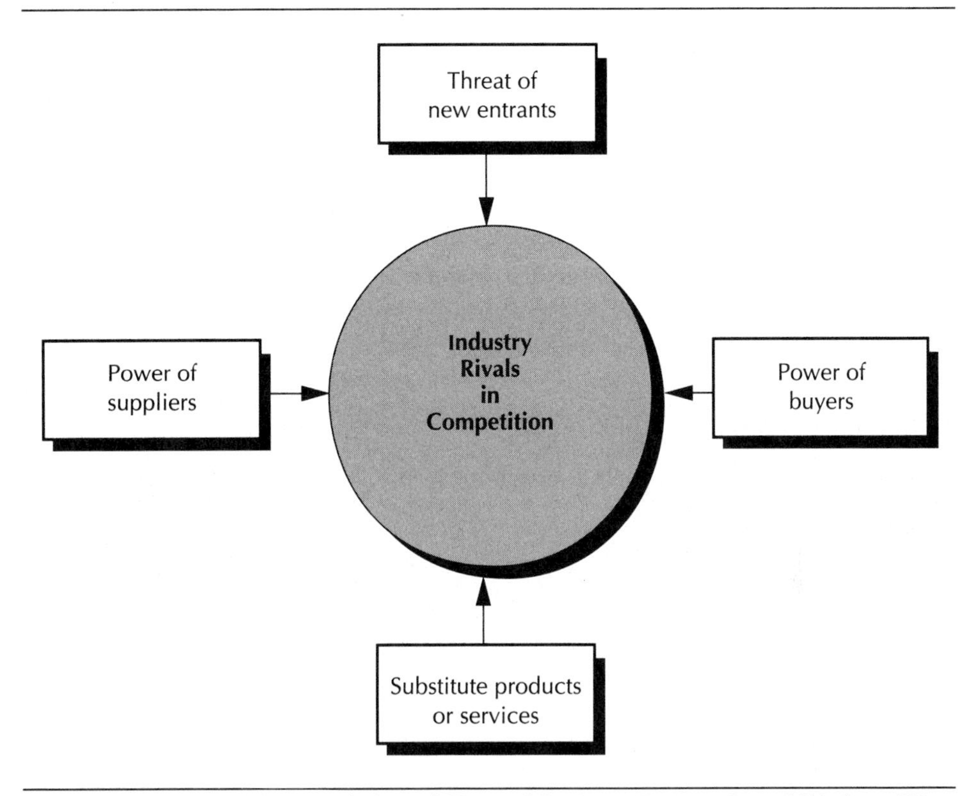

Adapted and reprinted by permission of *Harvard Business Review*. An exhibit from "How Competitive Forces Shape Strategy" by Michael E. Porter, March/April 1979.

FIGURE 2.1 Forces Governing Competition

TABLE 2.2 Strategic Thrusts

1. Differentiation
2. Cost
3. Innovation
4. Growth
5. Alliance
6. Time

The following paragraphs define the essential characteristics of these basic strategic thrusts.

1. *Differentiation.* The firm's products or services are distinguished from competitors' products or services, or a rival's differentiation is reduced. For instance, Automated Teller Machines (ATMs) distinguish the services of some financial institutions from others.
2. *Cost.* Advantage is attained by reducing costs to the firm, the firm's suppliers, or its customers, or by increasing costs to competing firms. Advanced order entry systems in the health care industry reduce ordering costs to both the customer and supplier, for example.
3. *Innovation.* Advantage is gained by introducing changes to the product or process, yielding fundamental changes to the manner in which the industry conducts its business. For example, brokerage firms have used information technology innovations, including telecommunications, to provide their clients with new products and services.
4. *Growth.* Advantage is secured by expansion, forward or backward integration, or by diversification in products or services. The *Wall Street Journal* and *USA Today* are two examples of daily newspapers that use telecommunication technology to seek and reach broad national markets, thus expanding their market share and stimulating growth.
5. *Alliance.* Advantage is attained by reaching agreements, forming joint ventures, or making strategic acquisitions. For example, the Business Vignette in Chapter 5 describes an interesting joint venture between Apple, IBM, and Motorola. Even large firms are using agreements and turning to joint ventures as a strategic thrust.
6. *Time.* Competitive advantage is secured by rapid response to changing market conditions or by supplying a more timely flow of products or services. Electronic design automation tools, computer-aided manufacturing systems, and the integration of the CAD/CAM systems and production logistics systems are thrusts aimed at increasing manufacturing's response to the marketplace.[9] According to Peter Keen, "innovation via technology requires resources of capital, technology, management, and time. Firms cannot buy time off the shelf. They need a catalyst to integrate the other resources in such a way as to 'buy' them the time they need."[10]

Information technology is an important ingredient in the competitive strategies of many firms. It is commonly used to shape strategic thrusts. The desire for strategic information systems stems from business managers' convictions that properly conceived systems can provide enormous advantages to their owners.

STRATEGIC SYSTEMS IN ACTION

The next three sections describe classic strategic information systems from the transportation, financial, and distribution industries. These systems were initiated more than 15 years ago and, after continuous improvements and enhancements, are still providing benefits to their owners. Some have changed considerably during their lifetimes as their firms responded to competitive actions or attempted to stay ahead of the competition. In many ways, these systems and the firms that own them are models for thousands of firms worldwide.

Airline Reservation Systems

In the late 1960s, American Airlines developed a rudimentary computerized reservation system called the Semi-Automated Business Research Environment, or Sabre. Sabre capitalized on the advanced third-generation computing equipment available at that time from IBM. American made an initial investment of $350 million in Sabre but did not achieve profitability on this investment until 1983.[11] Today, Sabre is the world's largest travel agency reservation system, serving 14,000 agents. The system enables agents to provide their customers with airline, hotel, and automobile reservations and other services. It is one of the most widely known and most valuable strategic systems in the world.

The Sabre system combines advanced computerized data processing and telecommunication technology. Reservations are taken by 8000 operators accepting calls at five regional sites and from 85,500 terminals in travel agency locations in 47 countries. The Sabre network consists of 45 high-capacity digital lines over which 470,000 airline reservations are booked daily.

During the 1970s, Sabre became an important marketing force in the industry and a valuable strategic asset for American Airlines. Travel agencies were considering the development of a system for their own use and, faced with this competitive threat, American decided to make Sabre available to them. Because of the high cost of reservation system development, many airlines joined the Sabre system and purchased reservation services from American.

To attain the dominance it has, American has continually updated and expanded the system and introduced new products and services. Sabre provides a basis for travel agent office automation, serves corporate customer's reservation needs, and makes available a host of travel-related services. American also utilizes the Sabre system internally in its advanced yield management system. Yield management is the process of allocating airplane seats to various fare classes to maximize the profit on each flight. Sabre databases are used to predict demand for each flight and to observe the rate at which seats are booked prior to flight departure. Based on this data, seats are dynamically allocated to full fare or discount fare passengers maximizing seat utilization and flight profitability.[12]

The introduction of EAASY Sabre for individual home use and Commercial Sabre for corporate use have been successful areas of growth for American. Only the schedules are available to these users; customers must make reservations through agents, thus preserving commissions for the agents. These new products stem the threat that new low-cost technology such as personal computers may provide competition and undermine Sabre's dominant position.

American has continually upgraded the system to provide end-users with increased flexibility. It implemented a feature named Calltrack which records all incoming calls to Sabre central. The calls are recorded by agency, caller name, and type of problem, thus providing American with valuable market and customer relations data. Other innovations such as the Frequent Flier program have helped the growth of Sabre and given American a competitive advantage by developing bargaining power and comparative efficiency.

Some companies have elected to develop their own reservation systems. Delta Airlines invested more than $120 million in its own reservation system, DatasII, and its software update DataStar but, after six years, has enrolled only 11 percent of travel agents nationwide. Delta is pushing its system by offering back-office capability for the travel agents through small networks of PCs and by providing stored video images of cruise ships, rental cars, and other travel products on hard disk. But the company admits that it is playing catch-up.

Later, in 1989, Delta purchased 40 percent of a firm known as Worldspan Travel Agency Information Services. Northwest owns 33.3 percent and TWA owns 26.6 percent of the firm, which combines Delta's DatasII with PARS previously owned by Northwest and TWA. The new firm, which combines the fourth and fifth largest reservation systems, hopes to capture 26 percent of the U.S. market.[13]

Many companies have chosen not to compete with Sabre but prefer to lease the service. Air France, KLM, and others are airline customers of American's reservation system. The system generated 5 percent of the gross revenue of the AMR Corporation, American's parent company, and accounted for 15 percent of its profits in 1990.

In the early 1970s United Airlines automated its in-house reservation system, and in 1976 the system was introduced as a travel agency system called Apollo. United spent $250 million to develop Apollo. It gives agents access to listings for all major airlines as well as for hotels, car rentals, and other travel-related services. The system was transferred to a separate business unit of United Airlines called the Apollo Services Division in 1976.

The objectives of the division are to add new, marketable services to the reservation system. One such service is an office automation system for travel agents that provides accounting, reporting, and other managerial features. This service has since been enhanced with a new modular Enterprise Agency Management System containing advanced features and functions. Clearly United intended to invest in Apollo to improve the quality of its information processing assets.

In 1987 the UAL Corporation decided to spin off a subsidiary centered around its computerized reservation system. This subsidiary became an independent affiliate of United Airlines and became known as COVIA Corporation. COVIA's mission is to enter new data processing business ventures and to capitalize on opportunities outside the travel industry through the use of its large worldwide network.

There have been many legal actions taken against United in connection with Apollo. These range from charges of monopolization of the reservation system market to charges of bias in the display and presentation of flight information. Many travel agencies have tried to break contracts with United to join other systems. For the most part, United has been successful in obtaining relief through the courts in these contract disputes. Competing firms have even offered to pay legal fees for agents who elect to break contracts with United in favor of their systems. Competitors are using every means available to combat the advantages of United and its system.

A market value of $1 billion was placed on the Apollo reservation system in 1988 when UAL Corporation sold 49.9 percent of COVIA for $499 million. The buyers were USAir, British Airways, and Swissair at 11.3 percent each, KLM Royal Dutch at 10 percent, and Alitalia at 6 percent. U.S. airlines sold $42 billion in tickets in 1988, 90 percent through travel agents. It is estimated that Sabre and Apollo have annual returns on investment of 50–100 percent.

In Swindon, England, a joint business venture composed of 10 airlines is developing an internationalized reservation system to offer a wide array of services. The joint venture, called Galileo, is owned by Aer Lingus, Alitalia, Austrian Airlines, British Airways, KLM, Olympic Airways, Sabena, Swissair, TAP Air Portugal, and the U.S. company, COVIA. The company plans to include most airlines worldwide, tens of car rental companies, thousands of hotels, major theaters, and sporting arenas. Office management services for travel agents will also be offered. A goal of the system is to give travel agents in Europe one user-friendly terminal to handle all their customers' needs. Clearly computerized reservation systems make up a large, dynamic business today.

Stock Brokerage

In 1977 the brokerage firm of Merrill Lynch announced the development of its Cash Management Account (CMA) at a New York press conference and began test marketing the product in Atlanta, Denver, and Columbus, OH. The CMA product consists of a combination checking account, debit card, and brokerage margin account supported by a computerized cash management system. The system provides customers with current information via phone and detailed printed reports at month end. Customer net cash balances are invested in one or more money market funds generating interest income for the client. Subscribers' expenditures are applied first against their net cash balance and, when depleted, against their lendable equity in the margin account. This innovative product involved an alliance with Banc One of Columbus, OH, which processes the checking activity for the CMA, thus preserving the separation of banking and brokerage as required by law.

In 1978 Merrill Lynch expanded the CMA to 38 offices in five states, and by 1980 the CMA became available in 39 states. By this time the number of accounts had grown to 186,000. To expand account growth, Merrill Lynch launched its first specialized version of the CMA designed for estate administrators. The CMA became available in all 50 states in 1981, by which time the number of accounts had passed the half million mark. The Professional Golfers Association adopted the CMA automatic transfer system in 1981 to manage funds and pay tournament winners. Additional features were under development to expand the range of services available to CMA clients.

The international CMA was launched in 1982, and the Working Capital Management Account, serving the needs of businesses and professional corporations, debuted in 1983. Accessing CMA reserves through cash machines became widespread in 1984. The Capital Builder Account, tailored for the needs of individual investors, became popular in 1985 and 1986. Additional features were under development—and required large investments at that time.

By 1987, the tenth anniversary of the CMA, 1.3 million accounts with $150 billion in assets were actively being served. Merrill Lynch introduced CMA enhancements including increased ATM access (24 hours a day at more than 22,000 locations) and the new CMA Premier Visa program. The Premier Visa program provides financial benefits and administrative features to the client and an additional $25 per year fee to Merrill Lynch per client.

Ten years after its debut, the CMA has been an unqualified success, and Merrill Lynch has derived a major competitive advantage from the program. The minimum balance required to open a CMA is $20,000, but the average balance is close to $100,000 for its 1.3 million clients. The minimum annual fee in 1988 was $65 per client. Fee income is substantially augmented by commissions on securities transactions, interest charges on debit balances, and service fees on the more than $28 billion in Merrill Lynch managed money market funds.

It was not until 1984 and later that the competition begin to employ similar information technology. Merrill Lynch carefully protected its position by securing a U.S. patent on the computer program and by defending its turf in court. In 1983 Merrill Lynch won a $1 million settlement from Dean Witter, its closest rival at that time.

The CMA patent application was filed on July 29, 1980, and was granted on August 24, 1982. The patent lists Thomas E. Musmanno as the inventor and is assigned to Merrill Lynch. It contains four drawings (flow charts) and six claims and is displayed in total on 11 pages. It represents an uncommon but important form of protection for a valuable strategic information asset.

Additional enhancements continue to be announced for the CMA implying continued investments by Merrill Lynch. By 1987, industry analysts estimated Merrill Lynch's technology budget to be nearly $1.5 billion, with several hundred million dollars dedicated to software development. Although Merrill Lynch's share is declining slowly, it holds approximately 50 percent of the market. With the CMA and other services, Merrill Lynch manages a total of $536 billion in client assets. The CMA exemplifies a brilliant combination of information technology and financial services employed for strategic competitive advantage.

But is Merrill Lynch secure in its position? Is it possible for competitors to provide more attractive offerings through the use of information technology?

The answers to these questions must await the market response to competitive offerings from Prudential, Smith Barney Shearson, Dean Witter, Olde & Co., and Fidelity. The Fidelity Ultra Service Account with its Fidelity On-line Xpress (FOX) computerized trading system promises significant competition for Merrill Lynch. FOX appeals to Merrill's CMA clients by offering a wide range of services to investors through their PCs at home. From a PC connected to the system through the phone network, investors are able to establish and manage many accounts, obtain the status of their accounts, access market reports or research information, obtain price information, and place orders to buy or sell securities. Some of the many features of the system are summarized in Table 2.3.

TABLE 2.3 FOX Features

1.	Real-time securities quotations
2.	Standard & Poors research reports
3.	Current financial and business news
4.	Order entry at reduced commissions
5.	Account management and pricing
6.	Portfolio management and tax accounting
7.	24 hour/365 day account access
8.	Password and PIN security features

Fidelity Ultra Service Account clients need a $10,000 opening balance and receive 100 free checks, a Visa Gold debit card, and can trade at a commission savings of 76 percent compared to full-service brokers. In contrast with others, Fidelity charges no annual maintenance fee. FOX allows investors to view their transactions and price their securities on their PCs at home or any other location convenient to the customer.

Using network, database, and personal computer technology now widely available to investors, Fidelity combines a series of strategic thrusts. It offers a differentiated product, utilizes innovative information technology, and provides convenient and timely services. Surely these are strong ingredients for success. But marketing capability, the implementation process, and customer acceptance are the ultimate determinants of success.

Distribution Services

Federal Express began the overnight package delivery business in 1973, processing 100 percent of the packages manually. Now, Federal Express is the leader in the U.S. overnight delivery market yet charges more for its services. Despite significant competition, Federal Express maintains its 40 percent share of the market due in large part to Cosmos, a database system that tracks all letters and packages that the company handles. Through Cosmos, the company can tell customers where their package is in 30 minutes or less, thereby easing customer fears of late delivery or lost parcels.

Much of Federal Express's growth and competitive advantage is due to innovation in the systems area.[14] Nine thousand of 14,000 couriers are online to local dispatching centers through console-mounted CRTs in their vans. The system supporting this capability is called Courier Dispatching. It helps drivers track packages and keeps them informed of schedules and locations while allowing the couriers to pick up new orders quickly.

Federal Express uses bar codes to track packages throughout the delivery system. Within two minutes, the scanned data is uploaded via telecommunications links into a central database where it is available to answer company or customer questions. Misdirected parcels are virtually nonexistent now and will become even rarer as more technology improvements are implemented. Packages are sorted by a scan-recognition system as they move at 500 feet per minute on conveyer belts. Federal Express processes more than 400,000 items in a two-hour window, and they believe these innovations are necessary to maintain their own service standards. Federal Express sorts packages and prepares bills in a paperless office where terminals display all necessary information.

The strategic advantage of Cosmos is that it differentiates Federal Express from its competitors, thereby providing substantial advantage to the company. Additional resources applied toward advanced features in Cosmos will enable the company to handle more volume and offer superior service in the future.

The examples we have studied relate directly to Porter's model of competition and to the theory of competitive thrusts. Telecommunication is a major ingredient of many important strategic systems. Many firms today are using this technology to gain competitive advantage. Some very successful strategic systems are composed mostly of telecommunications elements.[15]

Strategic Systems Update

Hoping to build on Sabre's success, AMR is looking for opportunities in the computer services business, perhaps bringing processing services to the healthcare or retail industries. But, efforts by AMR Information Services, Inc., to build Confirm, a reservation system for Marriott Corporation, ended in failure and the parties are in litigation over damages. AMR continues to make money on Sabre while losing large amounts on its core airline business. If AMR's officers had to choose between selling the airline or Sabre, they would probably sell the airline.

Europe's largest reservation system, Amadeus Global Travel Distribution SA jointly owned by Lufthansa, Air France, and Iberia Air Lines recently purchased System One, the reservation system of Continental Airlines.[16] The acquisition will give Amadeus entry into the U.S. market and will help System One expand globally. Software development for System One will be shared by Continental, Amadeus, and Electronic Data Systems which operates the system.

On another front, Southwest Airlines is using network technology to offer ticketless travel to more than 15,000 passengers daily.[17] When making reservations via phone or over the Internet, customers are given a confirmation code that is exchanged for a boarding pass at the gate. This saves the cost of producing the ticket ($15-$30 according to industry analysts) and eliminates the hassle at the ticket counter. The program took about four months of initial development to determine feasibility and is now being phased in throughout Southwest's system. In order to remain competitive, other national and international airlines are considering similar ticketless-travel systems.

Merrill Lynch continues to forge ahead with its CMA program, adding features and functions to help people manage their finances. In addition to brokerage, checking, Visa cards, and money market accounts, Merrill features direct deposit to CMA, bill paying services, automatic investing under dividend reinvestment plans, and statement coordination among various Merrill accounts. Through a series of subaccounts, the CMA statements provide savings and investment information on IRAs, college tuition accounts, retirement accounts, and others. The CMA is growing into a comprehensive, full-service vehicle for managing a family's financial affairs. Special features are available for trust and business accounts, too.

On February 1, 1995, more than 1,330,000 accounts were active with an average worth of nearly $204,000 each. Annual fees are now $100 per year. But for high-net-worth individuals, Merrill is proceeding with much more extensive services as discussed in the Business Vignette in Chapter 11.

Federal Express continues to expand its systems and automation and its worldwide business. Bar coding, scanning, and cellular links keep vital information current. When a FedEx courier picks up your shipment, he generates and scans the bar code for your package with his hand-held scanner. In the van, the courier inserts the scanner into a Digitally Assisted Dispatch System terminal which beams the information to Cosmos via an earth satellite. On average, various parts of the shipping system scan packages six times before delivery.

Customers can now link to FedEx's systems through FedEx Ship software at their offices. This feature enables customers to order and track shipments and to obtain billing information on-line. In addition to FedEx Ship tracking software, the company is introducing new systems automation for low-volume shippers, better systems supporting service agents as they deal with customers, and artificial intelligence applications to keep planes and vans on schedule during traffic tie-ups or bad weather. FedEx created a Business Logistics Services (BLS) division to engineer innovative logistics and information management solutions for firms that want to outsource their distribution facilities. Laura Ashley, a London-based retail chain, uses FedEx's BLS to supply its 540 stores in 28 countries, all within 24–48 hours of a shipment request.

In 1995, FedEx and its major competitor, UPS, implemented same day or next flight out service in the U.S. FedEx now services 95 percent of the U.S. population. Overseas it serves 191 foreign countries through 160 airports worldwide. The company is establishing an intra-Asian hub at Subic Bay in the Philippines to handle growing volumes in the Pacific Rim countries. FedEx operates 30,900 vans and over 450 aircraft. In 1992–1995, revenues are expected to grow 20 percent, cash flow 43 percent, and net income about 300 percent.[18]

WHERE ARE THE LEVERAGE POINTS?

Information systems theory and the examples of strategic systems implementation that we explored in this chapter validate both the theory and the practice of strategic systems. Some strategic systems are proprietary and are not discussed in the literature. Many other valuable systems, smaller and less obvious than those we discussed, are installed and operating successfully in firms worldwide. Thousands of firms own and operate information systems that provide advantage for them in their competitive spheres of operation. Thousands more are investing in strategic systems designed to capture similar advantages for their firms.

Opportunities

As the examples illustrate, there are numerous opportunities for leveraging the investment in information systems. These opportunities frequently consist of product or service offerings which have been differentiated from their competitors by the application of information technology. All the systems we studied exemplify this principle. The technology provides elements of superiority in the minds of the firm's customers through perceptions of high-quality products or services. Thus, technology applications inspire confidence and promote customer loyalty.d

Customer loyalty is achieved in other ways as well. Effective application of information technology can lower costs to the producer and to the consumer, thus improving the cost effectiveness of both. Significantly enriched services command higher prices; customers are willing to pay for higher quality. The leverage points are exploited for growth of revenue and profit for the producer of the product or service.

The examples illustrate how innovative application of information technology can be exploited to create value for the consumer. The development of innovative new products can create new markets. Demand is developed and satisfied to the benefit of producer and consumer. Some new products or services require the collaboration of several firms. Alliances must sometimes be formed to develop, distribute, or market the product or service and to capitalize on the technology most effectively.

Strategic systems are focused not only on the customer, but on supplier targets and competitors as well. Organizations that supply raw materials to the firm are affected by sophisticated systems that optimize buying strategies. And competitors feel the influence of systems used to design, develop, or manufacture superior competing products. Many firms use information technology to support or mold strategic thrusts. Effective utilization of information technology in the strategic competitive environment demonstrates high potential in many contemporary organizations.

The Importance of Technology

Technological advances and their introduction into business and industry worldwide constitute two of the principal drivers of competition. Advances in technology shape the products and services of the future and offer opportunity for innovative organizations to increase their value to the stream of economic activity. We have only to consider the advances in communication or transportation, or observe the revolution in chemicals or pharmaceuticals, or fathom the significance of information processing, to realize the importance of technological advances on international competition.

Information technology is particularly important because it is pervasive in the processes leading to advances in most other activities. All activities create, transport, disseminate, or use information. Advances in IT have a compounding effect on technological advances everywhere. Accordingly, information technology is exerting a remarkable and profound effect on worldwide competition.[19]

Advancing technology is important because it alters industry structure, shapes and molds competitive forces within and between companies and industries, and thereby changes forever the behavior patterns of billions of individuals. Technology for its own sake is not important. What is important is its dramatic impact on society as a whole.

The Time Dimension

Time is an asset and a source of competitive advantage. Firms must think about time resources as much as they do about capital, facilities, materials, technology, and management resources. The firm that views time as a valuable asset will strive to maximize this resource to its internal and external advantage.

The driving force behind the notion of time as a competitive factor is simply response to customers, markets, and changing market conditions. Responsiveness cuts across all functions of the firm, from product-requirements definition to installation and service. Timeliness involves suppliers and customers and impacts competitors. Thus time provides opportunities for competitive advantage. Within the firm, time is important in planning, implementing, and controlling; time factors are highly susceptible to manipulation with information technology. Although there are many examples, the notion of just-in-time (JIT) manufacturing, illustrates the value of the time dimension.

In Irvine, California, the McDonnell Douglas Computer Systems Company, a division of McDonnell Douglas Corporation, used the JIT approach to computer-coordinate the flow of parts and raw materials. The computer system required 111 new programs and 97 modifications to installed programs, for a total of 1900 person-hours of programming additions and changes.[20] Additional time was expended in planning, team meetings, and training, and some expenses were incurred for tags, labels, and other items. A thorough understanding of the flow of parts, products, and information through the company formed the basis for constructing the JIT approach. McDonnell Douglas reported impressive benefits from using the JIT approach. The resulting improvements are summarized in Table 2.4.

TABLE 2.4 JIT Benefits at McDonnell Douglas

Inventory reduction	38%
Work in process inventory reduction	40%
Printed circuit board assembly cycle time	80%
Rework reduction	40%
Improvement in inventory turns	100%
Quality-control process yield (now 99% perfect)	80%
Setup time reductions	50%

Using information technology in conjunction with parts-logistics control, attacks on setup time, design process improvements, and supplier logistics yielded substantial savings in time and money for this manufacturer of computer products. Other computer manufacturers agree. "We've ignored a critical success factor: speed," says John Young, CEO, Hewlett-Packard Corporation. "Our competitors abroad have turned new technologies into new products and processes more rapidly. And they've reaped the commercial rewards of the time-to-market race."[21] And in speaking about IBM, Louis

Gerstner, CEO, says that "One of the unusual things about this industry is that a disproportionate amount of the economic value occurs in the early stages of a product's life. That's when the margins are most significant. So there is real value to speed, to being first—perhaps more than in any other industry."[22]

The Strategic Value of Networks

Telecommunications technology speeds the flow of information between organizational entities bridging the gap in space and time. Given the pervasive role of information in business, telecommunications offers great potential for competitive advantage through reductions in time and mitigation of distance barriers. Multinational firms employ this technology as a condition of remaining in business, and they are continually searching for innovative applications.

Hammer and Mangurian developed a useful framework for visualizing the impact and value of communications technology. Figure 2.2 presents their framework.[23] By reducing the elapsed time of business processes, information in transit (float) is reduced and opportunities to improve business excellence are revealed. Errors are reduced and costs are lowered. Time-consuming processes reduce profitability and impair service while reducing delay times improves business efficiency. Information technology links between operating centers permit economies of scale and promptly disperse important business knowledge to where it is needed most. Business effectiveness and efficiency are increased as distance barriers are overcome.

Information technology is also a significant force for organizational change in business enterprises. It increases the span of communication and control thus permitting reductions in the levels of management. The need for filtering layers of middle managers is reduced. Electronic information exchange between organizations enhances management control and fosters closer ties to customers, suppliers, and business partners. Information technology permits alternative, restructured forms of business enterprise that respond effectively to competitive threats and help gain competitive advantage. IT is tremendously important in today's highly competitive global economy.

In addition to the strategic systems discussed earlier, E-mail, ATM machines, toll-free phone numbers, the Internet, Prodigy, and CNN are obvious examples of the impact of telecommunications technology. Additional examples will be developed throughout this text.

		VALUE		
		Efficiency	**Effectiveness**	**Innovation**
IMPACT	**Time**	Accelerate business process	Reduce information float	Create service excellence
	Geography	Recapture scale	Ensure global management control	Penetrate new markets
	Relationships	Bypass intermediaries	Replicate scarce knowledge	Build umbilical cords

Michael Hammer and Glenn Mangurian, "The Changing Value of Communications Technology."

FIGURE 2.2 Impact/Value Matrix

THE CHARACTERISTICS OF SIS

The strategic information systems studied in this text, and many which were not analyzed, have some common characteristics worth noting.

Organization and Environment

Strategic systems may alter the market environment and change the field in which the competitors are engaged. In the process, these systems will probably alter the internal environment or change the organizational structure. As the strategic system becomes effective, it influences the environment. In turn, the firm modifies its structure to accommodate the changed environment and the system itself. The metamorphosis of the competitive situation and the firm's response continues unless some form of stability becomes established among the competitors.

Financial Implications

Strategic systems require continued investments of resources to sustain their advantage. The owners of these systems find that competitors are always looking for ways to build systems for themselves in order to negate the owner's advantage. As the examples revealed, it is a challenging and never-ending task to stay ahead of the competition. Relatively secure positions in the brokerage industry, for example, are subject to threat by nimble, innovative competitors.

Some strategic systems produce revenue by themselves and become profit centers within the parent corporation. In the extreme case, the profit center grows into a major international business entity in its own right as the evolution of the COVIA corporation illustrates.

Additional Considerations

The owners of some successful strategic systems eventually become engaged in legal struggles with the competition. The legal questions involve competitive issues surrounding the appropriate use of the owner's business advantage. Some legal issues relate to protection of the competitive advantage. Over the years, the owners of major airline reservation systems have been in the courts litigating issues arising from the advantages derived from their systems.

THE STRATEGIST LOOKS INWARD

Many important systems originated as the idea of an individual or a small group who envisioned a way to capitalize on the emerging technology or to streamline some aspect of the firm. These individual or group insights became the catalyst that spawned the business opportunity.

Most strategic systems developed from analyzing the firm's internal functions. Efforts to automate (informate) the internal activities of the company more fully paid off later in competitive advantage. The Sabre system began life this way. A logical starting point to search for strategic opportunities is the firm's portfolio of application systems.

Many firms own application portfolios consisting of several thousand programs. Each of these programs is associated with and supports one or more business processes. Some of these applications may be candidates for strategic development. The thrusts that these applications may potentially contain range widely from cost reduction to innovative methods for attaining product or process superiority. These internal systems are found in marketing, development, manufacturing, sales, service, and administration. Superior market analysis tools coupled with automated design systems can significantly reduce the time and cost required to respond to changing customer requirements and market conditions. Automated manufacturing processes and sophisticated distribution systems speed the new products to the customer efficiently at reduced cost.

Information systems to help handle these basic tasks exist in most firms today. How can these systems be augmented or enhanced to improve the firm's posture in the marketplace? What new technology can be employed to improve these processes? What innovative actions will permit the firm to utilize internal resources to maximize its competitive position? These and other questions directed toward current applications form a basis from which to search for strategic opportunities internally.

EXTERNAL STRATEGIC THRUSTS

Another view useful in identifying potential strategic opportunities considers external factors. These factors include changing industry environment, recent actions of competitors, changing relations among suppliers, and potential business combinations among others. This view of the firm asks the questions: "What is happening external to the firm that may influence our opportunities to gain competitive advantage?" and "How can we capitalize on external factors through the use of information technology?"

The process used to answer these questions is different and more complicated than that used to view the firm introspectively. The individuals best suited to this task are found in the top positions in the company. They may be located in the marketing function responsible for competitive analysis or industry analysis, or they may be in product distribution with responsibility for ensuring timely and accurate dissemination of the firm's goods or services. The key senior people responsible for guiding the firm's long-term direction will have valuable insights. Their vision must be influenced by the information technologists who have their own vision of what is possible and feasible technologically. Potential opportunities will emerge from the intersection of these two visions.

Several cautions about strategic information systems are in order.[24] Strategic systems usually develop from deliberate attempts to improve or enhance current information systems. They do not begin life as separate and distinct systems from those in the applications portfolio. Most strategic systems do not emerge from specific attempts to meet corporate strategic objectives. Instead, they are the result of many incremental enhancements and sustained improvements to current systems: They do not evolve from radical changes to operational systems or from totally new systems.

Successful executives focus on improving corporate performance through constant attention to the many details of their businesses. They search for improvement through new technology, enhancements to systems and operational procedures, and modifications to the organization and its culture. The task of getting ahead of competition and of staying ahead is difficult. It requires adaptation to environmental changes and response to competitive forces. Successful executives know they cannot attain lasting competitive advantage from a few grand strokes.

INTEGRATING THE STRATEGIC VISION

Building on a model presented by Wiseman and MacMillan, the interrelationships of the strategic variables can be depicted graphically.[25] Figure 2.3 displays these relationships and illustrates the internal and external sources and uses of strategic systems. Time, the highly important strategic thrust, complements the other five thrusts in this model.

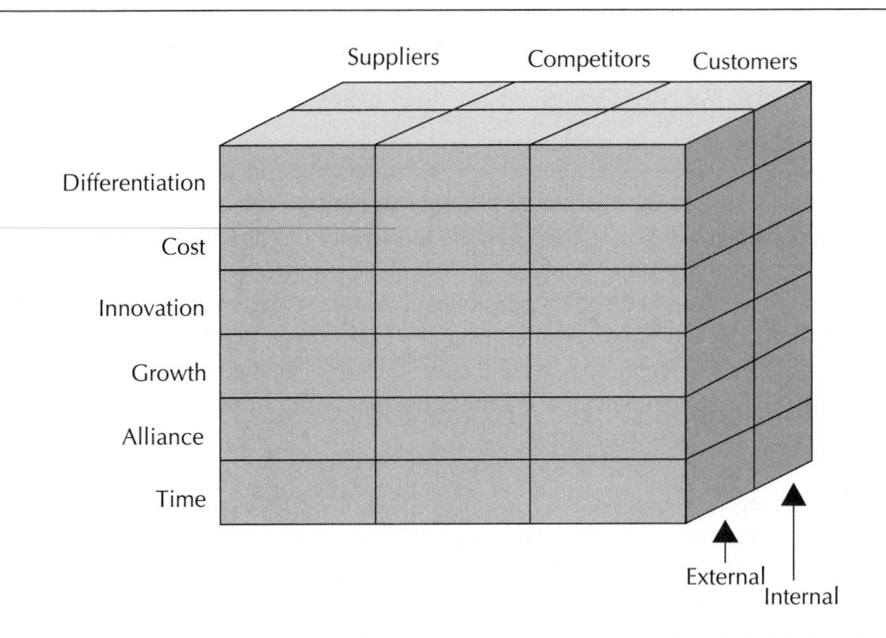

FIGURE 2.3 Integrated Model of Strategic Influences

Most strategic systems cover several areas in Figure 2.3. A system may utilize several of the strategic thrusts itemized down the left side of the diagram, or it may impact more than one of the groups illustrated across the top. For example, cost reductions and time savings obtained internally from the application of innovative technology may lead to growth in both revenue and profits affecting customers and competitors. Computer-aided design/computer-aided manufacturing systems have this potential. In the computer industry itself, innovative use of advanced technology in sophisticated electronic design automation systems enables firms to spawn new, high-quality products in ever shorter development life cycles. Such innovation significantly affects competitors and customers alike.

Systems change character over time as the firm exploits their potential. Some systems start life as an internal thrust and gain external importance to the firm. The Sabre system is one example. Throughout the life cycle of a strategic system, it may migrate through the model. It may play an important role for the firm in several arenas as its potential is developed via redirection and further investment. Change is a fact of competitive life, and successful strategic systems management involves anticipating, initiating, or reacting to change to obtain or sustain competitive advantage for the firm. Management of strategic systems is a major challenge for senior executives in nearly all organizations today.

SUMMARY

Information technology is becoming absolutely vital to the success of most modern-day firms in the industrialized world. Senior executives have many valid reasons to expect that information technology can provide advantage for their firms. They believe that innovative use of information technology promises improved corporate or organizational leadership. They are prepared to invest in the technology, and they expect their firms to attain a substantial return on this investment. Many realize this investment is not optional, but is required for the future prosperity of the organization. They also realize it is far easier to recognize strategic systems in action than it is to identify opportunities and capitalize on them.

Review Questions

1. Why are strategic concerns regarding information systems increasingly important today?
2. What distinguishes strategic information systems from other kinds of systems?
3. Explain Porter's model of forces governing competition.
4. What are the six basic strategic thrusts?
5. What are some examples of transaction processing systems that have strategic value to their owners?
6. What thrusts are employed by American Airlines in its reservation system?
7. As a competitive weapon, what distinguishes United's Apollo system from American's Sabre system?
8. In what ways did Merrill Lynch's CMA alter the environment in which brokerage firms were operating in 1977?
9. What protection is afforded Merrill Lynch as a result of the patent on the CMA program?
10. Using the information provided in the text and the model in Figure 2.2, trace the evolution of the Apollo system.
11. How would you describe the Cosmos system in terms of the Porter model?
12. What leverage can information technology provide business organizations?
13. What are some common characteristics of strategic systems?
14. What are the advantages and disadvantages of searching internally for sources of strategic systems?
15. What insights can the IT organization bring to the process of searching for strategic systems?

Discussion Questions

1. What is the main thrust of the strategy that Microsoft has been pursuing?
2. What appear to be the strengths of this strategy? Do you think Microsoft's strategy is proactive, or reactive, and why?
3. Discuss Moore's law in connection with Microsoft's strategy. What are the implications of Moore's law for IT and other functional managers in most firms today?
4. Research reveals that office automation has been declining in importance as an issue for IT executives. What might this mean in terms of strategic systems?
5. Discuss Porter's model as it applies to the computer industry today. How is the telecommunications industry different from the computer industry as viewed from the model?
6. The six strategic thrusts we presented are not mutually exclusive. Discuss the implications of this fact.
7. Deregulation of the airline industry has encouraged new entrants into the industry. Given the enormous advantage to the competitor who owns a reservation system, how can new entrants overcome the barriers to entry?
8. Discuss the role that IT managers play within the firm as the firm seeks to improve its competitive posture. What contributions can they make, and in what areas must they take the lead?
9. What additional actions might be taken by Merrill Lynch to capitalize further on its current position?
10. Discuss the importance of telecommunications to the strategy of Federal Express.
11. How might firms seeking international competitive advantage rely on information technology? What thrusts might they employ?

Assignments

1. Using the Wiseman book or another reference, select an additional example of a strategic system, and prepare an analysis of its characteristics. Be prepared to summarize your findings for the class.
2. Study the *Harvard Business Review* article by George Stalk, Jr., listed in the Bibliography, and prepare a report on why he thinks time is an important source of competitive advantage.

ENDNOTES

[1] *Value Line*, Dec. 9, 1994, 2122.

[2] Walter B. Wriston, *The Twilight of Sovereignty* (New York: Charles Scribner's Sons, 1991), 123.

[3] An information architecture describes the interconnections or configuration of information technology components such as hardware, software, and databases.

[4] Charles Wiseman, *Strategic Information Systems* (Homewood, IL: Irwin, 1988), 98.

[5] Peter G. W. Keen, *Every Manager's Guide To Information Technology* (Boston, MA: Harvard Business School Press, 1991), 74.

[6] Additional information on Porter's theories of competition can be found in the references at the end of this book. For a complete discussion see Michael E. Porter, *Competitive Advantage* (New York: Free Press, 1985).

[7] Charles Wiseman, *Strategic Information Systems* (Homewood, IL: Irwin, 1988).

[8] "Speed to Market," *Forbes*, May 28, 1990, 350. This special report to management by the Yankee Group presents a brief, succinct description of the strategic advantages derived from the time thrust. It contains six examples of firms which use or provide time advantages. Also, Roy Merrills, "How Northern Telecom Competes on Time," *Harvard Business Review*, July-August, 1989, 108.

[9] John Killkirk, "Firms Learn That Quick Development Means Big Profits," *USA Today*, November 22, 1989, 10B. Chrysler can create a 3-D computer model of a car body in about 10 seconds. It takes three to four years and costs Chrysler about $1 billion to bring a car to market, and computer modeling can reduce this time by up to one year.

[10] Peter Keen, "Vision and Revision," *CIO*, January/February, 1989, 9.

[11] Israel Borovits and Seev Neumann, "Airline Management and Information System at Arkia Israeli Airlines," *MIS Quarterly*, March 1988, 127. Using super minis and micros, this small airline built a high-function reservation system for only about $250,000!

[12] Dennis J. H. Kraft, Tae H. Oum, and Michael W. Tretheway, "Airline Seat Management," *Logistics and Transportation Review*, June, 1986, 115. American also analyzes yield management data on no-shows to overbook optimally. In 1988, American denied boarding to fewer passengers than any other of the nation's 12 largest airlines, according to an American Airlines newsletter.

[13] Alan J. Ryan, "Delta Lands Another CRS Partnership," *Computerworld*, February 12, 1990, 19.

[14] Federal Express more than tripled its revenue from 1980 to 1984, more than doubled it from 1984 to 1987, and more than doubled it again from 1987 to 1990. According to *Value Line*, March 24, 1995, Federal Express is expected to earn $355 million in 1996.

[15] W. Frank Cobbin, Jr., et al., "Establishing Telemarketing Leadership Through Information Management: Creative Concepts At AT&T American Transtech," *MIS Quarterly*, September, 1989, 361. This is an extremely interesting example of a strategic telecommunication system.

[16] "Amadeus Acquires Reservations System From Continental Air," *The Wall Street Journal*, April 28, 1995, B5.

[17] Ron Levine, "Southwest Soars With Ticketless Travel," *LAN Times*, April 24, 1995, 31.

[18] *Value Line*, December 23, 1994, 259.

[19] Michael Porter, "Technology And Competitive Advantage," *Journal Of Business Strategy*, Winter, 1985, 60.

[20] Lad Kuzela, "Efficiency—Just In Time," *Industry Week*, May 2, 1988, 2.

[21] "How Managers Can Succeed Through Speed," *Fortune*, February 13, 1989, 54.

[22] 1993 IBM Annual Report, 3.

[23] Michael Hammer and Glenn Mangurian, "The Changing Value of Communications Technology," *Sloan Management Review*, Winter 1987, 65.

[24] James C. Emery, "Misconceptions About Strategic Information Systems," *MIS Quarterly*, June 1990, vii.

[25] Charles Wiseman and Ian C. MacMillan, "Creating Competitive Weapons From Information Systems," *Journal of Business Strategy*," Fall, 1984, 42.

3 Developing the Firm's IT Strategy

A Business Vignette

Tough Newcomer—AT&T Strategy†

Three short years ago Alex Mandl was running a seemingly low-tech shipping company, Sea-Land Services Inc. Today he is a rising star at AT&T Corporation, overseeing the vast and most sophisticated communications network in the world. Employing a combination of workaholism and internal diplomacy, he has advanced from CFO to the most coveted slot next to chairman Allen—running the core long-distance group that churns out $42 billion in annual revenue. Mr. Mandl quickly forged a reputation for an obsessive focus on growth, relentless cost-cutting, split-second decision making and a brusque impatience with time-wasters.

Now Mandl faces the biggest challenge of his career: steering AT&T's flagship business into the 21st century amid tumultuous change in the telecommunications industry.[1] In defensive response, smaller rivals such as MCI and Sprint have lined up alliances with giant foreign backers. AT&T's offspring, the seven regional Bell phone companies, seem increasingly likely to win entry into the long-distance business. Legions of new wireless rivals loom on the horizon, and the advent of the "information highway" could bring myriad newcomers onto AT&T's telecom turf.

While colleagues viewed decreases in AT&T's onetime 100 percent share of the long-distance market as inevitable, Mandl has insisted on winning back customers. His lieutenants remember his first orders: "The erosion has to stop. Start taking back market share." When AT&T first approached McCaw Cellular Communications Inc., it proposed buying only a piece of the cellular-phone outfit to avoid arousing regulatory ire. Mandl successfully pushed to swallow the whole thing—an $11.5 billion bite.

Mr. Mandl's mantra comes down to this: "AT&T is the information superhighway," he declares. Forget the spate of unprofitable multimedia bets on everything from Hobbit chips to Eo communications. Mandl argues that the way for AT&T to thrive on future multimedia services—voice, data and video—is to exploit what no competitor has: its own global network.

† "Tough Newcomer," *The Wall Street Journal*, December 16, 1994, 1. Reprinted by permission of the *Wall Street Journal.*

AT&T's far-reaching tentacles hold 134 major switching centers, millions of miles of fiber-optic lines, thousands of miles of undersea cables, and an infrastructure of sophisticated billing systems and traffic-monitoring gear. Load up this speed-of-light highway with traffic from new kinds of customers, even rivals, Mandl argues, and AT&T can grab for growth rather than settle for stasis.

In this pursuit of what he calls "fat minutes," the AT&T network will carry more than routine calls to Grandma: computer files zapped among company facilities, video calls between lovers on different continents, joystick jocks in different states challenging each other on a game network, and thousands of other forms of digital fare.

The Mandl plan seems to have galvanized AT&T and surprised industry veterans with its clarity and simplicity. Most of them had taken AT&T's internal highway for granted.

The strategy contrasts sharply with the old AT&T, loath to let outsiders tap into its network and hobbled by a "not-invented-here" attitude. Now many of AT&T's new services will come via an ever-mushrooming number of alliances. Lotus Development Corporation, Xerox, Novell, and Intel have agreed to put their data and video traffic aboard AT&T's network.

More partnerships are in the works, producing "fat, sticky, addictive bits," says James Cosgrove, Mandl's multimedia-services manager. By 1997, AT&T hopes that the Lotus deal alone could add 10,000 corporate clients and $2 billion a year in revenue. Hewing to the old way of shunning such alliances "didn't make any sense," Mandl says, "when companies like Lotus and Intel already are huge players."

For the first time since 1984, AT&T is taking back market share from MCI and other rivals. Mandl's unit has more than tripled its growth rate to 8 percent fourth quarter 1994 from a year ago. The growth of long-distance usage at MCI, which had been 15.6 percent—three times as much as AT&T—has slowed to single digits. Sprint confirmed that its fourth-quarter 1994 traffic and earnings have fallen short of expectations.

Largely due to the "True Savings" program which gives callers discounts of 20 percent or better, AT&T has captured about one million new customers in seven months. Mr. Mandl expresses no reluctance to pound on a company one-sixth AT&T's size. "You cannot grow the business by standing still," he says. "You have to take business back."

In his first year at AT&T, Mandl chopped $1 billion in annual spending and imposed strict rules requiring that AT&T's business units earn enough profits to cover their own capital spending, marketing, taxes, and financing—or get axed. He also put in place a new plan tying managers' bonuses to escalating profit targets.

Still, it remains to be seen whether or not AT&T's latest gains are illusory. "Right now all they've done is put a Band-Aid on things. It's still just a discount plan," says Brian Adamik of the Yankee Group, a Boston research firm. MCI will respond soon and has hurt AT&T before, he says.

In 1995, AT&T is expected to earn $5.6 billion on revenues of $76.48 billion.[2]

INTRODUCTION

The process of searching for an area in which a strategic system may be developed is an important endeavor for the firm, and it requires considerable thought and research. However, strategic thinking covering all areas of importance to the firm is mandatory. If the firm is going to exploit information technology or other strategic opportunities for its own benefit, the firm's senior executives must apply a systematic, strategic vision in order to set organizational direction. The strategy must focus and coordinate the activity of the firm. A well-developed strategy ensures consistency of direction among the firm's units, and reduces uncertainty in shorter range decision making.[3]

If IT managers are to be successful in most firms today, consistent strategic planning must be a part of their routine business system. Strategy development promotes actions leading to sound planning, encourages organizational learning, and provides a process to ensure goal congruence. Elements of successful strategic planning include sound preparation and implementation, line manager involvement, correct definition of business units, action steps outlined in detail, and controls integrated into the plan.[4] These objectives are a critical success factor for IT managers. Strategic planning takes organization structure into account. Deciding whether to centralize or decentralize, for example, can be handled by thoughtful planning. This chapter develops the processes and techniques essential to successful information technology strategy development and strategic planning.

CONSIDERATIONS IN STRATEGY DEVELOPMENT

In the process of strategy development, thoughtful IT managers will perceive potential business opportunities and possible threats or pitfalls. To the extent possible, successful IT managers will attempt to redirect their efforts to maximize their advantages. The resulting proposed course of action may not capitalize on all opportunities available nor avoid every difficulty; but if optimally conceived, it will present a balanced approach to these conflicting influences. Since situations rarely remain static for a sustained period of time, thoughtful managers must reevaluate their situations periodically. And they must reassess and adjust their course of action in light of changing conditions.

The complete strategy expression must consider the elements discussed above, among others. Table 3.1 displays the ingredients of a strategy statement. The strategy statement sets forth objectives to be achieved and includes processes for its maintenance and use.

TABLE 3.1 Elements of a Strategy

1. Mission statement
2. Environmental assessment
3. Statement of objectives
4. Expression of strategy
5. Maintenance processes
6. Performance assessment

The first and most important step in formulating a strategy is to state the organization's mission and purpose. Developing a mission statement for the IT organization may be a challenging task, particularly on the first attempt. The mission statement should describe the organization's business and identify the customer base. The customers' needs and the organization's capabilities and resources must be considered. The IT mission should be stated in terms of meeting the needs of the customer and in terms of services the IT organization can provide. It is essential to define the IS markets within the firm itself.[5] In other words, the IT mission statement should be both reactive to customers and proactive toward customers. For example, the mission must include performing the transaction processing for the firm; but it must also include introducing new information technology for important new applications.

The process of visualizing and understanding the opportunities and the threats is called "analysis of the environment." Environmental analysis (or environmental scanning) attempts to account for the important trends impacting or likely to impact the organization. These trends may be political, economic, legal, technological, or organizational. The strategists must understand these trends in order to position their organizations optimally. Environmental scanning generates information that enables strategists to accomplish two tasks: to develop the objectives they intend to achieve, and to formulate the course of action, or strategy expression, that will guide them in achieving the objectives. The strategy will rely on assumptions where facts are unavailable, and it will account for the nature and degree of the risks involved in attaining the objectives. The strategy must include some degree of flexibility through the inclusion of options or alternatives. The statement of strategy includes the ingredients that lead to further planning and decision making.

Strategy maintenance is the process of reviewing the environment and reassessing the course of action in light of changing events. The extent to which strategy maintenance proceeds results directly from the environment's volatility. Businesses in relatively stable environments may experience long-lived strategies requiring infrequent maintenance activity. However, most firms and IT managers in today's economic and political environment find frequent strategy maintenance to be the norm.

The goals, the objectives, and the strategy statement form a useful document for the strategist to use in assessing the efficiency and effectiveness of the organization. Thus, after the fact, a strategy statement can also be useful in assessing organizational performance and evaluating the degree to which the organization achieved its goals and objectives.

Figure 3.1 portrays the relationship of these activities to each other and to strategic and operating plans. The planning activity will be discussed in the next chapter.

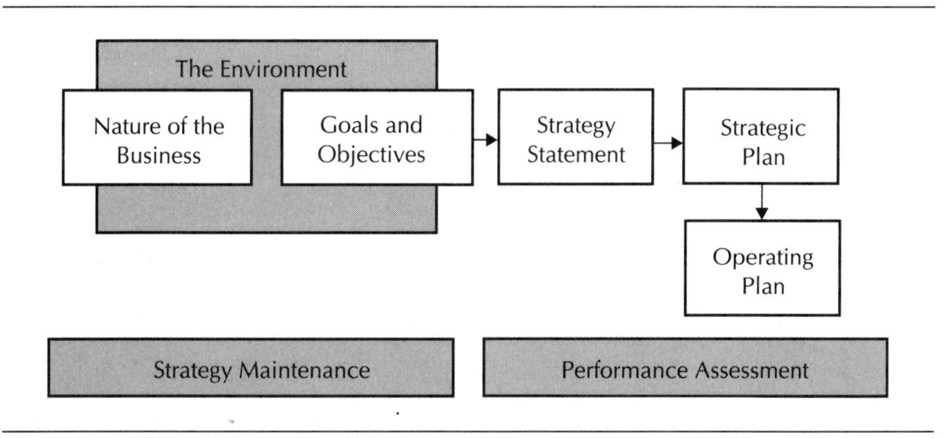

FIGURE 3.1 The Relationship of Strategy and Planning Elements

The basic elements essential to a complete strategy and planning process are the environmental assessment and nature of the business, goals and objectives to be achieved, strategy statement, maintenance activity, and performance assessment. The development of the strategic and operational plans flows from the strategy activity. Figure 3.1 suggests the relationship of these items to each other and to the maintenance and assessment activities.

A statement of strategy can be useful for an individual or for a group. As we consider strategies for groups, we need to be concerned with developing formats and procedures to improve communication among members of the group. Good communication increases the group's ability to develop and use strategies effectively. Solutions to the issues of information architecture, IS's role and contribution, and the use of information systems to integrate across functional lines hinge on the communication of strategic thrusts necessary to achieve strategy congruence within the firm. Succeeding sections of this chapter will deal with these aspects of strategy formulation.

STRATEGIES AND PLANS

The strategic plan for a firm combines the strategies for its business areas and for all of its functions. The combined detail supports its overall business strategic objectives. The strategic plan calls for detailed action during the next year or two. The detailed actions comprise the short-range or tactical plan. Therefore, business and functional strategies guide and coordinate the activities of the various functional and organizational groups toward agreed-upon objectives. This guidance and coordination is expressed in statements of strategy, strategic plans, and tactical and operational plans.

A strategy forms the basis for a plan, but it is not a plan in and of itself. The strategy spells out optimum actions required to achieve general objectives, but it does not provide detail sufficient to carry out the actions. Adding details to the strategy sufficient for implementation by the organization creates a plan from a strategy. This is true whether the strategy encompasses a two-year or a twenty-year period. Therefore, the task of planning is substantially different from the act of strategy development. One deals with stating optimal actions necessary to attain goals and objectives in the face of uncertainty; the other transforms the optimal actions into implementation tasks.

The processes of strategy development and strategic planning are essential to strategic management. Businesses have advanced their thinking about strategy from predicting the future to trying to create the future. Several phases of this development can be recognized. Phase 1 consists of basic financial planning which evolves to forecast-based planning in Phase 2. Phase 3 consists of externally oriented planning characterized by increased response to competitive pressures and evaluation of strategic alternatives. In turn, these form the basis for Phase 4, strategic management.[6] Strategic management means reorganizing and redeploying resources to attain competitive advantage. It also means implementing creative and flexible long- and short-range planning systems supporting resource deployment actions. Information

Technology managers play a vital role in this process because the technology they manage has considerable potential for initiating future change.

> According to George Sawyer, the plan for the management of opportunity deals somewhat with the present but more with the future because in the time required to fulfill any of its plans the business will have moved into that future. The first overview specification is for an attempt to project the evolution of this future, but management is not limited to this projection. Drucker speaks of 'creating the future,' meaning that management has a very real power to design the future and to make it come true.[7]

The fundamentals of planning for information technology will be developed in much greater detail in Chapter 4.

TYPES OF STRATEGY

Strategy expressions can be developed as stand-alone statements or as business or functional statements. The complete strategy statement for a function within the firm is the collection of stand-alone statements assembled with the functional statement. For instance, in the IT function, stand-alone strategies for improving office systems and for upgrading the mainframe computer form a part of the much larger strategy for the IT organization.

The statement of strategy for an IT function must contain the basic ingredients shown in Table 3.2. Although these elements are specifically applicable to the IT function, most of them apply to any function in the firm.

TABLE 3.2 Basic Elements of an IT Functional Strategy

Support to business objectives
Technical support
Organizational considerations
Budget and financial matters
Personnel considerations

A statement of strategy for a firm includes an overall long-range strategy to achieve the firm's objectives and more detailed strategies for each business and functional area. These detailed strategies support the firm's strategy. Thus, functional strategies will be developed for sales, marketing, manufacturing, IT, and other functions. Typically, the information technology functional strategy is assembled with strategies from other areas of the firm.

Not all firms have adopted the strategic process described above. The IT organization must work within the norms of the larger organization and it must account for the corporate culture. If the senior managers of the firm are committed to strategic development, middle and lower level managers will see the process as valid. They will be willing to invest their time in these activities. If the firm's planning process is forecast-based, then it will be difficult, perhaps futile, for the IT organization to shape the future in this firm.

Figure 3.2 illustrates the relationship of functional strategies to the business strategy for the firm. In a large firm, there will be more strategies than this diagram depicts.

THE FIRM'S BUSINESS STRATEGY		
IT Functional Strategy	Product Manager Business Strategy	Other Functional Strategies

FIGURE 3.2 The Firm's Business Strategy

The purpose of functional strategies is to support the firm's business goals and objectives. This is accomplished through development of functional goals and objectives that are congruent with those of the firm. Effective strategies provide internal consistency between IT goals and the firm's goals.[8] The alignment of IT goals with the firm's goals, or goal congruence, is a critical success factor for the IT organization. The management processes developed in this chapter are aimed at achieving this alignment.

For example, product managers in a firm are responsible for producing revenue, and their business strategy is part of the business strategy for the firm. Each function in the firm develops a functional strategy. The collection of all the functional strategies and business strategies describes the complete strategy for the firm. This detailed document outlines the strategic goals and objectives for the firm and shows how all the units within the firm contribute to the attainment of these objectives. The firm's overall business strategy must be premised on goal congruence in order for it to succeed.

In a large and complex business, the strategy is also complex and probably confidential. The strategy contains sensitive information relating to competition which the firm must carefully guard. Strategy documents are produced and used by people in the top echelons of the business who are responsible for guiding and directing the firm in the long term. In some firms, a special department is in charge of the strategy development process.

The personnel in this department assist the senior executives in utilizing the strategy. Strategy and plan congruence are critical success factors for executives in well-managed organizations!

Stand-alone Strategies

Occasionally, it is desirable or mandatory for a functional organization to develop a specific strategy for dealing with a unique opportunity or threat. Generally, the subject of this kind of strategy is an embryonic question of key potential significance to the function—one which has not been previously considered in detail. In many cases, these specific or stand-alone strategies will be developed outside the normal planning cycle in response to competitive or industry developments. In this sense, stand-alone strategies can be considered *ad hoc* actions to deal with currently emerging opportunities or threats. Once the firm accepts this strategy, it will be incorporated into the strategic plans of all the organizational units that it affects.

An example of a situation that might give rise to a stand-alone strategy for the IT function would be a vendor announcement of a new technology product. This new product might enable the function to further some organizational objective in the short term. In this case, a strategy would be developed to capitalize on the new opportunity. If accepted by the firm, the appropriate planning would begin.

Stand-alone strategies generally cease to exist as the subjects they deal with mature and become more specifically recognized in strategic plans of the firm or its constituent parts.

Business and Functional Strategies

Business and functional strategies form the backbone of the strategy development process for the firm. Each type of strategy responds to different kinds of opportunities or objectives. Business strategies have revenue and profit objectives for the firm. Mr. Mandl's strategy for maximizing AT&T's long-distance market share is an example of a business strategy. Functional strategies have goals that support the firm's business strategies. Unless the IT organization is revenue producing, and some are, its strategies will be limited to the functional type. That is, the IT function's strategy is in support of the firm's business objectives. If the IT function produces revenue for the firm, its strategy will also be part of the firm's business strategy.

For example, a business strategy for an IT organization that hopes to produce revenue could explore all aspects of exploiting an application in its portfolio. By making the program available to customers through new telecommunications technology, revenue and profit can be realized for the firm. A business strategy details how the organization hopes to accomplish these objectives.

In contrast, a functional strategy could address the introduction of new robotic systems in support of the manufacturing plant's cost and quality objectives. This item would appear in the technical support section of the functional strategy shown in Table 3.2.

Business strategies for the firm coordinate the functions of a business to the business objectives. Functional strategies, on the other hand, coordinate activities within or between functions. As an example, a functional strategy within the IT organization may be designed to support and augment the firm's business strategy for reducing product development cycles and increasing revenue in the near term. This functional strategy may contain many other supporting elements, as well.

The relationship of business strategies, functional strategies, and stand-alone strategies is shown in Figure 3.3. This illustration, emphasizing the IT organization, includes some revenue-producing activities. These are contained in the business portion of the strategy statement.

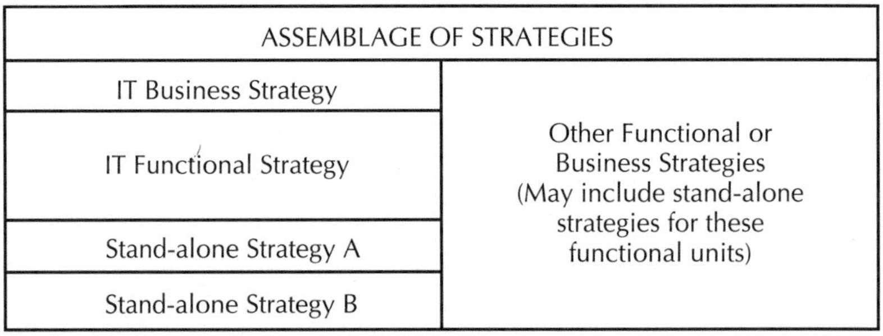

<table>
<tr><th colspan="2">ASSEMBLAGE OF STRATEGIES</th></tr>
<tr><td>IT Business Strategy</td><td rowspan="4">Other Functional or Business Strategies (May include stand-alone strategies for these functional units)</td></tr>
<tr><td>IT Functional Strategy</td></tr>
<tr><td>Stand-alone Strategy A</td></tr>
<tr><td>Stand-alone Strategy B</td></tr>
</table>

FIGURE 3.3 Assemblage of Strategies in a Firm

Well-managed IT organizations have an active functional strategy. Since the practical limitations on the allocation of functional resources must be allowed to influence business strategies, the process of developing and maintaining strategies and their associated plans inevitably becomes an iterative cycle bounded by resources of all kinds.

REQUIREMENTS OF A STRATEGY STATEMENT

Primarily, a statement of strategy is a vehicle for focusing management attention on strategic aspects of the firm's business. It is also a means of communicating to those who must review and approve the strategy and to those who use it to guide their actions. Additionally, the document must be available to those responsible for initiating adjustments to it. These adjustments take into account current input from the environment or business. The document must also be available to those who will use it to measure and evaluate the performance of the firm or its functions.

By itself, a statement of goals and objectives is insufficient to meet these needs. Information must be added regarding the environment, the basis on which the goals and objectives were selected, the assumptions on which they depend, the perceived risks, and the options that are available and reasonable. The firm's long-range plans also need this kind of information since both strategies and plans are snapshots of management's vision that direct the actions of the organization.

STRATEGY DOCUMENT OUTLINE

In order to present the required information in a coherent manner, an outline containing the main points of a strategy is usually developed, as displayed in Table 3.3. The subheadings in the strategy do not need to be addressed separately in sequence, but the strategy must be presented in a manner that leaves no doubt about whether a statement is an assumption, a course of action, a risk, or an alternative. The strategy must be a logical, coherent entity with the relevance of its parts clearly indicated.

TABLE 3.3 Strategy Outline

1. Nature of the business
2. The environment
3. Goals and objectives
4. Strategy ingredients

The section on the nature of the business must answer questions such as "What is the business we are concerned with?" and "What are the boundaries of this business?" In many instances the answers to these questions are obvious, and this section will require little development. However, the firm or organization must have a clear vision of its mission before the strategy-development process can proceed. Failure to understand the nature of the business in which the firm is engaged can be fatal.[9] Thus, the questions posed in this section are important and not trivial.

In this section, it is important to describe the IT business carefully. What purpose does the IT organization serve in the larger organization, and what type of organization is it? The roles of the IT organization can vary considerably. For instance, at Boeing, IT provides a capability to the main Boeing business, but it also operates as a service bureau for outside customers and generates revenue for the firm. In an insurance company, the IT group may provide major services for the company and act as a conduit through which all the firm's transactions flow. A construction company may use data processing for design or project planning purposes and spend less than one percent of revenue on IT activities. The IT organization must understand what business it's in and the role it plays within the firm.

The environment section of the strategy outline states what is known and assumed about the relevant and significant factors and trends surrounding the firm. IT managers must understand those factors impacting or influencing the firm's current or future behavior. As a test of relevance, the environment section should include those factors that can potentially influence the attainment of current goals and objectives. Another question that measures relevance is: "Can these factors force a change in our goals?" This question is especially important in cases where later readings on the environment show changes in the analysis of those factors.

Key environmental assumptions should be reviewed at the conclusion of the environmental section, because their credibility and consistency with each other need to be understood without ambiguity. Furthermore, tracking these assumptions will enable all levels of management to adjust and update them as part of maintaining the strategy.

What is the IT environment like? What are the current capabilities and what can they be in the future? IT should know the state of the application portfolio and what new technologies can be applied to the firm's business. Is the IT organization relatively mature and disciplined or is it underdeveloped, with relatively inexperienced people and an immature management system? What is the environment today, and what might it be in the future? These are some other key questions to address in the environment section of the strategy outline.

The goals and objectives section must state, in concrete terms, the ultimate objectives of the strategy. What IT capabilities are we trying to achieve, and what long-term objectives are we going to set for the organization? For instance, is our goal to enhance the application portfolio with several new strategic systems, or are we trying to link our business units with a new and modern telecommunication system? Perhaps organizational issues such as decentralization need to be addressed as goals and objectives.

W. R. King gives a methodology for deriving the IT goals and objectives from a firm's strategy. He begins the process by identifying the corporate strategy set and transforming the objectives into MIS objectives. For instance, if the corporate strategy calls for reductions in internal investments, then the IT organization might concentrate on inventory reductions through improved inventory management systems. Alternative means for reducing internal investments are developed and presented to management for decision-making purposes. King's methodology is complimentary to the approach presented in this text.[10]

An important test of a statement of strategy is whether the objectives are attainable and desirable given the goals and objectives of the larger organization, in conjunction with environmental considerations. The search for IT opportunities should be restricted to those which fall within the mission of the organization. Opportunities must serve the corporate purpose and must be aligned with corporate objectives. A second test is the degree to which the objectives can be used to measure and evaluate unit performance.

Goal setting requires some additional important considerations. Goals should be established within the resources of the firm and should have a reasonable chance of attainment. Managers must achieve a balance between easily attainable goals and challenging goals with respect to resources of all kinds. Goals should be explicitly stated and should be quantified whenever possible. Subsequent planning, measurement, and control are facilitated by goal clarity. It is preferable to have a small rather than a large number of goals. A small number of goals reduces the chances of ambiguity and conflicts and permits managers to direct resources more effectively. Finally, well-planned goals should be time-limited.[11]

The ingredients of an expression of strategy include a course of action and its accompanying and supporting factors. A summary of these items is presented in Table 3.4.

TABLE 3.4 Strategy Ingredients

1. Course of action
2. Assumptions
3. Risks
4. Options
5. Dependencies
6. Resource requirements
7. Financial projections
8. Alternatives

The strategy statement must describe the course of action that the firm will follow in attempting to achieve the objectives for the strategy. What steps will the organization take to achieve its goals and objectives? The steps should accomplish the following:

1. Lead to realizing the objectives
2. Be consistent with other long-range interests of the firm
3. Be preferred over alternative possible strategies

For example, if a goal of the IT organization is to improve the capabilities of its people, will it accomplish this through hiring, training, retraining, termination, or possibly some combination of these actions? The course of action spells out the steps to attain these goals.

What are the major assumptions on which the strategy is based? Assumptions that are inherent in or that exercise significant influence over the strategy are technical capabilities, functional support activities, and potential competitive reactions. The test of the assumptions is their credibility. The maintenance of the strategy requires the tracking of these assumptions. To continue with this example, one assumption might be that the current employees are trainable; that is, they possess the necessary background knowledge to facilitate further training.

Risk will always be present. In fact, risk should be one of the major parts of the IT functional strategy. Questions such as "What is the nature of the risks in the strategy?" and "What is their potential impact?" should be answered. In the example above, one risk of a strategy for hiring skilled employees might be that very few are available to the firm. In the event that some aspects of risk become significant, IT managers must review the optional courses of action to determine whether they offer reasonable insurance against risk.

Since no single course of action has 100 percent probability of success, it may be appropriate to improve the probability in some cases by building options or alternatives into the strategy. These options should take into account specific risks, assumptions, or dependencies that unduly depress the probability of success. IT managers should look for options that are available within this strategy. They should ascertain how long the options are valid and upon what considerations a selection should be made. Also, they should consider whether any of the options add some cost or expense and if so, how much. In the personnel example above, options have been identified. They include hiring, training, retraining, and termination.

In most firms, it is likely that any single strategy will depend heavily on other related strategies. For instance, in order for one technical strategy to work, it may depend on a capability or process generated by another technical strategy. "What are the key dependencies of this strategy?", "What is their nature?", and "In what ways are they significant to the strategy?" are some questions IT managers must answer.

The strategy must identify resources required to carry out the actions, and it must present financial projections of the revenue, cost, expense, profit, and capital required to implement the strategy. What resources are required to carry out this strategy? Are any unique resources required? Will hardware, software, or staff resources be available in the quantities and on the schedule required? These questions need answers in this part of the strategy statement.

Alternatives that were rejected in the selection of the strategy, and the reasons for rejecting them, should be retained for future reference.

The steps outlined above represent sound preparation and will lead to sound implementation if the planning process is successful. Action steps have been identified and details necessary for control can be developed.

STRATEGY DEVELOPMENT PROCESS

The Strategic Time Horizon

The development of strategy statements within a firm usually follows a schedule dictated by the firm's corporate planning director who is responsible for synchronizing strategy development events. This activity is usually an annual affair interspersed with planning, control, and measurement events.

A representative schedule of events depicting the strategy development process is shown in Figure 3.4.

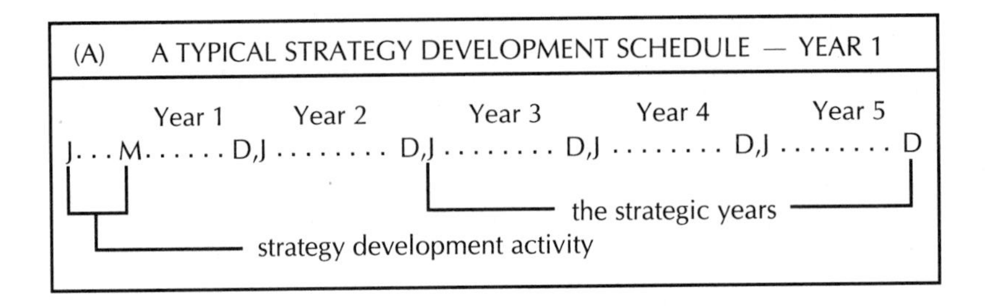

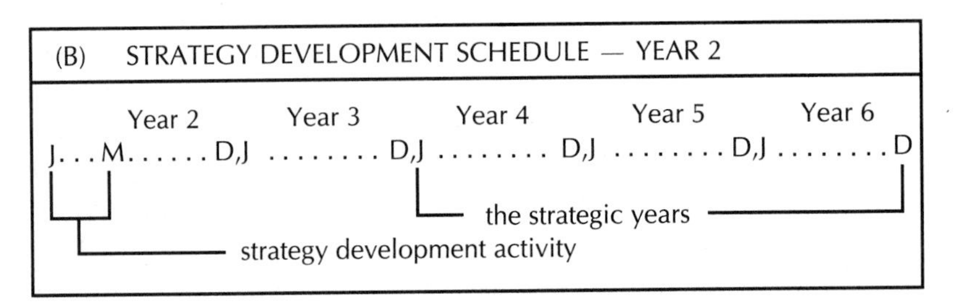

FIGURE 3.4 A Schedule for Strategy Development

Measured from the beginning of the current year, the strategic years extend from the beginning of Year 3 to the end of the strategic horizon, as shown in Part A of Figure 3.4. The strategic years may cover a period from 3 to 10 years or more, depending on the business or industry. In this example, the strategic activity occurs during the early months of Year 1 and considers the period of the strategic years.

The pattern is repeated annually as shown in part B of the figure. Each succeeding strategic activity removes from consideration the first year of the preceding period and adds one year in the future. The period of Year 1 and Year 2 is covered in the tactical or operating plan for the firm. This activity is discussed in the next chapter.

The schedule suggested above is most convenient for firms using the calendar year as the fiscal year for planning and reporting purposes. Organizations such as universities would adapt this process to the academic year, starting in September and ending in August. Other organizations, such as governmental agencies, may adjust this schedule for a fiscal year starting in July and ending in June.

Steps in Strategy Development

Effective strategy statements take time to develop properly. The strategist must take time to understand the environment and the area of opportunity or concern. The written statements should be concise and sharply focused. The statement of the opportunity or threat should highlight its relevance to the firm's future. If management believes that new hardware is required for the data center, for example, the statement should clearly indicate why this belief is reasonable. What substantiating evidence or trend information exists to support this belief? What evidence supports this concern?

The IT strategist should develop, in reasonable depth, a broad understanding of the future environment that influences the area to be studied. This environmental analysis might include an estimate of future computing costs, anticipated future computing loads, technical advances expected in telecommunications or processing capability, special future business conditions, and other relevant factors. Objectives are set and modified, if necessary, during the iterative strategy development process. Stand-alone strategies may need to identify several possible objectives. The selection of one occurs after testing the options for credibility and attainability. Alternative strategies should span the future environment. They must be clearly expressed so that readers of the statement can gain a thorough understanding of them.

In order to exercise a reasonable choice among the alternative strategies, selection criteria and the process for using them must be established. The criteria should measure the basic, long-range effects on the firm or the IT function. These effects may be measured in terms of profit and revenue, investment resources required, degree of risk, technological capability exploited, competitive reactions, and other factors important in each individual case.

The best strategy should be selected by using the criteria to measure the expected effect of each option. The selected strategy should then be developed in greater detail. Stand-alone strategies and functional strategies and their supporting data, reasoning, and other forms of evidence must be submitted to the appropriate executives for review and approval. The review and approval process enables and ensures congruence in planning. The process forces a review of the alignment of IT and corporate goals and keeps the executives of all functions informed of the important strategic directions. The issue of strategy integration across functions will arise naturally and will foster executive or organizational learning.

Upon approval by senior executives, these strategy statements will be incorporated into the strategic and tactical plans of the firm. The process of incorporation may require minor changes because this process may be iterative as well.

To be successful, the process of strategy development requires considerable effort and thought on the part of the firm's executives. This thoughtful activity focuses the best minds in the organization on the long-term health and welfare of the firm. Each function needs to be involved and the IT organization needs to be a full partner in all deliberations. The value of the resulting output is in direct proportion to the level of effort and the degree of cooperation among the senior executives. Strategy development processes and activities are vital to the firm that intends to remove IT strategic planning from its list of concerns.

STRATEGY AS A GUIDE TO ACTION

An IT strategy is developed with the explicit intent of guiding management in a course of action and appraising management's accomplishments in light of this guidance and direction. Various levels of management will take direction from strategies in planning and carrying out their activities. A standalone strategy may require significant effort in developing and carrying out new projects, as well as reshaping old ones. Reference to the strategy can serve as a guide to balance and direct the overall functional effort while taking advantage of breaks and unusual opportunities as they arise.

IT managers must exert strategic control; the strategy gives them the basis for doing this. Key dependencies and functional support activities, which have been identified in the strategies, must be tracked continuously. Significant departures from expectations should be analyzed to determine whether these variances are reason enough to review and revise the strategy. In particular, unfavorable changes in dependencies should be scrutinized carefully. This strategic control activity preserves the vitality and the viability of the strategy between planning periods. It also helps the organization maintain its strategic focus.

Periodically, the progress toward achieving the objectives of the strategy can be reviewed and appraised. The achievability of the objectives and the degree of stretch they require must be kept in mind. A moderate degree of accomplishment against a very difficult objective must be appraised differently from a high level of accomplishment against an objective easily attained.

THE STRATEGY MAINTENANCE PROCESS

The maintenance of a strategy is an essential step in its continued usefulness. Maintenance recognizes that a strategy is constantly subject to change. Frequently, change accounts for influences and factors beyond the firm's control. A complete statement of strategy requires careful documentation of the areas where change may take place. This documentation facilitates tracking actual developments against the assumptions, dependencies, and risks inherent in the strategy. The test here is whether the strategy has been documented in a way that clearly identifies the factors that need to be tracked.

Maintenance of the strategy is neither its implementation nor its use. Tracking will identify deviations from the strategy, which can be helpful in maintaining and implementing it. It is important to realize that strategy maintenance is a vital and essential part of the total strategy process. Tracking each of the key environmental and strategic assumptions, risks, and dependencies must be a part of the strategy process. The responsibility for tracking each area should be specifically assigned to someone in the organization. A diligent pattern of follow-up must be established in order for the strategy to succeed.

Periodically, or when tracking reveals a significant deviation, the entire strategy should be carefully reexamined and updated. The logical process for doing this is to follow the steps for developing a strategy, as previously described.

THE IT STRATEGY STATEMENT

The IT organization that follows the management process for strategy development indicated above will have met certain conditions necessary for success. The process establishes goal congruence, lays the foundation for planning, and carefully reveals the role and contribution of the IT organization. The strategy process is an important step in organizational learning, too.

The IT organization must develop and maintain a functional strategy that guides the actions of the IT function. These actions must support the goals and objectives of the firm and its constituent organizations. The IT function may also have stand-alone strategies. These strategies direct projects designed to enhance the function's objectives. The IT manager is responsible for developing and maintaining these strategies and for ensuring that they are synchronized with the firm's overall strategy process.

Strategic planning is one of the top issues confronting IT managers in most firms, and it is a critical success factor for them. Success in this activity paves the way for success in many other areas because sound and valid planning forms the base for most other management activities. Failure in this area is a prelude to failure elsewhere. Given the importance of this subject to the IT management team, what issues should be considered, and how should they be explored? Table 3.5 outlines some of the important issues for IT strategies. Minimally, the IT strategy statement should address these issues; there may be other issues specific to the firm that should be addressed as well.

TABLE 3.5 IT Strategy Issues

Business aspects
Technical issues
Organizational concerns
Financial matters
Personnel considerations

Business Aspects

The IT organization must maintain a keen awareness of the business goals and objectives of the firm and must develop strategies that support them. These business goals may include increased market share, improved customer service, lower production costs, or many other objectives of central importance to the corporation. CEOs expect IT to contribute to the firm's results in conformance with normal business practices. The strategy of the IT organization must understand, reflect, and support the firm's business goals.

Not-for-profit organizations, government agencies, and educational institutions among others have goals that may be different from those noted above. But in all cases and in all forms of enterprise, IT must support the parent organization's goals and objectives.

IT managers, interacting with the other senior executives, are key players in the development of the functional strategy. To the extent that the IT organization is involved in attaining the firm's objectives, IT managers must acknowledge their group's involvement in the function's strategy statement. They must ensure that the function's actions are in concert with the long-term goals and objectives of the firm. This involvement must be tested through an interactive review involving IT managers, their peers, and their superiors. Are the strategies congruent? Are the firm's dependencies on the IT organization satisfied? If the strategies are executed as written, will the objectives be met? The answers to these and other questions will establish the validity of the strategy process and help ensure its results.

Technical Issues

The IT manager provides leadership in using technology to attain advantage for the firm. The IT strategy is one place for this initiative to occur. The strategy should reveal the practical utilization of advanced technology in support of the firm's goals and objectives. This utilization must be consistent with reasonable risks and available or attainable resources. Additionally, the IT manager must ensure the technical vitality of the organization through development and implementation of current or advanced hardware, software, or telecommunication technology.

The IT functional strategy is the vehicle for paving the way toward improved technical health. It gives the organization the formal opportunity to establish objectives designed to maintain and enhance the technical health of the function and the firm. CEOs expect leadership from IT in discovering important technological applications. They expect IT to make efficient use of the firm's resources and to work within the norms of the organization.

Organizational Concerns

There are several reasons why organizational considerations are very important to the IT strategy. First, the introduction of information technology tends to have organizational consequences beyond the IT organization itself. These issues are difficult to envision and, in many cases, they are hard to resolve. Not all changes are looked upon favorably by everyone. And many changes, while important to senior executives, are resisted by nearly everyone else for a variety of reasons. The firm needs training and education in the subtleties associated with technology introduction. The IT manager needs to take the lead in satisfying these training and educational needs.

Further, the role of the IT organization and its contribution to the firm must not be taken for granted in the development of the IT strategy. Not everyone appreciates the IT role, and many managers in the firm do not fully understand the contribution, or potential contribution, of the IT function. Wise IT managers will take specific actions to champion their important role. The IT functional strategy will reflect these actions. Much more will be said about this in subsequent chapters.

Financial Matters

When the IT strategy is translated into strategic and tactical plans, it must satisfy the financial ground rules of the firm. In many cases, financial constraints limit the range of opportunities for the IT organization as they do for most other functions. These resource constraints force iteration in the process of developing the business strategy for the firm. These constraints also require successive revisions in the functional strategy for the IT organization. Wise IT managers guide their organizations in these matters in order to conserve energy and optimize results during the strategy development process.

Personnel Considerations

No functional strategy is complete without action plans that relate to the management task of recruiting, training, and retaining a base of skilled employees. These personnel considerations are intimately related to technical issues because strong technical people develop solid technical strategies while advanced technical strategies attract strong people. The IT functional strategy must develop these thrusts in coordination with one another.

IT managers must demonstrate productivity improvements in the IT organization. They must invest capital to increase their own productivity. IT must acquire the tools to assist the entire organization in improving productivity. In many ways, information technology is the productivity engine for the firm. The production sector of the business must reduce its operating costs through superior management aided by information technology.[12] Talented people, skilled managers, and advanced technical resources provide some necessary conditions for enabling major productivity improvements and achieving overall firm objectives.

SUMMARY

This chapter described the activities of strategy development from the perspective of information technology managers and their organizations. The responsibilities of IT managers were described in detail, and the processes were presented as they applied to the IT organization. These responsibilities include ensuring active participation of line managers throughout the firm. The IT strategy represents a collaborative effort. IT managers and line managers in all functions of the firm must share equally in these responsibilities because the only purpose of the IT strategy is to further the business interests of the entire organization.

If an organization is to successfully align IT objectives with corporate objectives, the managers must believe that the IT strategy is a strategy for the entire firm. The goal alignment issue is not present in firms where this belief is shared by members of the senior management team. It follows that senior managers throughout the firm will share joint responsibility for a sound IT strategy.

The development of strategy statements is a fundamental responsibility of management. Success in this endeavor requires significant time and energy. Coordination between the IT organization and the other functions is essential. Because information technology is pervasive in most firms, the strategy process is more difficult and more important for the IT organization than for its peer organizations. In many instances, the IT strategy unifies and integrates other functional thrusts. Thus, the IT manager's role is crucial in the strategy process of the firm. Senior managers in all functions touched by information technology share responsibility for the validity of the IT strategy. They must be active participants in its development.

Certo and Peter neatly summarize the important points of this chapter by stating that "strategic management is a continuous, iterative process aimed at keeping an organization as a whole, appropriately matched to its environment.

The process itself involves performing an environmental analysis, establishing organizational direction, formulating organizational strategy, implementing that strategy, and exerting strategic control. In addition, international operations and social responsibility may profoundly affect the organizational strategic management process. It is important that the major business functions within an organization—operations, finance, and marketing—be integrated with the strategic management process."[13]

Review Questions

1. What are the major ingredients of a complete statement of strategy?
2. Besides providing direction to the organization, what other purfff-poses does the strategy serve?
3. How do stand-alone strategies differ from other types of strategies?
4. Under what conditions can an IT organization produce both business and functional strategies?
5. For what purposes is a functional strategy developed, and how can it be used?
6. Toward what issues is the firm's business strategy pointed? What are the elements embodied in a firm's business strategy?
7. What is the relationship of functional strategies to the business strategy of the firm?
8. What is the relationship between strategies and plans?
9. Describe the relationship of a strategic plan to business strategies, functional strategies, and stand-alone strategies.
10. Where do tactical plans fit into this relationship?
11. What are the main elements of a strategy document?
12. What role does the environment statement play in the expression of strategy?
13. What are the connections between assumptions and risks? Why is it useful to have these items separated in the strategy statement?
14. The iterative process described in this chapter is useful in resolving dependencies, among other things. Why is this so?
15. What main issues should be addressed in the IT functional strategy?
16. The strategy development process permits executives to focus on goal incongruences. When and how does this happen?
17. Why do senior managers across the firm's various functions share responsibility for the IT strategy?

Discussion Questions

1. Discuss the main elements of AT&T's strategy discussed in the Business Vignette.
2. Discuss the role that alliances play in AT&T's strategy. Compare and contrast this stratagem with the pricing strategy in the True Savings program.
3. Give some examples of IT dependencies likely to be found in the strategy that cannot be resolved through the iterative process. How can the effects of dependencies external to the firm be mitigated?
4. Compare and contrast the actions of strategy maintenance and performance appraisal.
5. Identify the critical success factors facing the IT manager that can be addressed through the strategy development process outlined in this chapter. Establish your own opinion regarding the vital nature of strategic development.
6. An IT organization wants to develop a telecommunications capability to prepare for new services now being considered by the firm. What types of strategies will be useful in this connection, and how will they be used?
7. What are the consequences, in terms of strategy development, of the pervasive nature of information technology?
8. What are the characteristics of a high-quality IT strategy statement? Discuss these characteristics in terms of the process and the result.
9. What are the necessary conditions that must exist in the firm for the IT organization to achieve success in strategy development?
10. Managing expectations is a key IT issue. Discuss how the strategy development process assists in this task.

Assignments

1. Assume you are the manager of your university's information system organization. Your group maintains student records and the automated registration system. Develop an outline of a stand-alone strategy to incorporate the use of personal computers scattered across the campus into the system.
2. You are considering the purchase of a new, sophisticated personal computer. The computer is for use in your school activities and will be used in your job after graduation. Develop the elements of a strategy to acquire this computer.

ENDNOTES

[1] For an interesting insight into strategic planning at AT&T see Bernard H. Boar, *The Art of Strategic Planning for Information Technology* (New York: John Wiley & Sons, 1993). The author is an IT business strategist for AT&T.

[2] *The Value Line*, January 13, 1995, 745.

[3] Henry Mentzberg, "The Strategy Concept II: Another Look at Why Organizations Need Strategies," *California Management Review*, Fall, 1987, 25. According to Mentzberg, a good strategy also helps define the organization and gives it meaning and purpose.

[4] Daniel H. Gray, "Uses and Misuses of Strategic Planning," *Harvard Business Review*, January-February, 1986, 89.

[5] John C. Henderson and John G. Sifonis, "The Value of Strategic IS Planning: Understanding, Consistency, Validity, and IS Markets," *MIS Quarterly*, June, 1988, 187.

[6] Frederick W. Gluck, Stephen P. Kaufman, and A. Steven Walleck, "Strategic Management for Competitive Advantage," *Harvard Business Review*, July-August, 1980, 154. Strategic management is largely concerned with creating competitive advantage and shaping the future, claim the authors.

[7] George C. Sawyer, *Designing Strategy* (New York: John Wiley & Sons, 1986), 169.

[8] Henderson and Sifonis, 187.

[9] Theodore Levitt, "Marketing Myopia," *Harvard Business Review*, September-October, 1975, 26. This classic article describes the causes and results of failures. Using the marketing viewpoint, one can observe that the growth of end-user computing and the decentralization of information processing in some firms is partly the result of the centralized IT organization's product orientation, rather than customer orientation.

[10] W. R. King, "Strategic Planning for Management Information Systems," *MIS Quarterly*, March, 1978, 22.

[11] Richard T. Hise and Stephen W. McDaniel, *Cases in Marketing Strategy* (Columbus, OH: Charles E. Merrill Publishing Co., 1984), 130.

[12] Paul A. Strassmann, *Information Payoff: The Transformation of Work in the Electronic Age* (New York: The Free Press, 1985), 189.

[13] Samuel C. Certo and J. Paul Peter, *Strategic Management* (New York: Random House, 1988), 23.

4 Information Technology Planning

A Business Vignette

Team-Based Planning Aligns Goals and Objectives

Although corporate planning is a well-established activity, many firms have difficulty building plans that align goals and objectives among and between functions including IT. But not at the Indianapolis-based Citizens Gas & Coke utility where a series of management meetings devoted to IT planning are reminiscent of rural New England.[1] Information technology budgets are established similarly to the way a town or village might approve its budget at a series of town meetings. For example, in the 1995 budget planning sessions, managers approved an electronic mail and calendaring application but deferred a leading-edge CD-ROM-based customer service application until next year.

Driven by the impending deregulation of the utilities industry, a sense of urgency surrounds the corporate strategy process at Citizens Gas. The $350 million firm must maintain its services at the lowest possible cost in order to remain competitive in the emerging environment. Specifically, Citizens Gas is determined to keep its rates 15 percent lower than the average for comparable utilities. In order to meet this difficult goal, senior executives demand that Information Systems play a vital role in business activities and that users take ownership for their systems. To help meet these objectives, Citizens Gas established a process that integrates IS strategy with corporate strategy.

In order to begin its fiscal year the first of October with a comprehensive plan, the company's 14 business units meet in February, March, and June for three intensive sets of two-day planning sessions. Each business unit can sponsor IS projects and each has a vote on projects proposed by others. Business units propose IS projects that benefit the corporation and require a portion of the available resources. After weighing the benefits of each other's plans, they apportion the IS budget for the coming year. During these discussions, IS is represented by four managers and gets one of the 14 votes.

At these planning sessions, business unit representatives outline their visions of the future, present their strategies, and negotiate for a share of the budget. A consensus emerges on common directions and priorities as the attendees reach agreement on how the budget is apportioned. "This is a team-based process focused on customer satisfaction," Don Lindemen, Citizens Gas' president and CEO, reported to *Computerworld*.[2]

At the first meeting, departmental representatives attempt to agree on company direction. Following this discussion, each business unit develops short- and long-term objectives to share in the next meeting in March. Commonly, business units collaborate and present joint plans to the group. For example, IS and administration may combine expertise and resources to promote E-mail for the entire corporation. Their proposal would contain costs and benefits for Citizens Gas and would include a schedule for completion.

After the March meeting, the participants rank each proposal's value on a scale of one to five in preparation for the June meetings during which they decide funding.

Business units use three two-day meetings to establish objectives, strategy, and budgets for the next year.

FEBRUARY	Representatives from the 14 business units present vision, discuss strategy, and argue in support of their particular budget.
MARCH	Business units develop short- and long-term objectives and share them with other units. Negotiation and discussion are intense.
JUNE	Between the March and June meetings, each unit ranks the other group's proposals. The business units decide funding levels at the June meeting.
October 1	The agreed-upon budgets and plans take effect as the fiscal year begins.

"People do come in with their own agendas," Bob Steuber, Citizens Gas' Director of systems reported to *Computerworld*.[3] "But then that comes out in front of the whole group, and there's a feeling that you are sub-optimizing the system." However, Steuber continues "everyone comes out of the meetings understanding why certain decisions were made. We get a common vision out of these meetings," Steuber says. "If the strategies make the cut, there's more dedication to get them done because everyone knows we've all signed off on them."

INTRODUCTION

The preceding chapters concentrated on strategy development and the importance of strategic thinking for IT managers and senior executives in their firms. A model for developing several types of strategies useful to the IT organization, to the firm, and to its constituent functions was presented. These strategies and strategic statements express the preferred long-term direction for the organization. Based on a thoughtful analysis of the environment, they account for risks, dependencies, resources, and technical requirements and capabilities. These strategic statements are very valuable for the firm and its management because they lay the foundation for succeeding activity.[4] The strategies communicate and coordinate the efforts of the firm's executives and managers.

Strategic statements set the broad course of action for the firm and its functions. The strategy outlines the goals and objectives the organization intends to achieve and provides a general statement of how the firm expects to achieve them. It does not provide details. Strategy statements establish the destination and contain a general direction for reaching it. The activity of developing the detailed roadmap is planning. The specific details for reaching the goals and accomplishing the objectives are found in various plans. Formal strategy and planning documents are the vehicles for recording and communicating the firm's intended direction.

Planning begins with developing strategic plans from strategy statements, and it continues through tactical, or intermediate-range, planning to the development of operational plans and controls. This process forms a vital part of the management system for the successful IT organization. Successful IT planning is a combination of four related phases: a) agreement on the future: business vision and technology opportunity; b) development of the business ideas for information technology application; c) business and information technology planning for applications and architectures; and d) successful execution of the business and IT plans.[5]

The end-product of the planning process in this chapter is an organized framework and methodology from which the IT planning activity can take place. Planning is a critical success factor for IT managers and their organizations. It must be accomplished in an outstanding manner.

THE PLANNING HORIZON

The types of plans discussed in this chapter can be classified as strategic plans, tactical plans, or operational plans. Primarily, they are differentiated by the time span during which they are valid. This time span is called the planning horizon. A typical short-range planning horizon is 30 days, while the long-range planning horizon for most firms extends for years.

Strategic plans usually cover the period from two to five years or more into the future. They represent the long-term implementation of strategic thinking. Strategic plans contain specific details for achieving the organization's mission and for meeting its long-range goals and objectives. Strategic plans form the basis for short-range plans. Of course, all planning activity must support the strategic direction of the organization.

Tactical plans commonly address the period from three months or so to two years in the future. Tactical plans describe actions that begin the implementation of strategic plans. Operational plans consider the near term: from now to three months from now. Tactical plans bridge the period from the operational horizon to the strategic horizon.

Just as managers need stand-alone strategies to cover special business or functional situations, they also need plans to accompany these strategies. In some cases, it may be necessary for these plans to remain separate from the three types of plans discussed above. Usually, however, plans associated with stand-alone strategies are folded into the plan or plans covering the period addressed in the stand-alone strategy. At some high level in the firm, however, the resource portions of these separate plans are merged.

The extended period over which the operational, tactical, and strategic plans are effective is called the extended planning horizon. A diagrammatic depiction of a five-year extended planning horizon is presented in Figure 4.1.

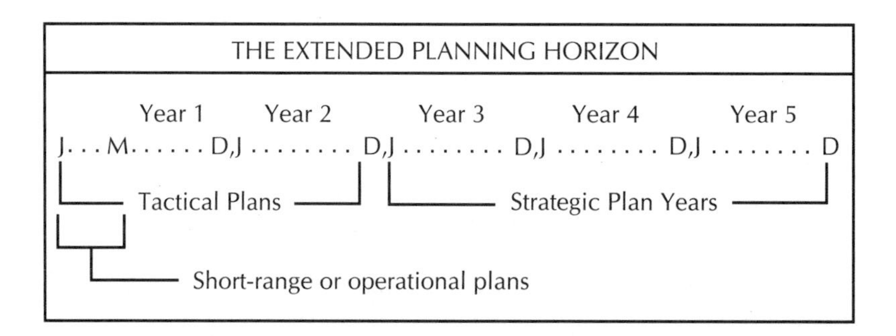

FIGURE 4.1 The Extended Planning Horizon

The extended planning horizon pictured in Figure 4.1 embraces five complete calendar years. It shows the time relationship among the plan types.

Operational planning is broad-based and contains a great deal of detail because it relates all the important firm activities happening in the very near term. Operational plans need to be regenerated frequently in response to changes in near-term business conditions. They need to bridge the gap from the present to the tactical time period. These plans contain significant amounts of detailed information. They are assimilated and implemented at fairly low levels in the organization. They give structure to the everyday operation of the firm for first-line managers and their teams. They are the basis for taking short-term action, and the reference points for measuring short-term results. Short-term variances result from comparing planned versus actual performance and often require correction.

Tactical or intermediate-term plans are less detailed and are effective during a longer time period than that of operational plans. Typically, these plans cover the current year plus the next year, and they provide overriding direction to operational planning. Ideally, the implementation of succeeding operational plans will lead to the successful implementation of the tactical plan, unless significant variances were generated during the implementation phase. Since activities seldom go as expected, the system must contend with variances in results from both operational and tactical plans.

Strategic plans extend from the end of the tactical period to the end of the extended planning horizon. Just as tactical plans give guidance to operational plans, strategic plans provide direction to tactical plans. The effective execution of tactical plans paves the way for accomplishing strategic objectives. Thus, implementation of the planning process results in a succession of overlapping plans moving through time.

The three timeframes must form a continuum over the extended planning horizon, and the three types of plans must form a unified picture of the future actions of the firm. The plans cover different time periods and contain differing amounts of detail. They portray a coordinated description of how the firm intends to carry out its strategies from now into the foreseeable future. The strategic plan supports the firm's strategy. The tactical plan provides shorter range support to the strategic plan. The operational plans give day-to-day or near-term meaning to the tactical plan. In the absence of any unanticipated changes in the environment during this entire period of time, and if the firm developed and carried out its plans flawlessly, then the firm attains its strategic objectives. Since these conditions are unrealistic, the true challenge lies in developing strategies and valid plans that account for anticipated environmental change, and revising the overall framework in response to unexpected events.

STRATEGIC PLANNING

Strategic planning converts long-range strategies into long-range plans. The process takes the information found in the strategy statement and adds detailed actions and elements of various resources required to attain the stated goals. These resources consist of people, money, facilities, and technical capabilities, blended together and working toward the objectives. The strategy contains statements about assumptions, risks, and dependencies.

Assumptions in the strategy must be converted to reality, and dependencies need to be accounted for in the plans. The risks noted in the strategy must be mitigated to the fullest extent possible by specific actions designed to contain and offset them. The process of planning must generate actions to accomplish the goals and objectives.

The strategic plan contains a detailed schedule of resource allocation designed to attain the strategic goals and objectives. The strategic plan for a firm is also a financial statement of projected revenues, expenses, costs, investments, and profits. The strategic IT plan also includes financial information, but this is usually limited to costs and expenses. The plan is detailed enough so that it can be tracked over its lifetime while comparing planned to actual performance. When the strategic plan has been translated into tactical and operational plans, the tracking takes place in the near term. The strategic plan translates strategic goals and objectives into strategic actions two to five years or more in the future. It is the foundation upon which the tactical plan is built.

For example, consider an IT group whose long-term goal is to improve market response times by deep automation of product design, development, and manufacturing. The strategy includes installing CAD/CAM systems for its design and development engineers and for its plant. The systems are to be effective in 48 months. The strategy statement contains the assumptions for using CAD/CAM and outlines dependencies and risks. Resistance to new systems, for example, might be a risk. The strategic plan adds details of schedule, actions, and resources: space requirements for the equipment, training for the users, capital resources, and operating expenses. The plan states that space will be leased, and that training will overcome any resistance to the new systems. The tactical plan will add more detail, particularly for the near term.

TACTICAL PLANS

Tactical plans generally cover the current year in detail and the following year in less detail. Management makes a commitment to implement tactical plans, and they are used to measure management performance; thus these plans are sometimes called measurement plans. Tactical plans form the basis for assessing the firm's performance, as well. Senior executives use tactical plans to measure unit performance. They also provide feedback for measuring the firm's progress toward attaining long-range goals. Tactical plans contain more detail and have a more direct bearing on near-term activities than strategic plans. They guide very short-range activities and link the near-term actions to long-range activities.

To continue with the CAD/CAM example, many activities will be planned for the first two years. Some of these are: selecting hardware and software, installing hardware and software systems, training users and IT people, converting from previous processes to new procedures, and setting up measurement systems. Questions, such as where and from whom the space will be leased and what type of training is required to overcome resistance to new technology must be answered. Managers must schedule these activities, assign human and material resources, and develop detailed budgets in order for the plans to be complete. Management will agree to the plan and will be measured on the degree of plan attainment. The tactical plan is the basis for day-to-day activity.

OPERATIONAL PLANS AND CONTROLS

Operational plans add detail to the very near-term activities of the firm and give them direction on a day-to-day or week-to-week basis. They are used by first-line managers or by nonmanagerial employees as they carry out their responsibilities. Operational plans usually require much analysis, in addition to that required for tactical plans, and they must contain control elements as well.[6] They show managers how well daily or weekly activities are proceeding.

The CAD/CAM installation will contain near-term tasks. These include: identification of hardware and software vendors, setting vendor criteria, developing space alternatives, and performing systems analysis of current procedures. At the same time, planning for the next short-term goals will begin, as will budget and cost-tracking tasks.

Planning Schedules

In most firms, planning is regularly scheduled and relates to the firm's fiscal calendar. Usually, the fiscal year and the calendar year are identical, and planning activity is seasonal. It is common and practical to develop tactical plans so they can be approved just prior to the beginning of the tactical period. For most firms operating on the calendar year, this means that the tactical plan for the next two years is developed and approved in the few months prior to the beginning of the new year. The new calendar year begins with an implementation of the new tactical plan for that year and the following year.

Citizens Gas' year starts in October, so planning begins in February for the next fiscal year. February, March, and June meetings lead to agreement on the funding levels for each business unit. Detailed budgets are developed and in place on October 1 to start the new fiscal year.

In a model planning calendar, the development of strategies and strategic plans takes place shortly after the beginning of the new year and is completed and approved around mid-year. When completed, it covers the period from 30 months in the future to the end of the extended planning horizon, perhaps five or more years hence. Figure 4.2 illustrates the second iteration of planning in relation to current plans. The new tactical plan incorporates the first year of the previous strategic plan, and the new strategic plan adds an additional future year.

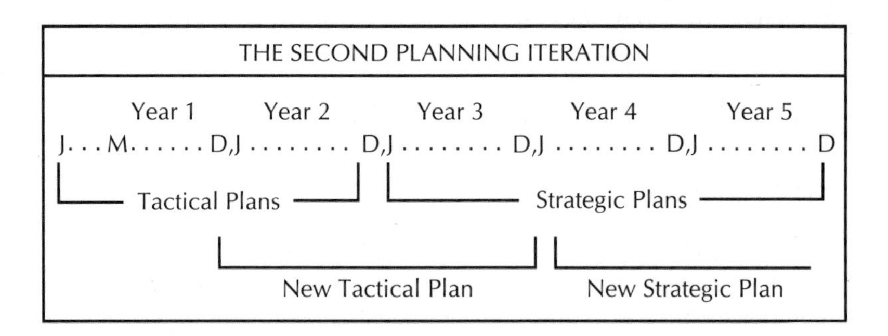

FIGURE 4.2 The Second Planning Iteration

For firms with a well-established planning cycle, planning does not start from the beginning with each iteration. As illustrated in Figure 4.2, the new tactical plan revises the second year of the previous plan and adds to it the first year of the previous strategic plan. The new strategic plan is a revision of the last years of the previous plan and adds a new year. The revisions reflect changes due to new or revised strategies, changing business conditions, or changes in the environment. Failure to achieve prior goals or objectives, or alterations related to risk management or dependency management, may also lead to revisions.

The entire planning process is cyclical and repetitive, with revised longer range plans becoming more near-term with each passing cycle. Usually, operational plans are not as tightly scheduled, and if the implementation of the tactical plan is proceeding smoothly, these plans may very well be of an *ad hoc* nature. The need for short-range plans depends on the circumstances and on the activities of the function and the firm.

The CAD/CAM example discussed earlier could proceed in a similar manner. Details of hardware, software, space, training, and many other items will be developed over time, as succeeding plan iterations and plan implementations occur. If all goes well, the systems will be operational as planned, and the firm will reap the benefits of shortened development and manufacturing cycles and increased responsiveness to changing market requirements. Ideally, the firm will achieve its strategic objective.

A PLANNING MODEL FOR IT MANAGEMENT

The discussion so far has related mostly to common planning elements. It is applicable to most functions in most firms. The planning horizon will vary between firms. The planning schedule could be altered considerably, depending on the industry and the desires of the firm. For example, electric utility companies usually have a long planning horizon, on the order of ten years or more, and firms in the forest products industry in the Pacific Northwest plan for 50 or more years in the future. Firms in highly volatile or highly competitive industries may benefit most from shorter range thinking. Firms on the edge of bankruptcy are interested only in survival. A year may be very long-range for a troubled firm.

The information technology function within the firm must adapt to internal and external forces and constraints in searching for a management system to govern its planning. The firm's planning system establishes many of the frameworks to which the IT organization must conform. The firm's planning department establishes the planning calendar and planning horizon, schedules the review procedures, and coordinates the approval processes. Usually, the planning department also prescribes plan formats and content outlines.[7]

The IT organization must have a view of future technology. Technology trend information is important because it permits IT to present credible technical detail pertinent to the extended planning horizon. Subsequent chapters will present methods through which IT can secure trend information. Some strategic plan elements can be specified in detail, while other items represent the manager's best judgment. In particular, long-range planning is a delicate blending of solid facts, intuition, and seasoned judgment. However, facts, if available, are always preferred in the planning process.

Information Technology managers, in particular CIOs, have responsibilities extending far beyond the boundaries of their line organizations. Organizational factors influence planning activity, and the CIO position is advantageous in this regard. Factors that can aid in planning are the perceived importance and status of the IT manager, the corporate culture, and overall management style. IT managers' strategies and plans must reflect the actions of their own organizations while including the information processing activities of the entire firm. Their view must encompass both their staff and line responsibilities. Their vision must include both technological and organizational factors. It must provide technical direction or at least generate practical technical alternatives. And it must consider organizational maturity and sophistication. The intelligent IT executive will not attempt to transform the firm from the contagion stage to maturity in one planning cycle.

Senior executives must include the entire organization's environment in their planning activity. Frequently, organizations undergo change and capture opportunities because of information technology. The corporate planning environment must anticipate and respond to IT-induced change.

The planning environment at Citizens Gas does this exceptionally well. The Business Vignette described how the company uses well-coordinated planning meetings to set expectations, obtain consensus, and establish unity between business functions. Upon completion of the meetings, goals, objectives, and plans for the units are congruent. Information Technology and the functions it supports obtain agreement from the CEO and other senior executives on the direction for the next year. And, in March, when everyone's plans have been in place for six months, the group begins the planning process again. Citizens Gas is an outstanding example of good corporate governance and planning.

Since planning considers constraints, ambiguities, facts, and uncertainties, what planning procedures might best serve the information technology function? What kinds of activities need to be planned? What tools and information will assist in the task of IT planning? A complete IT plan answers those questions. It is a detailed and thoughtful document describing the actions that will implement strategy over the planning period. It must address both the line and staff responsibilities of IT, and consists of the ingredients outlined in Table 4.1.

TABLE 4.1 IT Management Planning Model

Application considerations
Production operations
Resource plans
Administrative actions
Technology planning

Application Considerations

The applications portfolio is the complete set of applications programs used in conducting the firm's automated business functions. As we will learn later, the management of the applications assets is a very difficult task and the planning of this activity must be conducted skillfully. The plan for dealing with the application portfolio must include the selection of projects to be implemented during the plan period and the scheduling, control, and evaluation of these projects during implementation. The resources required for development, enhancement, maintenance, and implementation must be planned also.

Project selection describes which programs will receive maintenance or enhancement activity and what this activity will accomplish. Managers must know the schedule of this activity and the resources it will require. Applications that are not currently in the portfolio will be defined, and a plan for acquiring these programs is required.[8]

Identifying strategic systems to augment the portfolio is difficult. Programs may be acquired through purchase or they may be developed locally or by subcontractors. In all cases, the plan describes what will happen, when it will happen, how it will take place, and how much it will cost. The plan must justify the action, describe why the action is being taken, and identify who is responsible for taking the action.

Managers must know the answers to "who, what, when, where, why, and how" for application planning activity. If they have considered these fundamental items in their plans for the portfolio, they will have the basis for controlling implementation of the plan and for monitoring deviations from it. For completeness, effective plans must contain control items.

There are some major difficulties in obtaining planning direction for the applications portfolio.[9] These difficulties arise from competition for resources among work items. The resolution of these issues is a fundamental IT management task. A management system designed to resolve these difficulties is presented Chapter 8. The results of the process described there will comprise the elements for the applications portfolio plan.

Production Operations

Production operations consist of processes for running the firm's applications. Data is collected or developed, the applications are executed using current and possibly historical data, and the results are stored and/or distributed as planned. The processes may be continuous on-line systems, networked client/server systems, periodic, scheduled systems, or, more likely, some combination of these. In many cases, the production operation is highly dependent on advanced telecommunication systems. The strategic systems we studied in Chapter 2 are of this nature. In any case, the results generated by the applications are used by many members of the firm as they carry out their responsibilities. The results may be distributed to customers or suppliers as well. In most cases, the firm depends on these outputs for successful and efficient operation.

The IT management system must plan production operations to achieve satisfaction for customers both within the firm and outside it. Customers have expectations about the IT service they will receive, and IT must have a plan to satisfy these expectations. In production operations, the planning of service levels is very important.[10] The essential elements or disciplines for meeting service expectations consist of problem management, change management, recovery management, capacity planning, and network planning and management. Problem, change, and recovery management are elements for ensuring the attainment of service levels through operational management processes. Capacity planning and network planning ensure service levels through resource planning. Table 4.2 lists the planning elements of production operations.

TABLE 4.2 Production Operation Planning Elements

Service level planning
Problem management
Change management
Recovery management
Capacity planning
Network planning

Managing production operations is vital to the success of IT managers. Production operations management is a critical success factor for them and must be handled properly in both the short and long term. The basic considerations involving the elements of the production operations plan will be discussed in detail in Part IV of this text. This discussion includes the details of service-level planning and the operational disciplines of problem, change, and recovery management. Service levels are important elements of the IT plan. Problem, change, and recovery management are processes used to ensure that managers attain committed service levels.

Service-level attainment depends on adequate resources with which to process the applications. IT and functional user organizations must possess sufficient computing and network capacity to handle the application workload satisfactorily. The physical resources include computer processing units and all the peripheral equipment such as auxiliary storage and input/output devices. In distributed environments where end-user or client/server computing is well developed, individual workstations and associated equipment are important components of the system's capacity. Chapter 14 discusses performance analysis and capacity planning for the computing systems serving the applications.

Network planning and management are the subjects of Chapter 15. In the planning process, senior managers must consider vision, policy, and architecture.[11] Design and implementation of the network follows from the planning. As with the computer components of the system, performance analysis and capacity planning are required to meet service levels. The disciplines of problem, change, and recovery management apply to networks as well. Network capacity planning is an important element of the IT plan.

Resource Planning

The next element of the complete IT plan is the resource section. The resource items of the IT plan consist of money, equipment, people, and space. Technology is also a resource, but technology planning will be discussed separately. Table 4.3 indicates the elements commonly found in the resource section of the plan.

TABLE 4.3 Resource Elements

Resource Elements
Equipment plans
Space plans
People plans
Financial plans
Administrative actions

The IT plan describes the critical dependency on available resources throughout the planning horizon. A summary of the hardware and equipment required to operate the information processing activities of the firm is found in the resource section of the plan. Included in this section are the space requirements for housing the equipment and the special facilities for running the equipment. If the firm's production operation is growing, the space requirement can be substantial. The special facilities consist of heating, ventilation, air conditioning, electrical power, cabling conduits and cabinets, and perhaps raised floors. In many installations, the need for uninterruptible power sources to ensure reliable service is also a significant cost.

The plan must also include network equipment and services such as switches, routers, and perhaps dedicated leased lines. The financial obligations of the firm consist of purchase and lease costs and expenses for third-party services. The IT plan integrates the equipment and use charges with software and applications costs to form the complete telecommunication system financial plan.

People Plans

The next essential ingredient of the IT resource plan is the plan for people. By any measure, this is the most crucial element of the plan. People enable the firm and the IT organization to meet their goals and objectives. All the equipment, space, and money in the country is useless without skilled people. People management is a necessary condition for success, though not sufficient unto itself. The effective management of people resources is a critical success factor for IT managers. It is essential, therefore, that this portion of the plan is particularly well conceived and developed.

IT managers and administrators can provide much of the people-related information. Staffing, training, and retraining plans go hand in hand with the people-management activities of performance planning and assessment. Managers spend a great deal of time on these matters and must reflect their actions and anticipated actions in the plans for the organization.

The personnel plan includes requirements for people according to skill level and considers their deployment through the extended plan horizon. It must address attrition, hiring, training, and retraining. Sources of new employees are identified in the IT plan, and recruiting actions are formulated. For the people currently in the organization, managers propose individual people-development plans. Plans for people are also maintained for skill groups such as systems analysts, programmers, support specialists, telecommunication experts, and managers. The staffing plan must be consistent with the plans to accomplish technical and organizational objectives. Working with this detail, IT planners can calculate the total cost of the people resource by summing their salaries, benefits, training costs, and so on for the financial plan.

Financial Plans

The financial plan for the IT organization summarizes the equipment, space, people, and miscellaneous costs. Specific costs and expenses are segregated and identified by the firm's code of accounts. The timing of all expenditures reflects the rate at which resources will be acquired or expended. The timing must be consistent with work activities and completion of events. When the plan is approved, the budget for the planning years is constructed by funding the various accounts to the level requested in the plan. In almost all cases, the planning process is iterative, and intense discussion and negotiation determine the final budgeted amount. The early portion of the plan, typically the first year, represents a commitment. Management's performance is measured by comparing actual performance to the planned or expected performance. This comparison is part of the performance appraisal for IT managers.

Administrative Actions

Before, during, and after the planning cycle there are many administrative actions that assist in the planning process and in implementing the approved plan. Before planning begins, IT administrators can develop planning assumptions and provide planning ground rules. For example, keeping the number of programmers and analysts constant throughout the plan period may be a plan ground rule. For planning purposes, the firm might state that it expects to increase revenue by 10 percent per year. This is an example of a planning assumption. The list of ground rules and assumptions is established so that the resulting plan remains within the bounds of reason. Without a set of reasonable ground rules and assumptions, the plan process may iterate forever.

During the planning period, IT managers must coordinate the developing plan with the many organizations that it affects. This happens in a variety of ways, but the administrative function must ensure that communication is complete and unambiguous. Plan-review meetings should be held with managers from all the functional areas in the firm. For example, the plan must be reviewed with marketing managers to ensure that marketing requirements for information technology support are properly planned. Such communication meetings help ensure plan congruence and assist in organizational learning.[12] Through this process, the role and contribution of IT are clarified among the users. A well-coordinated plan review process like the one demonstrated at Citizens Gas mitigates several of the concerns that we studied in Chapter 1.

Many firms establish steering committees consisting of representatives from the functional areas in the firm. The purpose of such committees is to provide guidance to the IT planning and investment process. Effective steering committees increase the mutual understanding of users and IT personnel and tend to increase executive involvement in and understanding of IT matters.[13] Steering committees, consisting of senior executives from all parts of the business, provide valuable assistance in ensuring plan synchronization between IT and the rest of the firm. This is especially valuable for firms that have functionally decentralized IT operations.

Finally, managers and administrators are responsible for implementing the plan and for tracking activity within the organization to actions specified in the plan. Control of the organization is one of management's fundamental responsibilities. IT managers must install measurement and tracking mechanisms designed to ensure that the organization's activities are on target. The control functions must address the plan elements individually.

For example, actions affecting the application portfolio are routinely subject to project management control and monitoring. Control of the arrival and departure of equipment and facilities and the related space is essential. Production services must be measured and controlled according to the service-level agreements negotiated with the users. The firm's financial organization usually tracks the consumption and utilization of financial resources. Actual expenditures are compared with planned expenditures and variances in the accounts must be explained to the firm's controller and others. Schedules, budgets, and accomplishments are the principle control elements in IT plans.

Technology Planning

The IT organization has a continuing responsibility to monitor advances in technology and to keep the firm informed of progress in the field. Senior professionals throughout the IT organization must keep pace with state-of-the-art developments in their disciplines. Programming managers must track programming technology. System support personnel and telecommunications experts must maintain a current awareness of the technology and must improve their expertise. These experts provide technology assessments to the user organizations in the firm through the senior IT executive. Table 4.4 lists some of the many areas of technology that should be evaluated during the planning process.

TABLE 4.4 Technology Areas

Technology Areas
New processor developments
Advances in storage devices
Telecommunications systems
Operating systems
Communications software
Programming tools
Vendor application software
Systems management tools

Introduction of new technology leads to some predictable problems. Costs are usually underestimated and benefits frequently overestimated.[14] It is especially useful for the IT manager to fund a small advanced technology group whose purpose is to make assessments and provide models for utilization of the emerging technology within the firm. The ad tech group researches emerging technologies and evaluates new concepts. It establishes concept feasibility which begins the development and introduction of the new concepts. Feasible concepts are transferred to development and user departments for implementation and production. These activities assist in technology introduction and provide useful insights for planning purposes.

OTHER APPROACHES TO PLANNING

The first chapter presented two ideas fundamental to the management of information technology and particularly relevant to IT planning. These concepts are Nolan's stages of growth and Rockart's critical success factors. Both are important in the IT planning process. They add value to the processes outlined above and test the credibility of the results.

Stages of Growth

Nolan's theories, either the four or six stages, provide predictive insight to organizations within the firm and to the firm itself. To use the stage theory, management must assess the posture of the organization or the firm at the present time. Then plans can be based on an extrapolation of the observed trends. For example, if the firm is in the control stage, it would be wise to anticipate and plan for the integration phase followed by the data administration stage. If one function is in the data administration phase and others are in the integration phase, it is advisable to understand the reasons for this disparity. Perhaps the firm should concentrate on the laggards and bring them up to the level of the more advanced function. For competitive reasons, it may be useful to understand the position of other firms in the industry relative to stages of growth.

Before IT managers take action, they must communicate their assessments to the functional heads involved. There may be very good reasons for conditions being as they are. In any case, the decisions must be made by all interested parties, including the steering committee, not the IT organization alone. The analyses of the stage theory and its use in planning should be integrated into the more formal process outlined above.

Critical Success Factors

Critical success factors are excellent tools to use in conjunction with the formal planning process. They lend structure to planning and improve planning by focusing on important managerial issues.[15] They form an outstanding audit on the results of the process and ensure that necessary conditions for success are contained within the plan.

For instance, consider critical success factors 2a and 2d (Strategic and Competitive Issues), and 4b (Operational Items) identified in Chapter 1. If the IT plan is somewhat out of synch with the firm's plan, the resulting implementation will not totally support the firm's objectives. This unsatisfactory condition must be corrected. If the planning process develops unrealistic expectations or does not ensure realism in long-term expectations, the IT organization will suffer in later years. A third example from the production side of the organization is worth considering: If the IT plan does not contain enough resources to deliver service on time and within planned costs, the IT organization will experience short-term difficulties. The results of the planning process can be evaluated by reviewing the plan against the critical success factors for the IT organization and the firm. This evaluation can also help level expectations within the firm.

The first chapter presented critical success factors founded on issues prominent in the industry and on other factors of importance to the firm. Critical success factors are organized into long-range issues, intermediate-range concerns, short-range items, and business issues. This organization is designed to coincide with the planning timeframes discussed in this chapter. The IT plan should ensure that all necessary conditions for success are part of the plan.

It is extremely important to know the critical success factors of the senior IT executive's superior. It is even more important to know the critical success factors for the firm. If the plan prepared by the IT organization fails to account for these factors, the IT organization will not meet the necessary conditions for success.

Business Systems Planning

Another well-known approach to IT planning is the methodology of business systems planning, or BSP, developed by IBM.[16] BSP concentrates on the firm's data resources and strives to develop an information architecture supporting a coordinated view of the firm's major systems' data needs. The BSP process identifies the key activities of the firm and the systems and data which support these activities. The data is arranged in classes and an architecture is developed to relate the data classes to the firm's activities and to its information systems. In essence, BSP strives to model the business of the firm through its information resources. With BSP, the planning emphasis changes from the applications in the portfolio to the supporting databases.

The BSP process is very detailed and time consuming. It requires a bottom-up effort and tacitly assumes that a data architecture can be developed in one step. The process works best in centralized environments where the data, applications, and structure for processing are physically near one another. It is a useful approach in departments that operate departmental computing systems. The current trends toward decentralization, broad deployment of information technology, and increasing importance and value of the technology greatly complicate the task of IT planning. For these reasons, a variety of planning methods must be considered.

THE INTEGRATED APPROACH

This chapter has presented the important types of plans, plan ingredients, and some important planning approaches that IT organizations currently use. But other planning tools have been discussed in the literature.[17] Each of these approaches offer advantages under certain circumstances, and limitations or disadvantages under other conditions. But most methods are weak when information technology is critical to the firm, when the technology is widely dispersed, and when the organization is decentralized.

Cornelius Sullivan studied 37 major American companies to understand the effectiveness of their planning systems in relationship to factors indigenous to the firm.[18] Sullivan found two factors that correlate with planning effectiveness: infusion, or the degree to which information technology has penetrated the operation of the firm, and diffusion, the extent to which information technology is disseminated throughout the firm. Firms that considered themselves to be effective IT planners were tabulated in a matrix according to the type of planning system in use. Sullivan's conclusion are very revealing. Figure 4.3 presents them in matrix form.

Sullivan's research reveals that the stages-of-growth methodology is effective in firms in which the technology does not have a high impact and is concentrated. Most firms today will not find the stages-of-growth methodology to be effective as their primary planning mechanism. Critical success factors (CSF) were more important to firms in which the technology impact is moderate and the technology is dispersed. For many firms, the CSF methodology is a valuable adjunct to the formal planning mechanisms discussed earlier. BSP is more effective for firms in which information technology is centralized but is very important to the company. When it was developed, BSP was useful for centralized firms like IBM. It is much less important today.

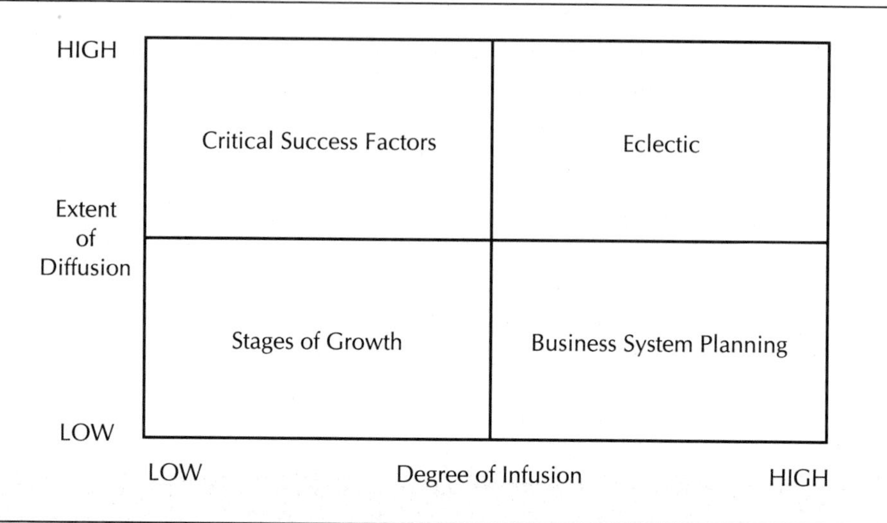

FIGURE 4.3 IT Planning Environments

Many of today's firms are in the upper right quadrant of Figure 4.3 or moving toward that position. For these firms, the planning process is more complex because the IT operations within the firm are very complex. Stages-of-growth concepts will be of limited value, critical success factors will continue to have importance, and concentration on information architecture will be valuable. As a planning mechanism, BSP will have limited utility. What planning methodology will best serve the firm in the Eclectic quadrant? How can IT executives plan effectively given the fast-paced transformations common among firms today?

Firms in the Eclectic quadrant, or moving toward it, face some predictable planning issues. Major changes are taking place in applications portfolio acquisition with the increases in purchased applications, joint development activity and alliances, and end-user and client/server application development. Organizations are being redesigned around information systems. Information technology managers and advocates are architects of the changing environment. Organizations are being reengineered to serve customers better and are looking for IT to support the new structures. Dramatic adoption of telecommunication technology and increasing reliance on networked systems greatly complicate information technology planning.

Michael Earl's studies in the United Kingdom indicate that planning approaches fall into five general categories.[19] The Technological approach attempts to develop an "information systems-oriented" architecture around which planning develops. This seems to be a variant of the BSP method but has some of the same difficulties. The Administrative method focuses on bottom-up resource allocation building off previous plans. In this method, strategy is deemphasized. The Method-Driven approach begins with previous IT plans and may consider recommendations made by consultants. In the final outcome, this method is likely to emphasize external factors. In the Business-Led approach, the firm assumes that business plans will eventually lead to IT plans. Unfortunately, results show that business plans are generally not specific enough for detailed IT planning.

Earl found that the fifth method, the Organizational approach, was consistently superior to the previous four. In this method, planning consists of frequent interchanges between the IT function and client functions to develop projects that lead to shared outcomes. Functional interactions on an as-needed basis supplement other long-range planning activities. This approach selects what is needed, when it is needed: It is indeed eclectic.

Successful IT managers must understand business trends in their firms and must plan to support their firm's business activity, sometimes in the face of considerable uncertainty. Understanding the management issues associated with the firm's application portfolio and managing the portfolio effectively is a critical goal. Supporting and managing user-driven computing and its telecommunications infrastructure are also critical activities. And establishing relationships with executives and client organizations to maintain flexibility and facilitate change efficiently is mandatory. These topics are all subjects addressed in subsequent chapters. In part, planning in this environment will always be eclectic.

MANAGEMENT FEEDBACK MECHANISMS

Throughout, this chapter mentions the need for control processes or management feedback mechanisms. Control consists of knowing who, what, when, where, why, and how for all essential activities of the organization. If managers know these answers, they are operating under control. If they don't, then they are operating out of control. An organization operating out of control causes very serious difficulties for the managers and for the firm. Control is vital for the success of IT organizations—not only for the organization itself, but for the entire firm. The basis of control is found in sound strategy development and planning.

The IT plan should provide data to gauge the answers to these questions. Control processes must be designed to compare the actual performance of the organization to the expected performance, as detailed in the plan. Chapter 3 discussed the notion of performance assessment in connection with the strategy development process. Plans describing how the strategy is to be implemented are used to gauge and assess actual performance. In other words, plans are the basis for assessing progress toward achieving strategic goals and objectives.

Figure 4.4 shows how these ideas fit together. Strategic processes lead to tactical processes which, in turn, are expressed in operational actions. Measurements lead to control assessments. These control assessments focus attention on tactical processes which, in turn, bring focus to strategic processes. Thus, the process is iterative, and strategies and plans are validated or altered with each planning cycle.

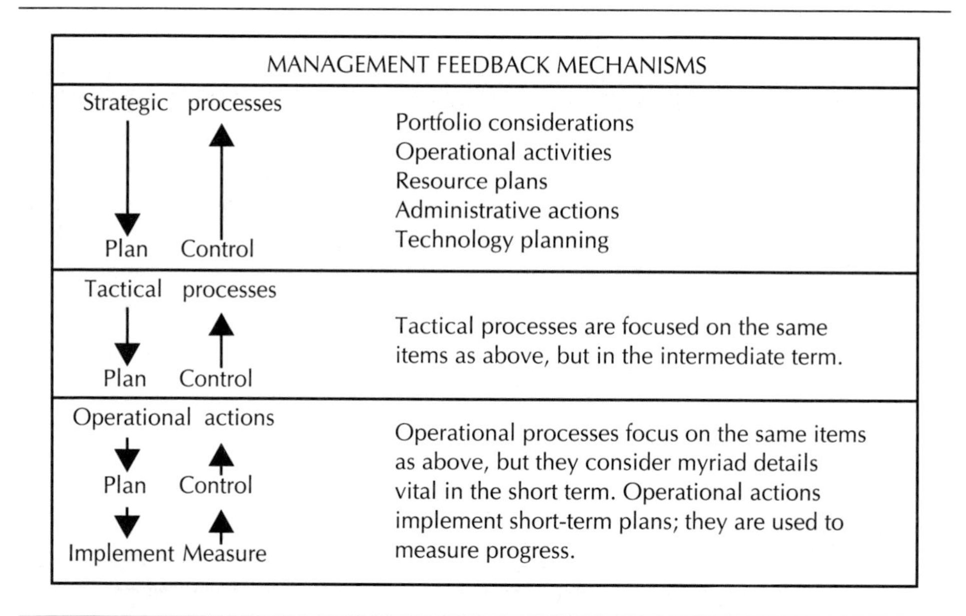

FIGURE 4.4 Management Feedback Mechanisms

SUMMARY

The task of planning for the IT organization is complex and best accomplished through systematic processes. The concepts of planning rely on separating the extended planning horizon into strategic, or long-term, tactical, or intermediate-term, and operational, or short-term components. The planning cycle is an annual event, thus, each cycle adds another year to the strategic period and each plan advances one year. This systematic process is synchronized with the plan calendar for the entire firm.

The IT organization's strategic plans are developed from IT strategies giving the strategic statements detail and specific implementation actions. Strategic plans usually cover the period from two years in the future to the end of the extended planning horizon. Tactical plans detail the actions to be taken for the next two years. They are derived from previous strategic plans and give implementation direction to previous strategic goals and objectives. Operational plans are the very near-term implementation of tactical planning.

Regardless of how they are developed, plans are only as good as their implementation and control mechanisms. Strategic and tactical processes and operational actions are related to measurement and control actions through management feedback mechanisms. Control is a fundamental management responsibility—especially important to IT managers. Measurements are also very important because they form the foundation for performance appraisal. Information technology planning is a critical success factor for IT managers. It is a necessary condition for their success.

Review Questions

1. What are the differences between the strategy and the strategic plan?
2. What do we mean by the extended planning horizon? Describe the planning horizons discussed in this chapter.
3. What is the relationship of strategic plans to tactical plans? How are operational plans related to tactical plans?
4. For the strategic plan, what resources must be blended together and planned?
5. What financial items are included in the firm's strategic plan?
6. Describe the purpose of tactical plans.
7. Who are the main users of operational plans? How does the firm's controller use operational and tactical plans?
8. Differentiate between the planning calendar and the planning cycle.
9. What are the elements of the IT organization's planning model?
10. What considerations surround the planning for the applications portfolio?

11. To achieve proper control, what questions need to be answered?
12. What are the elements of production operations planning?
13. Which elements ensure the attainment of service levels through management processes and which through resource planning?
14. What elements comprise the resource portion of the IT plan?
15. How is the financial portion of the IT plan constructed?
16. What does the term plan ground rules mean? How do plan assumptions differ from plan ground rules?
17. What role does technology play in planning for the IT organization?
18. What is the relationship between stages of growth and critical success factors given the planning process described in this chapter?
19. When does the business systems planning methodology become important in IT planning?
20. What does eclectic planning mean, and for what firms is it important?
21. What is the connection between operational plans and management control?

Discussion Questions

1. What are the favorable elements of the planning process at Citizens Gas as described in the Business Vignette? How many of the critical success factors presented in Chapter 1 are touched by the process at Citizens Gas?
2. Chapter 1 presented a list of critical success factors for the IT manager. Which of these are related to IT planning?
3. Using a mainframe installation as an example, discuss the continuum of timeframes inherent in IT planning. Describe the levels of detail in the plans throughout these timeframes.
4. Discuss the planning cycle and the planning horizon for a university operating on the academic year starting in September and ending in August.

5. Name several planning assumptions and ground rules that might apply to the university in Question 4.
6. What is the role of the firm's planning department in the plan development process?
7. What additions might be included in the planning model for an IT organization that provides services to customers outside the firm?
8. Discuss the relationship between the list of critical success factors and planning for production operations.
9. Equipment plans are often complicated by the need to include auxiliary items such as space, air conditioning, supplies, and maintenance. Discuss how the IT manager might determine an effective equipment plan.
10. What is the role of IT managers in the process of people planning? Discuss why this activity should take place on a continuing basis, not just at plan time.
11. Why is technology assessment vital to the IT organization? Describe how you think this assessment can be brought into the planning process.
12. Discuss some ways in which you think senior IT management should exercise control.

Assignments

1. Study BSP or one of the other planning techniques mentioned in this chapter. What are its advantages and disadvantages? How would it fit with the process discussed in this chapter?

2. A firm would like to install a more powerful, lower cost mainframe to replace its present mainframe. What are the kinds of items that must be planned to accomplish this objective?

ENDNOTES

[1] Leslie Goff, "I Walk Aligned," *Computerworld*, August 8, 1994, 68.

[2] See Note 1 above

[3] See Note 1 above

[4] There has been much criticism of strategic planning but most executives value it. See Daniel H. Gray, "Uses and Misuses of Strategic Planning," *Harvard Business Review*, January-February 1986, 89. This article describes some actions which overcome the difficulties of strategic IT planning.

[5] Marilyn Parker and Robert J. Benson, "Enterprisewide Information Management: State-of-the-Art Strategic Planning," *Journal of Information Systems Management*, Summer, 1989, 14.

[6] Albert R. Lederer and Vijay Sethi, "The Implementation of Strategic Information Systems Planning Methodologies," *MIS Quarterly*, September, 1988, 445.

[7] As in the previous chapter, this discussion assumes that the firm has some kind of formal planning process. The absence of a formal planning process increases the risk of plan incongruence and may lead to planning failures.

[8] Mick Rackoff, Charles Wiseman, and Walter A. Ullrich, "Information Systems for Competitive Advantage: Implementation of a Planning Process," *MIS Quarterly*, December, 1985, 285. Also see Albert L. Lederer and Vijay Sethi, "The Implementation of Strategic Information Systems Planning Methodologies," *MIS Quarterly*, September, 1988, 445.

[9] Martin Buss, "How To Rank Computer Projects," *Harvard Business Review*, January-February, 1983, 118. This article discusses the issue of who should make the decisions on applying resources to the portfolio. Buss endorses the formal planning process as the most effective method for prioritizing the firm's application projects.

[10] Edward A. Van Schaik, *A Management System for the Information Business: Organizational Analysis* (Englewood Cliffs, NJ: Prentice-Hall Inc., 1985). In particular see Chapter 7, Service-Level Management, pages 155-225. Also see John P. Singleton, Ephraim R. McLean and Edward N. Altman, "Measuring Information Systems Performance: Experience With the Management by Results System at Security Pacific Bank," *MIS Quarterly*, June, 1988, 325.

[11] Peter G. W. Keen, "Business Innovation Through Telecommunications," *Competing in Time: Using Telecommunications for Competitive Advantage* (New York: Ballinger Publishing Co., 1988), 5.

[12] See also John C. Henderson and John G. Sifonis, "The Value of Strategic IS Planning: Understanding, Consistency, Validity, and IS Markets," *MIS Quarterly*, June, 1988, 187.

[13] D. H. Drury, "An Evaluation of Data Processing Steering Committees," *MIS Quarterly*, December, 1984, 257. Drury found that approximately half of the firms polled had a committee which, on average, had existed for five years. The typical committee contained six members and was most effective when it held regular meetings and when agenda items originated outside the IT department.

[14] B. Gold, "Charting a Course to Superior Technology Evaluation," *Sloan Management Review*, Fall, 1989, 19.

[15] Andrew C. Boynton and Robert W. Zmud, "An Assessment of Critical Success Factors," *Sloan Management Review*, Summer, 1984, 17.

[16] *Business Systems Planning: Information Systems Planning Guide*, White Plains, NY: IBM, 1984.

[17] Albert L. Lederer and Vijay Sethi, "The Implementation of Strategic Information Systems Planning Methodologies," *MIS Quarterly*, September, 1988, 445.

[18] Cornelius H. Sullivan, "Systems Planning in the Information Age," *Sloan Management Review*, Winter, 1985, 4.

[19] Michael J. Earl, "Experiences in Strategic Informations Systems Planning," *MIS Quarterly*, March 1993, 1.

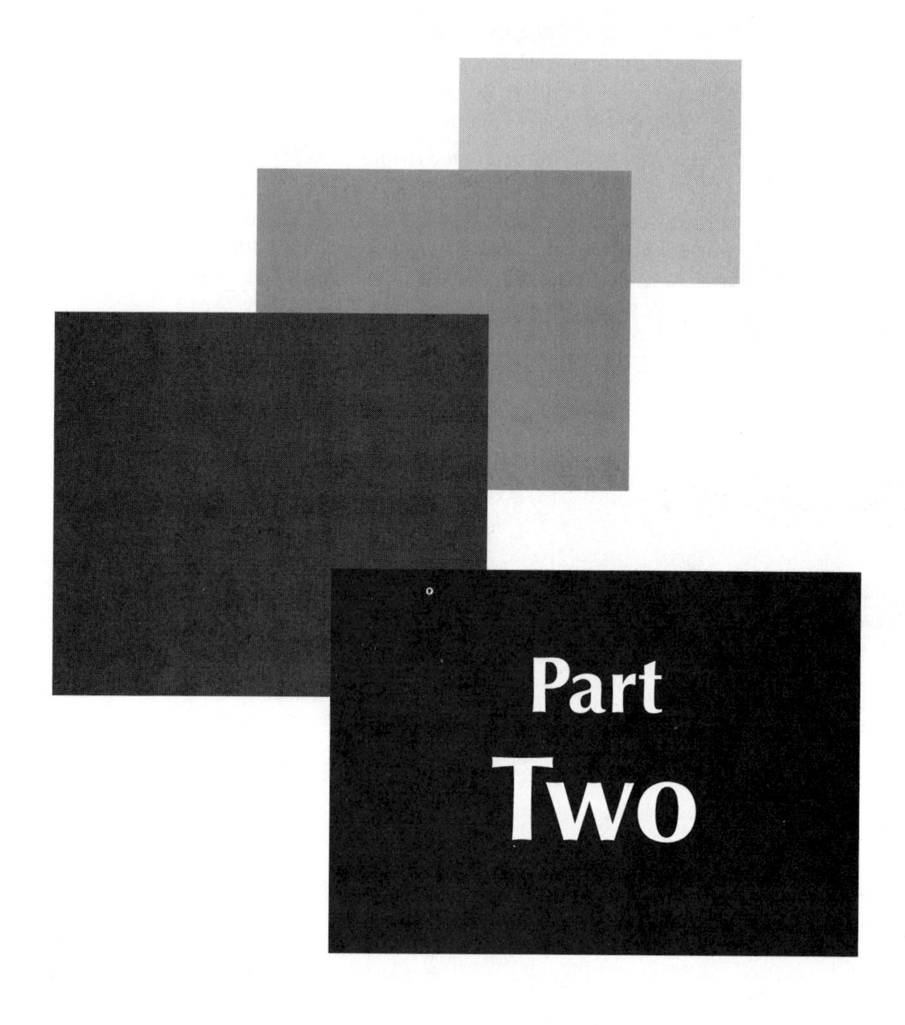
Part
Two

Technology and Industry Trends

Advances in computing hardware, operating system software, and storage technology create many opportunities for business managers. Simultaneous rapid advances in telecommunication systems technology allow managers to distribute data and data processing capability, integrate business systems, and leverage returns on their investments. Extensive deployment of personal workstations changes the work of employees and managers and permanently alters the firm's conduct of its affairs. Part Two chapters include Hardware and Software Trends, Modern Telecommunications Systems, and IT Industry Trends.

Successful information technology managers are highly cognizant of technology and industry trends. They use this trend information to prepare their people and organizations for the future.

5 Hardware and Software Trends

A Business Vignette

At Stake: Apple Computer's Survival

Apple Computer's survival was the topic of debate at the PC Outlook conference sponsored by Technologic Partners in December, 1993.[1] The debate came at a time when Apple's fortunes were bleak indeed. Gross profit margins dropped from 43 to 26 percent in 12 months, and 2,100 of 13,000 permanent employees had been laid off. Apple was in the midst of an executive transition as Michael Spindler replaced John Sculley as chief executive and Mike Markkula, Apple's cofounder, was selected to be Apple's Chairman. Could competition do anything Apple could do and do it just as well, or could Apple continue to develop easy-to-use, innovative PCs for its loyal followers? The issue was critical for Apple's survival.

The fortunes of Apple had turned negative just as they had for IBM earlier, and just as they would for Digital Equipment Corporation later. The makers of computers were in for troubled times during the early 1990s, but each for different reasons. In Apple's case, the fear was that the visionary spirit created by Steve Jobs, Mike Markkula, Stephen Wozniak, and John Sculley was gone. The appeal of Apple computers—ease of use, graphical interfaces, the mouse, stereo sound, and other features—was being imitated by many competitors. The poor marketplace showing of the Newton device reinforced the company's declining image. This small, hand-held personal assistant was touted as revolutionary technology. Apple boasted of its ability to read handwriting and to bring instant communication to its bearer. In fact, reading handwriting turned out to be much less successful than Apple anticipated, and Newton was poorly received.

Apple was a niche player from the beginning. Its systems were incompatible with those built from Intel parts, like IBM and all its clones. Apple relied on Motorola microprocessors, and its software and applications were not compatible with others. In addition, Apple refused to license its operating system to other manufactures thus preserving its uniqueness to the detriment of market share. To maintain market share, Apple needed to lead the field with innovations.

Feeling pressured, Apple decided to join forces with IBM and others. In July 1992, IBM and Apple signed a development agreement that led to two joint software projects: Kaleida Labs, Inc., and Taligent, Inc. In addition, joint development led to the PowerPC, a microprocessor designed by IBM and manufactured by Motorola and IBM, that is making its way into both IBM and Apple workstations. Apple is marketing a new version of Newton that stores handwritten notes without interpreting them—and the Newton architecture is being licensed to several

Japanese firms—a bold new step for Apple. In January 1994, Hewlett-Packard announced the purchase of 15 percent of Taligent and plans to use some of Taligent's object-technology in HP products. Like many other high tech firms, Apple is forming alliances to augment its strengths and to buttress its weaknesses.

Internally, Apple is beefing up management controls on advanced technology projects, shortening development cycles, and searching for technology to reinforce its stock in trade—user friendliness. As TVs and computers merge in the home entertainment market, Apple hopes to play a lead role with its Mac TV. The new management team at Apple is determined to take the lead once again, and to grow its cadre of loyal followers. Will Apple survive? Overwhelmingly, PC Outlook Conference attendees felt sure it would.

After peaking at $530.4 million in 1992, Apple's 1993 profit dropped to $285.6 million, declined further to $231.0 million in 1994, and is expected to rise to $575.0 million in 1995.[2]

INTRODUCTION

During the past 30 to 40 years, rapid technological innovations have fueled unprecedented advances in digital computing capability.[3] These innovations occurred in the fields of semiconductor technology, recording technology, telecommunications capability and, to a lesser extent, in the programming arena (Table 5.1). Taken together, these four technical activities form the foundation of the information technology revolution. Separately, each endeavor is subject to limitations imposed by the laws of physics, human or organizational limits, system or interrelational barriers, fabrication limits, and limitations reflecting economic reality. During the last three decades, practioners in these fields have marshaled intellectual, financial, and production resources to push practical limitations of these technologies toward their theoretical limits.

TABLE 5.1 Foundations for Information Technology Advances

Semiconductor technology
Recording technology
Telecommunications systems
Programming

In this endeavor, developers have been extraordinarily successful. Progress over the past three decades is measured in orders of magnitude (factors of 10). Reports indicate that technical capability has doubled in performance and cut cost in half over a period of a few years. Size, as measured by area or volume, has declined proportionately, and is one of the underlying contributors to performance increases and cost reductions. Only the ability to produce computer instructions—programming—has proceeded at a more reasonable pace.

Additionally, and again with the exception of programming, one can confidently predict that over the next five years or so, this rapid innovative pace will continue. We can be less certain about the practical utilization of these technologies within our social and business structures. Certainly, the results will be dramatic and highly meaningful for most of us and for the organizations in which we operate.

This chapter and the next will explore these trends and make careful projections concerning their future form and direction. In particular, the text will relate these trend data to the activities of our organizations and will extract meaning from them for ourselves as managers, for our employees and associates, and for our organizations. Successful managers remain alert to trend information; exceptional IT managers use leading technological indicators to prepare themselves and their firms for the future.

SEMICONDUCTOR TECHNOLOGY

The invention of the transistor in 1948 set in motion a stream of innovations and inventions that spawned the semiconductor industry.[4] The industry is developing many new products and is growing rapidly as a result. Advances in semiconductor technology have helped fuel the growth of the electronic data processing industry. Since 1948, the electronics industry has grown to become third largest in the U.S., behind automobiles and petroleum.[5] The U.S. computer industry creates annual worldwide revenues of about $150 billion.

Semiconductor technology is important to IT managers because it forms the foundation on which the information industry is built. Semiconductor chip technology makes today's microprocessors possible and permits system designers to pack ever more computational performance into systems of all sizes, at steadily decreasing unit costs. Advances in chip design fuel the growth of the information-handling industry and have created the information age as we know it. IT managers must have knowledge of the trends in this industry if they are to effectively perform their jobs.

SEMICONDUCTOR TECHNOLOGY TRENDS

Although the electronics industry is growing at an exceptional rate, growth depends upon the ever-decreasing size of the integrated circuit itself, and upon continued increases in the density of devices on the carrier. Industry growth rates are directly related to circuit density increases or growth in the number of devices per unit area of carrier. In addition to the increasing density of circuitry on the chip, the reliability of semiconductor devices is improving with time. Unit cost is rapidly declining. All of these trends are favorable for the ultimate consumer and for the semiconductor industry. Table 5.2 summarizes these trends.

TABLE 5.2 Semiconductor Technology Trends

Declining device size
Increasing density of devices on chips
Increased switching speeds
Expanded function per chip
Increasing reliability
Rapidly declining unit prices

Figure 5.1 displays the growth of circuit density versus time. Also shown are the density of integrated circuits on a chip as a function of time, from the late 1950s to the present, and projections into the year 2020.

Notice that the vertical scale is displayed in powers of ten, ranging from one to 10 billion circuits per chip. The number of circuits per chip during the 30-year period from 1960 to 1990 grew by seven orders of magnitude; from about three to 30 million during this time! Projections to the year 2020 range from slightly less than 1 billion to more than 10 billion circuits per chip, depending on assumptions regarding the processes required to build these superchips. In early 1995, papers describing experimental laboratory memory chips containing one billion bits, and experimental microprocessors capable of performing more than one billion instructions per second are being presented at technical conferences.[6]

The reasons behind the thrust to shrink the size of integrated circuits are to reduce overall cost and to minimize the switching time of the devices (the time needed for the device to go from one binary state to the other). Ever-decreasing circuit dimensions yield lower unit cost and higher unit performance, thus fueling continued semiconductor industry growth.

Present manufacturing technology produces circuit lines on a silicon chip that are fractions of 1 micron wide. (A micron is one ten-thousandth of a centimeter, or roughly one one-hundredth the width of a human hair.) Advanced photo lithographic techniques can reduce the line widths even further; in some cases down to 0.1 micron. Reducing the line dimensions from 1 micron to 0.5 microns quadruples the number of circuits per unit area and reduces signal travel distances, thus improving chip performance. Therefore, industry growth rates which depend on increasing performance and diminishing unit costs are inversely related to the size of the fundamental circuit.

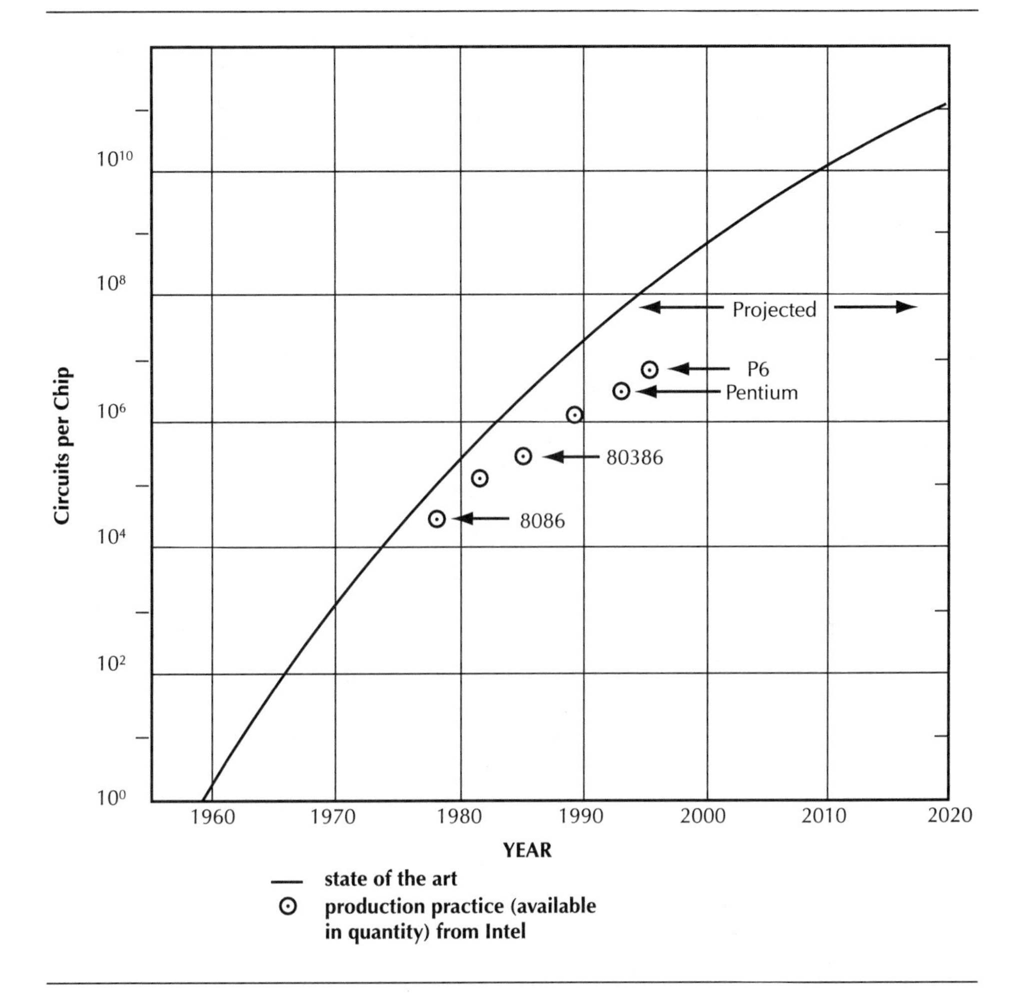

FIGURE 5.1 Semiconductor Chip Density vs. Time

Robert Keyes summarizes this concept, stating that:

> I am writing this article on a computer that contains some 10 million transistors, an astounding number of manufactured items for one person to own. Yet they cost less than the hard disk, the keyboard, the display, and the cabinet. Ten million staples, in contrast, would cost about as much as the entire computer. Transistors have become this cheap because during the past 40 years engineers have learned to etch ever more of them on a single wafer of silicon.[7]

Through the mid 1990s, 35 years since the commercial introduction of the single transistor chip, the number of transistors that can be fabricated on each chip has grown to more than 100 million—a factor of 10 to the 8th power and an average growth rate of 100 times each decade! While the performance increased by more than 100,000 times, the cost per chip has remained relatively unchanged. (Transistors on a chip can now be purchased for about 4000 per dollar.) The industry has maximized all the important factors in chip production. Unit size has declined substantially, unit performance has increased remarkably, and cost to the consumer has declined dramatically.

As the dimensions of silicon devices shrink and approach their minimum practical dimensions, scientists are searching for new technologies and materials to maintain the downscaling process.[8] One such material, gallium arsenide, possesses these physical properties. Gallium arsenide will potentially reduce device size by another factor of 100, increase switching speeds up to 200 billion times per second, and reduce cost per unit of function by orders of magnitude. Cray Computer Corporation is producing 3/16 inch square gallium arsenide chips for the Cray 4 supercomputer. Industrial, government, and academic laboratories around the world are researching this technology.

Intel's Pentium microprocessor chip is one example of the rapid pace of semiconductor development. First shipped in 1993, Pentium is the follow-up to the very popular 80486 chip and the latest in the 80x86 line. During development it was labeled P5 but the public thought it would be released as the 80586 chip. For legal reasons, Intel named the new chip Pentium so the name could be protected. Table 5.3 displays the advances in the Pentium family of microprocessor chips through 1994, and projected into 1996.[9]

TABLE 5.3 Technical Advances in the Pentium Microprocessor Chip

Internal Clock Rate (MHz)	Line Width (microns)	Availability
60	0.8	4th Quarter '93
66	0.8	1st Quarter '94
75	0.6	4th Quarter '94
90	0.6	3rd Quarter '94
100	0.6	4th Quarter '94
120	0.6	2nd Quarter '95
132	0.6	3rd Quarter '95
150	0.4	4th Quarter '95
200	0.4	2nd Quarter '96

The 75 MHz and faster chips differ from earlier versions by reduced line width and increased function. These faster chips contain an advanced controller function to accommodate a second processor on the motherboard. This enables symmetric multiprocessing, an advanced feature that

subsequent versions of advanced operating systems will support. The line width declines to 0.4 microns in the 150 and 200 MHz implementations, a four-fold increase in circuit density from early 1994 to late 1995, if analysts' projections are correct. The 200 MHz chip, called P6 internally at Intel, will execute instructions at more than 250 MIPS.

But Intel is not alone in the race to develop advanced performance microprocessor chips. In September 1994, Digital announced the next step in its Alpha chip series with a 300 MHz internal speed capable of processing 1.2 billion instructions per second.[10] This announcement places DEC in the lead for microprocessor speed, ahead of Intel's Pentium, IBM/Motorola/Apple's PowerPC, and SPARC from SUN Microsystems. Figure 5.2 displays the growth in processing speed over time for several popular microprocessor chips.

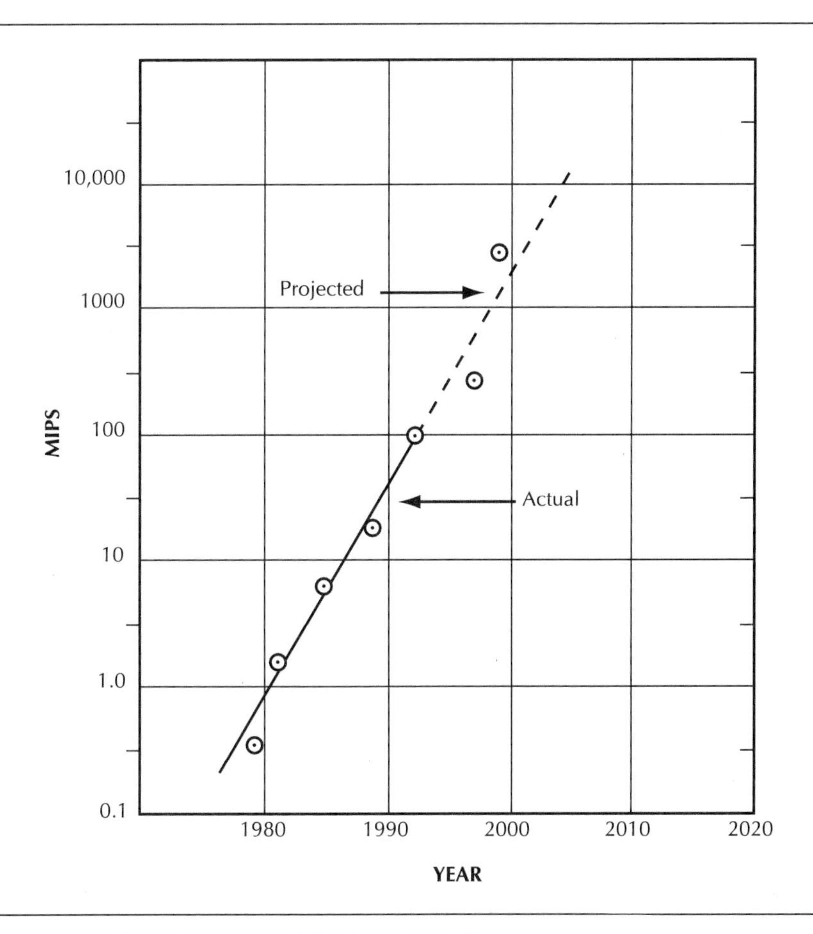

FIGURE 5.2 Microprocessor Chip Speed vs. Time

DEC's new chip is fabricated with 0.5 micron technology and could presumably be enhanced further with a move to 0.3 micron lines. Alpha is based on a 64-bit architecture and superscaler design. This gives Alpha one terrabyte of memory addressability and allows the initiation of four instructions per clock cycle. In addition, its internal microcode permits operating system independence, a major competitive advantage. Alpha is a superior design but its success ultimately depends on DEC's business skills.

Improvements in silicon chip design and manufacture continue in semiconductor laboratories around the world. For example, NEC Technologies is building a plant for 256 Mbyte memory chips, increasing storage by 16 times over the present advanced 16 Mbyte chip. By reducing line widths to 0.1 micron, experimental chips attain switching speeds of about 1/100th of a nanosecond. Achievements like this help close the gap between silicon and gallium arsenide. With extensions of existing technology, silicon will continue to be used for chip fabrication for the next decade.

The invention of the transistor, and its continued development over nearly 50 years has been enormously important in many ways, and its importance will continue to grow. Chip densities will continue to increase by orders of magnitude through advances in physics, metallurgy, and manufacturing tools and processes. We can plan on smaller, faster, lower cost, and more reliable memory and logic circuitry. IT systems, however, require more than logic and memory to function. Large-capacity storage devices are a vital and important adjunct to powerful computer chips. Fortunately, scientific advances abound in the field of recording technology, as well.

ADVANCES IN RECORDING TECHNOLOGY

We are witnessing very rapid progress in logic and memory circuit development and, fortunately for digital computing, recording technology progress is equally dramatic. Very high speed processors have enormous appetites for instructions and data, and they rely very heavily on high-performance auxiliary memory for system throughput. Approximately equal progress in both fields is essential to progress in systems development.

Magnetic Recording

Magnetic recording devices allow permanent storage of massive amounts of data, measured in gigabytes, at lower cost per byte than semiconductor devices. Access to the data is in the range of 10 milliseconds and the data transfer rates can be on the order of 16 megabytes per second. For example, the Western Digital AC31000 direct access storage device holds 1.083 gigabytes of data with a seek time of 10 milliseconds. The disk runs at 4495 RPM so the average rotational delay is 6.7 milliseconds. The manufacturer warrants the device for three years and advertises 250,000 hours mean time between failures. At a retail price of $489.00, the cost per megabyte of storage is $0.45. Higher performance drives are available at slightly higher costs per megabyte.

Magnetic recording technology is impressive. On a rigid disk, data is read or written by a read/write head flying over the surface on a cushion of air about 10 millionths of an inch thick. The disk under the head travels at a speed approaching 100 MPH. Manufacturing tools and processes required to construct rigid disks in the factory rival those found in the semiconductor plants. Tolerances on surface finish approach 1 millionth of an inch, and the purity of materials in the thin-film heads is similar to that of the silicon used in chip production. Disk manufacturing plants are highly capital intensive.

Advances in head design, recording materials, and production practices will ensure continued growth in capacity and performance of both rigid disks and magnetic tape devices.[11] As with silicon chip technology, advances in magnetic recording technology will migrate from the higher performance, higher capacity devices to the lower performance, more widely available devices. In addition, improvements in disk subsystem architecture will greatly increase storage system reliability.

Fault-Tolerant Storage Subsystems

The common practice of interconnecting large numbers of PCs into Local Area Network (LAN) configurations increases the need for reliable operations of critical network components. The failure of the server computer (the computer that coordinates network operations and acts as a data repository) or its storage devices may bring the entire LAN operation to a halt, disrupting operations, and perhaps causing lost revenue. Critical systems need to be fault tolerant. Fault-tolerant devices and networks are configured to reduce the impact of failure by introducing redundancy and backup mechanisms.

To reduce the exposure to failure in the storage subsystem, disk devices are organized to maintain some form of data redundancy. This concept is called RAID—Redundant Array of Inexpensive Disks. RAID is a set of standards (RAID 0 through RAID 6 plus some others) describing various types of redundancy that can be built into disk subsystems. For example, RAID 1 employs complete redundancy by writing each data set to two different disks. RAID 5 stores data on alternate drives in sector sized blocks with parity interleaved. RAID 6 is the same as RAID 5 but includes redundant disk controllers and power supplies. RAID 10 is a combination of RAID 1 and RAID 5.

Redundancy greatly improves the mean time between data loss (MTBDL) of a storage subsystem. For example, by replacing a four-disk non-RAID subsystem with a five-disk RAID 5 subsystem, MTBDL increases from about 38,000 hours to 48,875,000 hours, an improvement of about 1300 times.

The operating system considers the disk array to be a single unit but, in some systems, components can be replaced during system operation. For example, a diskette that may be showing signs of failure can be replaced with a new one without terminating system operation. This is called a hot replaceable or hot pluggable concept.

The notion of redundancy in storage systems can be extended to other system components such as power supplies, cabling, and controllers. Redundancy improves system reliability at an increased cost, but it may reduce slightly instantaneous system response. Technicians and managers must understand the trade-offs when making system decisions.

CD-ROM Storage

Magnetic recording will remain highly important for a long time to come, but optical devices are beginning to play an equally important role in multimedia and other applications. The ability to store video, audio, and text and to combine it in many useful patterns has numerous applications. Users have just begun to tap the potential of multimedia applications.

The compact disk, known mostly for its ability to record sound to near perfection, is also a powerful medium for storing computerized data. One disk, less than five inches in diameter, can hold 600 million bytes or characters of information, equivalent to about 30 sets of encyclopedias. If this isn't enough for an application, devices resembling jukeboxes are available for accessing many CDs from a single computer. About 4.7 million read-only disk drives are expected to be installed in 1995, up from 2.7 million in 1993.[12]

As with many new technologies, the standards-setting process lags far behind and there is a proliferation of vendor-defined hardware and software formats. Adopters of CD-ROM technology need to establish strategies that will permit future growth yet minimize costs associated with incompatible formats. Most organizations find CD-ROM storage attractive, and the applications are predicted to grow explosively in the future.

Numerous large databases, such as the 1990 census data, are being made available on CDs for computer users. Newspaper and magazine references; abstracts of academic research; legal and medical files; and very large databases of financial information are all available at a price that is declining with improvements in the technology and economies of scale in production and sales. This storehouse of information is now widely available and augments the printed media; it greatly improves the process of information retrieval for many purposes.

FROM MICROCOMPUTERS TO MAINFRAMES

The construction of computers from building blocks of logic and memory, on-line and off-line storage devices, and a variety of input and output devices spans a performance spectrum of several orders of magnitude. The spectrum is typically, but somewhat arbitrarily, divided into microcomputers, minicomputers, and mainframes. This division is based on central processing unit (CPU) performance measured in millions of instructions per second, or MIPS. The low end of the spectrum typically includes embedded micros and, on the high end, supercomputers. Embedded micros are small computing devices found in machine tools, automotive or aircraft control systems, or household appliances such as microwave ovens.

Mainframe computers did not attain one million instructions per second until around 1965, but they have steadily increased in performance at the rate of about one order of magnitude each 10 years since then. Prior to the advent of minicomputers in 1966, mainframes were our only CPUs. The idea of personal computers originated shortly after 1970, and it was at about this time that small computing devices were being embedded into equipment.

With CPUs intermediate in performance between mainframes and personal computers, microcomputers are a vital component of the computing spectrum. They serve small and medium businesses as central processors or as servers on departmental or enterprise networks.

WHAT'S HAPPENING WITH SUPERCOMPUTERS?

Supercomputers, the highest performance systems of all, are manufactured by firms in the U.S. and abroad to serve the needs of scientific laboratories, government agencies, and large commercial customers. The technology and architecture employed in these machines is leading-edge, and a precursor to tomorrow's more widely available machines.

With capabilities exceeding two hundred MIPS, uniprocessors are commonplace today, but many system designers are turning to multiprocessors to enhance performance. Multiprocessors are not a new phenomenon, having been in existence since the 1960s, but present-day processor technology is substantially different from that of two decades ago. Frequently, multiprocessors used in commercial applications consist of a front-end machine serving as a job scheduler for one or more back-end production machines. In some applications, the workload is divided by type between the processors; one processor handles communications and another performs batch production, for example. The trend of advanced technology, however, is toward multiprocessors having as many as tens of thousands of interconnected processors.

Since these large machines are typically used for scientific calculations, their performance is measured in floating-point operations per second, (flops). Floating point is a term used to describe the ability of the computer to keep track of the decimal point over many orders of magnitude while performing arithmetic calculations. A million flops is one mega-flop, or Mflop, and a billion flops is one giga-flop, or Gflop.

Supercomputers having a large number of parallel processors are capable of a trillion flops, or one Tflop. But even more powerful machines are being planned. Some plans contain 32 million processors. These machines, designed for image processing, execute the same instruction simultaneously in all processors. This architecture is called single instruction, multiple data.

Computer manufacturers and government agencies are discussing the development of ultra-fast computers with projected processing speed of 1 petaFLOP, or 1 million billion floating-point operations per second. Such performance can be achieved only by interconnecting large numbers of very high performance processors in a massively parallel organization. Presently, computer scientists are considering machine designs that have one million or more processor nodes each capable of executing instructions at the rate of 100 Mflops. Calculations indicate a computer of this capability would have more computing power than all the networked computers in the U.S. today.

But there are numerous major engineering and programming problems to be solved before the power of these computers can be realized. Operation of one million interconnected 100 Mflop processors implies massive internal data flows. Developing high-bandwidth internal communication channels among and between processors and to the huge main memory poses tremendous engineering challenges. The development of operating systems to manage these massively parallel machines and the construction of application software that takes advantage of this tremendous computer resource is also a formidable challenge. Machines of this type are ideal for scientific research.

But the usefulness of supercomputers and parallel processing extends beyond the bounds of scientific studies: Wal-Mart uses them to obtain vital information in its vast store of sales and purchasing data.[13] It uploads 20 million point-of-sale transactions from its 2100 stores daily, adding to its 6 terabyte database. Several thousand complex queries are issued daily against the database, two-thirds of which is stored for use by parallel-processing machines. Wal-Mart began moving to parallel processing in 1989 and credits its sophisticated information system with helping it master the details of sales and inventory management.

Advances in supercomputer technology show no signs of slowing and the technology and architecture embodied in them will migrate to a broader range of machines in the future. According to Seymour Cray, head of Cray Computer Company, Gflop capability will be available in desktop computers very soon.[14] The technology of computing is advancing rapidly and relentlessly from one end of the computing spectrum to the other.

THE MICROCOMPUTER REVOLUTION

The performance increase of personal computers since their introduction in the early 1970s has been the most rapid of all. Some PCs will soon be available with more than 200 MIPS capacity. Personal computer performance in 1995, as measured in MIPS, equaled that of mainframes widely available in 1990. Today's personal computer puts the power of yesterday's mainframe on the desks of millions of users. This trend has important social and business implications because it profoundly and irrevocably changes communication methods and the role of computing in modern society.

The exponential growth rate trends are sure to continue with Pentium, Alpha, their followers, and many other implementations. Sales figures indicate that the total number of PCs installed in 1995 will exceed 50 million, and that most of these will be part of a network. Currently there are about 30 personal computers per 100 people in the U.S. and the number is growing rapidly. The growth of PCs in both performance and sheer numbers will continue unabated into the 21st century.

Within the decade, 1000 MIPS desktop computers containing gigabytes of storage will be connected through high-speed data links to larger computers, giving users access to many more gigabytes of information. Ten years hence, office workers will have enormous computing and communicating power at their fingertips. The challenge for them, their managers, and their firms will be to use this capability advantageously. The field is ripe for inventors, innovators, and visionaries.

TRENDS IN SYSTEMS ARCHITECTURE

As we have seen, performance, capability, and function of computer systems continue to increase rapidly while unit size declines. Additionally, cost per unit of function is declining significantly. Given that we are confident of these trends, what alternatives are available to deploy this capability within the enterprise? What systems considerations or architectural innovations are being employed to translate continued hardware improvements into improvements in organizational effectiveness?

Client/Server Architecture

The abundance of low-cost, high-performance personal workstations, convenient and reliable means for interconnecting them, and the organizational need to use technology to empower workers is culminating in a strong trend toward moving processing power from centralized mainframe systems to local workstations. Organizations moved from distributed data processing models (departmental islands of computing using mid-sized computers) to end-user computing environments (individual islands sometimes linked to departmental or centralized systems). During this evolutionary period, the computing workplace was restructured as more function moved to the workplace, and as interconnectivity increased. This transformation has developed into a very popular architecture called client/server.

The first application of local networks to data processing consisted of workstation terminals (keyboard and display) connected to mainframe processors on which large data stores and application programs resided. Terminal operators entered or updated data in mainframe storage and reviewed the results of application processing on their displays. The application program included the format of the data presentation on the display terminal. In most cases, initiation of mainframe application processing was performed by computer operators working to a production schedule. In other cases, such as program development, operators entered source language statements at terminals, language compilers were initiated from the terminal, and the compiled results were displayed at the terminal from compiler-created data sets.

As their capabilities advanced, personal workstations were used to customize the presentations of mainframe application output and, for modest applications, to process the applications themselves. Voluminous data stores such as archival records, and sensitive information such as payroll records, remained securely protected on mainframe or departmental facilities. These larger CPUs provided file space and print facilities for numerous users at linked personal workstations. When used in this manner, the larger CPUs became known as file servers or print servers. Data and program storage, application processing, and control functions became rebalanced as application architecture migrated from traditional host-supported terminals through file and print server modes to client/server models of operation. The evolution of client/server architecture is shown in Figure 5.3.

In client/server mode, application data is entered into workstation storage and preliminary data processing takes place there perhaps using additional data from the server. Similar operations may be occurring simultaneously at several workstations throughout the business. Periodically, results of workstation processing are combined, and final or summary processing takes place at the server.

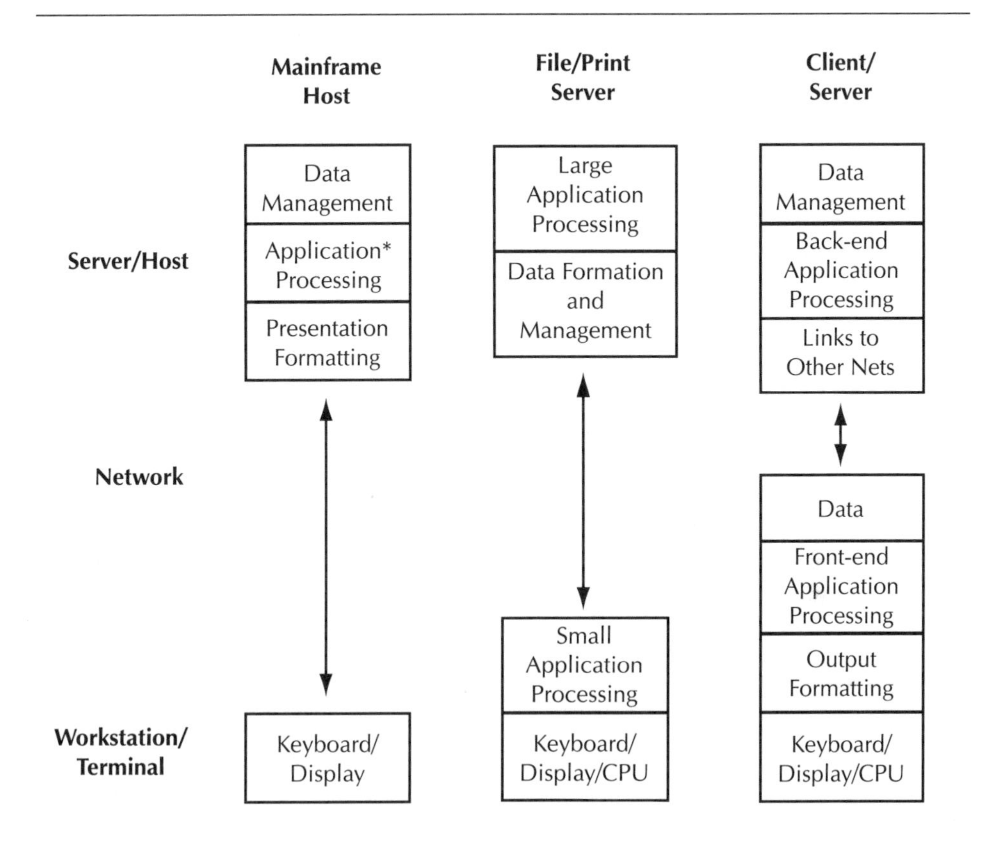

FIGURE 5.3 Evolution of Client/Server Architecture

For example, in an order processing operation, individual operators at workstations receive phone orders and enter them into order-entry applications. These workstation applications (front-end processing) prepare the screen, obtain customer data such as previous orders, credit limits, or product information, perform validity checks, and record the order after all verification is complete. Periodically, orders from all order-entry stations are collated for processing (back-end processing) on the host or server. The host application prepares picking information for the warehouse, shipping data for the distribution center, and performs many other administrative and logistical tasks.

To complete the example above, operators enter some order information from phone input and some from mail. In some cases, customers enter information directly if they are linked to the system through electronic data interchange. Extending the net to link customers electronically to the ordering system is a powerful way to enhance customer service and gain competitive advantage.

The order-entry example is typical among others that client/server architectures make possible. Financial operations such as payroll and accounts payable, engineering tasks such as product design or bill-of-material development, and many office and administrative tasks are performed more effectively in this mode. In fact, client/server facilities change the form of many traditional business processes. Employees are empowered with increased responsibilities and sophisticated data processing capabilities to carry them out. The process reduces information float, is less paper intensive, and more cost effective. Businesses, their customers, and their suppliers benefit from higher quality, lower cost operations.

Figure 5.4 displays a simple client/server architecture operating on a local area network. In many organizations, the server depicted in Figure 5.4 provides functions in addition to those shown here. For example, the server may manage a bulk printer for all workstation users or connect to a facsimile line for workstation initiated message transmission. The server may also act as a bridge or router/gateway connecting this LAN to other LANs or MANs that the organization uses. It may also connect to a large enterprise mainframe sharing processing and data storage functions in a hierarchical manner. The possibilities are almost limitless for enterprising network architects and business process engineers.

Because of the many important advantages to downsizing mainframe applications and converting them and their associated business processes to distributed processing modes of operation, businesses are making substantial investments in client/server implementations. Nearly half of application investment is now devoted to client/server operations; every major hardware and software vendor has introduced client/server products. Client/server operations have revolutionized the way information systems are implemented and managed with far-reaching ramifications for traditional business organizations. IS professionals and operations people have been profoundly affected by the wave of networked processing.

Client/servers represent a major step in the evolution of computing from stand-alone mainframes to fully integrated but widely dispersed facilities. They are the beginning of a more general environment of fully integrated network computing.

Research departments at computer firms, government agencies, and public and private research labs in this country and abroad are searching for system advances. Current trends include faster circuitry, increasing numbers of nodes in parallel architecture, improved networked computing environments, and revolutionary software and operating systems. The goal of the research is to achieve practical realization of inherently high-performance devices through new architectural and hardware designs. Research breakthroughs must occur in programming and operating systems in order to achieve full benefit of hardware and networking advances.

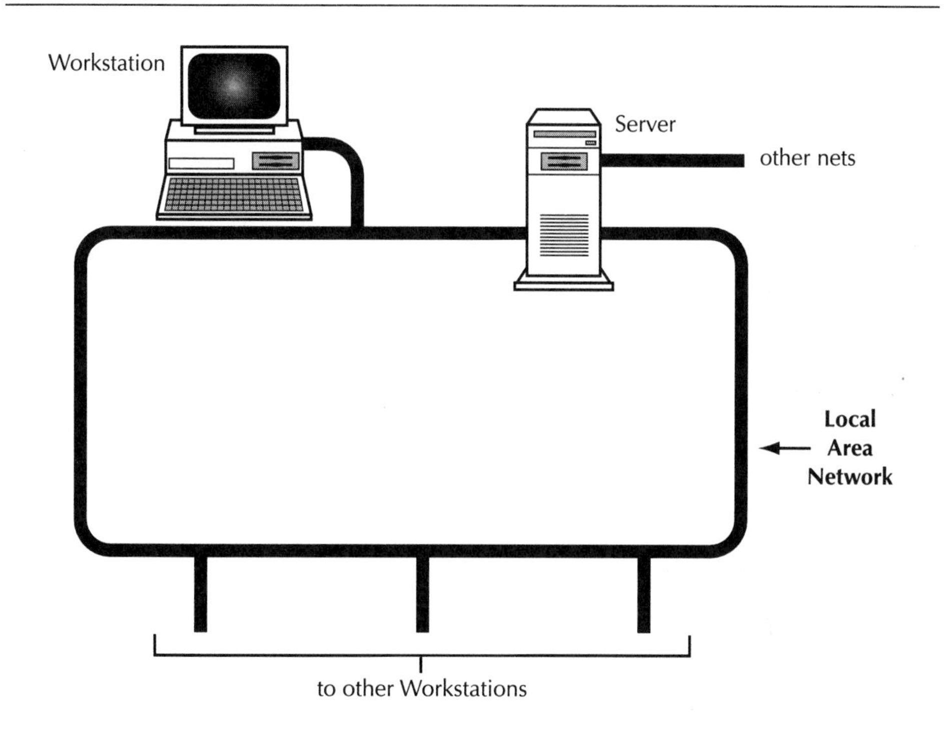

Workstation Functions
- Provide user interface
- Accept input data
- Format display output
- Perform data manipulation
- Send queries to server
- Perform front-end processing

Server Functions
- Perform back-end processing
- Do data management archiving
- Respond to user queries
- System administration

FIGURE 5.4 Client/Server Architecture

Parallel computer architecture raises very significant challenges for programmers and language designers. In the past, compiler designers focused their attention on the efficient execution of sequential operations. Conventional languages specified tasks that are carried out one at a time in a single thread mode, and brought together program instructions and data for discrete sequential operations. Faster operation required increases in the speed of purely sequential activities. Parallel computers, however, demand languages designed to attack problems in a multithread or simultaneous fashion. Now, new approaches to computer languages are needed in order to capture the benefits of parallel hardware.

Due to the large inventory of sequential programs and the ease of working in sequential mode, first attempts at parallel programming tried to convert sequential operations into simultaneous operations. For example, programs containing loops were examined for parallel operations with some success, particularly in scientific vector or matrix calculations. Computer programs have been written to search for parallelism in routinely sequential instructions. Then, these programs generate code that uses the implicit parallelism they have discovered. Other methods based on data flows have been implemented with similar success.

In single instruction, multiple data machines described earlier, the main memory is divided into increments, one for each processor (perhaps tens of thousands of increments). The data is distributed over the memory increments, one data element per increment. Then, as each processor carries out the same instruction in synchronized fashion, the data is transformed in all memory increments simultaneously. This parallel approach to problem solving works well on certain classes of problems. The computer requires a supervisor processor to coordinate the activities of all the other processors, and is itself controlled by a supervisor program.

Parallel-processing compilers are being developed for commercial applications. For example, IBM Clustered FORTRAN executes jobs across twelve processors at speeds up to ten times faster than would be possible with a single processor. Firms developing parallel processors and some that are using them are researching parallel processing actively; significant advances are expected.

The task of designing language translators and making parallel-processing compilers is formidable indeed, but the control programs needed to manage large parallel computers pose challenges that are even more demanding. Solutions to these challenges will permit users and their applications to capitalize on the enormous potential of parallel-hardware architectures. But, obviously, further innovation is required.

OPERATING SYSTEMS CONSIDERATIONS

An operating system is a program that interfaces between computer users and the computer hardware. The purpose of an operating system is to provide a friendly environment in which users may execute programs. Thus, the primary goal of an operating system is to make the computer system convenient to use. A secondary goal is to permit the computer hardware to be used efficiently.[15]

The Evolution of Operating Systems

The early use of computers involved a programmer or operator who wrote the program to solve a problem and operated the computer devices. This operation was called "hands-on." I/O devices were managed by the programmer/user by including device drivers, usually a small deck of punched cards, along with a deck of punched cards for the application program. The scheduling system usually consisted of a sign-up sheet, and set-up time played an important part in this type of operation. Some programs ran longer than anticipated, inconveniencing the person next in line, while others failed to run at all, causing idle time in the schedule.

As hardware and applications advanced, operating system improvements were needed to make the total system efficient and convenient to use. What are the stages through which operating systems advanced, and what can we expect from future operating systems? Table 5.4 displays the stages through which operating systems have progressed.

TABLE 5.4 Operating Systems Advances

Programmer/operator
System monitors
Uniprogramming operating systems
Multiprogramming operating systems
Multiprocessing operating systems

System monitors and permanent computer operators solved some of the problems discussed earlier. The computer operator batched similar jobs reducing set-up times and improving efficiency by managing the peripheral devices. The system monitor contained all the device drivers, compilers, assemblers, and the job control system. Small, peripheral computers were used to perform card-to-tape and tape-to-printer tasks thus making more efficient use of the CPU.

CPU efficiency was improved by multiprogramming application programs in main memory simultaneously. The operating system schedules the execution of these programs according to predetermined algorithms designed to minimize delays. The operating system contains the functions of memory management, task scheduling, main memory and data protection mechanisms, timer control, and others. Multiprogramming operating systems are extremely complex and require large amounts of time and resources to develop. Now, these operating systems are available for large personal computers.

Multiprocessor hardware configurations have been in existence for two decades or longer, and operating systems to manage them have evolved along with the hardware. These operating systems handle all the functions mentioned above but, in addition, they manage several interconnected multiprogramming CPUs. The degree of complexity in these systems qualifies them as one of the most complex advances that humans have undertaken. The complexity of operating systems supporting large parallel processors will probably increase by orders of magnitude in the future.

Contemporary Operating Systems

Contemporary operating systems contain extensive communications and telecommunications software that permit applications programs to utilize networks with ease. They also provide application programmers with data-management systems that handle and organize the large amounts of data associated with many modern applications. Most organizations have specialists who install and update the operating system. These skilled individuals, called systems programmers, assist applications programmers and users with the intricacies of the operating system.

Today's desktop computer usually uses one of the three most popular operating systems. Microsoft's DOS/Windows, used on IBM systems and on IBM compatibles, has the lion's share of the PC market. An improved version of Windows, called Windows 95, was introduced in 1995. Apple's System 7, used only on Apple Macintosh computers, has about 15 percent of the market. As noted in the Business Vignette, Apple refused to license its products to others. Now that Apple is changing its strategy on this point, its operating systems may gain popularity. IBM developed OS/2 as a competitor to Windows and DOS, but seems unable to gain ground against Microsoft. OS/2 is a more expensive system requiring more computer power. It is better suited to serving larger systems in sophisticated networked applications.

Network Operating Systems

Other operating systems have been developed to manage services on large computer networks. Windows NT, UNIX, and NextStep are in this category. This is an area of great importance given the trend toward network computing, so we can expect considerable development activity by computer suppliers.

In many instances, client/server operations provide opportunities for firms to become more efficient by improving productivity significantly. But the client/server architecture itself still offers many opportunities for increased efficiency. Generally, inefficiencies fall into the following four categories:

1. Available but inaccessible system resources
2. Processing or storage redundancy
3. Weak or ineffective system controls
4. Excessive need for manual operator intervention

In the client/server model, functions of an application are split into two parts. One part executes on the server and the other part on the personal workstation. Client workstations talk to servers only. In taking the next step toward the network computing model, processors are considered to be peers and each element can talk to all the others. Instruction execution, file storage, and database management resources are interchangeable among system elements. The goal of network computing generalizes the client/server model to provide users with easy access to all system resources required for their applications. This goal is easy to identify but extremely difficult to meet—in practice, it may be difficult to achieve totally.

By reducing or eliminating the inefficiencies in client/server configurations, well-developed network operating systems will improve the performance of networked systems. Application interfaces will be more consistent, easier to use, and will permit more flexible work flows and improved organizational effectiveness. Sophisticated operating systems for large computer systems evolved over many years, but the pace of evolution for small computer systems has been more rapid. The technologies of large systems are migrating to small systems and new technologies for interconnected personal workstations are developing rapidly, too. Alert IT managers will observe these trends and profit from them.

The software associated with future systems will be extremely complex and substantially different from software in common use today. System designers, and perhaps system programmers, must change their approach to data processing problems if they are to realize the full potential of hardware advances. Their new vision must include not only advanced hardware and telecommunications systems, but must embrace new methods of operation such as client/server and networked computing models. This new vision will make it possible to solve increasingly sophisticated problems; problems thought intractable in today's sequential world. The users of computing technology will benefit greatly from their vision and inventiveness.

COMMUNICATIONS TECHNOLOGY

The worldwide telecommunications industry has been capitalizing on the same technology as the computer industry, and for all the same reasons. Digitization of analog communication signals is permitting the industry to apply enormously productive digital technology effectively. New devices, new media such as satellites and fiber optics, and changes in the industry structure and environment such as the court-mandated breakup of AT&T, are all operating in concert. These and many more factors are bringing tremendous growth to the industry and benefiting users by delivering enhanced communications capability. Telecommunication advances will be explored in the next chapter.

TECHNOLOGY TRENDS

The technology of information handling is advancing very rapidly. IT managers must be alert to technology trends while preparing to take advantage of the trends for themselves and their organizations. Table 5.5 summarizes the technology trends.

TABLE 5.5 Information Technology Trends

1. Vast increases in computational power
2. Availability of huge data stores
3. Complex, easy-to-use operating systems
4. Numerous application packages
5. Extensive, integrated telecommunications networks
6. Increasing function at declining cost

The Meaning for Management

The economics of technology advances vastly increases the computational capability available to the firm. This capability is expressed in rapidly growing numbers of mainframes, mini- and microcomputers with improved performance compared with today's systems. Future managers will have orders of magnitude more computing capability than today's managers. They will be challenged continuously to achieve the full potential of this rapidly evolving capability.

The increased power of computer hardware will be accompanied by huge data stores containing all kinds of information. Much of this information will be generated within the firm, but increasing amounts will originate from public sources available to wide audiences. Access to these data stores will be rapid, easy, and inexpensive. The potential for information overload will increase.

Advanced hardware with great power and complexity will be made functional by enormously complicated operating systems. Operating systems will be designed specifically to make computing power available to a wide audience in an easy-to-use manner. These hardware and operating system advances will take place on systems of all sizes, from supercomputers and mainframes to small personal workstations. Extensive, fast, and easy-to-use telecommunications networks will enable parallel processing at the firm level.

For all the firm's employees, electronically interconnected computing devices will permit rapid communication and nearly simultaneous parallel problem-solving. In some firms, employees are beginning to engage in parallel processing with networks and client/server operations. Globally, firms are being linked to their customers and suppliers via electronic, interorganizational systems. Telecommunication will shrink time and distance barriers, enabling the efficient functioning of international alliances and global partnerships. Alternative forms of business organization will flourish in this environment.

For managers, these changes are significant. With relative accuracy, we can predict that in 10 years, the information technology required to solve business and other problems will exceed current capability by orders of magnitude. More important are the ways we will apply new technology. Here our vision is much less clear. As the editors of *Scientific American* said, "Imagine the citizens of the 18th century trying to envision the shape of the future that would include electrical power, telecommunications, jet transportation, and biotechnology. We who are alive at the end of the 20th century are having the same trouble discerning the impact of an evolutionary force that is reshaping our world: the fusion of computer and communication technology."[16] In the future, employees and managers will be challenged to capture the potential of advancing technology in ways we cannot envision now.

The potential benefits of information technology will not come without some eventual pitfalls. Management must not only harness this growing wave of potential, but must manage the inevitable personal and organizational problems riding in its wake. Recalling Drucker's comments that opened the first chapter, we realize that one of the challenges will be to accomplish the technology assimilation while facing managerial reductions and dislocations.

Skillful general managers will face rigorous challenges. They must develop strategies to introduce emerging technology. They must capture the advantages of the technology for their organizations while contending with a cornucopia of difficulties along the way.

The Meaning for Organizations

The first chapter emphasized that the conversion to knowledge-based organizations will cause major organizational changes. One of the most important of these will be the reduction of management positions by two-thirds, according to Drucker. Thirty-five years ago, some scholars thought the introduction of digital computers would generate widespread unemployment from the automation of thousands of routine tasks. Instead, the creation of new jobs and the upgrading of old ones offset employment declines. Now, the possibility exists that the anticipated job reduction will occur among managers. Ironically, these are the same people we expect to usher in the new knowledge-based age.

As in the past, perhaps we are being too pessimistic. Our knowledge-based society may once again create new and better jobs for those middle managers predicted to be displaced. In any case, our organizational lives will not remain static for long. Managers can look forward to evolutionary, if not revolutionary, changes in organizational structure and in the content of managerial roles. Adapting to the ever-shifting currents of change via structural adjustments will challenge future managers.

Other organizational changes are underway. Firms are creating alliances and forming joint ventures to exploit technology, while transforming themselves in other ways, too. For example, Baxter International and IBM joined forces in the health care field; Volvo's subsidiaries throughout Europe and North America are tightly coupled to headquarters in Sweden; and Corning's wheel and spoke structure is a radical departure from the norm of twenty years ago. These and other manifestations of information technology will be discussed in subsequent chapters of this text.

Rapid advances in technology are impacting organizations in yet another way. Alliances are forming among major corporations to speed the development and deployment of new technology by sharing financial and intellectual resources. For example, Pacific Telesis joined with the *Los Angeles Times* to bring shopping, information, and transaction services to consumers in Los Angeles via telephone. These partners are planning on-line applications for early implementation. There are many joint ventures between telephone and cable companies as they share their financial, intellectual, and physical resources in a race to bring advanced services to the public.

SUMMARY

The rapid technological advances experienced for nearly four decades will continue into the foreseeable future. These advances will require solutions to many technical problems in metallurgy, mechanics, systems architecture, programming systems architecture, and other scientific or engineering disciplines. Information processing technology will advance very rapidly, challenging our ability to utilize it fully. As is the case today, there will be technological solutions looking for problems. Management and enterprise-wide issues will continue to restrain our ability to capitalize fully on the rapidly emerging technology.

These technological advances bring huge opportunities for managers in all lines of work. These opportunities can be exploited by many managers; their employers will benefit from the exploitation of technology. The employment of sophisticated technology is not an optional activity. In our competitive global economy, failure to stay current technologically will be a high-risk option.

A Business Vignette

Trauma at IBM

"Is the environment today riskier than it used to be? You bet it is," said John Akers, Chairman and CEO of IBM. "Why? Technology is the key answer. Its pace is quickening, and any competitor who won't take the risks to be right on the curl of the wave won't survive for long." These profound and prophetic statements were made in 1988 by John F. Akers, Chairman and Chief Executive Officer of IBM.[17]

In 1989 and 1990, IBM's revenue continued its upward pace peaking at $68.9 billion in 1990 while net profit increased to $5.9 billion. But change had finally caught up with the computer giant. In the three year period ending in 1993, IBM lost a combined total of $15.9 billion! After peaking at 407,080 permanent employees in 1986, IBM shed 151,000 workers ending 1993 with 256,000 employees on the payroll. Among those leaving was John Akers, replaced by Louis Gerstner, Jr., in April 1993.

Commenting in the 1993 Annual Report on the cause of the precipitous decline of IBM's fortunes, Gerstner wrote:

> At the heart of the turmoil is one simple fact: IBM failed to keep pace with significant change in the industry. We have been too bureaucratic and too preoccupied with our own view of the world. We have been way too slow getting new things to the market. We had all this great technology coming out of our labs, but time and again someone else beat us to the marketplace. Once we got to the market, we usually had the better mousetrap. But one of the unusual things about this industry is that a disproportionate amount of the economic value occurs in the early stages of a product's life. That's when the margins are most significant. So there is real value to speed, to being first—perhaps more than in any other industry.[18]

After four or five difficult years, IBM appears to have turned the corner. In 1994, IBM earned $2.9 billion on revenues of $64 billion. But, like Apple and Digital, the company is profoundly different than its antecedent. Stunning developments from laboratories around the world have fueled industry dynamics that subsequently changed these prominent companies forever.

Review Questions

1. What are the technological foundations for advances in digital computers?
2. Which of these technologies has advanced at a moderate pace, and why?

3. What are today's observable trends in semiconductor technology development?
4. Why are semiconductor manufacturers striving to shrink the size of integrated circuits on chips, and how much smaller can they make them?
5. Why are advances in logic and memory technology and simultaneous advances in recording technology essential for progress in systems development?
6. Why are rigid-disk manufacturing plants similar in sophistication to semiconductor plants?
7. Why are optical recording devices for use in digital computing systems increasing in popularity?
8. Why are RAID devices becoming popular and on what principle do they rely for success?
9. What are the important trends in supercomputers?
10. What challenges do parallel processors pose to operating system developers?
11. How does the operation of modern-day PCs resemble the operation of mainframe computers 20 years ago? What are the implications for IT management?
12. How is a client/server system organized and how does it operate?
13. What is the relationship between communications technology and other parts of information technology?
14. Summarize the dominant trends in computer technology today.
15. What is the managerial significance of technology trends? How will these trends affect organizations?

Discussion Questions

1. Discuss Apple's strategy prior to 1993. Do you agree with Apple's strategy changes since 1993? What are the risks in changing or not changing strategic direction?
2. Discuss the business skills referred to in the following statement: Alpha is a superior design but its success ultimately depends on DEC's business skills.

3. The Intel 80386 cost about $100 million to develop. It's estimated that the follow-on product, the 80486, took about four years and $300 million to develop. What risks were involved in this development work, and how might these risks be quantified?
4. What is the annual compound growth rate in unit performance per dollar for a technology which doubles in unit performance and halves in cost over a four-year period?
5. Discuss the concepts of RAID storage and their application to system components other than storage subsystems.
6. Discuss the advantages and disadvantages of CD-ROM in comparison to hard drives, floppies, and magnetic tape.
7. Discuss how client/server operations might be extended to a firm's customers and suppliers in a catalog sales retail operation. What are the possibilities and limitations of this mode of operation and what might be its strategic significance?
8. Discuss why technology advances increase management risk. What actions might managers take to mitigate the risk factors?
9. What factors have encouraged the rise in alliances, consortia, and joint ventures? What are the international implications behind these movements?
10. Discuss the elements of risk and reward for corporations which are engaged in high-technology development.

Assignments

1. Research the amount of money that U.S. corporations spend on research and development activities in the areas of semiconductor technology, recording technology, telecommunications technology, and programming. Compare these amounts with the R&D activity in the pharmaceutical industry. What conclusions or inferences can you draw from the data you collected?
2. Obtain the descriptive details of one of the new operating systems mentioned in this chapter, and summarize them for the class. In particular, describe where this system would be used and why the developer expects it to succeed.

3. There are many alliances and joint ventures in process among IT companies today. Select one and discuss its goals, objectives, and prospects. Prepare a summary of your findings for the class.

ENDNOTES

[1] Julie Pitta, "Apple's Mr. Pragmatist," *Forbes Magazine*, March 28, 1994, 82.

[2] Revenue steadily increased from $6308.9 million in 1991 to an expected $10,500 million in 1995 but profit margins continued to drop through 1994. *The Value Line*, January, 27, 1995, 1074.

[3] During the 1940s and 1950s, analog computers were undergoing development and were thought to have great promise. The invention of the transistor and the development of high density silicon chips allowed digital computers to eclipse totally their analog counterparts.

[4] John Bardeen, Walter Brattain, and William Shockley invented the transistor at Bell Laboratories in 1948. In 1956, they received a Nobel prize for their invention.

[5] For an interesting and detailed discussion of the development of the semiconductor industry and microprocessors in particular see George Gilder, *Microcosm, The Quantum Revolution in Economics and Technology* (New York: Simon and Schuster, 1989).

[6] Gary Stix, "Toward 'Point One'," *Scientific American*, February 1995, 90.

[7] Robert W. Keyes, "The Future of the Transistor," *Scientific American*, June 1993, 70.

[8] Laboratory researchers are experimenting with dimensions near 30 Angstroms or about 0.003 microns where severe quantum effects make further downsizing very difficult. Robert W. Keyes, "The Future of the Transistor," *Scientific American*, June 1993, 70.

[9] From data provided by InfoCorp. as published in *Computer Shopper*, December 1994, 168.

[10] Charles Babcock, "Alpha May Not Win, But It's The Best," *Computerworld*, September 19, 1994, 6.

[11] IBM recently demonstrated a new world record for magnetic-data storage density of three billion bits per square inch, nearly five times the density of the most advanced disk available today. *The Wall Street Journal*, March 31, 1995, B12.

[12] Erica Schroeder, "Speed, Cost Propel CD ROM Into Offices," *PC Week*, February 22, 1993, 29.

[13] Charles Babcock, "Parallel Processing Mines Retail Data," *Computerworld*, September 26, 1994, 6.

[14] Interview with Seymour R. Cray in *Computerworld*, July 18, 1994, 20.

[15] James L. Peterson and Abraham Silberschatz, *Operating Systems Concepts*, 2nd Ed. (Reading, MA: Addison-Wesley Publishing Company, 1985).

[16] "The Computer in the 21st Century," *Scientific American*, Special Issue, Volume 6, Number 1, 1995, 4.

[17] Letter from Chairman and Chief Executive Officer, *Think*, Vol. 54, No. 3 (1988).

[18] Louis V. Gerstner, Jr., IBM Annual Report for 1993, 3.

6 Modern Telecommunications Systems

> The biggest demand pool in the world is for telephone service in Third World countries, including the former Soviet Bloc. There is no greater impediment to economic development than poor telephone service, and no greater spur than good telephone service. A telephone system is highly capital-intensive. But technologies that replace the "wiring" of traditional telephones with the "beaming" of cellular phones are radically reducing the capital investment needed. And once a telephone service is installed it begins to pay for itself fairly soon, especially if well maintained.[1]

INTRODUCTION

Telecommunications is defined as the science and technology of communication by electronic transmission of impulses through telegraphy, cable, telephony, radio, or television. Dictionary definitions make it clear that communication can occur either with or without physical media. Tele stems from the Greek word for distant; communicate has Latin origins meaning to impart. Thus we think of telecommunications as imparting information over a distance.

This chapter deals with electronic communication methods or electronic aids to human communication. In most instances, these electronic communication methods are tied together or linked into systems or networks.

Systems are groups of interrelated, interacting elements that perform input-process-output functions. In telecommunications systems, the type of interconnecting media is important (wires, cable, etc.) and the form or architecture of the interconnections is also important. The interconnecting media and its architecture is the network. The transmission medium carries the signals in the system. It may be wire, cable, or the atmosphere in the case of terrestrial microwave. The devices connected by the network define the network type; for example, telephone network, television network, or computer network. The telephone network is a critical, important, and complex telecommunications system.

THE TELEPHONE SYSTEM

The telephone system is characterized as switched, interactive, narrowband, and public. A switched network is one form of network architecture. Network architectures (the form in which the primary elements are interconnected) can be one of many types which traffic patterns, message types, frequency of use, and other factors dictate. Telephone traffic is usually sporadic and consists of two-way or interactive voice signals. It is not expected that all phones in a city will be in use simultaneously. It is expected, however, that any one phone may want to communicate with any other phone at any given time. These communication characteristics help determine the network architecture.

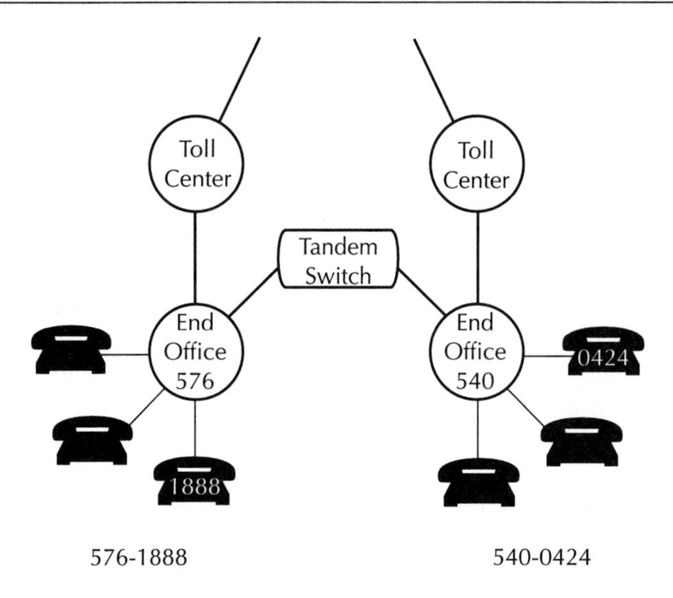

FIGURE 6.1 End Office Network Configuration

The part of the system nearest the user is organized or architected in a star pattern. Up to 10,000 access lines originate at the local exchange, fan out in a star-like pattern, and terminate at the residence or business phone. Figure 6.1 portrays this architecture.

The devices at the customer end of each access line are usually ordinary telephone instruments but may be fax machines, computers, or other terminal devices. Collectively, these devices are called customer premise equipment. The access lines, identified by the last four digits of the phone number, connect the customer premise equipment to the local exchange. The local exchange is identified to the network by the first three digits of the seven digit number (576 and 540 in Figure 6.1). Access lines are called the local loop or the twisted pair, referring to the pairs of copper wires that usually form the connection to the end office.

One purpose of the local exchange is to connect any one phone to any other phone on demand. To accomplish this, the exchange contains a switch that makes the desired connections between any two of the up to 10,000 access lines. Thus, the exchange forms a switched star network. It is very effective for traffic patterns in which most of the access lines are idle at any given time. At this level in the phone network, about 1500 of the 10,000 lines are busy at peak traffic times on average.

The local exchange is connected to other exchanges in the immediate area through tandem switches. Exchanges are all connected to toll centers. This arrangement or architecture is shown in Figure 6.2. The connections between the higher level centers (toll, primary, and sectional centers) are organized in a manner similar to a star network configuration but include additional direct toll lines between many of the adjacent centers. The links connecting the higher level centers to each other and to the end offices are high-capacity trunk lines capable of handling hundreds or thousands of individual calls simultaneously. This network in the U.S. (with some modifications) links about 140,000,000 individual access lines.

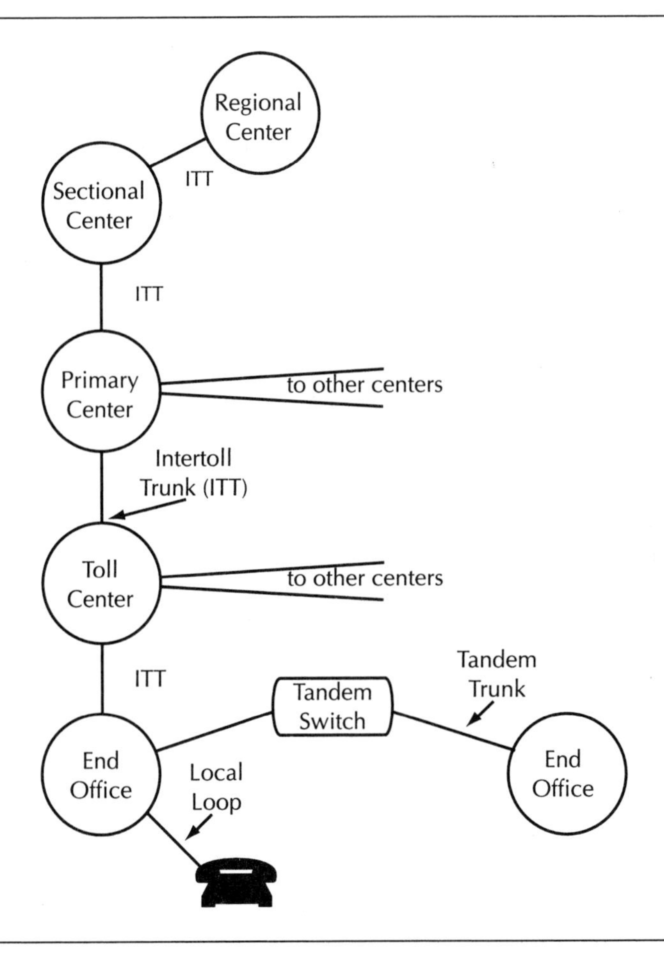

FIGURE 6.2 Bell System Switching Network

Operation of the switched phone network begins when the customer raises the instrument's handset initiating a connection through the local loop to the exchange. The dial tone (a 90 cycle per second note) signals the caller that connection to the exchange is complete and the switch is awaiting the destination phone number or address. The phone number sent to the exchange in the form of tones (pulse counts in older equipment) drives the switching equipment. Notice in Figure 6.2 that several switches may be involved in a long-distance call.

A long-distance call outside the local area is signalled by a leading 1. In this case, the caller's exchange routes the call to a toll center that can connect to the destination exchange. Referring to Figure 6.2, note that a primary or a sectional center may be involved in routing the call depending on the caller's location. Also, the routing may change between successive calls to the same number because of network loading, link failure, or degradation in toll line capacity. After the signal reaches the destination exchange, its switch connects with the desired access line. In 1951, Bell introduced direct distance dialing (calling outside the local area without operator assistance). This feature is now available to about 100 foreign countries. The word "dial" carries over from older equipment but seems to be a permanent part of our vocabulary. Although rather simple conceptually, the switched telephone network is complex operationally and relies on computers and high-technology switching equipment for effective operation.

Telephone Network Characteristics

In contrast with cable TV and broadcast radio or TV, where the communication flows in one direction only, the telephone system is bi-directional. Since the network provides bi-directional, on-demand communication, it is termed interactive. Various governmental agencies regulate phone companies as common carriers or public utilities and mandate that they make services available widely to the general public. The phone network is privately owned but the government considers it to be a public system.

In addition to being switched, interactive, and public, the phone system is narrowband. What is meant by narrowband and how does this distinguish the phone network from other networks and communications facilities? Narrowband is a term that describes the network's information carrying capacity or performance. To fully understand networks, we need to examine the signals they carry and their capacity to transmit information, or their performance.

Telephone Signals

Networks can carry two types of signals; analog and digital. Analog signals continuously vary in amplitude (signal strength) or frequency (signal pitch or tone). Voice and music signals are examples of analog signals: They vary continuously in amplitude and frequency. The purpose of the telephone instrument is to convert voice sounds (sound vibrations) into faithful electrical reproductions for transmission and to convert received electrical signals into voice sounds again. From end to end, the telephone network is designed to accomplish this conversion, transmission, and reconversion. Figure 6.3 shows analog signals that vary in amplitude and in frequency.

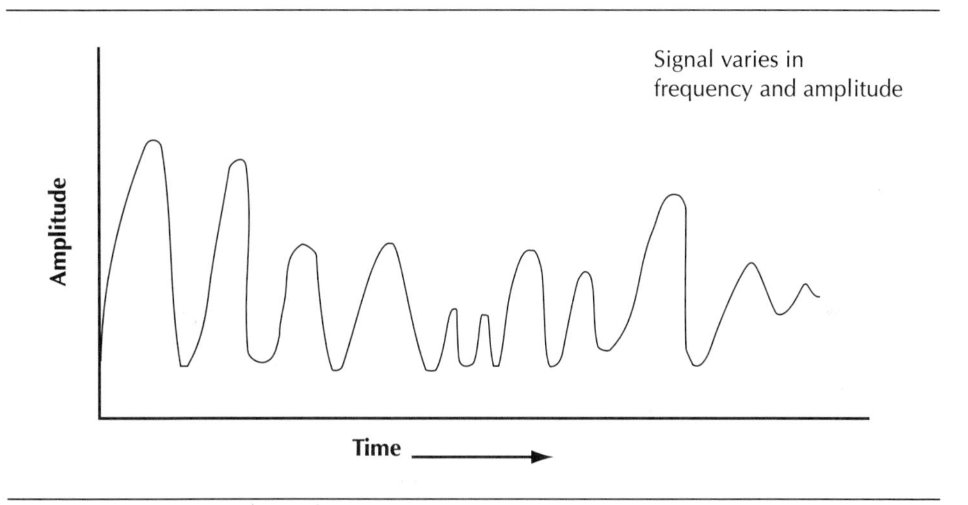

FIGURE 6.3 An Analog Signal

On the other hand, digital signals are discrete or discontinuous signals. A signal that can exist in only two states, plus and minus, for example, is a digital signal. Signals that can be transmitted in one of only two states are termed binary signals. Most naturally occurring events are analog. All digital events in information technology are binary, as shown in Figure 6.4.

Telephone circuits designed for voice transmission accommodate the range of voice frequencies found in normal speaking tones. Though the human ear can detect frequencies from 20 hertz to 14,000 hertz or so (hertz equals cycles per second, i.e., 1000 hertz equals 1000 cycles per second), normal voice communication takes place effectively in the range from 300 hertz to 3400 hertz or so. It is customary in voice telecommunications systems to allot 4000 hertz to one voice band or channel but to limit the signal transmission to the range 300 hertz to 3400 hertz. This provides some space (termed guardbands) on each side of the voice channel to prevent signal overlap with adjacent channels and to leave room for the phone system to transmit the signaling information necessary to operate the network.

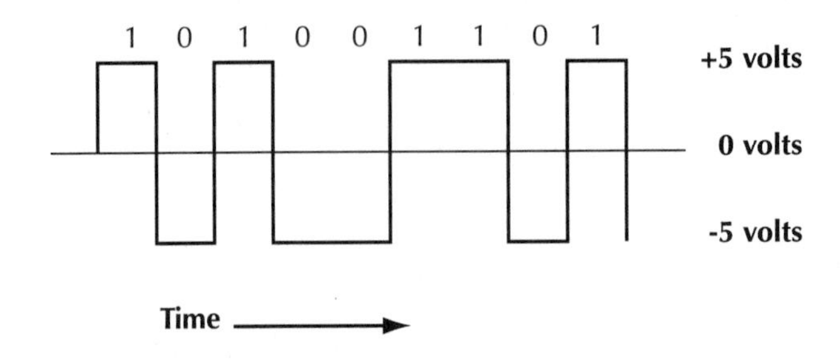

FIGURE 6.4 A Digital Signal

Multiplexing

The notion of guardbands and adjacent channels introduces the concept of multiplexing. Multiplexing is the subdivision of a transmission channel into two or more separate channels. One form of multiplexing is to divide the frequency range of the channel into narrow bands so that separate signals can be transmitted in each band independently. For example, let's assume there are two incoming voice signals, A and B, each limited to a frequency range of 300 hertz to 3400 hertz. One of the signals, B, is electronically added to a 4000-hertz tone and the resulting signal now varies between 4300 hertz and 7400 hertz. The frequency shifted signal B is combined with signal A and the two are sent down the line together. This example is shown in Figure 6.5. At the receiving end, the signals in each band are separated, and 4000 hertz is electronically subtracted from the higher frequency signal. At the receiving end, the result of these operations is the reproduction of the two original voice signals.

The importance of guardbands now becomes clearer. Between the multiplexed signals A and B, there is signal-free space of 900 hertz that prevents signal overlap or interference. The phone system uses this free space for transmitting call management (setup, maintenance, and termination) and network management information. Using guardbands for this purpose is called inband signaling.

The process described above can be repeated many times by combining incoming voice signals with tones at multiples of 4000 hertz and stacking them on the line for simultaneous transmission. At the receiving end, the signals are demultiplexed into individual voice signals each limited to 4000 hertz. The number of subdivisions or stacks in this method of combining and transmitting signals is limited by technical factors such as type and length of line, and line amplifier factors like linearity and distortion. This process, called frequency division multiplexing, is useful for transmitting analog signals.

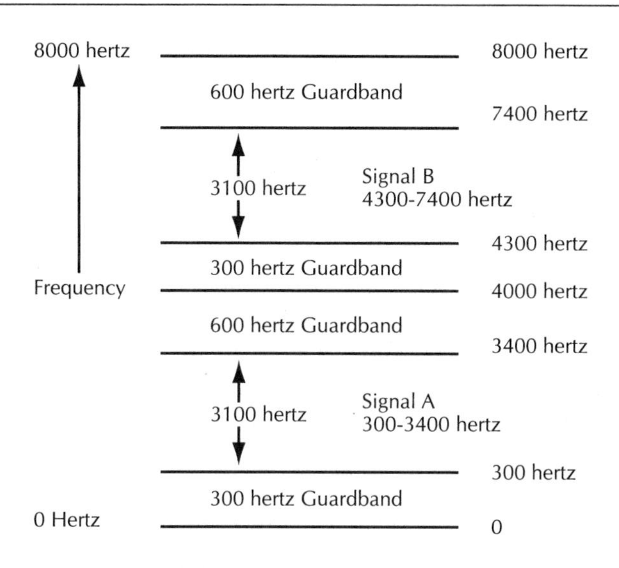

FIGURE 6.5 Two-Channel Voice Grade Line

Network Capacity and Performance

The 4000-hertz bandwidth of the normal voice grade channel compares to a bandwidth of about 500,000,000 hertz, or 500 megahertz, for coaxial cables used in cable TV, and to about 10,000 megahertz for microwave transmission. Coaxial cable and microwave transmissions are termed broadband. From the customer's equipment to the local exchange, the phone system is narrowband. Microwave or fiber optic trunk lines that carry hundreds of multiplexed conversations between toll switching centers are high bandwidth telephone network components.

Bandwidth measures the information carrying capacity of transmission lines or links; high-bandwidth links can transport more information per unit time than low-bandwidth links. In 1933, Harry Nyquist developed the theory of information transmission. Nyquist showed that the maximum number of discrete bits or digits of information that could be transmitted over a line of X bandwidth is 2X. This means that a voice grade line having a bandwidth of 4000 hertz can transmit 8000 bits per second. The inverse of this theory is that, if one wants to reproduce a signal varying at the rate of 4000 hertz by sampling it periodically, one needs to sample the signal 2X times or at least 8000 times per second. This is known as Nyquist's sampling theorem. These concepts are critical in converting analog signals to digital signals.

In 1948, Claude Shannon of Bell Labs refined Nyquist's theory to account for the average amount of inherent noise found on the transmission line. He showed that transmission speed in bits per second (bps) is related to bandwidth, signal power, and the amount of power in the noise. Shannon relates these variables according to the following equation:

$$bps = B \log(\text{base } 2)(1 + (S/N))$$

In this equation, bps is the transmission speed in bits per second, B is the transmission line bandwidth, S is the signal power in watts, and N is the signal power of the noise in watts. The ratio of signal to noise is on the order of 20 for many applications. For an S/N ratio of 20, a line having a bandwidth of 4000 hertz is theoretically capable of transmitting 12,178 bps. Note that these results differ from the Nyquist calculation. There are several technical reason for this difference but, in actual practice, the transmission capacity of a communication link is somewhat overstated by both these theoretical calculations.[2]

Digitizing Voice Signals

Information technology is the merger of communication and computer systems. To apply the power of digital computers to telephone systems, information must be converted from analog to digital format for processing. After processing, the signals must be restored to analog format for human use. Converting signals from analog to digital requires sampling techniques and relies on Nyquist's sampling formula.

Analog to digital conversion is a two-step process involving pulse amplitude modulation and pulse code modulation. Figure 6.6 describes the processes pictorially.

The top portion of Figure 6.6 shows an analog signal varying in amplitude over time quantized into eight discrete levels or pulse amplitudes. The bottom portion of Figure 6.6 shows the quantized signal being sampled at discrete time intervals. The sample heights are converted into binary digits; in this case of eight quantized levels, three bits are required to identify the signal amplitudes. The process of converting the samples into binary digits is called pulse code modulation.

Now, instead of using eight quantizing levels as in the example above, suppose that the number of quantizing levels is increased to 128. Increasing the number of quantizing levels improves the quantized representation of the analog signal and increases its fidelity. In this case, it takes seven binary positions to represent the sampled signal and, if one bit is added for control, the signal converts into eight bits or, in computer language, one byte. Further, using Nyquist's theorem, a 4000-hertz analog signal must be sampled 8000 times per second to be faithfully reproduced. This means that a digitized voice signal limited to a bandwidth of 4000 hertz converts into a bit stream of 8000 samples of eight bits each, or 64,000 bits per second.

1 The signal is first shaped so it occupies a discrete set of values.

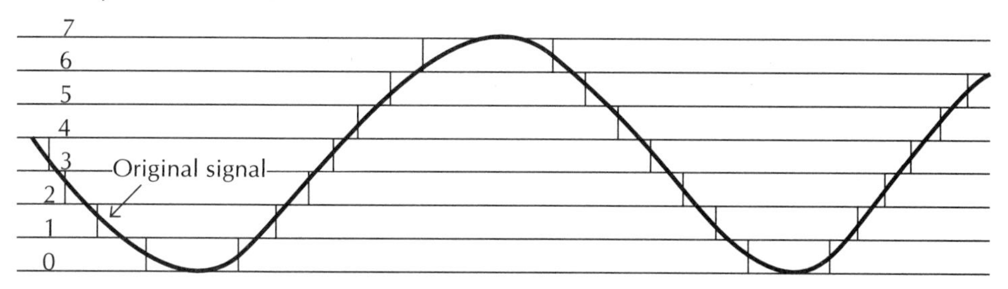

2 It is then sampled at regular intervals and the resulting signal coded for transmission.

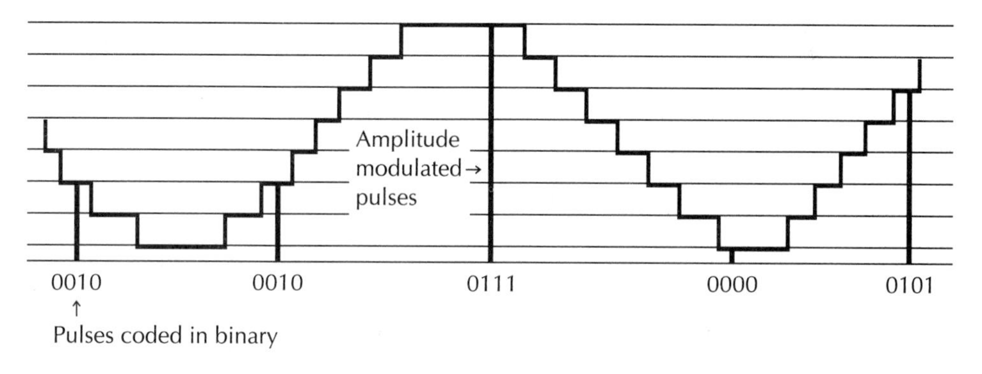

FIGURE 6.6 Pulse Amplitude and Pulse Code Modulation

This process is the basis for digitizing analog signals in the voice telecommunications system. Digitized signals are crucial to communications systems because they enable the application of powerful digital computers to communications problems. Switching is one such problem.

Telephone Switching Systems

One of the fundamental characteristics of the telephone system is its ability to interconnect any two phones within the network. To accomplish this, local exchanges and toll centers must be able to read addresses (phone numbers) and make the proper connections. Today, complex, high-technology switches in various network centers perform this task.

Electronic switches, built from solid-state logic and memory circuitry, represent the first step in integrating digital computer technology into the phone system. They were first installed in the Bell system network in 1965. During operation, electronic switches are controlled by a program stored in the switch's memory as in a digital computer. The program analyzes the address number, searches for a physical path to the address, and claims that path for the duration of the call. Of all the possible lines that go between the caller's exchange and the destination exchange, the switch seizes one for its exclusive use. This process of finding and claiming one of many lines is called space-division multiplexing.

Electronic switches replaced older mechanical equipment and represent a solid step forward in switching technology. Electronic switching gains efficiency by integrating computer technology into the telephone network. Further efficiencies can be attained by digitizing the voice signals and processing both the signalling information and the signals themselves in binary format using digital signal processors. Modern switching and transmission systems are entirely digital and employ another form of multiplexing called time-division multiplexing.

Time-Division Multiplexing

Time-division multiplexing is the process of successively allocating segments of time on a transmission channel to different users. In the phone system, this means that digitized information from many low-speed inputs is assembled in sequence for transmission over one high-speed digital line. One popular implementation is to interleave digitized signals from 24 voice grade lines for transmission over one line running at 1.544 megabits per second. This line is called a T-1 line and the assemblage of information is called the T-1 frame. The construction of the T-1 frame is illustrated in Figure 6.7.

Notice that the T-1 frame sequences 24 channels each containing eight bits of information plus one control bit per frame. The eight bits represent one sample of the original analog signal. Bytes of information from each of 24 channels are interleaved at about 5-microsecond intervals (one microsecond equals one millionth of a second), and the sequence is repeated every 125 microseconds. Thus, each analog signal is sampled once every 125 microseconds or 8000 times per second. The data transfer rate is 193 bits per frame multiplied by 8000 frames per second, or 1.544 Mbps. This is called the T-1 data rate. Within the phone network, multiples of T-1 lines are used.

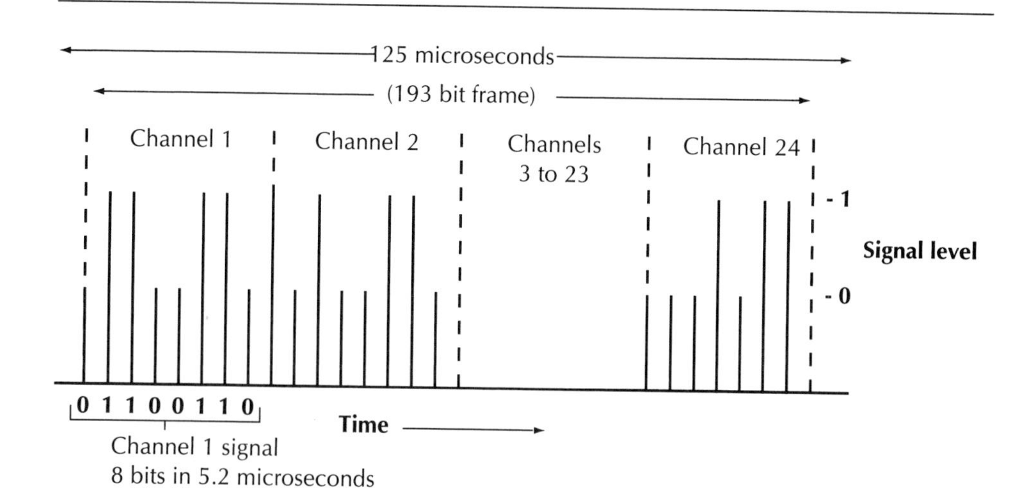

- Channels 2–24 each contain 8 bits.

  Therefore, 24 digitized signals are time multiplexed, each signal is sampled every 125 µsec, or 8000 times each second.

- Each frame contains 8 x 24 = 192 bits plus 1 frame alignment bit for a total of 193 bits per 125 µsec.

  Therefore, the data transfer rate is 193 x 8000 = 1.544 megabits per second.

FIGURE 6.7 Time-Division Multiplexing and the T-1 Frame

In digital systems, the digital signal processor manages the bit stream that includes network and signal information, acts on the data determining where the signal information should be sent, and routes the call interleaved with many others over a high-speed line. The switch is, in fact, a programmable digital computer. Because it can be programmed, the switch can implement functions such as call forwarding (when looking up a number it finds the alias), call waiting (upon detection of a busy line it signals the called party), or automatic number identification (the called party is sent the number of the calling party). It also stores information such as address called, and length of call for use in billing systems. Digital technology makes these and many other features possible.

The phone operations discussed above are premised on the idea that the switching system establishes the call routing and physical connections needed to complete the communication path, and that the call's route remains constant for its duration. This convention is called circuit switching. Circuit switching works quite well for voice communication because conversations take place at a rather constant rate and they utilize the circuits nearly fulltime (usually one party or the other is talking). With data communication, however, the transmission rate can vary widely and the communication can be sporadic or takes place in bursts. At times, the circuit can be fully loaded and at other times, it can be idle for prolonged periods. Another type of switching, called packet switching, helps improve efficiency in data transmission.

With packet switching, the data file to be transmitted is broken up into short packets, usually less than 1000 bytes, and prefixed with addressing and control information. Each packet is transmitted in turn, received by a forward node, stored briefly, then forwarded to the next node until it reaches its destination. Eventually, all the packets for a given message file are received at the destination device where they are assembled in sequence and the process terminates. This type of operation makes more efficient use of the network for bursty data but it requires more sophisticated switches and networks. There is no requirement for packets to take the same route through the net since the route is partly determined by instantaneous network traffic. Packet switching is important to advanced data transmission techniques.

Digitization of voice signals, multiplexing, programmable switches, and broad bandwidth lines have greatly improved the capability of the phone network. Broad bandwidth lines have been state-of-the-art between exchanges for a long time but now, phone companies are using new technology to increase the capacity of the line from the customer's equipment to the exchange.

If time-division multiplexing, circuit switching, and packet switching seem to require that a lot be accomplished in a very short time, recall that a modest size, general-purpose digital computer can process 100 million instructions per second. That is, in the five microseconds it takes to place one byte of information on the line, it can perform 500 add, subtract, or compare operations. And, computers specially designed for signal processing and switching operate even more efficiently than general-purpose ones. Processing binary signals, whether voice, data, or video, exemplifies the integration of computer and communications technologies.

Network Transmission Media

Media for signal transmission can be either conducting media or radiated media. Copper wires, such as those that connect the phone set to the wall outlet and to the local exchange, are conducting media. Conducting media imply direct physical connection between network components for signal transmission. On the other hand, radiated media are exemplified by broadcast radio transmission in which electromagnetic energy passes through the atmosphere or through space without a physical connection. Both media types are used in modern networks.

Today, most phone connections consist of four insulated wires, twisted around each other, within a jacket forming a cable. Twisted pairs grouped together into larger wire cables (a cable 3.35 inches in diameter carries 1200 copper pairs) are found today near exchanges where lines fan in from offices or residences. Frequency-division multiplexing can reduce the volume of wires by enabling one line to carry several conversations. However, copper wire limits high frequencies so coaxial cables are the next step forward.

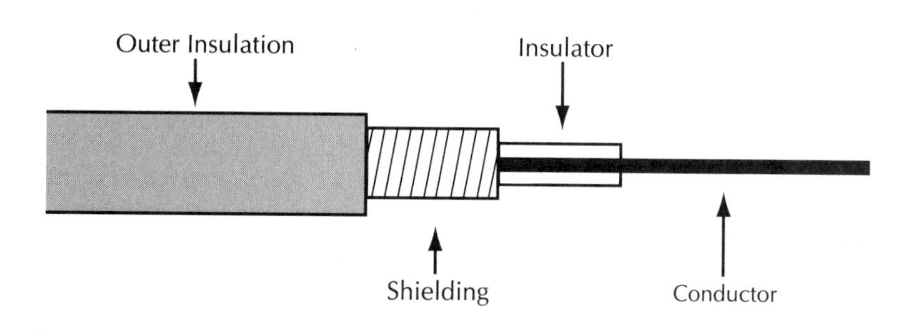

FIGURE 6.8 Coaxial Cable Construction

Coaxial cable has many telecommunications applications particularly in situations involving distances on the order of a few miles or so. Figure 6.8 shows the construction of a single conductor coaxial cable. It may range in diameter from one-fourth inch up to one inch. The outer shield offers protection from electrical interference and physical damage. Permanent taps are easily placed in the line thus aiding physical rearrangement of terminal equipment. Coaxial cable permits high data transmission rates, on the order of millions of bits per second, and is relatively immune to noise or interference. Its characteristics make it suitable for short range computer networks, telephone toll lines, or cable TV lines. Large diameter coaxial cables augmented with signal amplifiers can be used for longer distances in phone or TV networks.

Coaxial cables are very common in networks today. For many applications, they are ideal. But for some applications requiring very high bandwidth and low attenuation over long distances, cables made of optical fibers are more economical.

Optical fibers consist of a glass or plastic fiber surrounded by glass or plastic cladding or covering. The construction of the core fiber determines the type of transmission involved. Figure 6.9 shows an end view of a fiber optic cable and the three methods of transmission within the core. In the step index type, light signals are contained within the fiber by reflection off the walls as they progress down the cable. This was the earliest type of fiber used. The graded index fiber is constructed of core material having variations in the refractive index from the center of the core to the outside. Refraction of the light signal keeps it within the core. Single mode transmission uses a very thin core of glass (about 1/100th of a millimeter or less in diameter) to contain the signal. This technology is still under development but promises to be the most effective of the three types.

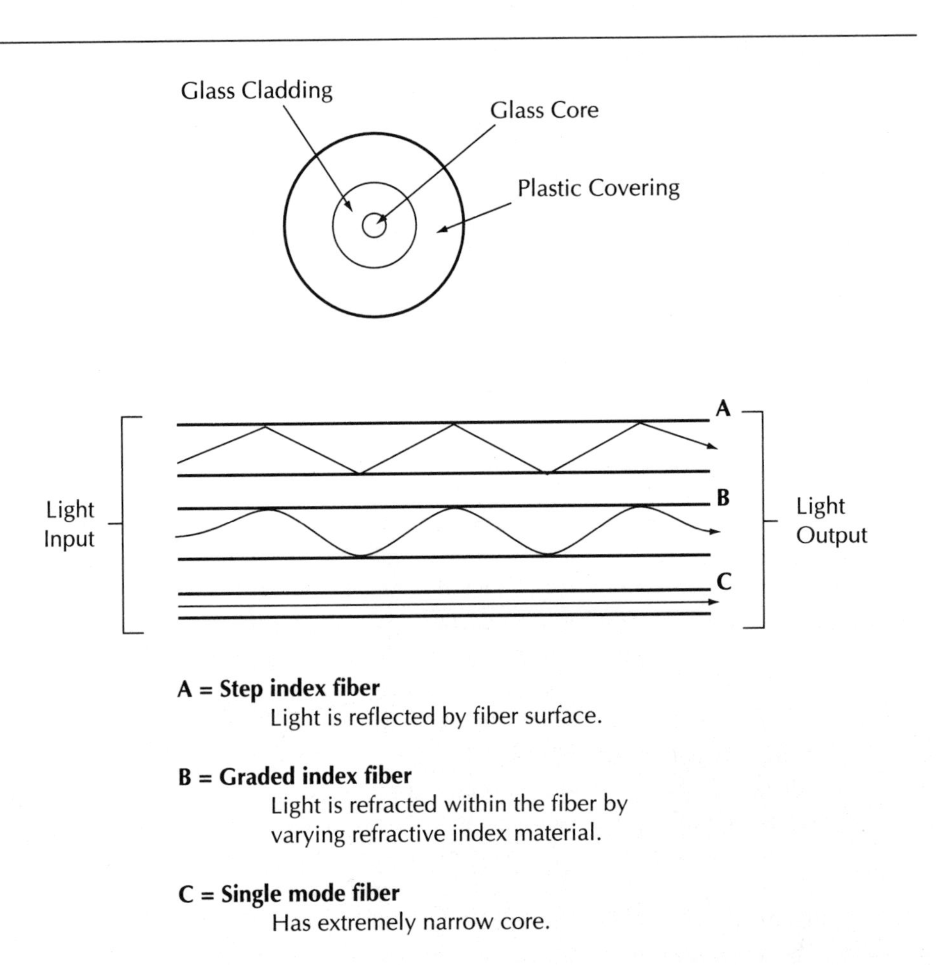

FIGURE 6.9 Construction of Optical Fiber Cable

Fiber optic cables offer many advantages over earlier cable types. They are low cost and have low attenuation making them suitable for long distances. Optical fibers have high bandwidth: They are capable of transmitting billions of bits per second.[3] They use light as the signal source; thus they are free of electrical interference. Fiber optic cables are difficult to tap and are very secure. However, taps and termination devices are relatively expensive so short fiber optic links are not economical for some applications. But because of their advantages, optical fibers are increasingly popular in voice, data, and video networks today.

Recall, however, that it's not necessary to have a physical connection such as wire or fiber to transmit information. Radio transmission, television, satellite communication, and even invisible infrared light (used in TV remote control devices) all are used extensively as communication media. The telephone network depends heavily on land-based microwave links and uses satellite microwave or fiber optic links for intercontinental traffic. Radiated media are very important in communications systems of all kinds including some portions of cable TV systems.

THE CABLE NETWORK

The cable network in the U.S. was initiated on a limited scale in 1949. Now, approximately 90 cable networks provide service to more than 60 percent of U.S. homes. Cable passes near another 30 percent of non-subscriber homes. Cable ranks third among wirelines behind power and phone in its penetration of households. Its growth has paralleled the growth of broadcast television in the U.S.

Cable is characterized as direct wireline (non-switched), generally not interactive, broadband, and private. It is expected that subscribers will select one of several tens of signals available on the line, and that all subscribers may want to use the line simultaneously. In contrast with the phone system's star architecture, the cable network architecture is tree and branch. All signals originate at the headend and are cablecast unidirectionally to subscriber equipment. Though the capacity of cable networks is high in comparison with phone networks, cable providers did not initially plan to use the network for bidirectional voice communication.

Cable Network Architecture and Headend

The point at which signals are placed on the local cable network is called the headend. The headend connects to one or more trunk lines that form the primary distribution network. Subscriber loops tap off the distribution trunks bringing the signals into individual homes or businesses. Thus, the network is called tree and branch. This arrangement is shown in Figure 6.10.

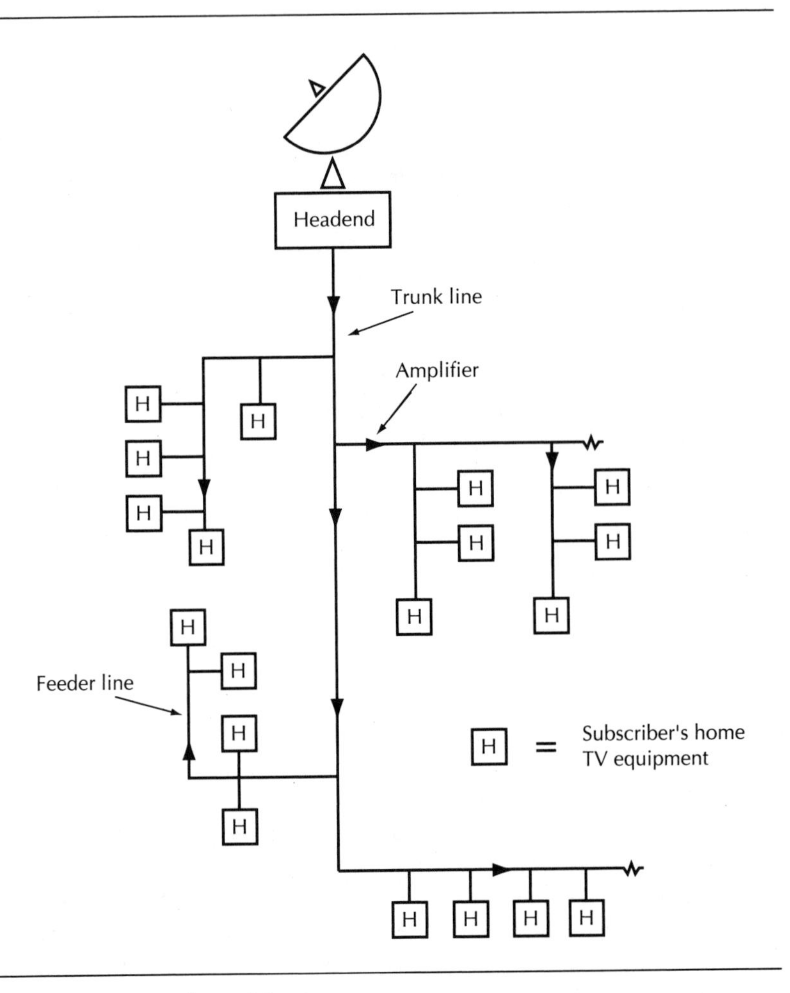

FIGURE 6.10 Cable Network Architecture

The headend of a cable network contains many types of sophisticated equipment needed to obtain signals from various sources and to place them on the cable for distribution. Source signals may be obtained at the headend from satellite or microwave transmissions via direct reception of radio or television broadcasts, or over landlines from other distribution sources. Headend sites include towers and antennas needed to secure signals.

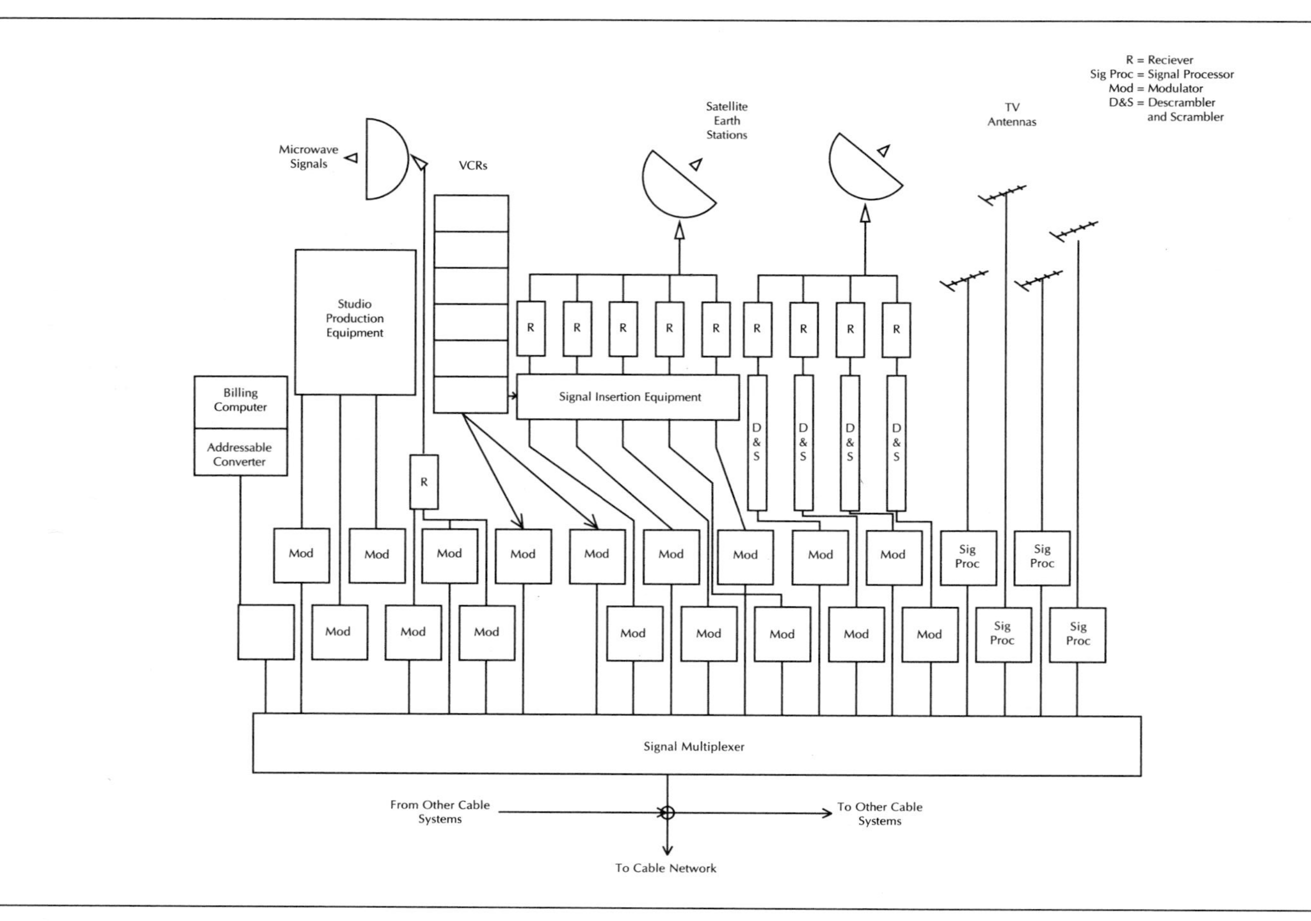

FIGURE 6.11 Cable Headend Configuration

The headend may also contain production equipment such as VCRs, video cameras, keyboards, control panels, and computers used to produce local formats and to control video signal distribution. Signal processors clean incoming signals of snow and hiss, filter interference created by unwanted signals, maintain signal strength, and, in some cases, convert the signal's frequency before passing it to the modulator. The modulator places the signal on the cable for transmission. In addition, the headend may transmit some signals to neighboring headends for retransmission to subscribers on adjacent cable networks. Figure 6.11 shows the configuration of headend equipment.

The headend also contains computer equipment for coding the signals so that only sites that have subscribed to the signal (have paid for it) are able to receive it. Converter equipment at the subscriber end decodes the signal thus enabling reception and viewing. Headend computers also handle billing.

The purpose of the cable headend is to acquire signals, maintain or enhance their quality, place them on the distribution trunk lines, and provide administrative network control. The headend is highly automated and intensely electronic, but it comprises only 6 percent of the network's costs. The remaining costs are in the distribution system.

Cable Distribution Systems

The cable network distribution system consists of trunk lines beginning at the headend, traveling through the service area, branching at various points, and terminating at the service area boundary. At points along the trunk lines, feeder lines branch out to individual subscriber locations. The trunk and feeder line system is shown in Figure 6.12. In a typical cable network, 19 percent of the cost resides in the trunk system and 75 percent in the feeder system (the last mile) linking sites to trunk lines.

The trunk line is a coaxial cable about three-quarters of an inch in diameter designed to carry up to 75 analog video signals with minimum signal loss or attenuation. Even with cables of this size, amplifiers and gain control loops must be placed along the trunk every one-quarter to one-half mile to maintain uniform signal strength and quality. As shown in Figure 6.12, at points along the line, taps into the trunk permit connections to the feeder system.

The feeder system is constructed of smaller cable (three-eights to one-half inch in diameter) that transmits quality signals over the shorter distances from trunks to individual homes or businesses. The feeder system contains line splitters (multiline connectors), taps for home drop lines, and feeder amplifiers, if feeders extend beyond 300–400 feet or so. In addition to video signals, trunk and feeder lines must carry electrical power for the trunk and feeder amplifiers and maintain interfaces for use by line technicians.

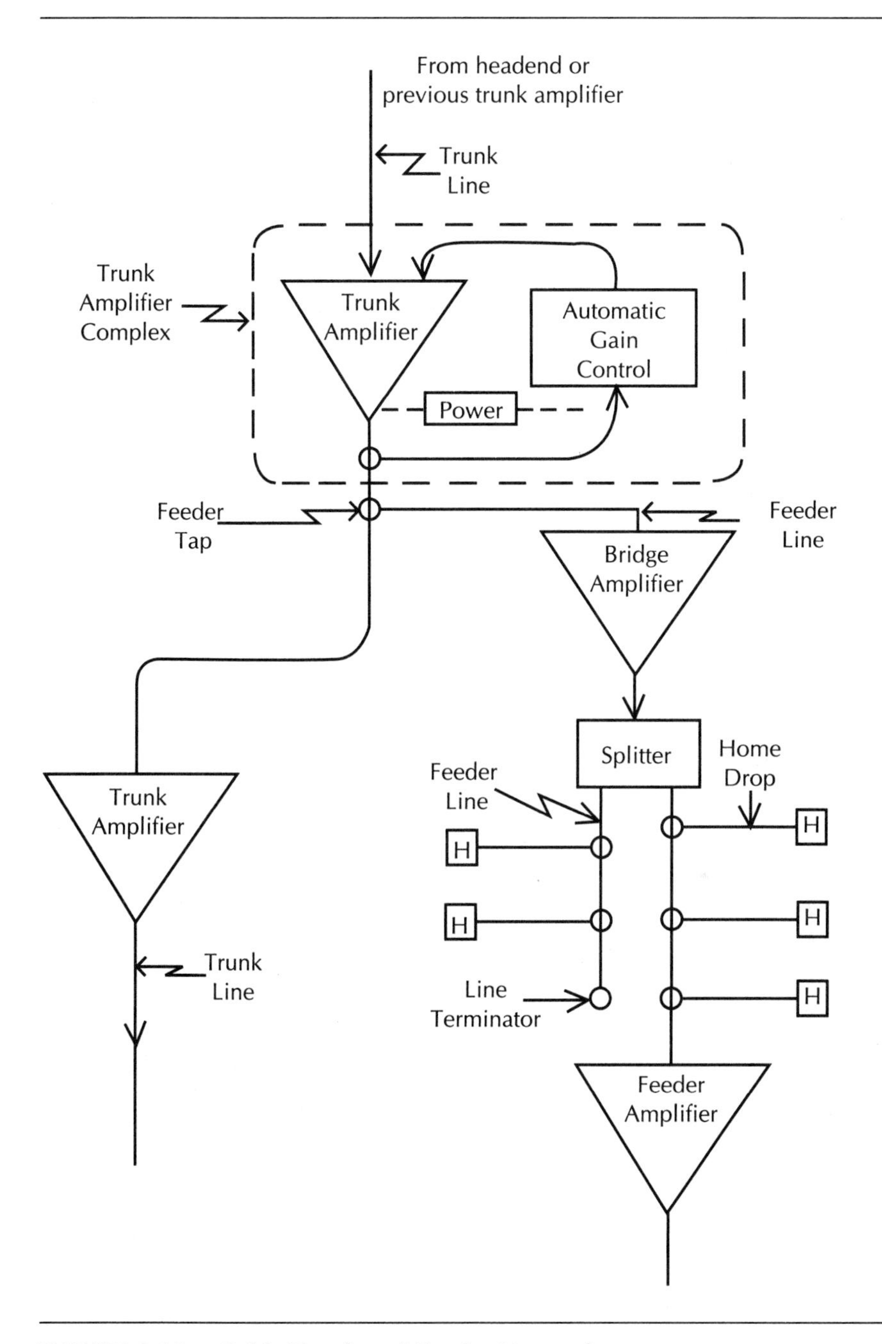

FIGURE 6.12 Cable Trunk and Feeder Network

The cable network becomes more complex as operators attempt to improve system reliability, security, and function or performance. In some instances, the network consists of dual cables to increase system capacity and, in some instances, operators have added cable to provide special functions for individual companies or institutions. For example, to assist telecommuters, Digital Equipment Company links some employee homes to their office locations. Other firms use cable for teleconferencing—maintaining interactive video between conference rooms in individual office buildings. Some cable systems include devices within the feeder system to maintain signal security and prevent unauthorized use. Driven by many forces and responding to many opportunities, the cable network is rapidly becoming complex.

Cable services provide one-way transmission of video programming (cablecasting) that requires end-user selection or intervention to obtain the desired programming. Defined and delimited this way, cable services are not regulated as common carriers or public utilities but must be franchised by local governments.[4] Thus, cable networks are deemed private, not public, utilities. However, cable systems come under governmental surveillance at several levels and are attracting increased attention as they depart from delivering basic cable services.

Cable System Capacity

Cable networks are broadband. In contrast with the 4000 hertz telephone local loop, cable feeders carry many frequency-division multiplexed channels each with a bandwidth of approximately six megahertz. The information carrying capacity of a single feeder is 50 thousand times that of the phone access line as currently configured. However, new developments in digital signal processing will allow feeder cables to carry even more video signals. These developments in cable engineering, along with improvements to the telephone network, will also increase the capacity of local phone access lines, perhaps to the extent of enabling the phone network to carry digital TV signals!

Coaxial cable TV channel capacity is currently limited by the frequency response of the signal amplifiers. The frequency spectrum found on modern trunk lines extends from 54 megahertz to about 550 megahertz with several small gaps to allow for system control, digital information, and FM broadcast signals. Dividing this spectrum into the 6 megahertz (including video guardbands) subdivisions required for each channel permits 80 signal channels. In practice, most systems contain 50 to 54 channels. New technology in the form of fiber optic links and digital signal compression may add 450 additional channels to the cable network. In addition, cable operators are anticipating interactive, multimedia applications.

Integrated Services Digital Network

Integrated Services Digital Network (ISDN) technology integrates digitized voice and data for transmission over the copper circuitry and other links currently installed in the phone system. ISDN eliminates the analog portion of today's networks and makes the network digital from end to end. It offers enhanced communications services and high-speed data transmission over one network element.[5] These services are now becoming available to subscribers who desire them.

There are two classes of service offered by ISDN: the basic rate and the primary rate. A depiction of these services is shown in Figure 6.13. The basic rate, called 2B+D, consists of two 64K bps B channels and one 16K bps D channel. The B channels can transmit either digitized voice or data, while the D channel is used for signal control and routing information. An additional 48K bps is also available for other purposes.

The primary rate service, named 23B+D, consists of 23 64K bps B channels and one 64K bps D channel. The primary rate supports 1.544 Mbps bidirectional transmission. The B channels can be used for a mixture of voice and/or data communication needs. ISDN is useful for interconnecting data processing facilities within the firm, for connecting sites within the firm for voice communication, or for integrating voice, facsimile, and data transmission. Transmission capability much beyond 1.544 Mbps, called B-ISDN, is also being implemented.

Since the 1960s, local phone companies have been upgrading their switches to digital systems, but the circuitry to the individual phones remains analog in nature. The phone network is inefficient because it contains this mixture of analog and digital technology. Being totally digital, ISDN removes this inefficiency. For example, ISDN equipment is able to communicate with the central switch during routine calls using the D channel, thus permitting performance improvements and paving the way for new and different functions.

Since adoption of ISDN in the U.S. must proceed despite a huge investment in copper and analog circuitry, and the relatively effective phone system must remain operable during the transition, there are obvious financial barriers to a speedy implementation. Some countries with less well-developed phone systems, or with systems in need of replacement, may be motivated to implement ISDN on an accelerated schedule. Parts of Europe are adopting ISDN faster than the U.S.

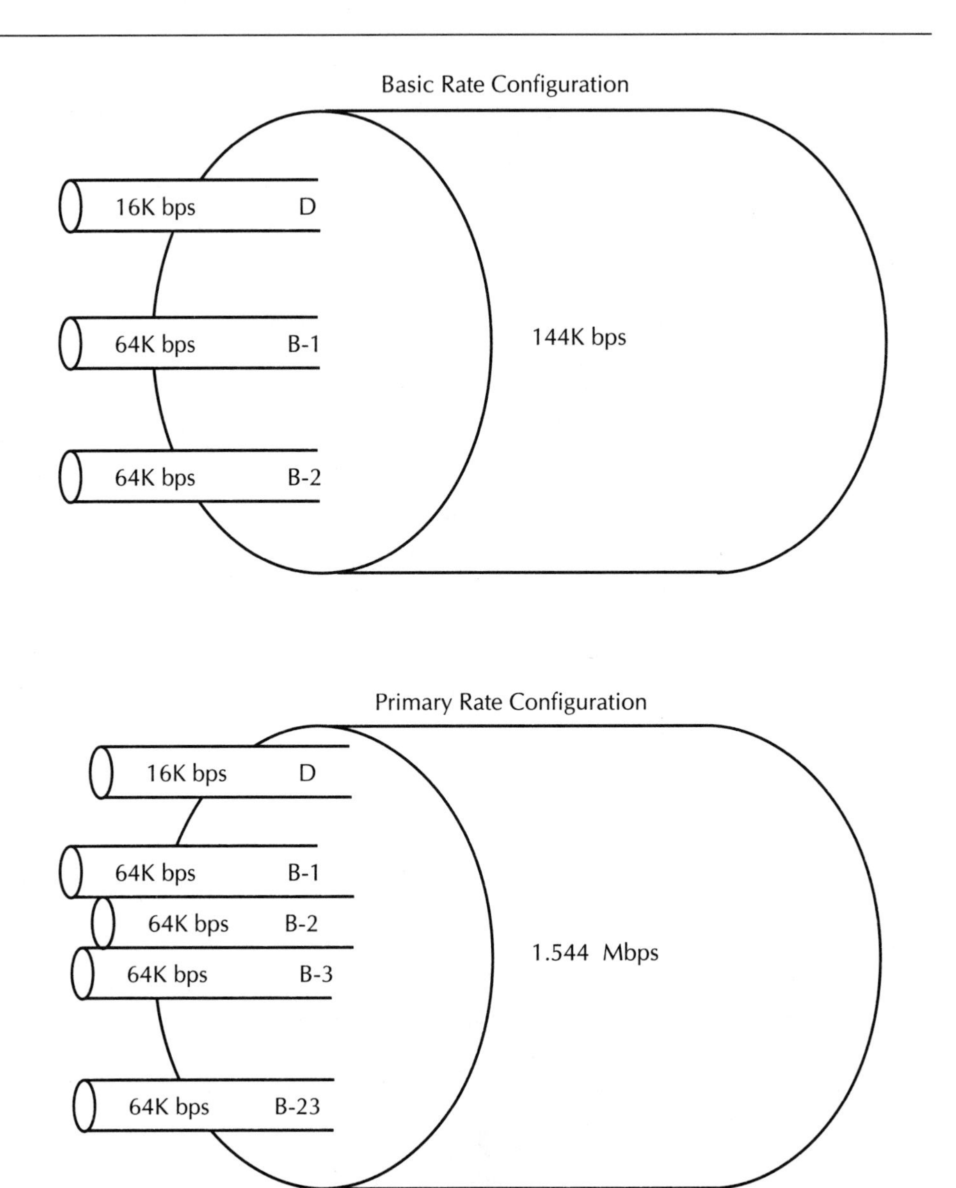

FIGURE 6.13 ISDN Basic and Primary Rate Interfaces

T-1 Service

In response to its new competitive environment, AT&T made T-1 service available to the public in September 1983. It had been implemented internally within the phone system for many years to connect exchanges, but had not been made available commercially prior to 1983. A T-1 link can be constructed of copper wires enhanced to handle digital signals at rates exceeding 1.5 Mbps, or from other media. The link can handle up to 24 simultaneous phone conversations at approximately four times the cost of an ordinary line. Many firms, among them American Airlines and Merrill Lynch, have been able to capitalize on this technology, and its use is expected to expand as many more firms do so in the future.

Now, extensions in T-1 services include very much higher bandwidths than originally proposed. Table 6.1 displays the current structure of extended service in North America. In addition, there is widespread interest in fractional T-1 service which permits customers to use a portion of a T-1 link.

Telecommunication equipment suppliers build and market multiplexers for handling one or many T-1 lines. These devices manage the traffic, detect and isolate defective lines, and provide redundancy, thereby improving reliability. Multiplexers are becoming more sophisticated as the functional capability of their logic and memory circuity continues to grow as it has in computer technology. This sophistication improves throughput because it makes voice compression possible. It also increases reliability through error detection, correction, and redundancy techniques; and it increases customer satisfaction. These functional improvements are increasing the demand for telecommunication services. More than one-half of the locations owned by U.S. Fortune 1000 companies now employ T-1 technology.

TABLE 6.1 T-1 Services

Digital Facilities	Transmission Rate	Number of T-1 Equivalents
T-1	1.544 Mbps	1
T-1C	3.125 Mbps	2
T-2	6.312 Mbps	4
T-3	44.746 Mbps	28
T-4	274.176 Mbps	168

T-1 technology is not confined to the private sector. The Federal Reserve System is consolidating 12 separate networks into a new network based on fractional and full T-1 capability called Fednet. Fednet consolidates the Fed's 12 data processing centers into three, and permits depository institutions with similar equipment to connect to the Fed system. It's expected that the network will carry more than $1.25 trillion in daily transactions.[6]

VOICE SIGNAL COMPRESSION

Is it possible to increase the number of phone conversations beyond 24 over a T-1 line while maintaining satisfactory quality? Is there some way to increase the effective capacity of a line for digitized voice information?

As it turns out, the traditional sampling rate of 64,000 bits per second is overly conservative for voice conversations and that 32,000 bps is quite satisfactory. The technique that uses this sampling rate is known as Adaptive Differential Pulse Code Modulation (ADPCM). Decreasing the bit rate by 50 percent through ADPCM technology increases the number of voice messages carried on a T-1 line from 24 to 48. Other more sophisticated speech compression techniques that encode voice information based on speech characteristics have also been developed. One such technique developed by GTE yields an increase in the effective T-1 capacity by a factor of two, raising to 96 the number of simultaneous voice conversations—a factor of four improvement.

Other techniques that detect short, silent pauses between words and sentences offer additional room for significant improvement in throughput. One such algorithm, called Time Assignment Speech Interpolation (TASI), can add another factor of two or three to the throughput of a T-1 line. Under optimum conditions using current technology, up to 288 voice messages can be transmitted concurrently on a single T-1 line! Digital compression techniques are implemented in computer-like digital signal processors. Digital compression greatly increases the capacity of the phone network and gives scientists great incentive to develop compression technology for video signals.

Today's standard analog television broadcast contains about 100 million bits of information per second. By digitizing the analog signal, digital compression techniques can reduce the bandwidth needed to transmit the signal by a factor of ten without noticeable signal degeneration. Some techniques mathematically smooth color details not noticeable to the human eye while others eliminate information that has not changed from frame to frame. Video signal compression requires specially designed, high-speed processors. The notion that video signals can be compressed by a factor of ten quickly led to the well-known belief that 500 channel TV is in the near future.

Several important organizations develop and set standards that coordinate the details of data communication networks. Founded in 1946, the International Standards Organization (ISO) has been very effective in establishing standards on a broad range of topics. Mostly, its membership consists of standards organizations representing participating nations operating on a voluntary, nontreaty basis. The ISO has authored more than 5000 standards designed to facilitate international commerce and economic activity. One of its most important efforts has been in data communications where its Open System Interconnect model is one of the most influential standards.

The Open System Interconnect Model

The Open System Interconnect model (OSI) creates an open or nonproprietary data communications architecture intended to promote end-to-end compatibility between communications subsystems. The model is extremely complex and detailed requiring many thousands of pages to describe. It has been under development for nearly two decades by individuals representing nearly every facet of the communications industry. The OSI model represents an important, critical development in data communications technology.

The model describes how the network operates and the protocols it uses. It consists of a seven-layered architecture in which each layer views the one below it as if it were the actual path to the other communication system. The model achieves compatibility between communication systems independent of specific application coding. The effect of layering is to achieve modularity of communications systems through carefully defined rules and terminology. In addition to accomplishing this important task, the model has had a major influence on our perceptions of network communications. The layered OSI model is shown in Figure 6.14.

In the OSI model, data flows from one user's application program through the model's seven layers to the network media itself. At the terminating end, the data leaves the network media and moves through seven layers to the receiving application. The path through the model is bidirectional. Each layer specifies or provides services and has at least one protocol or definition of how the applications establish communication, exchange messages, or transmit data. The layered architecture, which defines divisions of responsibility, permits changes within one layer without disrupting the entire model. Thus, OSI modularity greatly reduces complexity. OSI is extremely important because it provides an organized, logical solution to a very large communications problem.

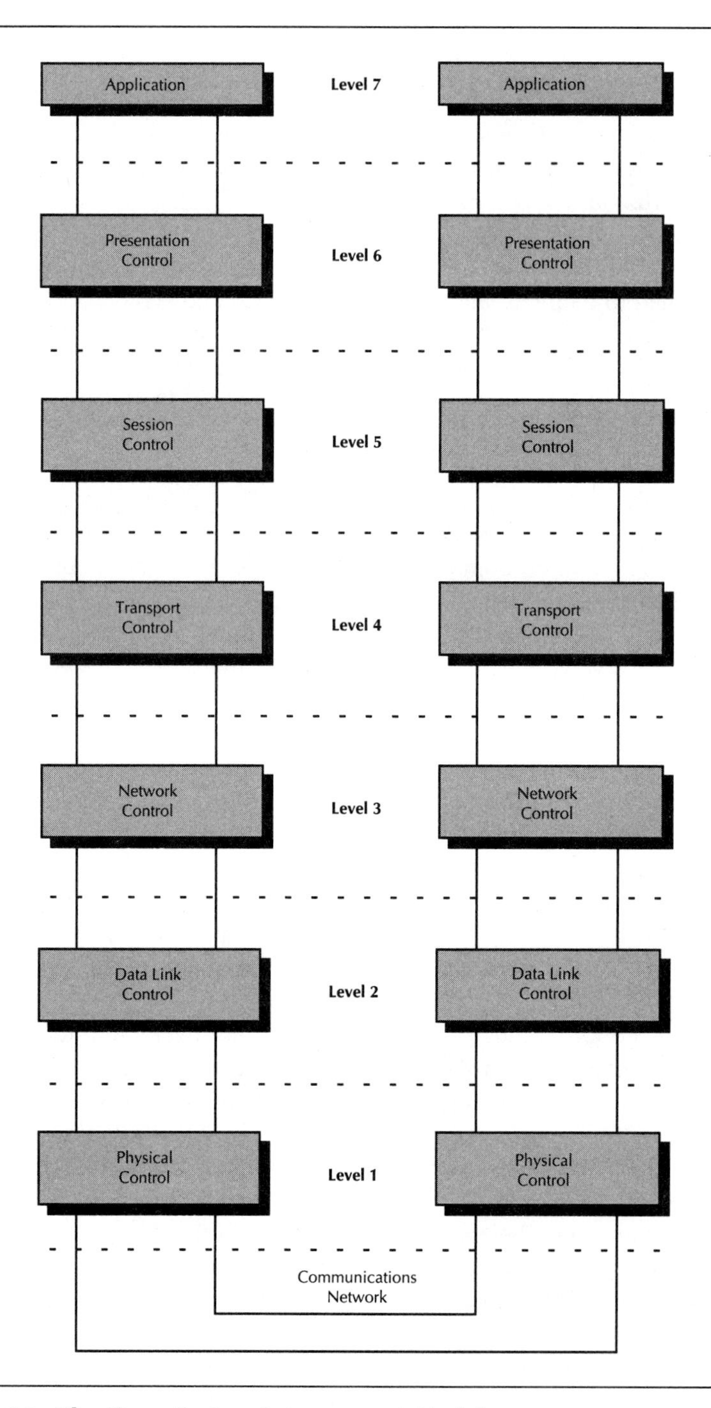

FIGURE 6.14 The Open System Interconnect Model

Communication Between Networks

Devices, and the protocols and formats that they support, may differ from one user network to another. These differences must be resolved for communication to proceed. For example, users of one local area network (LAN) with a specific communication data format may need to communicate with users on another LAN having a different data format. All the facilities to handle these incompatibilities are provided by the lowest three layers in the OSI model. Thus, devices operating within the physical, data link, and network layers of the OSI model facilitate communication between networks.

Extending a single network, or communicating between stations on different networks, requires specialized facilities operating at layers 1, 2, or 3 in the OSI model. One such facility is a signal repeater. The purpose of the repeater is to retime or reshape pulses thus mitigating the effects of signal attenuation or distortion and extending the network's length. Thus, a repeater is placed at intervals along the line to forward the signal after improving its quality. In local telephone and cable networks, repeaters are inserted every mile or so to restore signal quality. Repeaters on the communication line operate at the physical layer and are called layer 1 relays.

Many times it's necessary to connect two networks that use identical data communication formats or protocols, such as two LANs. In these instances, individuals using devices on one LAN want to communicate with individuals on an adjacent LAN. The two LANs must be electrically connected to communicate. For this to happen, knowledge of device addresses must be shared between the networks and disciplines must be followed to prevent disruption to network operations. The facility to accomplish this is called a network bridge.

A bridge is a programmable facility that connects two LANs operating with identical protocols. The bridge resolves sending and receiving addresses and passes packets of data between the LANs. The bridge does not modify the packet and does not add anything to it. Information on a network communication line is intercepted by a bridge that passes it along if the addresses are on the same LAN, or routes the information upward through its data link layer to the connecting LAN if the addresses are for different LANs. Thus bridges are layer 2 facilities.

A router/gateway is a device used to connect two networks that may or may not have similar protocols. If the networks are similar, the device is usually called a router. If they are not similar, the device is usually termed a gateway. (Definitions in the literature are not consistent on this point.) The gateway employs an internet protocol or, in other words, it has knowledge of each network's operating protocol and can translate messages from one protocol to another. It also resolves addresses and establishes routes or paths.

During operation, routers/gateways determine if the message is intended for a user on the same LAN. If so, it merely passes the information down the communication line. If the message is for a user on a LAN operating with a protocol different from the protocol of the sending LAN, the gateway makes the necessary internet conversion and sends the information down the line of the receiver's LAN. In the process, the gateway resolves addresses within its data link layer. The function of the router/gateway embodies that of a bridge but also includes the much more complex functions of protocol conversion.

A collection of communication networks connected together with routers/gateways is called an internet. An internet with a complex topology presents a significant challenge to routers and gateways, and their capabilities must be quite sophisticated. Figure 6.15 depicts one such network.

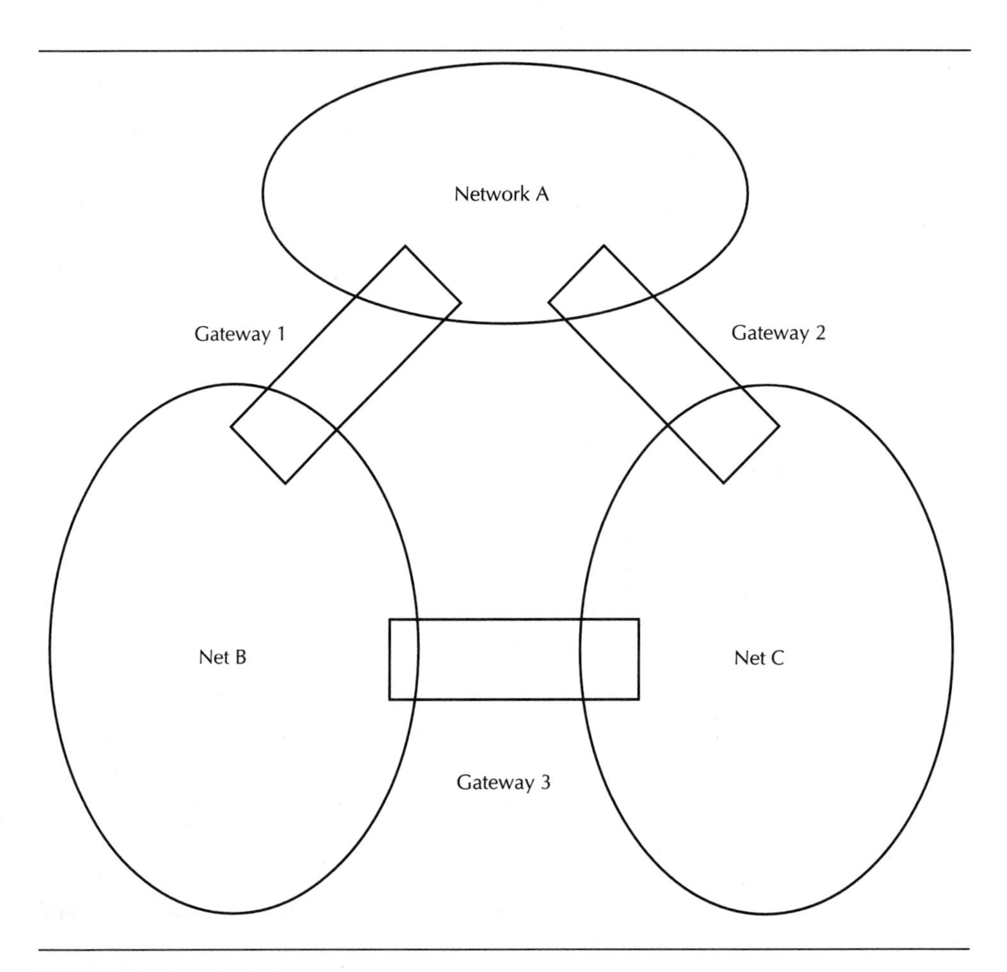

FIGURE 6.15 An Internet

As Figure 6.15 suggests, multiple paths can exist in the internet and the gateways must resolve path routing. For example, a user on net A may communicate with a user on net B through gateway 1. Under some circumstances such as traffic congestion or equipment failure, traffic from net A destined for net B may get routed through net C. Since the nets comprising the internet usually contain different protocols, gateways must be able to resolve routing problems and protocol differences. These complexities, along with the usual problems of network and human errors and the need to satisfy accounting, security, and a host of other demands, make internet operations very complex. It's this complexity that the OSI architecture organizes so effectively.

Vendor Network Architectures

Several organizations have developed network architectures that have gained popularity and widespread adoption. For example, the U.S. Defense Department sponsored Transmission Control Protocol/Internet Protocol (TCP/IP) while Digital Equipment Company and IBM initiated Digital Network Architecture (DNA) and Systems Network Architecture (SNA), respectively.

IBM introduced SNA in 1973 and has secured a large user base since then. SNA is an IBM proprietary product and is considered a *de facto* standard for IBM customers worldwide. SNA is a hierarchical, layered model supporting distributed mainframes, token-ring local area networks, and peer-to-peer communication. The layered architecture is similar to that of the OSI model. DNA is DEC's proprietary network standard and, like SNA, it is a layered architecture roughly comparable to ISO/OSI. DNA supports distributed, interconnected DEC minicomputers and terminals attached to minis. DNA is popular for local area networks operating within a building or group of buildings.

TCP/IP is a set of protocols developed by the U.S. Defense Department for its ARPAnet network and, therefore, it is vendor-independent. Many vendors support TCP/IP. UNIX and ethernet environments use TCP/IP, and IBM, DEC, and Apple Computer support TCP/IP products. TCP/IP is widely used as an approach toward network interconnectivity. In addition to TCP/IP, most vendor architectures (there are many others not discussed here) are migrating slowly toward the OSI standards.

The demand for voice and data communications in government operations and in commercial businesses continues to grow rapidly as communication costs drop, and as organizations learn to apply new technology advantageously. Data traffic on U.S. telecommunications networks is growing about 20 percent per year and voice traffic at about 5–6 percent per year.

The result of these growth rate differentials is that data accounts for about half of the telecommunications traffic. Lower costs and increased availability of data communications components drive some of the increased demand, but new applications and new forms of enterprise to capture the technological advantages are the primary driving forces.

Options available to network implementors have also increased as network standards such as OSI, ISDN, and others evolved. Many other options for network planners are being developed thus spuring growth in local and wide-area nets for voice, data, and video communication. New architectures, powerful network operating systems, and sophisticated network management systems are required to attain full advantage of the network and computing technology.

ADVANCED TRANSPORT TECHNOLOGIES

High-speed networks with data rates of 45 Mbps or higher frequently employ microwave or fiber optic transmission technologies. For reasons of cost, size, weight, and flexibility, and because many nets are being installed in large buildings in large cities, most high-bandwidth networks are being constructed with fiber optic media.

Fiber Distributed Data Interface

One important standard for high-speed LANs using fiber optic technology is Fiber Distributed Data Interface (FDDI). The FDDI standard was developed by the American National Standards Institute and is designed for firms that need flexible, high-performance networks. FDDI performance is defined as 100 megabits per second. The standard supports 500 LAN stations extended over a maximum fiber length of 200 kilometers. Station spacing must not exceed 2 kilometers. FDDI conforms to the OSI model.

FDDI is designed to accommodate future growth in the number of network stations and in traffic loads resulting from client/server operations and complex applications such as computer-aided design and graphics manipulation. By employing fiber optic technology, FDDI transport rates provide substantial improvements over typical 4 to 16 Mbps LAN implementations; thus, FDDI is useful for linking together lower speed LANs to configure internets.

One of the basic concepts of the FDDI standard is dual counter-rotating rings. Figure 6.16 shows this configuration with data flowing in opposite directions in the primary and secondary or backup ring. The dual ring concept aids in network initiation and reconfiguration and provides backup and recovery capability in case of ring failure. The dual ring maintains network operations if a device on the network fails or a fiber fault occurs. In case of failure, the LAN path is restored using the secondary ring, which is normally idle. This redundancy increases the fault tolerance of FDDI networks.

FDDI networks can be configured in several ways to support high-speed workgroups, interconnect separate LANs, or provide a backbone net for multiple buildings or campus networks. In Europe, for example, the Italian government installed an FDDI backbone network connecting over 300 toll stations on its highway system. The system gathers traffic statistics, updates weather and road conditions, and manages toll collection. This state-of-the-art system cost about $35 million to install.

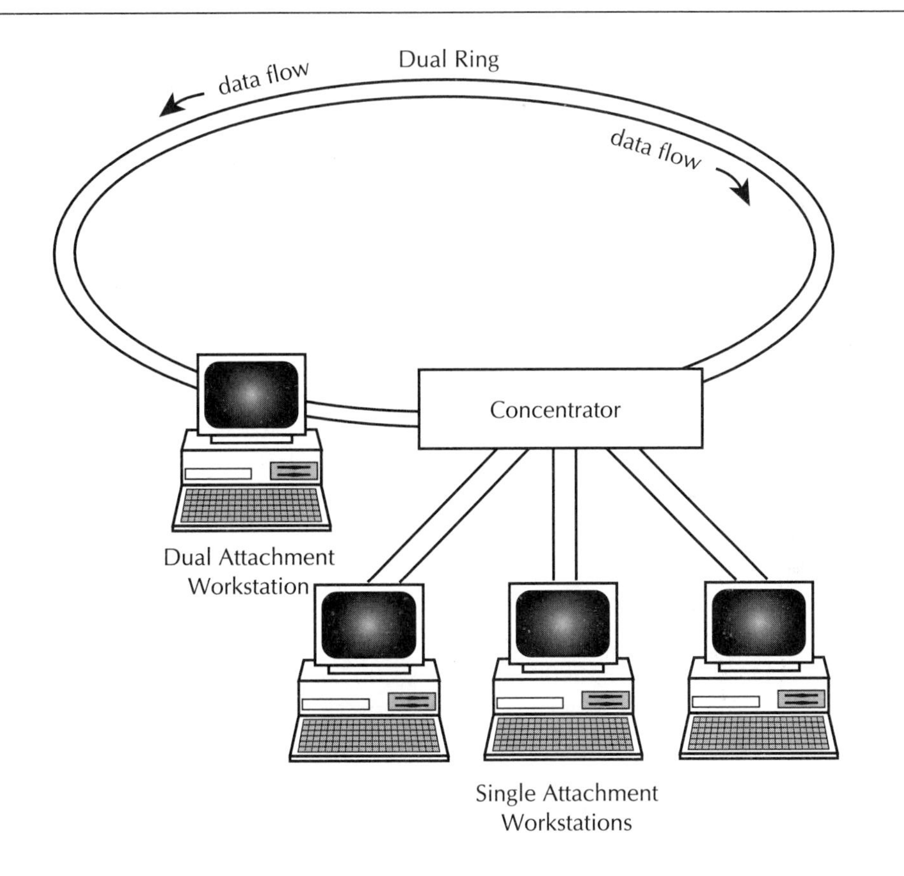

FIGURE 6.16 FDDI Network Configuration

Metropolitan Area Nets (MAN) will benefit considerably from FDDI and its extensions. FDDI and its enhancements FDDI-II, FDDI Follow-On, and its kin, Distributed Queue Dual Bus (DQCB) will increase network throughput, improve reliability, and form the base for growth in fiber optic networks.

Synchronized Optical Network

Synchronized Optical Network, or SONET, is another standard specifically designed for optical fiber media. It is a hierarchical standard starting at a base of approximately 51.84 megabits per second. This rate is called Synchronous Transport Level 1 (STS-1). The STS-1 data rate is multiplexed to higher transmissions and is defined to a rate of 2.4 gigabits per second. Actually, the multiplexing is defined by STS-n, where n takes on values of 1, 3, 9, 12, 18, 24, 36, and 48 only. The STS hierarchy is shown in Table 6.2.

TABLE 6.2 STS Line Rates

STS Level	Line Rates in Mbps
STS-1	51.84
STS-3	155.42
STS-9	466.56
STS-12	622.08
STS-18	933.12
STS-24	1244.16
STS-36	1866.24
STS-48	2488.32

The STS-1 SONET signal is a 9 by 90 byte data block that repeats every 125 microseconds, or 8000 times per second. The block contains 6480 bits (9 x 90 x 8); thus, the data rate is 51.84 Mbps (8000 x 6480). Extensions to the data rates above STS-48 are expected to yield transmission speeds greater than 10 gigabits per second.

Rapid advances in SONET technology are expected during the next decade in conjunction with advances in digital switch interfaces, optoelectronics switches, and photonic switches. These and other developments are likely to make SONET technology the preferred choice for high-speed data transmission in the next 20 years.

Other Advanced Transport Technologies

Most of the world's communication technology is either circuit switched, as in the narrowband phone system, or packet switched for data traffic. The main inefficiency of circuit switched voice or fax is that dedicated bandwidth is wasted during breaks in the voice or fax communication. Most voice communication is on the order of one word per second, slow in comparison to the capability of the dedicated line. Packet switching buffers data into packets before transmission and shares one circuit among several data sources. Thus, packet switching uses available bandwidth more fully than circuit switching. Slight delays caused by circuit sharing are immaterial in data transmission but unacceptable in voice communication. Advanced transport technologies using broadband switched nets are able to use packets for virtually all types of communication.

Two advanced transport technologies are frame switching (frame relay) and cell switching (asynchronous transfer mode or ATM). Both technologies use a star-wired network with dedicated lines connected to a switch at the center (like the phone net). Thus, ports can communicate with each other at full line speed. The difference between ATM and Frame Relay lies in the construction of the information packet, as shown in Figure 6.17.

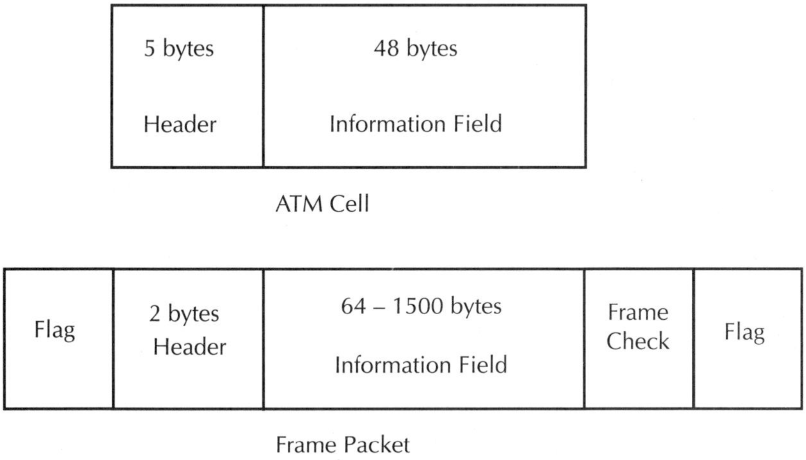

5 bytes Header	48 bytes Information Field

ATM Cell

Flag	2 bytes Header	64 – 1500 bytes Information Field	Frame Check	Flag

Frame Packet

FIGURE 6.17 ATM Cells and Frame Relay Packets

The frame contains an information field of variable length and packet data that ensures integrity of the information. The information field can vary from 64 bytes to 1500 bytes in length. The ATM cell structure is a fixed 53-byte length and assumes network integrity to ensure information transmission accuracy (a good assumption for advanced networks). With frame switching, switch management is also different. Here, the circuit is established for each packet. But with ATM, the initiating station requests the destination and the bandwidth. The switch establishes dedicated bandwidth to the message for its duration.

ATM equipment merges all forms of information (voice, data, video, etc.) for transmission to the switch where it is dispatched to its destination, which also contains ATM equipment. Since the packets are small and fixed in length, the traffic is very predictable. Voice data can be given a fixed number of cells to ensure high quality and the remaining time allotted to data traffic. As this method suggests, ATM is a sophisticated form of time-division multiplexing occurring over broadband lines. Switched multi-megabyte services such as ATM can deliver data rates exceeding 100 megabytes per second, limited by line, switch, and equipment bandwidth. ATM seems to be the technology of the future. Widespread implementation is on the rise as facilities and equipment costs decline.[7]

Desktop communication capability is growing very rapidly, too. Transmission speed at the desktop was about 1.0 Mbps in 1980, 50.0 Mbps in 1990, and projected to be 10.0 Gbps by the year 2000. This phenomenon is providing extreme challenges to microprocessor developers. Having increased very dramatically, PC processor speeds serve very well for dedicated computing but, as the PC evolves into the center of communication, bandwidth capability becomes increasingly important. By the year 2000, PCs will be processing several hundred MIPS *and* handling data traffic at the rate of gigabytes per second. Enormous computation and communication capability will be available to office workers in the near future.

WIRELESS SYSTEMS

In 1991, George Gilder predicted:

> What currently goes through wires, chiefly voice, will move to the air; what currently goes through the air, chiefly video, will move to wires. The phone will become wireless, as mobile as a watch and as personal as a wallet; computer video will run over fiber-optic cables in a switched digital system as convenient as the telephone is today.[8]

Radio and wireless communication, rapidly developing since 1920 or so, play an important role today in AM, FM, and TV broadcast systems and in a wide variety of communications for governments, military organizations, private corporations, and others. But in the next 10 years, wireless communications networks will link workers in offices, distribution centers, and field locations with voice, fax, data, and video, providing abundant information for the most demanding 21st century jobs. Through wireless transmission, everyone who wants to can be on-line, everywhere, all the time. The transitions occurring between forms of communication will profoundly change how the world communicates.

As many observers have noted, these communication transitions are occurring with revolutionary speed. Cellular voice and data services and mobile data networks are growing 20 to 30 percent annually, and the market is just opening in many places. In the U.S., mobile data networks, the convergence of wireless communications and portable computing, are growing rapidly. For example, a survey by the Yankee Group found that 12 percent of IS and telecom managers were presently using wireless products and that another 20 percent were in the initial stages of implementation.[9] Another 32 percent were considering acquiring wireless equipment. Advancing technology, declining costs, and immediate benefits are driving increased customer demand.

Land-Based Systems

Current cellular technology is based on a standard transmission protocol for analog phones called Advanced Mobile Phone Services (AMPS). Developed by Bell Labs, AMPS is being replaced by digital technologies that promise to bring many efficiencies to the cellular network just as digitization made possible many advances in wired systems. Tiny microprocessors within cellular phones will perform digitization and multiplexing and will control the radio operation, too. Future users will be able to send voice, fax, and data from a single cellular device. For the next several years, however, analog systems will coexist with digital systems. Eventually, digital systems will replace them.

Future digital cellular systems will be designed as a dense network of low-power, broadband radio transmitters and receivers. The architecture of the cellular network is portrayed in Figure 6.18.

The operation of the network depends on a combination of radio and landline communication. As shown in Figure 6.18, a call originates from a cellular phone and is received by the transmitter/receiver located in the cell site. Each cell or geographic region contains one transmitter/receiver that controls traffic within the cell and incoming and outgoing traffic. The message moves from the cell site via wires or radio to the Mobile Telephone Switching Office (MTSO) to the local exchange, where it proceeds as a conventional call. If the call is to another cellular phone in the area, the MTSO broadcasts the number to all its cell sites. Upon confirming the location of the called phone, the MTSO completes the communication between the two local cells.

If either the originating caller or the called party is traveling, the transmitter/receiver detects a weak signal and hands off the call to the transmitter/receiver in an adjacent cell. If a caller leaves the local service area for another, the call is completed through another MTSO. This is called roaming. Computers and computerized digital switches control and monitor the operation of this complex, high-tech network. Microchip technology makes cellular systems possible.

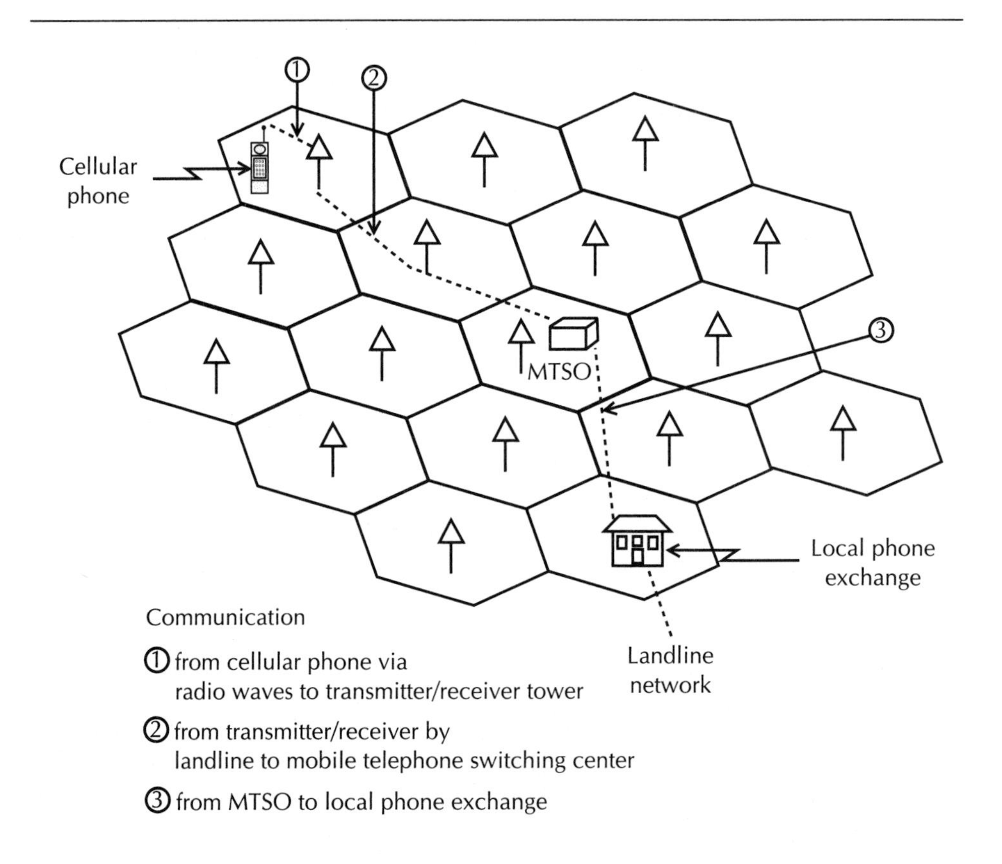

FIGURE 6.18 Cellular Network Architecture

In contrast with standard AM and FM broadcast stations which increase their service area by increasing their signal strength, cellular systems depend on low-power devices that are light weight and easily transported. Thus, the service area, or cell, must be relatively small and the transmitter/receiver network relatively dense. To understand how this works, refer to Shannon's formula discussed earlier in this chapter:

signal rate(bps) = Bandwidth(B) x log(base 2)(1 + (S/N))

If the signal power (S) equals the noise power (N), then

log(base 2)(1 + 1/1) = log(base 2)(2) = 0.693

or

signal rate(bps) = bandwidth(B) x 0.693

or

signal rate is directly proportional to the bandwidth.

As the signal power declines below the noise power, signal rate can be maintained by increasing the signal bandwidth according to Shannon's theory. Thus, small, light-weight, very low-powered cellular phones can communicate effectively in the presence of noise if sufficient bandwidth is allocated to the transmission. Cellular systems rely on a rather wide portion of the radio spectrum and a dense network of transmitter/receivers.

Several new digital technologies are replacing the analog AMPS system. Cellular Digital Packet Data (CDPD) places digital message packets in unused or underused portions of the radio spectrum or during conversational pauses. This speeds up data transmission by a factor of about eight over the speed of conventional cellular transmission.

Another method, Code Division Multiple Access (CDMA), tags each message with a code and spreads it over a wide spectrum. With CDMA, more low-power messages (calls) can occupy the wide channel. Some claim that CDMA offers nearly a 20-fold improvement over conventional AMPS. Another system based on time-division multiplexing called Time Division Multiple Access (TDMA) divides the channel into time slots and multiplexes 3 to 7 conversations in a manner similar to time-division multiplexing used in wired systems. TDMA is not as effective as CDMA, but its deployment is more advanced.

Cellular systems have great potential but are limited to land areas and are less economic in sparsely settled regions. To overcome these limitations, several firms are planning or building cellular systems that replace stationary cells with mobile cells controlled by low-earth-orbit satellites. The system is a truly global cellular network, but the technology is exotic, expensive, and risky.

Satellite Cellular Networks

The concept of a global cellular network is based on precisely locating numerous low-orbit satellites so that any point on earth can communicate with at least one satellite at all times. The satellites can communicate with each other and with ground stations so that any two people with suitable cellular phones can communicate, regardless of their locations. The concept includes an electronic, international directory containing a unique phone number for each cellular phone in the system. In principle, an individual on a trans-Pacific cruise using personal communication equipment can maintain phone contact with the home office 24 hours a day using voice, fax, or even video communication. At least eight separate systems aimed at capturing this business opportunity are in the conceptual or developmental stages.

One system, conceived by Motorola in 1987 and estimated to cost $3.4 billion, uses 77 satellites in polar orbit 500 miles above the earth, and 20 ground stations.[10] Satellites at this altitude are moving about 4.5 miles per second relative to a fixed point on the earth's surface and have a viewing area (cell size) of about 400 miles in diameter. Thus, an individual using the system for more than a brief period will need the services of several satellites. The system automatically transfers an individual's message to the appropriate satellite to maintain the communication path. For long-distance communication, several satellites in turn may serve the call originator, while several others may serve the recipient. The system switches and processes calls in space, but one or more ground stations may also be involved.

Transferring traffic from satellite to satellite is akin to transferring traffic from one land-based cellular station to another, as mobile callers move from one station's jurisdiction to another's. For obvious reasons, the satellite process is much more complex.

The Motorola system, termed Iridium (the element iridium has atomic number 77), is considerably different from current communications satellite systems. Traditional communications satellites operate in geosynchronous orbit about 22,300 miles above the earth. At this altitude, the satellite remains nearly stationary over a fixed point on land where it can view about one-fourth of the earth's surface. Because of the distance from satellite to earth, electromagnetic transmission to the satellite and back is delayed about one-fourth second. Thus, cellular phones require bulky battery packs. In contrast, the Iridium system suffers imperceptible communication delays once the call is established and transmission can be maintained with small cellular phones.

The real issues facing the Iridium project and others like it center around demand for the system and the developer's ability to market it at its anticipated costs. In Motorola's case, the phone device is expected to cost about $2500 per unit. Monthly service fees are projected at $50, and a five-minute call will be about $15. In comparison, land-based cellular service in the U.S. averages about 50 cents per minute, monthly charges are $20-30, and the instrument costs on the order of $100 or so. Advocates claim that demand in lesser developed countries and in sparsely settled regions of the world, combined with business and personal traffic, will make the systems profitable. By the year 2000, developers hope to capture 1–2 percent of the estimated market of 100 million cellular users and establish profitability.

For some people, these projects indicate communication overkill. Some believe brilliant technology is deceiving developers who are overlooking lower tech and lower cost alternatives. Regardless of the merits of the arguments on either side, satellite cellular systems, land-based cellular networks, land and undersea cable systems, and conventional satellites are greatly increasing global communication capability.

SUMMARY

Global telecommunications is one of the important results of advanced information technology. Semiconductor technology and programming are used extensively in the development and implementation of telecommunications equipment. Telecommunications hardware and software are being integrated with data processing equipment; and voice, image, and data communications are being integrated on a large scale.

Hardware and software that serve customer needs are available for communication between individuals and businesses with LANs or internets using wireline or wireless technology. Most business organizations in the U.S. find these capabilities attractive and have adopted them. On a larger scale, telecommunications systems serve national or international needs. Multinational firms, and firms engaged in international commerce, are rapidly implementing these technologies. Telecommunications technology and its uses in business and industry are critically important to information technology management.

Review Questions

1. Describe the conditions that are making the phone industry a growth industry, according to Peter Drucker.
2. What determines the architecture of a network system?
3. What are the characteristics of the telephone system in the U.S.?
4. What are the major elements of U.S. telephone network systems?
5. Describe the function of the local exchange. Why is the local access line called the bottleneck in the phone system?
6. What forms of media are found in a telephone network?
7. Describe the various forms of multiplexing discussed in this chapter. Which forms do not require digital signals?
8. From Shannon's formula, what factors affect signal transmission rate?
9. What are the characteristics of the cable TV network in the U.S.?
10. What is T-1 service?
11. Describe some conditions that make voice signal compression feasible. What additional factors apply to video signals?
12. What are the advantages and disadvantages of ISDN?
13. Why is the International Standards Organization model so important?
14. What are the main features of the OSI model?
15. Distinguish between repeaters, bridges, and gateways. What are some of the uses for these devices?
16. What is the theoretical ATM packet rate on an STS-1 SONET line? What are some of the implications of this rate?
17. Differentiate between AMPS and CDPD. How does TDMA differ from CDMA?
18. Describe the operation of the Iridium satellite system.

Discussion Questions

1. Describe the operation of the phone network in a call going from New York to Los Angeles. What possible forms of media may be involved in this call?
2. Discuss the steps required in making the phone system a digital network end-to-end.
3. Discuss the process for digitizing voice signals. What are the prerequisites for signal digitization and how does it enable multiplexing, switching, and advanced transmission techniques?
4. Discuss the advantages and disadvantages of coaxial cable and fiber optic cable. Based on what you have read or heard, which media will become dominant, and why?
5. Discuss the similarities and differences between the phone network and the cable network in the U.S. today. What modifications to each can be made to make them more alike?
6. Assume you are the CIO of a U.S. firm. Discuss the possible uses of the TV cable network in your business.
7. ISDN has been implemented rather slowly in the U.S., according to some. Discuss the reasons why this might be so. What are some of the consequences of this for IT management?
8. Discuss the difficulties for CIOs and their firms in making network choices from today's menu. What management principles are useful in resolving these difficulties?
9. Compare and contrast the land-based cellular system that is being deployed with the developing Iridium system.
10. George Gilder (see endnote 8) said "The new rule of radio is the shorter the transmission path, the better the system. Like transistors on semiconductor chips, transmitters are more efficient the more closely they are packed together." Discuss the significance of this statement.

Assignments

1. Visit a local telephone switching center in your area and investigate the types of switches in use. Write a two-page report describing the center.
2. Using library references to obtain a detailed description of cable system headends and assess the difficulties involved in providing video on demand.
3. Obtain a description of the standards pertaining to one type of LAN network and summarize its main features on one or two pages.
4. Write a brief description of one of the systems competing with Motorola's Iridium network. What are the relative advantages and disadvantages of each?

ENDNOTES

[1] Peter F. Drucker, "Where the New Markets Are," *Wall Street Journal*, April 9, 1992, A14.

[2] From the previous discussion one notes that the information carrying capacity of coaxial cable is about 50,000 times that of the local access line. The capacity of a microwave link is about 20 times that of coaxial cable. Thus the phone network, limited by the bandwidth of the access line, is termed narrowband.

[3] In 1991 Hitachi achieved a data transmission rate of 40 gigabits per second (40,000,000,000 bps) using an optical fiber—the practical equivalent of about 600,000 simultaneous phone conversations.

[4] The Cable Communications Policy Act of 1984 amends the Communications Act of 1934 to provide a national policy regarding cable television. The Act outlines its purposes, specifies channel use and ownership restrictions, provides for franchising and regulation, and deals with issues of privacy, consumer protection, coordination of authorities, and other administrative details. It was approved on October 30, 1984.

[5] James C. Brancheau and J. D. Naumann, "A Managers Guide to Integrated Services Digital Network," *Data Base*, Spring, 1987.

[6] "Fed Uses F-T1 to Reach Every Corner of USA," *COMMUNICATIONS NEWS*, April 1993, 11.

[7] Based on a survey of 350 sites in Oct. 1994 by the Business Research Group, 10 percent planned LAN implementations and 21 percent planned WAN implementations of ATM. Stephen Klett, Jr., "Users Anticipate Benefits From ATM," *Computerworld*, Nov. 14, 1994, 2.

[8] George Gilder, "Into the Telecosm," *Harvard Business Review*, March-April 1991, 154.

[9] The Yankee Group, "Wireless Communications," White Paper supplement to *Computerworld*, November 14, 1994.

[10] "Phone Calls From Anywhere," *The Gazette Telegraph*, August 30, 1992, E1.

7 IT Industry Trends

A Business Vignette

Prelude to Divestiture

Formed in 1885 from the union of the Bell Telephone Company and Western Electric, the American Telephone and Telegraph Company grew its phone system to 36 million installed phones and captured more than 84 percent of the U.S. market prior to 1949. It survived many technical, legal, and business challenges, but its success caught the attention of the U.S. Department of Justice which filed a civil anti-trust suit against AT&T and Western Electric in 1949.[1]

Claiming that AT&T and Western Electric conspired to monopolize the telephone equipment market, the Department of Justice wanted the companies to restructure into separately owned entities, and asked that the Bell operating companies obtain competitive bids from all manufacturers on equipment purchases. It also placed other demands on the companies. In 1956, during the Eisenhower administration, the Department of Justice and the defendants signed a consent decree ending the litigation. The decree limited the defendants to furnishing regulated communication services, required Western Electric to confine its production to phone equipment, and mandated that the defendants license their patents to other firms. However, the Bell system remained integrated and regulated.

But at that time, the telecommunications industry and its technology were experiencing rapid changes. Cable systems, radio, and microwave were growing explosively; the Bell system came under increasing pressure from the Federal Communications Commission (FCC) and other agencies. Several firms developed equipment, such as answering machines, for attachment to the phone instruments or the network but, for years, AT&T successfully prohibited their attachment arguing that such devices could harm the network. Finally in 1968, after much maneuvering before the FCC and in the court system, AT&T was directed to allow attachment of such devices unless they were proven to be harmful. This event, known as the *Carterphone* decision, opened the door to competitors who offered numerous attractive devices; many firms entered the interconnect business as a result of this decision.

Technology was fostering competition in many other ways, too. In 1963, the firm now known as MCI sought permission to establish a microwave link between St. Louis and Chicago. The service intended to compete with AT&T by lowering costs and providing special features. Over many objections, the FCC ruled in favor of MCI in 1969 and allowed the company to establish its microwave link. And in 1962, Congress passed the Communications Satellite Act establishing COMSAT as the agent for the U.S. in international satellite communications. AT&T had launched and was operating the Telstar I satellite, but was unable to monopolize the satellite transmission business. During the 1960s and early 1970s, many inquiries and legal actions were taken against AT&T. Primarily, these resulted from growing competition, technology advances in computing and data transmission, and concern for the negative effects of regulation on service to users.

In 1974, the Department of Justice filed an anti-trust suit demanding the separation of local from long-distance facilities, and requiring that service providers and equipment manufacturers be separate and independent of each other. After more than seven years of debate, the litigants settled the anti-trust case on January 8, 1982, by agreeing to a Modified Final Judgment (see the Chapter-end Appendix for an overview). About 100 years after its beginning, the U.S. telephone industry was permanently changed in one of the most significant business events of all time. On January 1, 1984, the agreement commenced: Thus began a new era in U.S. telecommunications.

INTRODUCTION

The information technology industry is large, very complex, and growing rapidly. It is now one of the world's largest industries with 1993 revenues of about $380 billion. The U.S. electronic computing industry has grown to several hundred billions of dollars in the short span of barely 50 years, and now involves thousands of firms employing upwards of one million individuals.[2] Some firms develop and manufacture components such as semiconductor chips or disk storage devices. Others assemble and market systems constructed from various components purchased from others or manufactured themselves. Many firms develop the tens of thousands of application programs that make the hardware useful. All firms are attempting to capitalize on the technology; some by inventing it and others by developing its usefulness.

The U.S. telecommunications industry is much older and about as large as its computer counterpart. It, too, is developing and capitalizing on new technology. Some of the breakthroughs depend on inserting semiconductor and computing technology in communications devices such as switches; others stem from developments in new media or devices such as optical fiber or satellites. In addition to developing and implementing new technology and providing a rapidly growing array of new services, the industry is responding to important regulatory and environmental changes. Developments in the telecommunications industry here and abroad are dramatic and highly significant.

THE SEMICONDUCTOR INDUSTRY

Semiconductor devices are the fundamental building blocks of information age products. More than 100 companies compete for a share of the $60 billion worldwide semiconductor market. Most outputs of the semiconductor industry are commodity products such as memory or microprocessor chips, chips for consumer electronics or industrial products, or devices for automotive, communications, or military applications. Product development and manufacture is very expensive. This is a highly capital-intensive industry.

Major companies around the world participate in the semiconductor business, but Japanese firms hold about 43 percent of the market. North American companies captured about 39 percent of the market in 1993, while European and other firms secure the remainder. The world's largest suppliers of semiconductor products are presented in Table 7.1

TABLE 7.1 The World's Largest Semiconductors Suppliers

1.	Intel	U.S.	6.	Texas Instruments	U.S.
2.	NEC	Japan	7.	Fujitsu	Japan
3.	Toshiba	Japan	8.	Mitsubishi	Japan
4.	Motorola	U.S.	9.	Samsung	Korea
5.	Hitachi	Japan	10.	Matsushita	Japan

Because the industry is capital intensive, the market is dominated by large companies. (Intel is expected to generate about $15.3 billion in 1995 revenue.) However, many small firms have found a niche making speciality products.[3] In addition to the firms listed above, many manufacturers also build semiconductors for their own consumption. Among others, IBM, Hewlett-Packard, and Digital Equipment Corporation produce $4 billion, $475 million, and $240 million, respectively, for internal consumption. AT&T's Microelectronics division produces chips for its computer and telecommunications products, and sells to others, too. Many firms share design and manufacturing with larger companies.

Worldwide consumption of semiconductor products is primarily for computer/datacommunications systems (45 percent) and consumer products (23 percent). Communications systems consume 13 percent of the output, industrial products 12 percent, and the remaining 7 percent goes to automobile and other products. Tiny semiconductor devices are rapidly becoming ubiquitous in modern society.

The electronic computer industry, barely 50 years old, is now one of the world's largest. Hardware, software, and service products are rapidly diffusing throughout industrialized nations, permeating offices, businesses, and individual residences. About 70 percent of the market is in the U.S. and Western Europe; but four of the top eight information technology suppliers are Japanese firms.

Although, the first computers went to government agencies, businesses quickly invested in them and now, industry growth is being driven by desktop and mobile PCs for businesses and individuals. The U.S. currently has 32 PCs installed per 100 citizens. Modem sales are growing at an average annual rate of 17 percent worldwide (22 percent in Europe), indicating rapid growth in connectivity.[4]

As the industry grows and matures, companies within it prosper unevenly; some of the early starters departed rather quickly, General Electric and RCA, for example. Others with more staying power grew rapidly. Some companies struggled, then merged and ultimately survived. For example, Unisys was formed from the merger of Burroughs and Sperry Corporation in 1986. Still others with seemingly winning formulas eventually were overcome by agile competitors, weak internal strategies, and the rapidly changing environment. Digital Equipment Company and IBM, the largest in the industry, were humbled in the early 1990s but are presently on the mend. Thousands of smaller firms prosper by developing and marketing an almost endless stream of hardware or software products; numerous firms provide specialized information systems services from consulting to disaster recovery to systems integration.

Hardware Suppliers

Mobile personal computing, with growth rates estimated to be 25-40 percent for the next several years, is the most rapidly growing portion of the computer industry. Market leaders in portable PCs, notebooks, and subnotebooks include Toshiba, Compaq, Apple, and IBM, but many others compete in this market, too.[5] The largest supplier of notebooks, Compaq (currently 100th on Fortune 500 list), has about a 14 percent market share. But, more than 30 percent of all notebooks are sold by firms other than the top ten suppliers. The rapid growth in mobile computing foretells similar growth in wireless data communication and networking.

Desktop computers and workstations connected to servers are becoming the standard in modern firms. Most businesses have invested heavily in personal computers for their employees and many are networking them to servers and to enterprise systems in client/server configurations. Many firms are active in this market. The leaders are Apple, IBM, Compaq, NEC, and Dell; the top ten firms hold about 50 percent of the market. The workstation/server market is more concentrated: Sun Micro Systems captures about 40 percent, Hewlett-Packard has 20 percent, and Digital Equipment Corporation (DEC) claims nearly 11 percent of the business.[6] Among midrange systems, usually networked and often serving client systems, IBM, Hewlett-Packard, and Digital garner nearly three-fourths of market.

The market for large-scale, enterprise systems is shared by IBM at 53 percent, Unisys 20 percent, Amdahl 11 percent, with 16 percent supplied by others. Mainframe activity is declining as businesses decentralize their computing and rebalance with personal and networked computing. In addition to processors, many of these hardware suppliers also market peripherals such as printers, storage systems, video display devices, tape storage, and CD-ROMs.

The Software Business

The software business has grown rapidly with many firms competing for software sales. The top 100 independent software vendors generate about $20 billion in annual revenue and employ upward of 100 thousand people around the world.[7] In addition, hardware vendors such as IBM (about $11 billion in annual software sales), DEC, and others support their products with software.

Microsoft, an industry leader in software, has grown to 250th position among the Fortune 500 by successfully capturing a large share of PC operating system software sales. MS-DOS led the market supporting Intel microprocessors in PCs but has been supplanted by Microsoft Windows, the current de facto industry standard. Windows NT and future products indicate that Microsoft will retain its dominance for some time to come. Interestingly, while noted for its operating systems, Microsoft garners two-thirds of its revenue from applications software such as its popular word processor, MS-Word. As with past market leaders, Microsoft has been investigated by the Department of Justice for possible anti-trust violations.

Many other firms supply important software. Among the largest, Computer Associates, Oracle, Novell, and Lotus Development Corporation are noted for their many general business applications, database systems, networking systems, and spreadsheets.

Industry Dynamics

The computer industry is in a state of flux as some of the major players, notably IBM, DEC, and Apple, adjust their strategies to cope with competition, changing market conditions, and technology advances. Mergers, acquisitions, and joint ventures are common among the smaller hardware, software, and peripheral manufacturers as they attempt to marshal resources and marketing clout to ward off larger competitors. But the phenomena is found among the larger players, too. After decades of priding itself on self-sufficiency, IBM is building alliances by the thousands to strengthen its position. After years of going it alone, Apple is also participating in joint ventures and licensing its software to others. The trend is unmistakable. The business is so dynamic that nearly all firms are seeking safety by joining with others.

The trend is not confined to the U.S. but includes major firms in countries around the world. Walter Wriston puts it best in the following passage:

> During the 1980s, high-technology firms, such as Siemens (Germany), Philips (Benelux), GGE, Bull, and Thomson (France), Olivetti (Italy), AT&T, IBM, Control Data, Fujitsu, Toshiba, and NEC (Japan), each forged numerous foreign alliances; some of them formed dozens. A chart of Siemens's international cooperative agreements is a genealogist's delight, including, among others, Ericsson, Toshiba, Fujitsu, Fuji, GTE, Corning Glass, Intel, Xerox, KTM, Philips, B. E., GEC, Thomson, Microsoft, and World Logic Systems. IBM has so many alliances in Japan that there is a Japanese book on the subject called *IBM's Alliance Strategy in Japan*.[8]

But dynamicism is not confined to the semiconductor and computer industries; the telecommunications industry here and abroad is experiencing considerable turbulence, too. As semiconductor logic and memory chips are combined with mass storage and software to build consumer products and computers, telecommunications engineers are using these same building blocks to fabricate switches and other devices to control transmission media. In effect, they're constructing the global information infrastructure.

THE INFORMATION INFRASTRUCTURE

In the U.S., the information infrastructure consists of many elements including broadcast radio and TV, publications such as newspapers and periodicals, cable and telephone wireline and wireless networks, the Internet, and many other networks owned or operated by businesses and government agencies. The collection of these, excluding broadcast and publications, forms the National Information Infrastructure or the Information Superhighway.

Regardless of the terminology, phone, cable networks, and the Internet are rapidly expanding to provide electronic access to vast communications systems linking individuals here and abroad. This communication revolution defines the information age.

The companies that operate the phone and cable systems are investing large sums to fabricate the infrastructure. To understand the factors driving these investments, Figure 7.1 considers the critical contrasts between these networks in transmission bandwidth and switching capability.

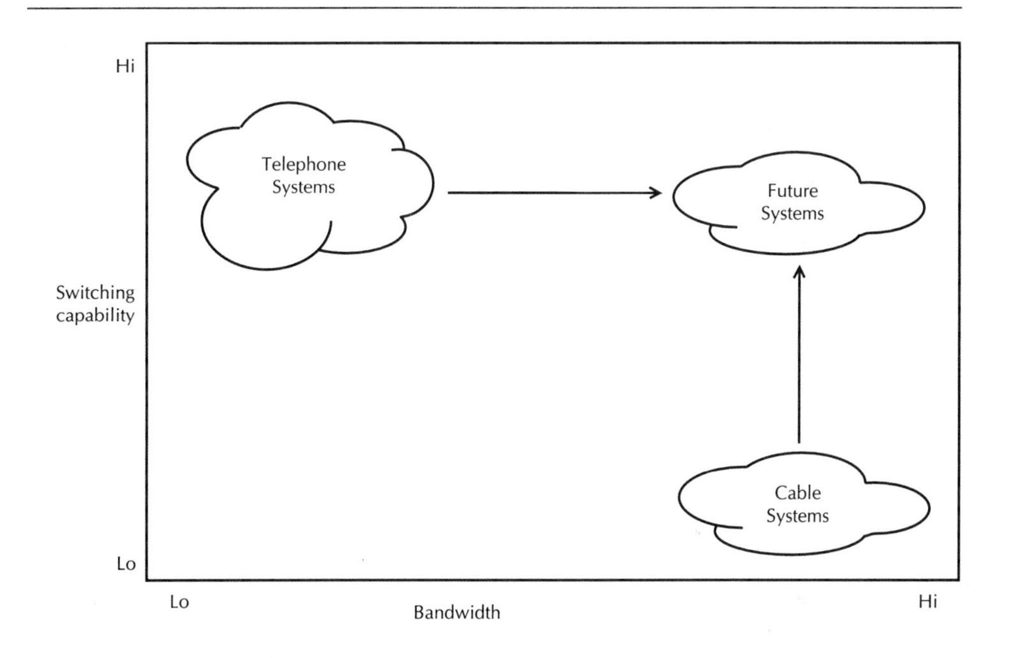

FIGURE 7.1 Switching vs. Bandwidth

The switched phone network permits individuals to reach businesses and 93 percent of U.S. households through direct distance dialing. Rapid, trouble-free switching and broad deployment are the most important strengths of the phone network. However, the bandwidth of the local loop is extremely narrow; inadequate for delivering hi-fi sound or video.

In contrast, cable delivers video over wideband links with broad deployment in urban areas. But without switching capability and two-way communication, cable is unable to deliver narrowband voice messages between customers over its broadband system. In moving toward a single broadband national network, cable networks must install broadband switching, and phone systems must greatly increase the effective bandwidth of the local loop in their switched networks. Figure 7.1 depicts industry positions and trends as we move toward the network of the future.

In practice, phone and cable firms are developing their networks and collaborating to build the infrastructure. They are using telephone, video, and computer technologies to provide businesses and individual consumers with many new options to meet their information needs. Technology advances and business ventures will help realize the network of the future, but other conditions such as consumer protection must be addressed as well.

THE TELECOMMUNICATIONS SERVICES INDUSTRY

The conclusion of the anti-trust case against AT&T has had a profound influence on today's U.S. telecommunication services industry. The Court ruled that the American telecommunications monopoly had to divest its local telephone companies and compete in the long-distance and data communications markets. The seven operating companies created from the breakup of AT&T, and many non-Bell companies, now serve the telecommunications needs of their respective geographical areas. AT&T and a host of smaller companies provide long-distance service in the U.S. and beyond.

Local Service Providers

Following the divestiture agreement of January 8, 1982, AT&T created seven regional Bell operating companies (RBOCs) from the 22 wholly owned subsidiary telephone companies. AT&T also transferred its cellular facilities to the RBOCs along with a one-seventh interest in Bell Communications Research Corporation (Bellcore).

At the time of divestiture, Bell companies served approximately 87 million access lines, while independent companies owned about 22 million. Bell Atlantic and Ameritech each served approximately 40 percent more access lines than US West or Southwestern Bell. Though geographically much smaller than US West, whose service territory included a total of 14 Western states, Bell Atlantic and Ameritech are located in areas of higher population giving them the leading market share.

Activities of the RBOCs are carefully controlled according to the provisions of the Modified Final Judgment (MFJ). They are required to provide local service to all interexchange carriers equal in all respects to the service provided to AT&T or its affiliates. The MFJ specifically prohibits discrimination between AT&T and all other persons or firms providing similar services. Additionally, it prohibits the RBOCs from providing long-distance services, manufacturing telecommunications products or customer premises equipment, or "provide(ing) any other product or service, except exchange telecommunications and exchange access service, that is not a natural monopoly service actually regulated by tariff."[9]

Currently, more than 200 million telephones and other communications devices obtain local service from the seven RBOCs serving about 116 million lines; from GTE with 17.4 million lines, or from more than 1300 smaller companies. There are nearly 100 tiny companies that own fewer than 1000 access lines each, mostly in less densely populated rural areas. Altogether, these local telephone companies serve nearly 154 million access lines. About 25 companies account for 90 percent of the local and long-distance traffic. In addition, the largest local companies also provide wireless services in their operating areas.

Business Developments

Local operating companies have invested more than $275 billion in plant and equipment and employ 585 thousand workers in serving U.S. businesses and households. In the ten years following divestiture, the RBOCs have grown their access lines by more than 30 percent. The seven RBOCs and GTE are expected to achieve $113 billion in revenue in 1995 with a net profit of nearly $14 billion on their investments. They are planning additional investments of nearly $100 billion over the next five years, modernizing and expanding their capabilities as part of the information superhighway expansion.

The local telecommunications service industry is changing rapidly. As former wholly owned markets begin to face competition, cable operators, competitive access providers (CAPs), and long-distance companies are struggling to obtain a piece of the local phone market. Cable operators are hoping to obtain a share of the residential business market and CAPs are wooing the local companies' high-volume business customers. In addition, the local companies are attempting to enter the cable business and are trying to persuade the courts to let them enter the long-distance business.

In order to protect their markets, local companies are reengineering their operations to reduce costs and are offering new services to retain customers. The "majors" are rapidly expanding their wireless operations and are introducing interactive multimedia networks, frequently in joint ventures or alliances with others. The majors are planning to enter the promising, potentially huge personal communications services (PCS) market, too. In addition, most players are expanding their overseas operations and exploiting their strengths in areas having high growth potential and less competition.

To illustrate these trends, the following paragraphs provide some individual detail on BellSouth (BLE), GTE, US West (USW), and other market leaders.

In 1994, BellSouth added 887,400 access lines in Alabama, Florida, Georgia, Kentucky, Louisiana, Mississippi, North Carolina, South Carolina, and Tennessee. Now serving 20.2 million lines, BLE is the largest of the U.S. local phone companies by this measure. Wireless and international activities now account for 12 percent of its revenue, and network and related services for 72 percent. Cellular service is one of BLE's most rapidly growing markets, now exceeding two million customers. About one in 18 of BLE's customers is a cellular phone user.

BellSouth is participating in a joint venture with the Walt Disney Company, Ameritech, and SBC Communications (formerly Southwestern Bell) to develop, market, and deliver innovative video programming to consumers. The venture expects to invest $500 million in video markets during the next five years. In North Carolina, BLE is investing in the state's Information Highway now serving 60 locations with video conferencing, medical applications, distance learning, and high-speed data transfer. Overseas, BellSouth operates or shares operations in a total of 16 countries throughout Latin America, Asia, Western Europe and the Middle East, and the Pacific Rim. These operations include cellular service, long-distance service, private data networks, and mobile data nets. BellSouth also formed alliances with two Chinese companies to develop advanced telecommunications technologies for rapidly expanding markets in China. BellSouth is positioning itself to grow its markets in the U.S. by offering advanced services, and to expand rapidly in selected areas around the globe.

GTE grew its access lines by 4.9 percent in 1994 bringing its U.S. total to 17.4 million and its worldwide total to 22.9 million. The addition of 754,000 mobile-cellular customers in 1994 increased GTE's base to more than 2.3 million in the U.S. In contrast to individual RBOCs, GTE provides wireline or wireless service in 33 of the 50 states: GTE has a wireline presence in 164 metropolitan areas.

GTE is the fourth-largest telecommunications company in the world, the largest U.S.-based local phone company, and the second-largest cellular provider in the U.S. Wireless growth has doubled in the past two years and is expected to continue at a similar rate during the next several years.

Overseas, GTE has operated the Dominican Republic's phone system for 40 years and also manages Venezuela's system; both systems are wireline and wireless. In cooperation with others, GTE provides cellular service to Argentina and plans to expand operations in Latin America where it has an established presence. In Canada, GTE owns a majority share of BC TEL and operates the system serving British Columbia and Vancouver, and the network serving the province of Quebec.

Among many other initiatives, GTE is expanding rapidly into the data services business. In 1994, the company introduced its World Class Network, built on SONET (synchronous optical network) and ATM (asynchronous transfer mode) switches. Chapter 6 describes the details of these technologies. By yearend 1995, GTE will have made this service available in 14 major large business markets. The service is expected to grow to 275 SONET rings by 1998, from 125 planned for 1996.

US West has taken a somewhat different approach to market expansion than either BellSouth or GTE. In 1994, in addition to serving 14.3 million access lines and nearly one million cellular subscribers, USW purchased Wometco Cable Corporation and Georgia Cable Holdings for $1.2 billion. These acquisitions accompanied a $2.5 billion investment for 25.5 percent of Time Warner Entertainment Company. US West's investments in these firms indicates their belief that cable in BellSouth's region may permit future phone competition with BLE and that content (programming) will be profitable, as cable expands toward 500 channels. Clearly, USW is making long-term investments.

USW was the first to announce plans for multimedia networks in its service area. The first trial is underway in Omaha, and nine more are planned for later in this decade. In the wireless market, USW and AirTouch Communications, the wireless operation spun off by Pacific Telesis, have reached an agreement to combine their domestic cellular operations. They also agreed to join Bell Atlantic and NYNEX to seek licenses to operate PCS networks in major U.S. markets. USW has been aggressive in acquiring positions in cable, wireless, content, and in U.S. and European markets, while launching many other initiatives.

Overseas, USW holds 50 percent of TeleWest Communications plc, a U.K. cable television/telephone business, and 50 percent of Mercury One-2-One, a personal communications joint services venture also in the U.K. Both are growing rapidly. USW also has wireless joint venture operation in the U.K., Hungary, the Czech Republic, Slovakia, and Russia, and holds interests in other European cable television ventures.

Many of the RBOCs are considering separating their traditional wireline operations from their wireless and cable operations in order to gain more flexibility in these new endeavors. In April 1994, Pacific Telesis spun off its domestic and international cellular, paging, and other wireless operations in a stock distribution creating a new company, AirTouch. Pacific Telesis intends to expand its services with the next generation of wireless, the personal communications services (PCS) to allow customer to place and receive calls from office, home, or on the move. In April 1994, US West confirmed its intent to create two new classes of stock separating its services into basic wireline phone and all other. BellSouth and Bell Atlantic are reported to be considering similar moves.

Cable systems and programming have captured the attention of the RBOCs and others as they struggle to maintain a competitive posture in the exploding telecommunications marketplace. In addition to the US West purchases, SBC Communications, BellSouth, and Ameritech have a joint venture with the Walt Disney Company. Bell Atlantic, NYNEX, and Pacific Telesis initiated a joint production venture using Creative Artists Agency as a consultant. And Sprint is part of a joint venture with Tele-Communications Inc., Comcast Corporation, and Cox Communications. As the information infrastructure develops, these and many other business ventures will play increasingly important roles.

Local and Long-Distance Considerations

Since 1984, the MFJ restricts the RBOC companies from entering the long-distance market and stipulates that they must provide long-distance companies equal access to their local facilities. This means that long-distance messages, using higher level exchanges depicted in Figure 6.2, will pass from the local company to the long-distance company then to a local company again as the call is completed.

The service areas of each of the seven Regional Bell Operating Companies (RBOCs) are divided into Local Access Transport Areas (LATAs) within which the local company provides service. There are approximately 200 LATAs in the U.S. centered around metropolitan areas. Some LATAs cover a lot of territory in sparsely populated states, but others are geographically smaller in more densely populated areas. For example, Idaho, Maine, Nevada, New Hampshire, New Mexico, South Dakota, Utah, Vermont, and Wyoming are single LATA states. California and Florida each have ten LATAs and Texas has 15.

Local companies serving the LATA may charge a toll for some calls, but regulations require that all interLATA traffic be transferred to a long-distance carrier. As an example, consider the cities of Denver and Colorado Springs, each in separate LATAs within Colorado. Denver customers can make local calls at no extra charge above their monthly bill within a defined area, but for calls within the LATA beyond this defined area, they must pay toll charges. InterLATA calls, such as from Denver to Colorado Springs, must be handled by one of the long-distance carriers. The long-distance carrier's fees for these calls are collected by US West. The long-distance carrier also pays a fee to access the local network.

The situation is more complicated in some regions of the country where independent local companies operate in a single community interconnected with RBOC and long-distance equipment.[10] RBOCs must coexist with many large and small local suppliers. Thus, non-Bell companies are an important part of the U.S. phone system, too.

LONG-DISTANCE SERVICE PROVIDERS

Long-distance service in the U.S. is provided by AT&T with 59.6 percent market share, MCI (17.8 percent), Sprint (9.6 percent), and about 790 smaller companies that share 13.4 percent of the market. In the past eight years, the annual growth in long-distance service has averaged 10.5 percent as compared with about 3 percent growth in local access lines. During the last decade, the number of local companies has steadily declined as industry consolidation takes place, but since divestiture in 1984, the number of long-distance providers has grown from 108 to 790. Further growth is expected through at least 1998, according to industry estimates.

Developments in the Long-Distance Market

Competition in the long-distance market is keen, not only in the U.S. but around the globe, as service providers form partnerships and alliances and make large investments in technology and facilities. With the lion's share of the market, AT&T is unwilling to allow its position to erode. It has embarked on an ambitious strategy, briefly described in the Chapter 3 Business Vignette, to build on its strengths here and abroad.

AT&T is capitalizing on the merger of computer and communication technology by making continued investments in its chip division AT&T Microelectronics, AT&T's Bell Labs, and by its $8 billion purchase of NCR Corporation in 1992. It is entering the terminal business by developing improved cordless phones, TV converter boxes to handle 500 cable channels, and a wireless communicating personal computer. At Bell Labs, where UNIX and C++ were developed, 60 percent of its 18,000 professionals are development programmers.

Since the RBOCs and other local phone companies are trying to enter the long-distance market and will probably succeed in doing so, AT&T is entering portions of the local market through its 1994 purchase of McCaw Cellular, the largest cellular provider in the U.S., for about $11.5 billion. By entering the cellular business, AT&T will gain revenue from local and long-distance traffic while avoiding some of the more than $10 billion it paid to local providers for local access.

Overseas, AT&T has established a long list of partnerships with foreign providers and is expanding its network to connect more capacity to them. In 1992, AT&T and its international partners opened service on their new fiber optic cable linking the U.S. to Germany and the Netherlands.[11] This addition to the trans-Atlantic cable system, called TAT-10, cost about $300 million and can handle 80,000 simultaneous phone conversations. Cables with 10 times the capacity of TAT-10 are expected later this decade.

Laying underseas cable is big business for AT&T which operates six cable-laying ships worldwide. Other underseas cable operators include British, French, and Japanese interests and Russia which operates 14 cable ships. The rapidly expanding underseas cable network now claims more traffic than international satellites carry. Competitive pressures from the cable network are being felt by Intelsat, the international satellite consortium sponsored by the UN. Cables also threaten satellite phone systems under development by Motorola and others.

As construction on the information superhighway expands, AT&T will gain revenue from increased communication and from its equipment division, which sells transmission systems, switches, wireless products, and cable systems. Products and systems sales are now about 27 percent of its revenue and will grow from contracts with China, Saudi Arabia, and many of the U.S.-based local phone companies. AT&T is a strong beneficiary of the growth in global telecommunications.

MCI Communications, the second largest long-distance carrier, serves the U.S. and overseas markets with high-quality voice networks. MCI operates more than 3550 million circuit miles of microwave and fiber optic network providing toll free (800) service, worldwide direct dialing, and other functions.

MCI is entering the local phone market with its MCI Metro Plan: a five year, $2 billion investment in a fiber rings and switching system in 20 major U.S. cities. Eventually, the plan will extend to 200 urban markets. Like AT&T, MCI hopes to reduce local access fees and handle business phone traffic, whether local or long-distance. With this initiative, MCI joins Teleport Communications Group and Metropolitan Fiber Systems as competitive access providers to local companies in urban areas.

Sprint exemplifies some of the important changes taking place in the telecommunications industry. Sprint's long-distance division serves about eight million customers in the U.S. and provides voice service to all of the world's direct-dial countries, more than 290 countries and locations. Its U.S. network was the first to be 100 percent digital fiber optic; Sprint takes pride in the quality of its transmission service.

Through acquisitions in the local access industry (Centel Corporation being the largest), Sprint now provides local service to 6.4 million local lines in 19 states. Local service is concentrated in Washington, Kansas, Missouri, Indiana, Ohio, the Carolinas, and Florida. Sprint also has a significant cellular presence in these states and in New Mexico, Minnesota, Illinois, and Iowa.

In October 1994, Sprint announced plans for a joint venture with three of the largest cable companies in the U.S.: Tele-Communications Inc., Comcast Corporation, and Cox Communications. This corporate alliance intends to combine local telephone, long-distance, and wireless service with cable service into a single product for consumers and businesses. This combination will have direct access to about one-third of U.S. residences and is expected to grow as other providers join the venture. In a move overseas, Sprint and Europe's two largest telecommunications entities, France Telecom and Deutsche Telekom, are working on an agreement to form a large international enterprise. Sprint is not to be outdone by its larger competitors. Sprint's revenues will climb to about $14 billion in 1995. In its markets, Sprint is intent on competing as a full-service provider.

In another interesting development, British Telecommunications, hoping to enlarge its role in the international market, paid $4.3 billion for a 20 percent share of MCI Communications. The company also plans to enter the Italian and Swedish markets. BT already has alliances with phone companies in Norway, Finland, Denmark, Australia, and Japan.[12] MCI acquired BT North America and Tymnet as part of the deal and will invest more than $1 billion developing and marketing global network services. British Telecom is the fourth largest global carrier and MCI the sixth largest.

CELLULAR AND WIRELESS OPERATORS

Current regulations require at least two cellular providers in each metropolitan and rural area, one of which is usually the local wireline operator. Each of the RBOCs, except Pacific Telesis provides cellular service in addition to a large number of other firms. Table 7.2 displays the top 10 cellular companies as measured by service area population equivalents or POPs.[13]

Capital investments made by cellular providers exceed $14 billion or about $875 per subscriber. Wireless data services are rapidly expanding, too. The major companies, RAM Mobile Data, a joint venture between BellSouth and RAM Broadcasting, and Ardis, a joint venture between IBM and Motorola, serve more than 500,000 customers. This will be a high growth area in the near future.

In order to improve cellular operations for subscribers as they travel (roam), 15 providers are developing MobilLink, a seamless service that covers 89 percent of the U.S. and 81 percent of Canada. Cellular companies know that ease-of-use is important in the cellular business, as it is in computing.

TABLE 7.2 Top Ten Cellular Providers

Company	POP Equivalents (millions)
McCaw Cellular (part of AT&T)	66.0
GTE Mobilenet/Contel Cellular	53.0
LIN Broadcasting	52.7
BellSouth Cellular	36.4
AirTouch	35.9
Southwestern Bell Mobile Systems	30.9
Bell Atlantic	25.9
Ameritech Cellular Services	24.6
United States Cellular	23.7
Sprint Cellular	20.2

Satellite Cellular Networks

Tempted by the potential of a global cellular network and by Motorola's advances in this area, at least eight separate systems aimed at capturing this business opportunity are in the conceptual or developmental stages. Together, these systems, if completed as planned, will add 1200 low-earth-orbit (LEO) satellites to the approximately 200 already in geosynchronous orbit.[14]

Many communications satellites have been parked in geosynchronous orbit about 22,300 miles above the earth. Because of the long distance involved, electromagnetic transmission to the satellite and back is delayed about one-fourth second. The power demands are relatively high, too. LEO systems overcome these difficulties but add enormous technical complexity.

In addition to Iridium from Motorola, Bill Gates (Microsoft) and Craig McCaw (McCaw Cellular) are proposing a system called Teledesic. Loral and Qualcomm are developing Globalstar. Meanwhile, GM-Hughes, now the leader in geosyschronous orbit satellites, intends to expand its coverage by adding up to 17 more satellites. Competition among these and others will be keen if all parties continue to forge ahead. These firms will face numerous difficulties as they work to implement their systems.

Some of the obstacles these systems must overcome are displayed in Table 7.3.

TABLE 7.3 Obstacles Facing LEO Satellite Systems

Obstacle	Related Factors
Construction financing	$2 billion up to $9 billion.
User terminal cost	$750 up to $2000.
User operating cost	Up to $200 per hour.
Political	Obtaining electromagnetic spectrum. International regulations.
Technical decisions	Spectrum sharing; modulation type.
Ground facilities	Facilities sharing agreements.

Note: Predicted operating costs are highly dependent on volume of traffic, traffic type, and other financial and technical factors.

In addition to the technical, political, and financial difficulties inherent in satellite systems development, marketing uncertainties and competition stalk the players. If all current players ultimately launch satellite systems, analysts believe some are destined to fail. Motorola hopes to improve service and reduce costs by reducing the number of satellites from 77 to 66. In addition, by completing call switching among satellites, ground stations are eliminated. TRW's 12 satellite system called Odyssey and Loral's 48 satellite system both depend on a combination of satellites, ground stations, and existing networks to provide service.[15] The developers believe that re-tooling some current facilities will help lower development and operating costs. Funding for several of the projects seems assured and development is proceeding accordingly.

Fearing that Europe will be left behind in the race to develop low-earth-orbit satellite communications systems, the European Community Commission is calling for industry and government to coordinate research and technical standards. The ECC intends to stimulate European interest in these systems before the field is captured by companies from other nations.[16]

THE INTERNET

One of the most significant telecommunications developments in the U.S. is a network of computer networks, the Internet, born in 1973. The Internet evolved from the government's Advanced Research Projects Agency network (ARPANET) and now consists of about 100,000 connected networks. Attached to Internet are more than 5 million computers. The Internet reaches 94 countries and is growing at the incredible rate of about 10 percent per month; by the year 2000, Internet expects to have a staggering 180 million computers attached to its network system. To many, Internet *is* the information superhighway.

Using virtually all available modes of data transportation, Internet users can search databases, share research, transfer files, or chat about almost anything they desire. Thousands of groups use the net to communicate about common interests ranging from solid research to absolute fantasy.

Commercial activity on the Internet is growing, but there are many difficulties inherent in commercial expansion. Commercial exploitation of the net is deplored by some, while serious technical obstacles, including insufficient security, remain challenging. Nonetheless, commercial enterprises (.com domains) now total 1.5 million, as of January, 1995, and are growing rapidly.

Commercially available on-line information services are attracting customers at the rate of about 10,000 per day, according to some reports. The combination of Compuserve, America Online, Prodigy, Delphi, Genie, Apple's Eworld, and smaller services such as AT&T's Interchange and Microsoft's Network will exceed ten million subscribers shortly. Prodigy, a joint venture of IBM and Sears, Roebuck is supporting ISDN in cooperation with BellSouth, NYNEX, and Pacific Telesis Group. The service is intended to open the path to multimedia services and to provide faster data access. As the infrastructure is developed, exploitation of the superhighway is destined to grow.

REGULATION

One of the most critical issues facing the telecommunications industry is that of regulatory reform. The Modified Final Judgment, effective since January 8, 1982, and numerous rulings since then govern the actions of the RBOCs. The Cable Communications Policy Act regulates many issues facing the cable industry in the U.S. In addition, states and localities regulate various aspects of telecommunications services on a smaller, more regional scale. In the era of exploding technological growth, regulation presents challenging issues for Congress, the Courts, the Department of Justice, and both federal and state regulatory agencies.

The large phone companies generally favor regulatory reform in the belief that the current environment limits competition and consumer choice. They would like to see restrictions lifted on equipment manufacturing, cable television programming content, and long-distance service. The long-distance providers would like to retain their market share but have the freedom to enter the local market. At present, their entre into the long-distance market now is through wireless services. Competitive access providers would like to have physical access to RBOC central offices to connect and operate equipment within them. All are calling for a level playing field, although, arguably, the playing field remains poorly defined.

While the majors jockey for position, consumers expect reasonable services at fair prices; some expect preferential treatment for various public interest groups. Many issues will need to be resolved before a balanced position can be achieved.

The traditional cost-based rate of return is being replaced by various alternative forms of regulation designed to increase pricing flexibility and provide incentives for increased productivity. For example, SBC Communication's intrastate activities in Kansas and Missouri are operating under incentive regulation plans, but Arkansas and Oklahoma are regulated under traditional rate-of-return concepts. California has taken the lead in these initiatives. For example, the California Public Utilities Commission (CPUC) permits GTE to retain any earnings up to 15.5 percent return on investment, but requires GTE to refund all earnings above the 15.5 percent cap. The CPUC also allows other companies to offer local toll service in GTE's and Pacific Bell's service territory in preparation for full competition beginning in 1997.

Ameritech has actively sought to obtain access to the long-distance market in its five-state service area in exchange for opening its local service to competition. The Department of Justice has recommended to Judge Harold Greene that Ameritech be granted permission to offer limited long-distance service. Some believe that the market is likely to open in a test mode before restrictions are lifted nationwide.[17] Congress may enact legislation governing how and when local telephone companies, long-distance carriers, and cable TV operators can enter each other's markets.

Bell Atlantic, serving the Mid-Atlantic states, was the first RBOC to obtain permission to offer video services in its area. The company plans to spend $11 billion to upgrade its network for video transmission.[18] In early 1995, Judge Greene ruled that Bell Atlantic could offer video services nationwide. This ruling fueled the belief that other RBOCs will soon obtain permission to offer similar services. To capture these opportunities and others, the industry continues to invest heavily in the information infrastructure.

THE MEANING FOR BUSINESS MANAGERS

Although the entire information technology industry is undergoing rapid and profound change, networking advances are outpacing other changes, as their business impact will attest. Powerful technology, created from semiconductor and computer building blocks, is dramatically increasing the ability of individuals and firms to transport voice, data, and video information on a global scale. By exploiting new fiber, wireless, and satellite technology, telecommunications firms are investing large sums of money in their race to build a global information infrastructure.

Developing the global infrastructure is a massive effort that ushers in dramatic challenges for private industry and government agencies alike. Businesses are rising to the challenge by marshaling resources capital, people, and technology to extend current facilities and to provide new services and capabilities. Several billion individuals, now without phone service, will prospectively attain it through business and technology innovation. Another billion customers, or so, now served by wireline and wireless, stand to benefit from the business and technological innovations of phone, cable, and content providers. Here and around the globe, the information infrastructure is being constructed at breakneck speed.

Individuals and organizations are rapidly exploiting these new telecommunications developments. Around the globe, small individual computers and large computing systems are "wired," or interconnected, permitting nearly instantaneous interchange of important personal or corporate information. As Walter Wriston points out, "borders are not boundaries."[19] Networking is the key to exploiting information; modern information systems are increasingly telecommunications-based.

Businesses are increasing their interaction with technology providers, not only by building on their routine service, but in more fundamental ways as well. As businesses examine their strategies by asking "what business are we in?" and establish core competencies to exploit, frequently they conclude that parts of their information system are outside the core. Outsourcing has been widely discussed in the literature, sometimes with negative connotations, but the trends are unmistakable. In many cases, computer utilities are far superior to captive operations; in many instances, application development firms offer superior products and services. Many firms are finding that their networks are better operated and managed by telecommunication outsourcing firm's professionals.

Consequently, organizations are forging partnerships with the firms discussed in this chapter, and many others too numerous to mention. These partnerships develop and operate applications, manage system operations, and manage and operate important networks. Increasingly, service firms are absorbing noncore functions. Consulting firms, disaster recovery operators, IT training organizations, and others are thriving because they offer superior products and services. As organizations mature in their use of information technology, the desire for and the tendency toward informations systems self-sufficiency will continue to diminish.

For managers in typical IT organizations, these trends mean change on a grand scale. In effect, IT organizations and their employees are being held to increasingly high standards of performance as comparisons are made between them and competing suppliers. For many critical functions, their performance will be unchallenged, particularly if they pay attention to the firm's core mission. For other functions, IT managers will seek alliances with outside service providers, negotiate the arrangements, manage the transition, and supervise the delivery of services. For IT managers and their employees, accomplishing all of these formidable goals will be a major accomplishment.

Thus, IT industry trends have profound implications for professional managers everywhere. Wise managers will understand these trends and their implications and will take advantage of them. Unwise managers, trapped in the past, will resist and, eventually, succumb to forces beyond their control.

Review Questions

1. What factors contributed to the Bell System breakup in 1984?
2. What is the Carterphone decision? Why is it important?
3. Using figures from Intel, describe why the semiconductor industry is considered to be capital intensive.
4. Where are semiconductor products consumed? Where would one find semiconductors in a private residence?
5. There are many different parts to the computer industry. Name as many as you can and identify the significance of each.
6. What are some of the business ramifications of the rapid growth in mobile computing?
7. Why are alliances and joint ventures so popular in the computer industry?
8. What elements comprise the U.S. information infrastructure?
9. Describe the essential technological differences between telephone companies and cable companies.

10. What are the main provisions of the Modified Final Judgment?
11. What forms of competition exist for the RBOCs? How will this change in the future?
12. What is BellSouth's growth strategy? How does this differ from US West's strategy?
13. What is a LATA? What is its significance to the telephone industry?
14. Define POP, as the term is used in the cellular phone business. Which telecommunications firm owns the most POPs?
15. Who are some of the key players in the LEO satellite business? Explain why the Iridium project needs about 70 satellites.
16. Describe some of the obstacles that LEO satellite cellular systems must overcome.
17. Which firms offer individual access to on-line network services? What is the Internet's relationship to these commercial network service providers?
18. What role do regulatory agencies play in the telecommunications industry?
19. Describe how the industry trends discussed in this chapter will impact upon IT managers.

Discussion Questions

1. Discuss the role of technology in the breakup of the Bell System.
2. Discuss the significance of the Carterphone decision as it applies to today's telecommunications infrastructure.
3. Consider a semiconductor firm that must invest more than $1 billion to produce commodity products that sell for ever declining prices. Analyze the associated risk factors. What actions can be taken to minimize these risks?
4. Discuss the business significance of the fact that there are many suppliers of mobile and desktop computers. How does this fact relate to the strategies of the larger computer companies?

5. Many large computer manufacturers produce software for their products, yet the growth in independent software developers has been extraordinary. Discuss the reasons for this, noting in particular the reasons for Microsoft's phenomenal rise to prominence.
6. Using Figure 7.1 as a reference, discuss the actions that telecommunications firms can take to build the future network.
7. Discuss the elements of GTE's growth strategy.
8. Since 1984, the number of local phone suppliers has declined, but the number of long-distance providers has grown. Discuss the reasons for these trends and explain what you think will happen in the future.
9. Discuss the advantages and disadvantages of underseas fiber cable as compared with satellites for international communication.
10. Outline the elements of Sprint's strategy. What might it do to enhance its strategy?
11. Consider AT&T's purchase of McCaw Cellular and discuss its significance to the telecommunications industry.
12. Itemize the advantages and disadvantages of LEO satellites versus geosynchronous satellites for cellular phone communications.
13. Discuss the Internet and commercial network service providers in relation to Figure 7.1.
14. Government agencies must balance the interests of many parties in fashioning regulations. Identify the parties the agencies must deal with in the telecommunications field? Discuss the divergent interests among these groups.
15. Some analysts believe that regulatory actions always lag technology developments. Assuming this is true, do you think this is an advantage or disadvantage for the public? Discuss your reasoning.
16. What actions can IT managers take to cope with the rapid and tumultuous changes occurring in industry today. In preparing your thoughts, consider material you learned so far in this text.
17. Discuss the main points of the MFJ section I, AT&T Reorganization, found in this chapter's Appendix.

Assignments

1. From annual reports or other industry information, sketch the strategy of one of the major telephone or cable firms. On what technical, business, or regulatory developments is the firm hoping to capitalize? What risks are involved in the strategy?
2. Through library research, determine the agencies and organizations that have regulatory authority in your state. Describe several state regulations and the processes involved in establishing and enforcing them.
3. Write a two-page report on telecommunications developments in Europe, including privatization actions, regulatory changes, and the development of alliances and joint ventures.

APPENDIX

The MFJ describes the agreement between the American Telephone and Telegraph Company and the Department of Justice reached on January 8, 1982, in one of the most significant business events in U.S. history. This Appendix includes the introduction and section I, AT&T Reorganization, of this document.

MODIFICATION OF FINAL JUDGMENT[20]

Plaintiff, United States of America, having its complaint herein on January 14, 1949; the defendants having appeared and filed their answer to such complaint denying the substantive allegations thereof; the parties, by their attorneys, having severally consented to a Final Judgment which was entered by the Court on January 24, 1956, and the parties having subsequently agreed that modification of such Final Judgment is required by the technological, economic and regulatory changes which have occurred since the entry of such Final Judgment; Upon joint motion of the parties and after hearing by the Court, it is hereby ORDERED, ADJUDGED, AND DECREED that the Final Judgment entered on January 24, 1956, is hereby vacated in its entirety and replaced by the following items and provisions:

I

AT&T Reorganization

A. Not later than six months after the effective date of this Modification of Final Judgment, Defendant AT&T shall submit to the Department of Justice for its approval, and thereafter implement, a plan of reorganization. Such plan shall provide for the completion, within 18 months after the effective date of this Modification of Final Judgment, of the following steps:

1. The transfer from AT&T and its affiliates to the Bell Operating Companies (BOCs), or to a new entity subsequently to be separated from AT&T and to be owned by the BOCs, of sufficient facilities, personnel, systems and rights to technical information to permit the BOCs to perform, independently of AT&T, exchange telecommunications and exchange access functions, including the procurement for, and engineering, marketing and management of, those functions, and sufficient to enable the BOCs to meet the equal exchange access requirements of Appendix B; (Author's note: Appendix B describes in considerable detail how the BOCs must provide equal, non-discriminatory access to any and all interexchange carriers.)
2. The separation within the BOCs of all facilities, personnel and books of account between those relating to the exchange telecommunications or exchange access functions and those relating to other functions (including the provision of interexchange switching and transmission and the provision of customer premises equipment to the public); provided that there shall be no joint ownership of facilities, but appropriate provision may be made for sharing, through leasing or otherwise, of multifunction facilities so long as the separated portion of each BOC is ensured control over the exchange telecommunications and exchange access functions;
3. The termination of the License Contracts between AT&T and the BOCs and other subsidiaries and the Standard Supply Contract between Western Electric and the BOCs and other subsidiaries, and
4. The transfer of ownership of the separated portions of the BOCs providing local exchange and exchange access services from AT&T by means of a spin-off of stock of the separated BOCs to the shareholders of AT&T, or by other disposition; provided that nothing in this Modification of Final Judgment shall require or prohibit the consolidation of the ownership of the BOCs into any particular number of entities.

B. Notwithstanding separation of ownership, the BOCs may support and share the costs of a centralized organization for the provision of engineering, administrative and other services which can most efficiently be provided on a centralized basis. The BOCs shall provide, through a centralized organization, a single point of contact for coordination of BOCs to meet the requirements of national security and emergency preparedness.

C. Until September 1, 1987, AT&T, Western Electric, and the Bell Telephone Laboratories, shall, upon order of any BOC, provide on a priority basis all research, development, manufacturing, and other support services to enable the BOCs to fulfill the requirements of this Modification of Final Judgment. AT&T and its affiliates shall take no action that interferes with the BOC's requirements of nondiscrimination established by section II. (Author's note: Section II places requirements on the BOCs to provide equal access to all interexchange carriers and prohibits other activities such as manufacture of equipment and provision of interexchange services or information services.)

D. After the reorganization specified in Paragraph A(4), AT&T shall not acquire stock or assets of any BOC.

Following this, the remainder of the document describes the judgment's many provisions. It requires the operating companies to provide service to all interexchange carriers and specifies that the Court retain jurisdiction for subsequent modification or enforcement of the judgment. U.S. District Court Judge Harold H. Greene, who presided over the anti-trust proceedings, continues to retain jurisdiction over subsequent events in this matter.

ENDNOTES

[1] For a detailed history and analysis of the U.S. telephone industry, see Leonard S. Hyman, Richard C. Toole, and Rosemary M. Avellis, *The New Telecommunications Industry: Evolution and Organization* (Arlington, VA: Public Utilities Reports, Inc., 1987), Volumes I and II.

[2] The contract to build the ENIAC computer is dated June 5, 1943, marking the beginning of the computer industry.

[3] Intel has invested about $200,000 per employee and spends about $33,000 per employee annually on research and development. In 1993 Intel started construction on a wafer manufacturing plant expected to cost $1 billion.

[4] James O. Jackson, "It's A Wired, Wired World," *Time*, Special Issue, Spring 1995, 80.

[5] From S & P *Industry Survey*, Vol. 1, No. 1, 1995, C77.

[6] See Note 5 above, C81.

[7] *InformationWeek* May 31, 1993, 38.

[8] Walter B. Wriston, *The Twilight of Sovereignty* (New York: Charles Scribner's Sons, 1992), 86.

[9] Modification of Final Judgment, Section II, Paragraph D3. United States District Court for the District of Columbia: United States of America, Plaintiff, v. American Telephone and Telegraph Company, Western Electric Company, Inc. and Bell Telephone Laboratories, Inc., Defendants. Civil Action No. 74-1698.

[10] Smaller than the seven RBOCs and GTE, 15 local companies such as Southern New England Telecom and Pacific Telecom serve 50,000 to 5 million customers.

[11] "AT&T Racing Rivals to Lay Undersea Lines," *Wall Street Journal*, August 14, 1992, B1.

[12] "BT to Announce Expansion Plans For Italy, Sweden," *Wall Street Journal,* April 12, 1994, B3.

[13] POPs are the number of people in a service area divided by the company's ownership percentage of that area.

[14] George Gilder, "Telecosm: Ethersphere," *Forbes ASAP*, October 10, 1994, 132.

[15] "Phone Space Race Has Fortune at Stake," *Wall Street Journal*, January 18, 1993, B1.

[16] Richard L. Hudson, "EC Commission Gives Wake-Up Call For Satellite Phone," *Wall Street Journal*, April 16, 1993, A5B.

[17] Ellis Booker, "Long-Distance Market Poised to Open Up," *Computerworld*, April 10, 1995, 8.

[18] "Bell Atlantic Likely To Be First Bell Entering Cable Arena," *Houston Chronicle*, March 20, 1995, 4B.

[19]See Note 8 above, 129.

[20] See Note 9 above.

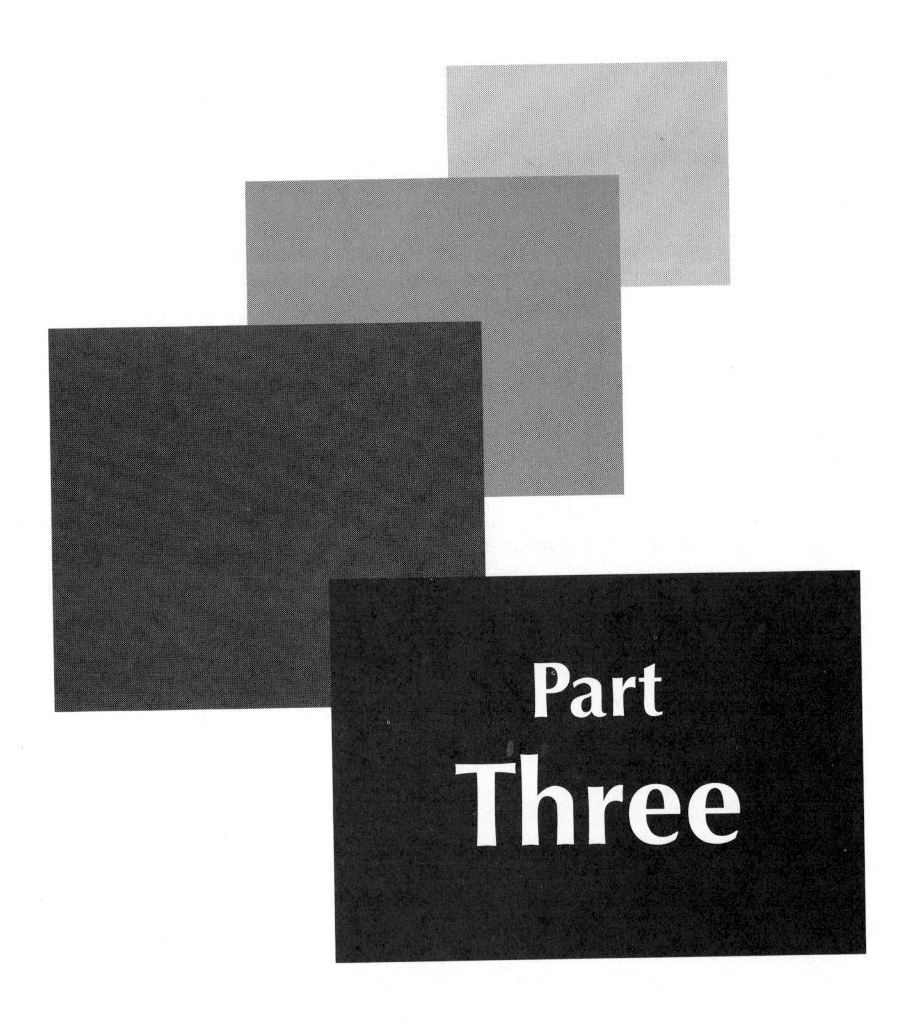
Part
Three

Managing Application Portfolio Resources

The application portfolio and related databases encompass large and critical firm resources. These resources are growing in value; they form a base from which the firm can gain strategic competitive advantage. The firm must add value to these assets and mitigate the effects of technical and business obsolescence. IT professionals and skilled users collaborate in this activity. This part addresses the management issues associated with the application portfolio and data resources. Chapters include The Application Portfolio Resource, Managing Application Development, Alternatives to Traditional Development, and Managing Client/Server Implementation.

Managing application and data resources is a challenging task. Successful IT managers consider this activity a cornerstone of their mission.

8 The Application Portfolio Resource

A Business Vignette

Building Big Systems—Managing Huge Risks

The information management industry has created another industry in its midst: rescues. Under the direction of corporate consultants and IM gurus, specialized shops have made millions in the last few years—$10s of millions per year for one accounting firm, according to a story in *Business Week*.[1] These shops catch and tame runaways: systems-development projects that get out of control.

Businesses and government agencies in the U.S. have many tales to tell of the horrors that can occur when the risks of large-system development are not managed or understood properly. Two out of every eight large software systems under development are canceled; most exceed their schedules by factors of two or more; and most large systems do not function as intended or are not used at all.[2] Software development remains a cottage-industry craft while society rushes at full speed into the information age. The following discussion highlights some of these problems:

1. Systems must connect with other systems, especially in businesses that have large databases that many applications and users access. In earlier days of small systems, developers could get their arms around individual programs. They built stand-alones with some degree of success. Now, however, different skills and techniques are required for developing networked systems that talk to each other.

   For example, the Department of Motor Vehicles in California intended to allow patrons to renew licenses at one-stop kiosks conveniently located throughout the state. This meant merging the vehicle and driver registration systems—a seemingly easy task. When the schedule expanded and the cost rose by more than six times original estimates, the DMV canceled the seven-year project after spending more than $44 million.

2. When developers need business, or organizations want to expand, they tend to oversell what they can deliver. The programmers and analysts who do the grunt work of building the applications may be overworked, understaffed, and demotivated. But their executives will promise unrealistic performance, including low cost, short development time, high reliability, and ease of use. What happens when the programmers can't or won't deliver? In some cases, organizations suffer severe embarrassment; some spend large, unplanned amounts of money; some incur other liabilities. In companies where reliability

of data is especially important—health care firms, for instance—lost or incorrect data can lead to loss of subscribers, bring hardships and danger to those subscribers, and result in business losses to the firm itself.

3. The more complex the application, the longer development will take, and the more it will cost. These factors leverage each other; over a period of several years, the basic cost of development will continue to grow. Therefore, a project that takes four years to complete instead of two will be billing far higher developer rates as the years roll by. Delays mean more cost. For example, *Business Week* cites an Allstate Insurance project begun in 1982, and scheduled for completion in 1987, with an $8 million projected cost. By 1988 the cost had risen to $100 million, and completion was pushed out to 1993.[3] Ultimately, the project was completed at about 14 times the original cost estimate.

How can these problems be solved? There are no easy answers, but experience indicates that the risk of failure rises rapidly as system size increases. Unfortunately, development departments frequently fail to produce sound estimates; failing to estimate the complexity of the project, they err on the low side. After that, things only get worse. But several things can and should be done to minimize the risks.

First, according to the *Business Week* article,[4] the decision-makers and those paying the bills must get deeply involved, must understand the project in detail, and must stay involved in the process from beginning to end. Second, the project must be reviewed frequently—continuous review and on-going audits are not unreasonable for large, high-risk projects. At each review, progress reports, budget items, and all performance criteria must be analyzed for variances from the plan. Even small plan changes usually spell big trouble for large systems.

Third, if outside contractors are used, firms must find some way to guarantee the high standards and abilities of the people who will do the design and the programming for the system. This may require the hiring firm to specify qualified individuals by name in the contract.

Fourth, the people who will use the system must be part of the development effort. In effect, the project needs a constant series of user trials; it will not suffice to wait until the end of a project to bring the users on board. A surprising number of development projects have failed to involve users; in those cases, transaction problems, culture problems, and lack of general acceptance have resulted. On the other hand, knowledgeable users can alert the developers to possible conflicts in transactions; they can provide information about how data flows and is used in the organization and they can champion the system throughout the company during the trial and introduction phases.

But, there needs to be a fundamental shift in software development and management from cottage-industry craftsmanship to professional engineering. Most likely, this cultural change will require decades to develop. Shifts in university computer science curricula, software engineering certification, and the practice of management in industrial software development will be necessary. Meanwhile, risk aversion in applications development is a prudent course of action.

INTRODUCTION

The previous section discussed systems used for the delivery of information and the industry that provides them. It concentrated on advances in semiconductor and recording technology, operating systems, and hardware and operating system architecture. The strong, dominant trend of rapidly declining unit costs and ever-increasing functions is the basis for the rapid infusion of new technology into nearly every niche of our society.

Continuing to support dramatic hardware advances is the inexorable growth of software, such as operating systems, telecommunications programs, and utility programs. Application programs developed within the firm are also growing, adding to the total already in use. It is estimated that there are 100 billion lines of COBOL in use today that cost about $2 trillion to produce.[5] A proliferation of commercially developed programs supplement them. Typically, these application programs are business utilities, such as payroll, general ledger, billing, and inventory control programs. They are designed for applications across a broad spectrum of user organizations, and they support some of the fundamental activities of the firm. Recently, many of these application programs have been made available for implementation on personal computers. This software, and the hardware it supports, comprise the delivery system base for the firm.

The remaining ingredients of the information system for the firm are the large databases used in concert with the application portfolio. Typically, this vital information, on which the firm thrives, is unique to the firm. These program and data resources represent an enormous effort invested over a long period of time. The applications and the data are operationally and strategically vital to the firm. Therefore, managing the firm's software and data resources is critical.

Depreciation and Obsolescence

The application portfolio and related databases are growing in magnitude and importance over time. The firm must continue to invest in and apply resources to these assets to prevent depreciation and obsolescence. Depreciation results from the accumulation of functional inadequacies due to gradually changing business conditions. For instance, the work-in-process inventory system designed for a labor-intensive process will lose effectiveness as the labor-intensive processes are replaced by robotics. Obsolescence results from the introduction of business changes that reduce the appropriateness or value of current applications. For example, the introduction of labor accounting terminals on a plant floor will render the previous labor accounting system largely obsolete.

Maintenance and Enhancement

In contrast to money invested in hardware technology, costs associated with maintaining, enhancing, and improving the portfolio are mostly related to personnel. They are rising over time. New tools to improve the individual productivity of systems developers are being developed. But, for the bulk of the applications written in third-generation languages, no major cost breakthroughs are anticipated. Additionally, and again in contrast to hardware, there is no easy, low-cost way to move from an older generation of application software to a new and modern generation.

Maintenance and enhancement of the 100 billion lines of COBOL installed in U.S. businesses is estimated to cost about $30 billion annually. A typical Fortune 1000 company maintains about 35 million lines of code, according to Keen.[6] Clearly, maintaining and enhancing computer applications is an expensive proposition.

But the portfolio is valuable because the worth of the firm's information system is gauged by the total system function performed for the firm. The applications codify the rules and procedures used in the firm's business processes and describe the culture of the business—*the way we do things around here.* They represent long-term knowledge of the firm's operations. To a considerable extent, the functions that the application programs provide represent the yardstick by which the firm's information system is measured; the firm's application portfolio is as valuable as its ability to meet corporate objectives.

DATA RESOURCES

The databases that accompany the application programs are among the firm's essential assets. Over time, they grow in size and importance to the firm. These databases are the result of continued accumulation of information, and they represent an investment of resources. The firm's databases form the foundation upon which future advances in applications technology can proceed. These data systems provide the lifeblood that sustains the day-to-day operation of the firm. These data resources and their inherent database management systems, coupled with the hardware on which they reside, are a vital and critical adjunct to the application portfolio.

The data resource may also lose value as the applications themselves depreciate. The database management system, which may be satisfactory for the current applications, may constrain modernization of the portfolio and make enhancements much more difficult. In addition, large databases are probably involved in numerous interactions among and between applications. This database coupling greatly complicates the revitalization or enhancement of the application programs themselves.

Historically, in some firms, the database resource became dispersed widely as the firm expanded geographically. In most firms, local dispersion of the data accompanied the introduction of individual workstations. The information architecture for the firm is an important issue, as noted in the first chapter. In all cases, the intrinsic value of the data demands that effective protection mechanisms be installed. The data resource is vulnerable to loss and destruction, and it suffers from systemic forms of obsolescence.

APPLICATIONS AS DEPRECIATING ASSETS

In actual practice, applications pose a management dilemma for the firm. The application portfolio represents a large resource base on which future applications can be developed. At the same time, improvements and enhancements to the applications are mandated by ever-changing business conditions. This implies depreciation or even obsolescence. Increasing amounts of time and money are required to capitalize on the stream of opportunities available to the firm through its application programs. Resources are required to avoid the negative consequences of obsolescence and depreciation. Since the continuing demand for resources to fuel the growing asset base of applications shows no signs of diminishing, most firms face difficult and important decisions regarding resource allocation.

In most organizations, the applications programming department faces a large workload backlog. In many firms, the backlog ranges from one to four years in length. In firms where client application development is occurring, they too may be experiencing backlogs. The backlog stems from the organization's need to keep the business competitive via new applications and the requirements to keep the current application portfolio functionally modern. In addition, the process for maintaining and enhancing applications is inherently inefficient; for reasons discussed later, programmer productivity in this activity tends to be low. Program development organizations typically struggle to manage the backlog effectively. In attempting to do something for everyone, they perform a less-than-satisfactory job for the organization as a whole. Well-managed organizations can successfully overcome many of these difficulties.

The Programming Backlog

The following formula defines the application programming backlog:

$$\text{Backlog} = \frac{\text{Work to be accomplished in person-months}}{\text{Number of persons to do the work}}$$

For instance, if the organization has two programming tasks to perform for a total of 36 person-months of work, and has three programmers to do the work, then the backlog is 12 months. Some assumptions are made in calculating the backlog this way. One assumption is that there is no wasted time due to skill imbalances during the development cycle. This is most likely to be true for large development organizations. Another assumption is that none of the code is reusable, so it must be uniquely developed for each application. This is less likely to be true in large organizations.

In addition to the identified backlog, there frequently exists an unidentified or "invisible" backlog, which occurs when the identified backlog is large. An invisible backlog develops because departments in need of programming work do not make known their requirements. They are reluctant to add to an already long list of outstanding work requirements. The true backlog facing application developers is the sum of the identified and the unidentified outstanding work. In many firms, the identified backlog is on the order of two to three years or more. The invisible backlog may be very large indeed, depending on the dynamics of the firm.

The Need to Prioritize

For many firms, the alternative of applying more resources to application development is not reasonable. Good programmers are difficult to obtain, they are expensive, and many firms are reluctant to make a long-term investment in programmer development. Newly acquired programmers are not immediately effective because they need to learn the culture and business aspects of the organization. For other reasons, too, adding additional programmers may actually reduce output.[7] Therefore, the crux of the issue is the difficult challenge of prioritizing the application development resources.

Programming prioritization is important for other reasons, also. The application portfolio is a strategic resource. It has been developed over a long time, and it will remain important in the future. Thus, expenditure of application development resources demands high-level consideration. In addition, the portfolio is a source of management expectations. IT managers need a thoughtful process for managing these strategic resources and associated expectations over the long term.

For these reasons, a development plan is needed. The plan must describe the application of available resources, application programmers, and available financial support to manage the tasks that the enterprise wants to accomplish. Figure 8.1 shows the process in which the organization is involved and portrays several alternatives for accomplishing these goals.

The application portfolio in Figure 8.1 consists of existing applications owned by the firm and some required applications that do not exist in the portfolio. The portfolio analysis process described below provides an organized and businesslike approach to decision making. The first level of decision making selects among the choices of maintenance, enhancement, acquisition, or termination. Additional decisions between abandonment or replacement, as well as alternatives to acquisition, are displayed. The process depicted in Figure 8.1 and discussed below provides a rational approach toward maximizing the value of the portfolio. Upon completion of the process, resources are identified for optimizing the separate tasks of maintenance, enhancement, or acquisition.

ENHANCEMENT AND MAINTENANCE CONSIDERATIONS

In most firms, it is common for programming resources to be deployed in the maintenance and enhancement of existing applications.[8] Generally, fewer resources are available for new development. The 100 billion lines of COBOL in use today are mostly in programs that are depreciating because of new and changing business conditions. This huge program asset base requires a continuous investment of resources to prevent depreciation and obsolescence. Dealing effectively with program depreciation leads to important positive outcomes for the firm.

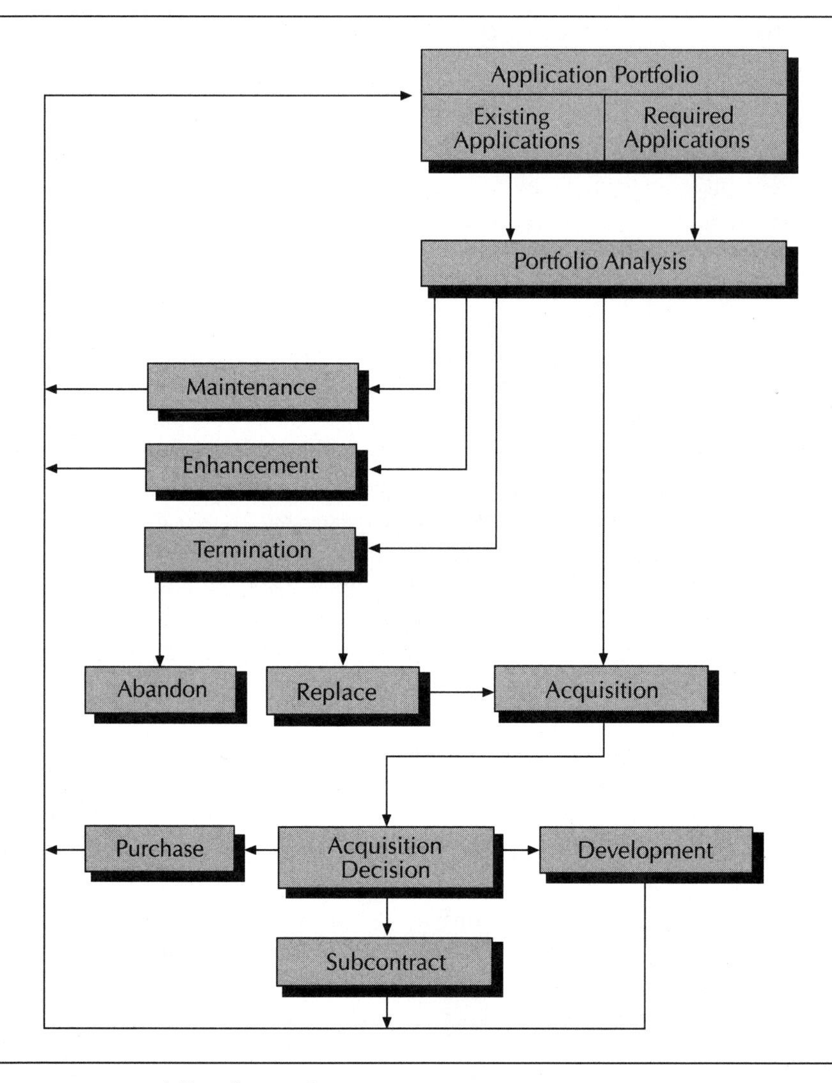

FIGURE 8.1 Portfolio Alternatives

Vendors have produced several generations of hardware over the years, each generation having significant cost and performance advantages over its predecessor. The hardware vendors have helped ease the transition from one generation to the next for the application programmer. Computing systems are created to make the old programs relatively easy to migrate to the new hardware. While this upgrading of the hardware has ushered in cost and performance advantages, the applications supporting the firm have been much more difficult to upgrade. For many firms, the result is that very old application programs are running on very modern hardware. For example, the FBI's National

Crime Information Center runs 900,000 transactions per day. The program was written in the 1960s in System/360 assembly language. The FBI's payroll and personnel system, written in COBOL, has been enhanced and maintained for more than twenty years.[9] Businesses have many examples like these. It is relatively easy to upgrade the CPU or to add new input or output devices; it is much more difficult to upgrade the application software.

The High Cost of Enhancement

Many firms feel trapped by the past. Their applications require significant modernization, yet the resources to do the job are not available. The expenditure of money on hardware brings early returns, but the cost to correct the deficiencies in the application portfolio remain extremely high. What are the principle factors behind the situation as it exists today?

There are a number of reasons why application maintenance and enhancement require large investments of time and money. They are summarized in Table 8.1.

TABLE 8.1 Why Application Maintenance Costs Are High

Obsolete programming techniques were originally used.
Documentation is obsolete or absent.
Many uncoordinated modifications have been made.
Old versions of languages were used.
Languages were mixed within the program.
Unskilled programmers made enhancements.
Architecture changes are required.
File structures need major changes.

The old programs in the portfolio were built with programming techniques that are now considered unsatisfactory. The techniques that were used were not designed for ease of modification. Many old programs aren't well structured. They probably contain strings of code, arranged in a convoluted and unstructured manner. These old programs have had a continuous series of enhancements over the years, probably performed by many different programmers. Because of the enhancements and the enhancement process involved, the program documentation is generally poor or absent. In some cases, more time is spent trying to understand the program than fixing the programming problem.[10]

Many of the older programs are written in earlier versions of a current programming language and may even contain embedded assembly language code. Modification to these applications is an error-prone process. Modification introduces risk into the operation of the programs. Most programmers consider this low productivity activity to be undesirable and perhaps less "glamorous" than new program development.

In many firms today, enhancement and maintenance programming is frequently considered good training. Therefore, it is assigned to the least experienced programmers in the organization. Because of their inexperience, junior programmers may produce less efficient code, and the code contains more errors. Skilled, senior programmers usually prefer new development work. Attracting good people to perform maintenance is becoming increasingly difficult. In some firms, individuals are reluctant to take responsibility for program maintenance.

Older programs generally require major architectural improvements. For example, legacy mainframe systems operating in sequential batch-processing mode must now must be converted to client/server mode of operation.[11] Perhaps the original input media was punched cards, and now graphical user input must replace card-image input. The implementation of a new database management system may be required, impacting data that currently exists on sequential tape files. Changes such as these alter the architecture and foundation of the applications and require massive amounts of effort. Architectural alterations may demand more resources than were invested during the initial development cycle.

In many instances, the files associated with each application are unique. They may consist of sequential tape files copied to disk. This means that a change in the application's function may dictate large and expensive changes to the database, or to other programs that use the database. In total, these enhancements require large investments. Because of the complexities involved, modifications and enhancements to the portfolio proceed at a slow pace. For the same reasons, the maintenance and enhancement process is inherently error prone and risky.

Trends in Resource Application

The conditions described above result in the continued expenditure of large sums of money, increased dissatisfaction with the program development department, and likely deterioration of application quality. Figure 8.2 portrays the deployment of funds over time in maintenance, enhancement, and new development. This situation is typical of many organizations today.

Figure 8.2 shows that the typical firm is increasing its total expenditures on programming. The increase is due to escalating costs per programmer, additions to the programming staff, or both. However, the resources devoted to new development are declining over time because of the demands for changes and alterations to current applications. Maintenance and enhancement activity consumes increasing fractions of the available resources. It may account for up to 80 percent of the total planned expense.

There are some reasons to be optimistic about finding solutions to the applications problem. However, the current situation in business and industry is far from satisfactory. Most large firms have several thousand applications in the portfolio. Even small to medium-sized companies will own several hundred to a thousand or more programs. These applications may be recent additions to the portfolio, or they may be ten or more years old. Occasionally, an application will be found that was designed for now-obsolete hardware. Such applications may require some form of hardware simulator to run. In some cases, the program cannot be modified because the documentation or skills are no longer available.

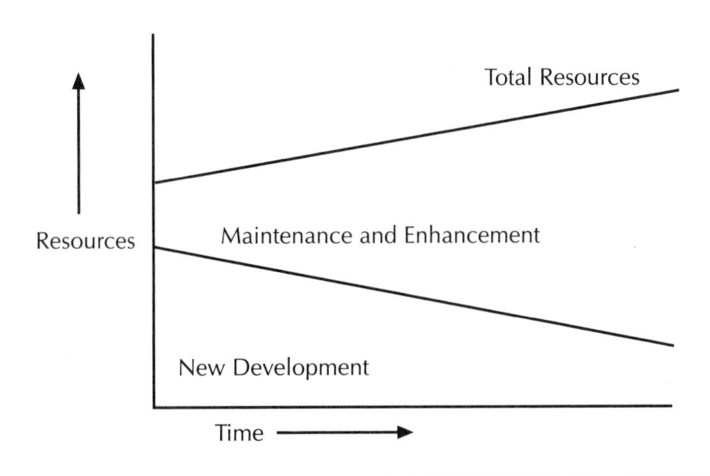

FIGURE 8.2 Typical Expenditures on the Portfolio

Full-time maintenance programmers usually support many applications. For example, one maintenance programmer may be responsible for financial applications and another for marketing applications. Several of them may support manufacturing applications, and others respond to work requests from various departments. As problems arise, or when enhancements are required, the maintenance programmer moves from program to program. Since there is a large backlog of work, the programmer may take short cuts and ignore some important aspects of the work. In the interest of pleasing the greatest number of clients, maintenance programmers frequently fail to document changes completely. Many totally ignore documentation, further complicating future maintenance.

The term *maintenance* itself takes on several meanings in this situation. Some effort is devoted to repairs. These repairs consist of fixing bugs in new programs and finding and repairing bugs in enhancements to old programs. As the portfolio ages and enhancements grow, this effort increases in magnitude. Frequently, additional effort is required to interface or bridge enhanced programs to old databases. These bridge programs add to the portfolio size and also require enhancement as other programs or databases change. Bridge programs may also introduce errors, further complicating the situation. Maintenance may also include minor rewrites. In some cases, it becomes easier to rebuild portions of the application than to work with old code. Application maintenance is highly inefficient. Firms engaged in this activity are on a course depicted in Figure 8.2.

For all the reasons discussed above, software maintenance is not considered to be highly desirable work; programmers usually prefer other assignments. However, the prevention or deferral of obsolescence to the application portfolio is very important to the firm. Poor quality workmanship or untimely maintenance and enhancement can severely reduce the firm's ability to conduct its affairs effectively.

In many firms, the situation is not improving. Maintenance and enhancement consume more than 50 percent of the programming budget. Programmers continually add to the amount of usable code by 10 or more percent per year. Typically, the people working on these applications were not involved in the original development activity. Many of the difficulties noted above continue to plague maintenance programmers today.

TYPICAL AD HOC PROCESSES

Many firms lack an organized and disciplined approach to this thorny problem and rely on one or more reactive methods for prioritizing the application programming team's work. Table 8.2 itemizes some common, default approaches to programming prioritization.

TABLE 8.2 Common Approaches to Prioritizing Programming Resources

"The squeaky wheel gets the grease."
React to perceived threat of failure.
Recover from embarrassing situations.
Adjust to threats from competition.

In the first instance, managers throughout the firm contend for application development resources by making personal appeals to application development managers or directly to programmers. The success of the appeals depends on the manager's degree of persuasion, the relative position of the person appealing, or the degree of implied or actual threat. Seasoned managers are more likely to succeed than new managers. Friendship frequently plays a role. A system of score-keeping (who owes what to whom) may develop. The process is rich with emotion, high on anxiety, and very low in objectivity. In an attempt to calm troubled waters, IT managers frequently try to do something for everyone. This usually raises expectations, generates overcommitment, and yields unsatisfactory results. Effective IT managers must use disciplined processes successfully in these situations.

In spite of these difficulties, the "squeaky wheel" theory is rather popular. Most managers believe they can negotiate a better deal for their organizations than their peer managers can. Old timers are especially reluctant to give up their favored position. They tend to resist a more rational process. In some firms, the squeaky wheel is the corporate culture; in others it is the default management system. Firms with this culture or this management system are characterized by a lack of teamwork and frequent finger pointing.

In a reactive environment, priorities are frequently readjusted to avert difficulties. When problems are recognized within the squeaky wheel environment, corrective action usually follows promptly. Unfortunately, the firm often rushes from one disaster to the next in a futile attempt to keep all systems from collapsing at once. The inevitable result is a terrible waste of resources, generally unsatisfactory performance in the long run, and the conclusion by even the most casual observers that management does not know what is going on.

Since the process described above is often not successful, embarrassing situations are likely to develop for the firm. Payroll processing goes astray for all employees to observe firsthand, incorrect or duplicate billings are sent to customers, or, worse yet, collectibles are not received. Occasionally, they are received, but improperly recorded. These situations require immediate action. Resources are deployed from around the firm to correct the problems post haste.

When activities within the firm appear to be on a steady course, competitors may marshal their IT forces for offensive action. The firm's detection of these competitive actions elicits a prompt reaction. Again, resources are deployed to contend with these threats and to ward off their consequences, if possible. When survival of the firm is at stake, all other activities have lower priority.

These reactionary actions and out-of-control situations are not desirable, however popular they may be. Is it possible to develop a process that better serves the firm? What management tools and techniques are available to assist in prioritizing the resources? Given the limited and constrained resources available to most firms, what approach is preferred under these circumstances? Obtaining sound answers to these questions must be a high priority for business managers because major consequences result from the actions taken or omitted under these circumstances.

SUPERIOR PORTFOLIO MANAGEMENT

The following discussion presents an approach that allows the firm's senior managers to combine their strengths with those of the IT organization to yield a preferred course of action. The methodology focuses on business results. It presents managers with alternatives to consider. The prioritization process requires significant communication among and between the various players. It causes them to view the problem from the level of the senior people in the organization. This approach demands a general management perspective because it focuses on considerations fundamental to the firm. For these and other reasons, this approach is likely to achieve superior results.

Several factors are important in prioritizing the backlog. Among these are the firm's business objectives and the financial and other benefits derived from the applications. Frequently, the applications have important, intangible benefits. Some, in fact, may be leading-edge technology and have technical significance. Usually, there are relationships among and between these factors. For example, a common situation is an application that is required to meet business objectives but does not show a positive financial return. The benefits may be great, but intangible. Likewise, an investment to attain technological leadership may be valuable for competitive reasons.

The questions that these considerations pose make resource allocation decisions difficult. IT clients alone do not have enough information to allocate resources. A combination of IT managers and client managers may lack the vision that top executives have. To resolve these issues, many firms utilize a steering committee of top executives to prioritize the application development backlog. If the steering committee does not approach prioritization in a systematic manner, but acts informally and with a political orientation, the results are likely to be mediocre. If the committee incorporates the advice and counsel of senior executives as part of the firm's formal strategy and planning process, the results are likely to be superior. This text strongly favors the planning approach.[12]

How is superior portfolio management accomplished in practice? What steps can the IT organization take during the strategy and planning process to prioritize the programming work? What is an appropriate management system to carry out these difficult and important tasks? The following analyses present an organized and disciplined methodology to resolve these issues.

Satisfaction Analysis

The first step in this disciplined approach is to perform an analysis of satisfaction on each application in the portfolio. Satisfaction ratings for each application are obtained from the client organization and from the IT organization. The satisfaction analysis focuses on attributes important to each of these organizations and quantifies emotional perceptions. Typical factors used in the satisfaction analysis for the IT organization and the client organizations are shown in Table 8.3.

TABLE 8.3 Factors Used in Satisfaction Analysis

Client Factors	IT Factors
Sound function	Good documentation
Easy to use	Modern language
Good client documentation	Ease of operation
Healthy cost/benefits	Trouble free
Sound architecture	Well architected

This list is not exhaustive, but it illustrates most of the attributes contributing to satisfaction. In specific cases, there may be firm-dependent attributes that should be included.

The applications in the portfolio are scored from 0 to 10 by each organization, then sorted in several different ways. The scorings or ratings are used first in an attempt to reach consensus on which applications will or will not receive additional funding. Agreed upon applications that will receive no additional investment are removed from further consideration.

Plotting the results of this type of analysis in a diagram as displayed in Figure 8.3 can reveal some interesting results. This figure portrays the analysis conducted by the application users and by the IT organization. The applications have been evaluated with a closed list of attributes, such as shown in Table 8.3. Each application is plotted to show its score by both groups. The results for three programs are shown in Figure 8.3. For discussion purposes, the programs are identified as A, B, and C.

Newly developed and recently installed programs generally reside in the upper right quadrant of Figure 8.3. One such program is identified by A. It is not uncommon to find new programs, such as program A, scoring lower than 10-10. Because of trade-offs made during development and changing business conditions during the elapsed development time, newly developed applications are not completely satisfying. Frequently, requirements become clear only after the program has been implemented!

Many applications, perhaps those currently being enhanced, will be found in the lower left quadrant of Figure 8.3. An example of this type of program is B. In many instances, the bulk of the enhancement resource will be devoted to programs having low satisfaction to both the IT organization and users. Given the information available at this point in our example, we cannot yet judge the effectiveness of this course of action.

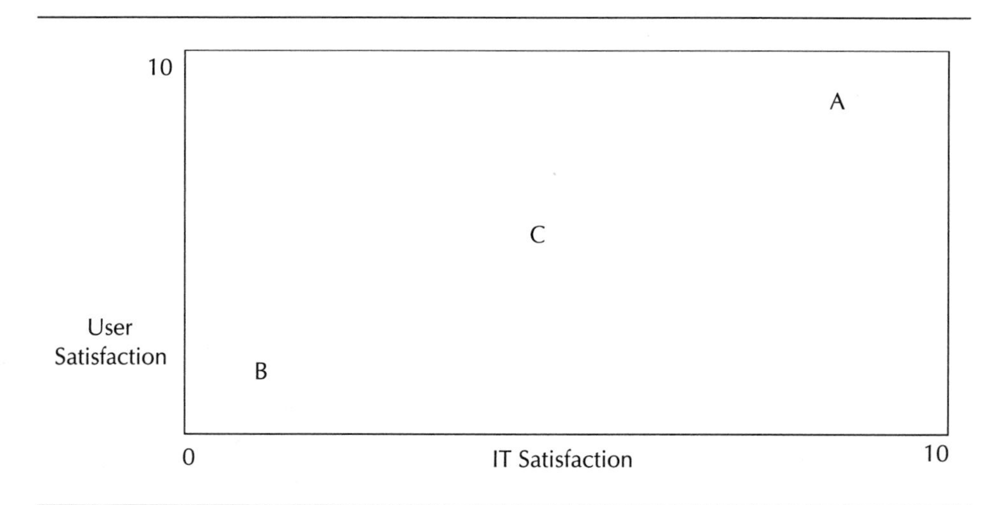

FIGURE 8.3 A Graphical View of Some Satisfaction Analysis Results

Program C is an average program in the portfolio. It is not very satisfactory to the clients nor the IT organization. Programs of this type probably make up the bulk of the applications in most firms. When all the results of this analysis are plotted, they tend to fall along the diagonal shown in the diagram. It is unusual for an application to be very satisfactory to the client organization while also very unsatisfactory to IT. The converse is also true. Generally, difficulties with applications affect both groups.

As it now stands, this analysis provides insufficient information for deciding how to prioritize the backlog. One missing ingredient is the short-term versus the long-term perspective on the value of each application. The next step supplies this information.

Strategic and Operational Factors

To delineate more clearly which of these applications merits additional resources, managers must obtain insights from a different perspective. The organization must understand the strategic or operational value of the applications. The importance of the applications should be considered from both the short-range and the long-range perspective. The firm's managers must evaluate both the strategic and operational value of each application. As in the previous analysis, to accomplish this the client-managers score the applications on their strategic importance and operational importance using the 0–10 scale. Figure 8.4 portrays the results of the strategic versus operational analysis for selected programs.

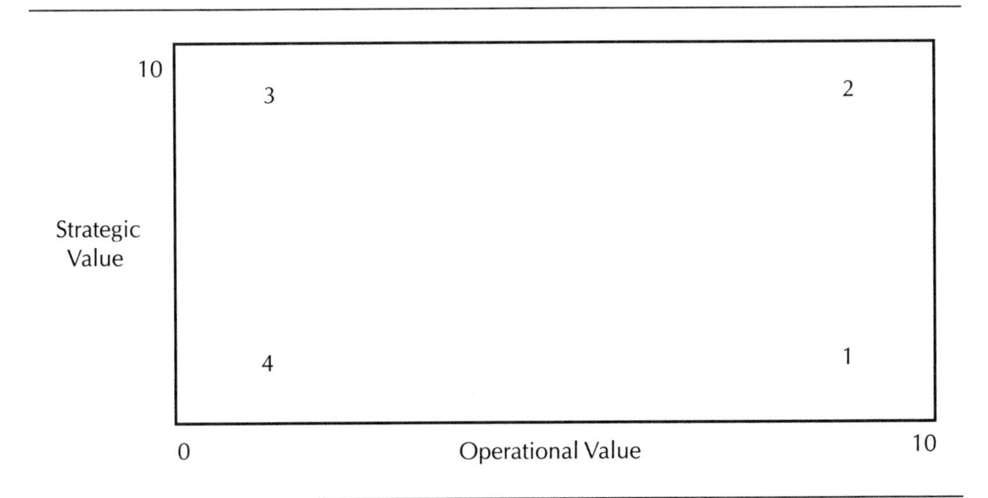

FIGURE 8.4 Strategic vs. Operational Value

Some examples of applications that illustrate the insights gained in this analysis are identified in Figure 8.4. The examples have been selected to show how this analysis forces differentiation between applications in the organization.

Typical programs indicated by 1 are applications such as payroll, accounts payable, or student accounting. Operationally, these applications are very important to the firm, but they generally provide little or no strategic advantage. Number 2 identifies programs such as the Merrill Lynch Cash Management Account, the Sabre reservation system, or a seismic data processor for an oil exploration company. These programs are important both operationally and strategically. Applications such as these are vital to the firm on a daily basis. They provide strategic competitive advantage in addition to important firm functions.

Number 3 indicates other programs such as the CMA in 1977, computer-aided instruction programs for elementary schools, or the next generation vector processor. These applications have relatively low operational importance to the firm at this time, but they have high potential long-range strategic advantage. Programs in the Number 4 position are relatively unimportant strategically and operationally. Examples of these applications are office furniture inventory programs and one-time statistical analysis programs. These applications are needed, but they are not highly important to the firm's mission.

Application owners do most of the evaluation, taking into account strategic goals and objectives of the firm. IT input should be obtained during the process, especially in relation to anticipated technology improvements. This input is especially useful for emerging applications, or for those applications that are candidates for major enhancements. The list of programs included in this analysis should not be restricted to those currently in the portfolio. It must include new requirements presently in the backlog, as well.

Again, the scoring scale is 0-10, but the plotted results will show considerable scatter as compared with the previous diagram. This is normal and desirable because it allows for discrimination among the candidate programs. When this analysis is used in conjunction with the previous analysis, a clearer picture of the candidates for resource deployment becomes available. Either analysis taken by itself provides some useful information but is not sufficient for making decisions. Together they build a base for the next step: cost-benefit analysis.

Costs and Benefits

The additional element, which completes the prioritizing process, is an analysis of the cost and benefits of the actions proposed for each application. The analysis must include the cost of failure to do the work and the benefits of intangible results.[13] In most cases, it is not necessary to have highly refined data to reach final conclusions. Most analyses can proceed if the costs and benefits are predictable to within 20 percent or so. Additional refinement may be required in the event of a close decision. The time value of money must be incorporated into this analysis to ensure financial integrity. Some examples of various analysis methods can be found in texts on financial accounting.[14]

Benefits information is provided by client managers while development and implementation costs are provided by IT. The benefits may be purely financial, or they may be largely intangible. If a program later becomes a candidate for resource expenditures, the firm's controller should "book" the benefits and the costs. That is, the client organization commits expected benefits and IT commits the costs and expenses. The controller tracks both. This process ensures a higher degree of data integrity than other methods. It places commitment responsibility on the organizations where it belongs.

After completion of the scoring and the cost-benefits analysis, a tabulation of all the applications can be very revealing. Table 8.4 illustrates some of the conditions that may be found in a large portfolio of applications.

TABLE 8.4 Compilation of Program Ratings

Program Name	User Sat.	IT Sat.	Strategic Importance	Tactical Importance	Costs	Benefits
prgm 1	8	8	1	8	10K	14K
prgm 2	2	3	1	3	30K	40K
prgm 3	7	9	8	4	100K	200K
prgm 5						
prgm 6						
etc						
"						
"						
prgm 2000	...					

Three examples of the analysis results have been listed. Program 1 is a recently completed application with fairly high tactical importance to the firm. Additional investment in Program 1 will yield a modest return. Program 2 is an old program that is not very satisfactory to either the user community or to IT. Although this program is not strategically or tactically important, modest financial benefits will occur if additional expenditures are applied to it.

Program 3 is a relatively new program for which a great deal of strategic importance is anticipated. Both the client group and the IT organization are quite satisfied with this program as it currently exists. Additional investments are required to achieve the relatively high future payoff anticipated for this application. In a large portfolio, there are many individual differences between applications which permit decision makers to discriminate between candidates for resource allocation.

Sufficient information is now available for making decisions. The decision makers are the firm's senior functional managers and their superiors.

The appropriate time to conduct this activity is after the strategy has been established and just prior to the beginning of the planning cycle for the firm.

Using all the factors brought out by this analysis, the management team prioritizes the list of programs from high to low.[15] Their judgment and experience are combined with the data to reorder the list until consensus is reached or a decision is made. This process puts decision making at the management level, where it belongs, and reduces overcommitment. The list of work items is closed when the sum of resources required equals that available to the firm. This may mean, for example, that the first 400 applications of 2000 receive resources, and the bottom 1600 remain at *status quo*. This process points out what will and will not be done, and the firm agrees on an achievable plan.

Additional Important Factors

Given the analysis outlined above, the firm's senior decision makers are in a good position to allocate critical resources confidently. The IT organization should direct the analysis process. The IT organization must obtain the client's view of the applications in the portfolio. It must also take the lead in developing the discussion preceding the decision to deploy resources. IT managers are responsible for bringing a completed analysis to senior executives.

Each senior member of the review team will obtain a clearer understanding of the firm's business needs. If the proper environment has been established by the top individual, the needs of the firm will be well served. If this process is followed rigorously, it eliminates other less-disciplined approaches, such as the squeaking wheel method. This approach puts the decision making where it belongs, at the top of the organization. It eliminates many small *ad hoc* deals between individuals at lower levels in the organization that could reduce the firm's effectiveness. This methodology sheds light on the complex problem of resource allocation. It leads to a much better-informed organization because it fosters organizational learning, and it prepares the organization to use information technology advantageously.

Usually, required resources far exceed available resources. As illustrated in Figure 8.1, if money is in greater supply than people, the alternative of purchasing applications or subcontracting program development may appear attractive. Although each alternative raises some additional issues, many organizations prefer these choices to the option of hiring and training more analysts and programmers.

An additional alternative, not shown on Figure 8.1, is to utilize a service bureau to fulfill certain data processing needs. Information of the type displayed in Figure 8.4 will be most helpful in deciding whether to buy the service, buy the application, subcontract the development, or perform the development in house. Sometimes, routine tasks such as payroll processing can be performed best at a service bureau. Many other routine data-handling activities are service-bureau candidates also.

Strategically valuable applications, or those that utilize proprietary processes or data, are best completed within the firm. This local development is likely to be completed by the resident application programming department. In some cases, the applications can be developed by the clients themselves. In many cases, clients can perform many maintenance or enhancement changes. This activity, termed *end-user computing*, can be very effective. It is discussed more fully in Chapter 11.

When the prioritization process is complete, IT managers have all the necessary information to complete the applications program portion of the IT plan. The management systems for prioritizing and for IT planning are essential portions of strategic application management. The firm's tactical plans contain cost and benefit information developed during this process. This tracking information becomes part of the performance measurement of the managers who own or control the applications. In a similar manner, the expected applications costs are committed and tracked. IT development managers are measured on this commitment. The firm's normal cost-accounting system is used for measurement purposes. This process incorporates strategy development, strategic planning, and operational control elements. It ensures a high degree of integrity, secures commitment, and places responsibility where it belongs.

Finally, the process introduced above offers one major advantage. The disciplined process, involving the firm's senior executives, provides a splendid opportunity to identify and fund new strategic systems. Using the framework developed in Chapter 2, executives can begin the task of identifying strategic opportunities. At this time, they have a great deal of valuable information on the portfolio, and the environment is favorable. This is the proper time to search for strategic opportunities. Rackoff and colleagues have documented a structured methodology for accomplishing this task.[16]

The prioritization methodology presented in this chapter provides an ideal background from which to begin the search for strategic information systems opportunities. The right people are present, the timing is right, and a decision-making process has been established. New strategic opportunities can be introduced and resources allocated to their implementation using this prioritization scheme. All conditions necessary for sound decision making are present.

MANAGING DATA RESOURCES

Data resources accompanying the applications, the database management system, and certain hardware items, must form an integral part of the prioritization process. Applications enhancements, conversions, or rebuilding activities will require resource expenditures on the database, in many cases. In some instances, database conversions may be important enough in their own right to be considered as separate items in the methodology. That is, a database conversion or reorganization activity should be considered for prioritization in the same way that the firm evaluates applications.

In many cases, database considerations and resources will bear on the sequence (and perhaps on the priority) of the actions planned for the applications. An information architecture will begin to emerge from this work, if the firm does not already have one in place. In the final analysis, the interplay among and between the applications and data must be clearly understood. Data sources within the firm must be well-defined for the process to be effective.

Application specialists in the IT organization understand the technical nature of the database system and the interactions among applications. IT professionals need to develop the architectural considerations of the portfolio for the firm's senior managers. This element of executive education must be conducted thoroughly and skillfully to improve the executive decision-making process. This activity provides another opportunity for the IT executives to obtain agreement on direction and expectations among their peers and with their superiors. It's an opportunity that should not be lost.

WHY THIS PROCESS IS VALUABLE

The methodology developed in this chapter organizes the decision-making process and provides a framework for judging how best to deploy scarce resources. This methodology causes decision makers to focus on issues fundamental to the well-being of the entire organization. It discourages parochialism. The process puts decision making at the senior executive level where it belongs. It serves to integrate the IT function into the fabric of the business. Executives at many levels in the business will become educated in information technology issues during this process. The results are embedded into the strategizing, planning, and control operations of the firm.

The management system for dealing with the application portfolio attacks some of the critical issues discussed in Chapter 1. It offers the opportunity to advance organizational learning, and helps define IT's role and contribution. It focuses on data as a corporate resource. This process must be used in conjunction with strategy and plan development for maximum effectiveness. This management system forces congruence between the firm's strategic plan and the strategic plan for IT regarding the application resources. Finally, the process ensures realistic expectations regarding the portfolio. All senior executives know what to expect from IT; and IT knows unambiguously what it must deliver. Many critical issues faced by the IT organization and its clients are best attacked by employing these processes effectively.

SUMMARY

The application portfolio and related databases form a large and important resource for the firm. These resources are frequently accompanied by significant problems and unique opportunities. They require senior management attention, and they relate to several critical success factors for IT managers. IT managers must handle issues presented by the application portfolio skillfully.

There are many advantages to the methodology presented in this chapter. It puts decision making in the hands of those most responsible and most capable of making well-informed decisions about important issues. It makes IT managers jointly responsible for the results, and forces proper accountability throughout many management levels. This methodology eliminates many other less-effective methods for prioritizing. The development of unrealistic expectations regarding the applications is less likely using this process. This management system causes decision makers to consider alternatives not likely to be revealed by other, less well-disciplined methods.

Managers must have knowledge of some fundamental attributes about the applications and about their requested enhancements. This knowledge may not be readily available, but informed decisions cannot be made without it. If the firm is not accustomed to handling resource questions in a disciplined manner, this approach will require some executive education. This prioritization process may be useful in other situations within the firm, as well. Lastly, this process yields substantial educational benefits regarding information technology for many of the participants. For all of these reasons, the IT organization and its management team will be well served by this management approach.

Review Questions

1. What are the ingredients of a firm's information system?
2. Identify the intangible assets of the firm's information system.
3. What is the difference between depreciation and obsolescence as applied to application programs?
4. Why are the firm's databases growing in magnitude over time? Why is the database management system an important asset?
5. Distinguish between *maintenance* and *enhancement* as these terms apply to application programs.
6. The terms *depreciation* and *obsolescence* are used in connection with programs. How do these terms compare with depreciation and obsolescence of more tangible assets?
7. Define application backlog.
8. What is the invisible backlog, and under what conditions is it likely to be large?
9. What courses of action for a program can result from the application of the process discussed in this chapter?
10. Describe the dilemma posed by the application backlog.
11. Why does maintaining and enhancing the application portfolio consume large amounts of resources?
12. Data files are frequently shared among application programs. What additional complications to the enhancement process are caused by this fact?
13. What are the essential elements in the methodology presented in this chapter?
14. Why is a plot of the satisfaction analysis (as shown in Figure 8.3) likely to display scatter along the diagonal? Why is it unlikely that any application will be rated 10-10?
15. How should the IT organization participate in the strategic/operational analysis?
16. Why are costs and benefits important to the process?

17. How does the process outlined in this chapter focus executive attention on a variety of alternatives?
18. Under what conditions may the firm's databases restrict the choice of options, or otherwise impact the decision-making process?

Discussion Questions

1. Discuss the causes of the computer runaways in the Business Vignette. What additional causes may be present?
2. Using data presented earlier in this text and appropriate estimates, establish a relative value for the application programs and data resources in a typical firm. Be prepared to discuss your results.
3. Discuss the pros and cons of using application maintenance and enhancement as a training vehicle for junior programmers.
4. What additional attributes might be useful to include in the satisfaction analysis? Under what circumstances might additional attributes be required?
5. Using information from Chapter 2, trace the evolution of Merrill Lynch's CMA program in Figure 8.4 as you believe it occurred.
6. How does the application backlog relate to the issue of expectations discussed in Chapter 1?
7. Discuss the problems that will arise if satisfaction analysis alone is used to prioritize the backlog.
8. Discuss the issues associated with compiling the cost and benefit data for the applications. Discuss the effect on decision making of intangible costs and benefits.
9. How can the time value of money be incorporated into the cost-benefit analysis?
10. Discuss how the management process described in this chapter helps the firm find areas in which strategic systems may be found.
11. Some people believe the methodology presented in this chapter is too time consuming and bureaucratic. What are the alternatives? Compare and contrast them to the methodology discussed here.
12. Discuss the relationship between this methodology and the task of reducing or removing the issues found in Chapter 1.

Assignments

1. Read the material referenced in Note 13. Itemize the various costs and benefits Yourdon discusses. Summarize the writer's thoughts on strategic benefits.
2. Study the Rackoff article in Note 16 and prepare a report on the details of the SIS planning process at GTE.
3. What additional factors are important for firms developing application programs for sale? In these cases, how would the processes in this chapter need to be modified?

ENDNOTES

[1] Jeffrey Rothfeder, "Its Late, Costly, Incompetent—But Try Firing a Computer System," *Business Week*, November 7, 1988, 164.

[2] W. Wayt Gibbs, "Software's Chronic Crisis," *Scientific American*, September 1994, 86.

[3] See Note 1 above.

[4] See Note 1 above.

[5] Peter G. W. Keen, *Every Manager's Guide to Information Technology* (Cambridge, MA: Harvard Business School Press, 1991), 59.

[6] See Note 5 above.

[7] Frederick P. Brooks, Jr., *The Mythical Man-Month* (Reading, MA: Addison-Wesley Publishing Company, 1975), 16.

[8] Scott D. Palmer, "Software Maintenance," *Federal Computer Week*, December 5, 1988, 26. The General Accounting Office estimates that maintaining software costs the federal government 40 to 60 percent of its software budget. International Data Corp. found that enhancement maintenance consumes 50 percent of the maintenance budget in the federal sector. Fixing bugs requires 24 percent, and adapting programs to new hardware or system environments consumes 26 percent of federal maintenance expenditures.

[9] Palmer, 26.

[10] Robert L. Glass, "Help! My Software Maintenance is out of Control," *Computerworld*, February 12, 1990, 87.

[11] Legacy systems are "systems that have evolved over many years and are considered irreplaceable, either because re-implementing their function is considered to be too expensive, or because they are trusted by users." Edward Yourdon, *Decline & Fall of the American Programmer* (Englewood Cliffs, NJ: Yourdon Press, 1993), 238.

[12] Michael R. Mainelle and David R. Miller, "Strategic Planning for IS at British Rail," *Long Range Planning*, August 1988, 65. The method used at British Rail parallels that developed in this chapter.

[13] Edward Yourdon, *Modern Structured Analysis* (Englewood Cliffs, NJ: Yourdon Press, 1989). This book contains a thorough discussion of cost-benefits analysis in Appendix C.

[14] For example, see Carl L. Moore, Robert K. Jaedicke, and Lane K. Anderson, *Managerial Accounting*, 6th Ed. (Cincinnati, OH: South-Western Publishing Co., 1984), 347. Also Yourdon, 510.

[15] Eric Clemons says, "Even when it is not possible to compute explicit, precise values associated with embarking on strategic programs, it may be possible to estimate, with enough accuracy, to rank alternatives." "Evaluation of Strategic Investments in Information Technology," *Communications of the ACM*, January 1991, Volume 34, Number 1.

[16] Nick Rackoff, Charles Wiseman, and Walter A. Ullrich, "Information Systems for Competitive Advantage: Implementation of a Planning Process," *MIS Quarterly*, December 1985, 285. The planning process includes instructing executives on competitive strategy and SIS; applying SIS concepts to actual cases; and reviewing the firm's competitive position. This is followed by brainstorming and discussing SIS opportunities, evaluating the opportunities, and developing detail for strategic systems planning.

9 Managing Application Development

A Business Vignette

Building Better Systems[1]

It shouldn't take long for a visitor to Perdue Farms Incorporated to realize that Bob Cook appreciates the people who work for him. If the customer-satisfaction scores posted on the walls escape notice, the numerous "Associate of the Month" certificates are easy to spot.

"Having satisfied associates (programmers and analysts) is the key to delivering quality products and having satisfied customers," said MIS Director Cook. As evidence, he points to the successes by his development staff since IS began emphasizing communication and employee development. Since 1989, the average time spent on maintenance has been shaved from 146 to 52 hours per week. Today, associates spend 94 percent of billable hours building new systems or improving the existing ones deemed most valuable to business. As a result, a six-year project backlog has been eliminated. The average program development time has fallen from 60 hours to 16, while average cost per program has dropped from $1,950 to $568. These impressive figures add up to a 300 percent increase in associate productivity and an enviable systems development ROI of three dollars of benefit for each dollar invested, according to the company's calculations.

Perdue Farms Inc. is a $1.3 billion integrated poultry producer with 12,500 employees located in Salisbury, Md. In 1988, Perdue, the fourth-largest integrated poultry company in the U.S., responded to a disastrous decentralization effort by reorganizing the data center and rethinking its attitude toward both associates and users. The new approach emphasizes communication goals up front and publicizes results after the fact. In addition, associate's training has increased to five weeks per year. As a result of these changes, Perdue has slashed its IS turnover rate to eight percent, down from 30 percent in 1988, according to Cook. Management and process changes have contributed more to the department's accomplishments than new technology, says Cook. "We're interested in proven technologies," he said. "Being a generation behind is acceptable."

IS has certainly come a long way since 1988. That was the first year Perdue found itself operating at a loss, caused in part by the decentralization begun in the early '80s. At that time, one data center was spun off into three separate centers in an effort to push decisions down through the corporate ranks and provide more autonomy. But when the strategy that was supposed to control costs actually drove them up instead, company executives responded quickly by reconsolidating and downsizing.

"We needed to focus on basic things, like reacting quickly to business changes and meeting on-time delivery," said George Reiswig, now Perdue's CFO. We were an industry leader in terms of marketing, sales, and logistics. We wanted to create an MIS department which would be consistent with our corporate objectives and culture." IS had to ensure that its services were valuable and relevant to the business units while convincing customers that they had a vested interest in development projects from the start.

Toward these ends, a monthly chargeback system was implemented in 1988, and a formalized mission statement and list of Critical Success Factors were incorporated into the systems-development methodology. One major tenet of the methodology is that automation is never the first step in a project. Rather, it occurs only after superfluous business processes are eliminated, and necessary ones simplified (business process improvement).

The CSFs for each project include senior-management sponsorship; limited project size, duration, and scope; precise definition of requirements; and continuous involvement of both the systems staff and the customer. Planning is a collaborative process and each project's requirements are stated up front in customer/supplier agreements. IS spends a lot of time identifying its customers, finding out what they do, and figuring out how technology can help them.

"MIS spent a lot of time in our area identifying needs," one user said. "They performed a cost-justification and payback analysis and suggested a presentation system for use with our outside customers. Formerly, we did one yearly presentation for the top 20 percent of our customers, but now we give presentations to nearly all on a quarterly basis. This is typical of IS and user activity in many Perdue departments.

Keeping users informed about plans, goals, and activities is another important part of IS managers' jobs. "We work closely with senior line managers," Cook said. "We issue monthly status reports which I look on as advertising." As a result, users have a better understanding of the traditionally mystery-shrouded IS function. "Knowing about MIS makes you have a buy-in to what they're doing," one user stated.

Perdue's long-range, strategic-planning efforts present one of the IS organization's greatest challenges. "Historically, Perdue wasn't good at strategic planning," says Milton Shupe, Perdue's IS strategic planner. As part of the effort to change and enlarge the IS role, Shupe and his fellow IS managers have undergone education in strategic planning. "We've always been great executors," Shupe said. "Now we're learning to become great planners."

INTRODUCTION

The previous chapter concentrated on the management of the application portfolio and related databases. The central problem with the portfolio is that of prioritizing scarce resources toward application maintenance, enhancement, or enlargement. This book emphasizes a management system to use with strategizing and planning systems developed earlier. This approach prioritizes resources: It focuses on doing the right things. The current chapter focuses on doing things right.

This chapter emphasizes application project management. It concentrates on the task of developing applications, one of several alternatives for application acquisition. For many applications, the preferred option is to perform maintenance, enhancement, or development activity within the organization. This option is preferred for large, unique applications or for important, strategic applications. Applications that require specialized, proprietary knowledge or those that depend on exclusive, restricted databases must be maintained by the firm itself. Directing application development is a critical success factor for managers.

Technical considerations are important in application development: Management considerations are critical. Weak or ineffective development managers are the most frequent and most expensive source of project difficulty. This chapter emphasizes application project management. It concentrates on details important to application systems. The purpose of this emphasis is to develop a framework and a process for delivering application systems on time, within budget, and to the client's satisfaction. This chapter relies on traditional approaches to application development as a means for exploring these management issues.

THE CHALLENGES OF APPLICATION DEVELOPMENT

Application development continues to be a significant concern for most firms today. In fact, the difficulty of building applications seems to be increasing. For many firms, some of which were highlighted in the previous chapter, application development is a very traumatic experience for developers and clients. The trauma is not confined to the private sector. At a major university, the introduction of a telephone registration system, and its subsequent failure, created major embarrassment for the administration. The breakdown caused an estimated 5,700 students to wait in line for up to ten hours to add classes. The administration attributed the failure to a series of minor glitches that came together to create a gridlock. "We have just experienced the pain of new technology," commented the embarrassed president.

The problem seems to spare no one. California's $1.2 billion information technology budget is under legislative review as a result of claimed mismanagement of its largest projects including the failed DMV system. Amid much fingerpointing, the FAA is scrapping portions of its air traffic control system after costs, first estimated at $2.6 billion, are now expected to reach nearly $7 billion. At Denver's new airport, the $190 million baggage handling system didn't work, and the resulting delays attracted national attention. One hundred networked computers control a system of tens of bar-code scanners, hundreds of radio receivers, and thousands of electric eyes that collectively manage 4000 carts delivering luggage to 20 different airlines. Unfortunately, software problems cost Denver over $1 million per day in interest and other costs as the airport remained idle. At one time or another, reliable firms like IBM, Lotus Development, Ashton-Tate, and Microsoft have all fallen victim to program development problems.

Reasons for Development Difficulties

Program development is one of the most difficult tasks for IT managers. The difficulties seem to fall into one of two categories: those associated with programming itself, and those stemming from the firm or its management. Problems stemming from the programming discipline or from the program characteristics are:

1. Large program size
2. Increased program complexity
3. Measurement weaknesses
4. Weak theoretical foundations

Over time, programming itself has become more difficult for several reasons. Many of the small, easily written programs were completed years ago; but frequently, current applications are large and complex. For example, the first accounting programs or early billing systems consisted of perhaps ten or twenty thousand lines of code. Their cost was several hundred thousand dollars or less. The student registration system referred to earlier was under development for five years, the Allstate system is a $100 million project, and the space shuttle project contains 25.6 million lines of code and represents a $1.2 billion investment. It's estimated that a space station project would require more than 75 million computer instructions. Here are some figures that suggest the size, in lines of code (LOC), and cost of current application systems.[2]

• Lotus 1-2-3, version 3	400,000 LOC	$22 million
• Citibank Teller machine	780,000 LOC	$13 million
• Supermarket checkout scanner	90,000 LOC	$3 million

Like those noted in the previous section, most current systems are built around complex telecommunication nets. They interact frequently with many users in highly visible ways. Some are founded on older, less complicated, but relatively inflexible systems. Many important systems today are larger and more complex than previous systems by orders of magnitude. While applications continue to increase in size and complexity, tools and techniques to build these difficult applications have not improved in capability and ease of use at the same rate.

Because software development involves intellectual processes, software is very flexible; it is very easy to change a line of code. It may be very difficult to understand the consequences of the change completely, however, particularly if the line of code is part of a very large program.[3] Large programs, therefore, tend to be rather inflexible and difficult to change. Advances in the theoretical foundations of software development, and in tools for development, lag behind the demand for applications by a large measure. Software development tools can be very worthwhile, but their adoption and use is still somewhat limited.

Software development largely remains an individual craft lacking in solid measurement systems and based on weak theoretical foundations. Consequently, large system development is difficult to understand and to predict, and nearly impossible to control. Professional standards in software engineering have yet to be established. Until they have been clearly defined, large development projects proceed in the face of high risk.

The firm and its management are also part of the program development problem in many cases. Some firm-based sources of difficulties are:

1. Environmental factors
2. Inadequate development tools
3. Improperly skilled developers
4. Failure to use improved techniques
5. Weak management control systems

In addition to shortened development cycles, firms expect increases in development productivity and improved application quality. Senior executives want to capture the benefits of new or improved applications sooner; consequently, they expect development managers to create applications in less time. Executives demand productivity improvements from their organizations, and they expect developers to build applications at reduced cost. They also expect higher quality applications. IT executives are searching for controllable development processes so that the results are more predictable, and complexities can be managed effectively. Lastly, continuing maintenance expenses for error correction, enhancements, and changes must be minimized.

These executive expectations of application development appear reasonable. Unfortunately, there usually is a large gap between these expectations and program managers' ability to deliver. The gap results from management's failure to reach agreement on the firm's capability versus its expectations. In other words, discontinuities in the environment set the stage for application development failures.

Program development tools and techniques are less than state-of-the-art in most firms today. Many firms are looking for the one new technique or tool that will improve their development productivity but fail to understand that progress must be made across a broad front. For example, object-oriented analysis and development, now the latest technique, is important but will not by itself improve productivity if programmers are overworked, unmotivated, and otherwise badly managed. As Yourdon states, "There is no one single bullet. But taken together, perhaps a collection of small silver pellets will help to slay the werewolves of software development quality and productivity."[4] Improved management systems is one of the silver pellets.

Most management systems for application development utilize the notion that systems have life cycles. Embryonic concepts for new applications emerge; the ideas are developed; and systems are designed and implemented. They are maintained and enhanced, and are ultimately replaced. This systems-development life cycle concept is the basis for studying the management issues and considerations in the traditional approach to application development. The life cycle approach, sometimes called the waterfall method of systems development, is widely used, and its use is expected to continue.[5] The model was conceived to bring order to complex activities and to provide the basis for constructive management intervention in systems development. This chapter outlines the management aspects of this subject, beginning with a discussion of the ingredients of the life cycle approach and culminating in some broad considerations of the overall process. The management issues surrounding the traditional approach to systems development are explored in detail.

THE TRADITIONAL LIFE CYCLE APPROACH

The life cycle approach to system development divides this complex task into phases, each culminating in a management review. There are several valid reasons for using a partitioned or phased approach to development.

Complex development activities are more easily understood and controlled in small increments. Skill requirements vary considerably over the life of the project, and various skill groups are best managed if the project is subdivided. In addition, interactions between the development and client functions during development are much easier to manage incrementally. Finally, managers must evaluate progress and make decisions on an interim basis for maximum effectiveness. Figure 9.1 illustrates the waterfall life cycle and the associated phase reviews.

Note that each phase has distinct characteristics; each contains unique development activities. Each phase has management activity, but they tend to be similar in nature from phase to phase. Typically, some non-management activities within the life cycle span several phases. This segmentation and overlapping will become clear in the following discussion.

The life cycle approach to systems development usually divides the project into five or more phases. The specific approach selected depends on the size of the project and the management system unique to the firm.[6] Table 9.1 outlines the phases as they are frequently employed in practice and as they are discussed in this chapter.

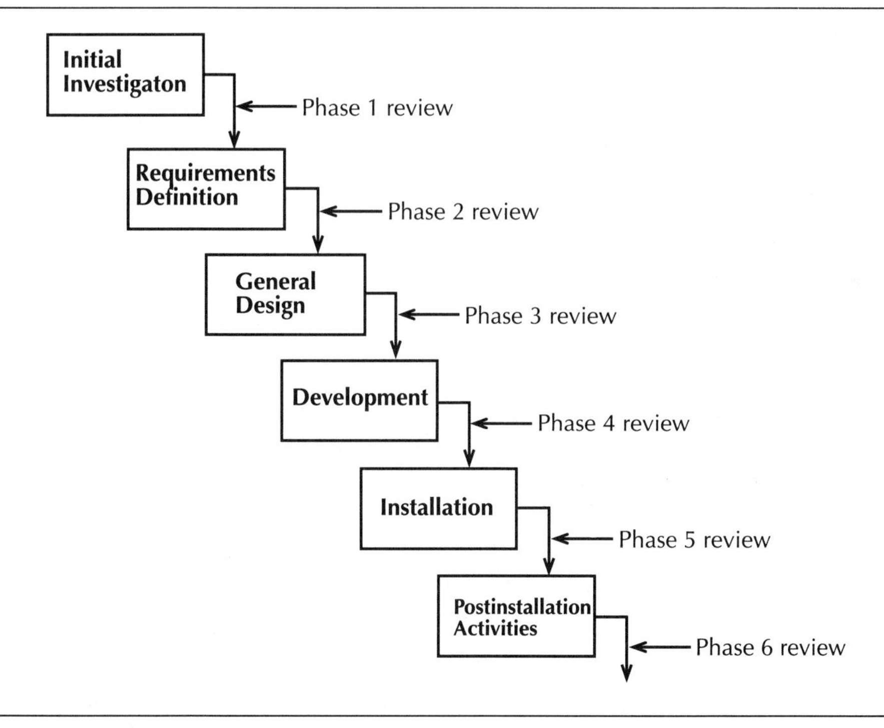

FIGURE 9.1 The Waterfall Life Cycle

TABLE 9.1 Phases in the Development Life Cycle

1. Initial investigation
2. Requirements definition
3. General design
4. Development
5. Installation
6. Postinstallation activities

Each of the phases in Table 9.1 requires both unique and routine management activities. Specific types of management information are required during the phases and at the end of each phase. The activities and information requirements for each phase will be presented in detail.

The techniques of application development are widely discussed in the literature. Many good texts discuss systems analysis, systems design, and system implementation. There is much less information on *management systems* for systems design, development, and installation. This chapter concentrates extensively on the management of systems analysis, design, and installation activities because, as in most information technology endeavors, the ability to manage the work effectively is critically important.

APPLICATION PROJECT MANAGEMENT

The management of application development projects contains many elements common to the management of other types of projects, but some elements are unique to programming projects. The success of application development managers, especially on large and complex programs, is relatively low. And, as pointed out in the Chapter 8 Business Vignette, there have been some magnificent failures. It appears that there are important and significant differences between application development and other types of projects. These differences are critical for the management system employed by application developers.

Table 9.2 lists elements of the application project management process. The items in the table are necessary ingredients for success in managing application development. Development managers utilize these six elements as a framework for decision making. Although adopting these project management notions will not guarantee success in application development, not employing these techniques will greatly increase the probability of failure. Collectively, these actions and controls form the basis of a management system for application development.

TABLE 9.2 Elements of the Application Project Management Process

1. Business case development
2. Phase review process
3. Managing the reviews
4. Resource allocation and control
5. Risk analysis
6. Risk reduction actions

BUSINESS CASE DEVELOPMENT

The business case for an application programming project is intended to itemize the investment resources and estimate the return on those investments. It illustrates and clarifies the costs and benefits of the project. The firm can expect to incur these costs and enjoy these benefits when the application is complete. The business case must address both tangible and intangible costs and benefits, providing a basis of comparison with the alternatives. One alternative must be the business-as-usual case. The tangible costs and benefits of the selected alternative, the remaining alternatives, and the business-as-usual case will be evaluated using one or more traditional financial analysis tools. In many cases, intangible factors are very important.[7]

The department manager who owns the application is responsible for providing the business case. The IT development manager must propose the most cost-effective solution to the business problem, and this solution is used by the owner to articulate the business case. The business case is revised during the life cycle. It is an important constituent of the data on which the outcome of the phase review depends. The purpose of the business case analysis is to give executives vital information with which to make important decisions.

The first step in preparing the business case is to establish the objectives that the development activity is designed to meet. These objectives provide the base for all future analyses; they are the foundation for subsequent decision making. They include a statement of the problem this application is targeted to solve. The objectives may describe an opportunity on which the organization intends to capitalize. The issues may be operational, tactical, or strategic, and the goals of the system may yield tangible or intangible benefits. The second step is to perform an analysis of these benefits and the related costs.

The cost/benefit analysis must be performed for each alternative solution to the problem or opportunity and must also be completed for the business-as-usual case. It is important to understand the current situation in detail.

Will the current system be viable five years hence? Will competition render the current approach obsolete in the future and, if so, what will the consequences be? Against this background, the decision to proceed with alternative courses of action will be judged.

The analysis of the current system may not seem difficult, but it is substantially complicated if one includes the costs associated with lost opportunities or the cost of less than satisfactory customer service, for example. These items must be evaluated in both the current solution and the proposed alternatives, if sound decisions are to result.

The financial evaluations must cover an extended period so that costs associated with the operation of the proposed solution can be assessed. This may mean that the analysis involves five years or more and that the time value of money becomes important. This also means that the analysis is subject to a wide variety of uncertainties and may require substantial amounts of judgment for interpretation. Under these circumstances, it may be best to make a range of projections of both costs and benefits. The decision makers can then make assessments considering the expected probabilities over the range of values. In addition, the economic assessment must include an evaluation of all the tangible and intangible costs and benefits accumulated across *all* functions affected by the application, over the appropriate period of time.

The estimate of development costs includes people-related expenses such as salary and benefits, hiring and training, occupancy and overhead costs, and costs associated with development activity. The development activity costs include hardware or computing costs, travel, and other items directly related to the development tasks. These costs will be found in the IT organization and in the client organizations. They must be summarized for the period starting at project inception and ending at application installation.

Operating costs in all affected functions include people-related costs from the time of implementation to the end of the calculation period. Additionally, all costs associated with hardware and telecommunication systems, costs of purchased services and equipment, and all other cost items related to the operation of the system must be included. If the system is installed in phases, cost estimates need to reflect this phased approach accurately.

Upon completion of the cost analyses, it is possible to calculate the return on investment. The return on investment may be calculated via the payback method, which tells when the operating benefits exceed the development costs, the net present value method, or the internal rate of return method.

The net present value (NPV) method recognizes the time value of money over the life of the project. The NPV method accounts for the fact that money received today is worth more than money received at some future time, and that costs incurred today are more expensive than future costs. For most development projects, the NPV is the sum of the discounted net cash flows over the life of the project. Usually, the discount percentage is related directly to the firm's cost of capital. It may also be a function of time.

Internal rate of return (IRR) also considers the time value of money and the time span of the project. The IRR calculation yields a percentage rate that the firm receives for the money spent on the project. People from finance and accounting take part in these financial analyses to provide guidance in the details of the analysis. This gives confidence to the financial executives that appropriate financial considerations have been applied. Large projects require the concurrence of the Controller in any case. Details of these financial-analysis techniques can be found in most texts on financial management.[8]

If the analyses have been performed consistently and the considerations are mostly financial, the selection among the alternatives, including the business-as-usual case, will be straightforward. The alternative that yields the fastest payback, has the highest NPV, or yields the most positive IRR should be selected. However, in actual practice, this situation almost never occurs because generally the intangible factors are very important and more difficult to quantify.

When purely financial considerations dominate the business case, some additional factors should be taken into account. For example, the payback method favors operational systems over tactical systems, and tactical systems over strategic systems. The payback method will almost always exclude long-range investments; the use of this method tends to encourage short-range thinking. On the other hand, if the development costs are incurred over long periods of time, there are difficulties, as well. There is a risk that technical obsolescence, changing business conditions, changes in management objectives, or other factors may make the system less desirable when ultimately installed than when originally conceived. A balance among these conflicting ideas can be achieved only through experience and judgment.

Other nonfinancial factors may weigh heavily in the decision to develop an application. For example, executive information systems may be a good investment because they improve decision making, some systems are helpful in satisfying government or legal requirements, others may be important as vehicles to introduce or evaluate new technology or new ways of doing business.[9] In each of these cases, there is a risk in doing the project *and* in *not* doing it. The risks and benefits are usually evaluated without relying primarily on financial considerations.

The most common situation is one in which intangible benefits are very important or where long-range implications weigh heavily in the decision making. Strategic systems, those applications most highly sought by executive management, have both of these conditions by definition.[10] In these circumstances, the value of the strategic planning process and the resource allocation methodologies discussed earlier are very valuable. These processes and the thoughtfulness and carefully considered judgments made by senior executives over sustained periods of time all contribute to increasing the probability of success in what is a risky endeavor.

Firms that embark on large, complex application development projects incur risk. If the firm proceeds with the development project without the preparation embodied in the management systems and disciplines discussed earlier, it is incurring needless risk. The executives, their organizations, and the firm itself accepts this risk. Misunderstanding the risk, or thoughtless acceptance of it, lie at the heart of the difficulties found in most application project runaways.

THE PHASE REVIEW PROCESS

Management control of important activities in the application development process is best accomplished when the work is divided into logical segments. The primary purpose of the phased, life cycle approach is to ensure commitment and to understand and control the risk elements in the project.[11] The management control aspects of each phase have four dimensions: scope, content, resources, and schedules. The detailed activities of the phase and the information required for the phase review reveal these dimensions. Each phase depends on other phases of the project; the management process leads naturally from one phase to the next throughout the life cycle.

The phase review process is management oriented. It is a means for managers to inspect the progress to this point and to examine plans for the future.[12] It focuses on decision making. The decision to continue, continue with modifications, or terminate is an essential part of the process. The decision usually is conditional, and it will be reviewed again during the next phase. That is, a subsequent review may uncover facts that lead to a different decision from the one made during a previous review.

Typically, phase reviews are conducted during meetings but may be handled via correspondence in unusual circumstances. Each review should be documented in a phase review report that forms a part of the project's permanent record. The documentation presented in the phase review process should develop from the workings of the management system employed to manage the project.

The management system must be well-defined, organized, structured, and consistent across the phases of a project, and between projects. It must ensure that the phase activity includes the ingredients highlighted in Table 9.3.

TABLE 9.3 Ingredients of a Phase Review

Ingredients of a Phase Review
Project description
Well-defined goals, objectives, and benefits
Budgets and staffing plans
Specific tasks planned vs.accomplished
Risk assessment
Process to track plans vs. actual accomplishments
Asset protection and business controls plans
Client concurrence with objectives and plans

Phase reviews should produce documents and results useful for implementing subsequent phases. Implementation proceeds from one phase to the next, provided that the review results in a favorable outcome. The application system owner must concur that completed work meets specifications. Participants in the next phase must agree that they have all the input necessary to continue.

It may appear that implementation progresses from phase to phase in a step-wise manner. Frequently, however, especially in large or complex applications, there is concurrent activity within and between phases. This does not eliminate the need for valid checkpoints that determine project status. Checkpoints required for tracking, measuring, or authorizing project continuation are more important when the project involves parallel activity.

It may not be possible to resolve all the issues at a given phase review. The decision to continue, continue with alterations, or terminate is a management judgment based on the circumstances and the information available at that time. The application owner may elect to proceed with the project despite an unresolved issue. However, this option should include an assessment of the associated risk. All open issues must be resolved promptly to ensure project integrity.

Phase Review Objectives

The objective of the phase review process is to provide the greatest probability of project success. This means that phase reviews must measure the accomplishment of agreed-upon objectives within the planned time and at the planned cost. They must provide a tool to assess activity status and to develop alternate action plans, if required. They must provide a vehicle for reviewing plans and objectives for subsequent phases. All persons affected by the design, schedules, cost, data, and operating requirements of the project must have the opportunity to evaluate the status of the project at specific times during the process. The phase review process documents the decision reached by all affected managers.

Timing of Phase Reviews

Phase reviews should take place at the completion of each phase in the system development life cycle. Successful completion of the review indicates successful completion of the phase. When this process is applied to maintenance or enhancement activities that qualify as significant projects, phases and corresponding schedules must be identified. For large projects, with phases extending beyond six months or so, interim reviews should be held. Interim reviews should concentrate on plans versus actual accomplishments and expenditures, and on the estimates to complete the phase.

If the system is large and complex, it is probably composed of multiple, concurrently developed sub-systems. Phase reviews should be conducted for each sub-system. In addition, the entire project should undergo a comprehensive review at least once every six months. The purpose of this review is to coordinate the sub-system reviews and to assess overall project status. Each review must establish, alter, or confirm schedules for subsequent phase reviews.

Phase Review Contents

The activities of Phase 1, initial investigation, begin with an idea for a new application or for an enhancement to an existing application. These embryonic concepts usually (but not always) arise in the application-using department. Strategic systems, for example, may originate through the strategizing or planning processes, as discussed earlier. Systems analysts perform a preliminary review of the existing system and define the new system requirements. They develop a preliminary concept for the system and generate alternatives. Next they evaluate feasibility and develop plans and schedules for the Phase 1 review. Table 9.4 itemizes the management information requirements for the Phase 1 review.

Phase 2, requirements definition, consists of modeling the existing physical system and deriving a logical equivalent to which the new system requirements are added. This results in a logical model of the new system. The global technical design is created from the model. Updated costs and benefits are developed for the project and system control and auditability requirements are established.[13] System performance criteria are established at this time. The Phase 3 plan is complete, and the Phase 2 review is scheduled. The management information requirements for the Phase 2 review are shown in Table 9.5.

TABLE 9.4 Phase 1 Review — Management Information Requirements

Statement of need and estimate of benefits
Schedule and cost commitments for Phase 2
Preliminary project schedule
Preliminary total resource requirements
Project dependencies
Analysis of risk
Project scope
Plans for Phase 2

TABLE 9.5 Phase 2 Review — Management Information Requirements

Documented statement of requirements
Refined benefits commitment
Schedule and cost commitments for Phase 3
Refined project schedule
Refined total resource requirements
Updated analysis of dependencies
Analysis of risk
Project requirements and scope
Plans for Phase 3

Activities of Phase 3, general design, consist of developing external and internal system specifications. The system software specifications are refined, and utility program requirements are specified. At this time, hardware requirements are completed, and system architecture definitions are finalized. During Phase 3, system control and auditability requirements are completed, and user documentation and training are planned. The plans for Phase 4 are developed, and the Phase 3 review is scheduled. The management information requirements for Phase 3 are shown in Table 9.6.

TABLE 9.6 Phase 3 Review — Management Information Requirements

Finalized general design
Final benefits commitment
Schedule and cost commitments for Phase 4
Committed project schedules through Phase 5
Committed costs through Phase 5
Resolution of remaining dependencies
Analysis of risk
Preliminary user documentation
Preliminary installation plan
Plans for Phase 4

During Phase 4, development, the activity consists largely of program design, building, and unit test. File and data conversion strategies are developed during Phase 4, and program modules are written. Program module testing is completed during this phase, and system installation is planned. The development team begins user training and develops client documentation. Plans for Phase 5 are developed, and the Phase 4 review is scheduled. Phase 4 management information requirements are presented in Table 9.7.

TABLE 9.7 Phase 4 Review — Management Information Requirements

Finalized installation plan
Satisfactory completion of program test
Schedule and cost commitments for Phase 5
Committed project schedules through Phase 5
Reaffirmed commitment to system benefits
Commitment to system operational costs
Analysis of risk
Final installation plan
Plans for Phase 5

The activities of Phase 5, installation, are critical. During Phase 5, user training and user documentation are completed. Acceptance testing is passed during this phase, and file conversion is completed. The system is installed, and operation is ready to begin. The Phase 5 review is scheduled. Upon satisfactory completion of Phase 5, the strategy for beginning the use of the new system and phasing out the old system is implemented. The management information requirements for Phase 5 are shown in Table 9.8.

TABLE 9.8 Phase 5 Review — Management Information Requirements

Satisfactory completion of system test
Final user documentation
User acceptance document signed
Finalized application business case
Reaffirmed commitment to system benefits
Commitment to system operational costs
Analysis of risk
Plans for Phase 6

Phase 6, postinstallation activities, consists mainly of the operation and maintenance of the new system. This is the time to evaluate the effectiveness of the life cycle management system and to review the management techniques applied during the development process. The system is evaluated against the original specifications and objectives. Managers conduct an analysis of programming and implementation effectiveness and review requested enhancements.

The Phase 6 review evaluates the results of previous phase reviews and makes plans for incorporating new knowledge into subsequent projects. This is the application of business process improvement techniques to application development. It is a critical activity for healthy development organizations. Phase 6 is important to IT application development management because it is introspective and reinforces sound management techniques. The IT organization learns through experience, just as individuals do. The Phase 6 review provides checks among the strategy, the various plans, and the actual implementation. These checks are an essential part of management control.

This outline implies a significant amount of management information. This information enables managers to assess the progress of the programming project and make judgments regarding its future. Without information of this type, and a setting in which to evaluate it, managers accept unnecessary risk and may face the prospect of operating out of control. Phase 6 is mainly a review of the quality of the management process, and an application of business process improvement techniques to application development. This introspective review assists management in refining the process for the organization's future use.

The Participants

The owner of the application system and data, key client managers, and representatives of all functions affected by or affecting the project must participate in and evaluate the results of phase reviews. Representatives of other senior executives should participate in cases where the project is of significant potential benefit to the firm. Usually, it is not difficult to gain the attention of executives when large amounts of money are involved. IT managers or their representatives normally orchestrate the review process.

Phases for Large Projects

Exceptionally large or complex projects may require special attention that is best gained through modifications to the phase review schedule. The terms "large" and "complex" are relative to the size and skills of the organization. The decision to invoke special treatment must be individually determined for each organization. If the proposed project is the largest or most complex that the firm has ever considered, then it must receive special handling. Other projects may qualify for special treatment, as well. Special consideration consists of additions to the phased process which are illustrated in Table 9.9. For major projects, the phases consist of more events.

TABLE 9.9 Expanded Phase Review Process

Phase 1	Initial Investigation
Phase 2	Requirements Definition
Phase 3	General Design
Phase 3A	Detailed External Design
Phase 3B	Detailed Internal Design
Phase 4	Development
Phase 4A	Detailed Program Design
Phase 4B	Program and Test
Phase 5	Installation
Phase 6	Postinstallation

The modified phased approach presented in Table 9.9 divides the design and development phases discussed earlier, creating a project with eight phases. The review that concludes Phases 3A and 3B evaluates the detailed external and internal design. Likewise, the program design can be evaluated separately from the programming and testing as accomplished in Phases 4A and 4B. This additional perspective greatly assists in maintaining project control. Further subdivisions can be made if conditions warrant. In addition, project managers must define appropriate intermediate checkpoints within each phase to assure proper controls.

Phase reviews are essential in controlling project activities. They relate directly to the system development life cycle, and facilitate thoughtful, organized decision making. Managing phase reviews is a critical IT activity.

MANAGING THE REVIEW PROCESS

The review process is management oriented and leads to sound decision making. Each phase review must always contain unambiguous documentation with which all parties concur. At a minimum, the documentation must describe the scope, content, resources, and schedules of the work effort. In addition to these items, it is advisable to produce a very clear statement of the assumptions and dependencies involved in each phase and in the complete plan. At some point in the project schedule, prior to successful project completion, the assumptions must turn into facts and the dependencies must be formalized explicitly.

The management team must devote every effort to resolving issues and nonconcurrences prior to or during the phase review. The emphasis during the review should be on the items of scope, content, resources, and schedules. Changes to any of these major items should receive special attention. Managers who search for the underlying reasons for changes and variances at each step are less likely to get into trouble later on.

A management system for issue tracking and reporting must be developed. Each issue must be assigned to an individual for prompt resolution within an agreed-upon schedule. Subsequent phase reviews must document the fact that all open issues have been resolved satisfactorily. Unresolved issues inevitably lead to an unsatisfactory review; they can become sufficient cause to terminate the project.

The results of the phase review must be documented and placed in the project's management file. A summary of the information should be prepared and distributed promptly to all concerned parties and to appropriate members of the management team.

RESOURCE ALLOCATION AND CONTROL

Project planning describes the ebb and flow of skills from Phase 1 through Phase 6. Analyst and client activity is high during the beginning of the project and is high again just before and during installation. Programming activity is low at project definition. It peaks during implementation and tapers off after installation. Similarly, computer operators, technical writers, database administrators, trainers, and others all have patterns of deployment over the life cycle. Managers need to track and control these resources just as they do physical or monetary resources. Life cycle resource management and control are fundamental to the success of application program development.

Information on the deployment of resources by skill type is a part of the application development plan. The resource plan must be available during the life of the project; it is fleshed out in additional detail at each phase review. During each phase, the management team must track resources applied by skill type versus the plan. For example, during Phase 3, analyst effort declines and programmer activity increases as the application moves toward implementation. Deviations from the plan during this phase in these skill areas are a signal for management to attack the underlying cause. Simply analyzing total resources will not reveal the degree of detail that management needs.

Stephen Keider identifies six tasks that are most likely to cause failure if mismanaged, and presents seven early danger signs.[14] Through interviews, he learned what 100 MIS professionals thought was the single most important cause of system failure. The results are depicted in Table 9.10.

TABLE 9.10 Major Causes of System Failures

Reason for Failure	Number of Responses
Lack of project plan	23
Inadequate definition of the project scope	22
Lack of communication with end users	14
Insufficient personnel resources and associated training	11
Lack of communication within the project team	8
Inaccurate estimate	8
Miscellaneous	14

Some projects fail because of technology or design problems, but to a great extent, the project manager can control the difficulties itemized above.

RISK ANALYSIS

An essential and vital part of application project management and of each phase review is the analysis of risk. Software projects fail for a variety of reasons, but most of the reasons for failure can be traced to inadequacies in the project management system. Management system failures can be greatly reduced by using the principles developed in this chapter. But there is one additional technique that will significantly improve the project's chances for success. This additional element is the analysis of risk.[15]

In the application development process, there are analogies to the familiar economic concepts of leading, coincident, and lagging indicators. Some project managers are relatively uninformed and are observing lagging indicators. For example, some development managers will find out one day that the program they developed has failed because their clients have never embraced the program, or have ceased using it because it lacks function. These managers will discover the failure after the fact.

Most programming managers are better informed. They have an exhaustive list of project metrics that they track on a continuous basis. These indicators relate to budget, schedule, function, or to a host of other relevant items. The managers' very careful monitoring of these indicators allows them to know the very moment when their project fails. They are the first to know because they are employing coincident indicators. This is certainly better than using lagging indicators, but it is not nearly good enough. The idea, after all, is to succeed, not fail. What is needed is a set of leading indicators, indicators to alert managers in advance that their projects are headed for trouble unless they take action. The analysis of risk provides these indicators.

What is the analysis of risk? How will this analysis work to provide early warning for project managers? How will project managers be alerted to impending difficulties? To answer these questions, managers need to search for the sources of programming project difficulties. Next, they must develop a set of quantified measures that describe the extent of risk to which they are vulnerable in each of these areas. Additionally, managers must track these risk measures to discern their trends over the life of the project.

The major sources of risk for an application programming project can be grouped into five categories and assigned approximate weight corresponding to amount of risk contribution. Table 9.11 lists these sources of risk and their approximate weights.

TABLE 9.11 Sources of Risk

Source	Weight
1. Client activity	20
2. Programming and management skill	10
3. Application characteristics	30
4. Project importance and commitment	20
5. Hardware requirements	10
6. System software requirements	10

Each category can be further subdivided into quantifiable and measurable items that collectively provide a basis for the assessment of project risk. First, let's discuss the ingredients of these six categories and then develop a rationale for quantifying them.

Active client involvement in any programming project is vital for the success of that project. If support from the client community is weak or missing, the project is in jeopardy. Client activity can be measured by considering the quality and quantity of the requirements definition and the depth and breadth of their involvement in the pre-requirements activity. Client involvement throughout the post-requirements activity, and their knowledge level of the proposed system and its relationship to their other systems, are good measures of client activity. The extent of training efforts and the success of these efforts in training the application users in the new system are also valuable measures.

The knowledge, skill, and experience levels of the system implementors and of the project manager are extremely important for project success. Having insufficient numbers of implementors, or having implementors with insufficient skills to perform the tasks, indicates possible corrective action. For instance, if the system requires new telecommunication software, and if the technicians are not sufficiently experienced or trained in this area, management should be concerned.

A third and obvious area of concern is the nature of the project itself. The items of importance are the size or scope of the system under development, its duration and complexity, and the anticipated project output or deliverables. The project logistics and the extent to which development takes place over a wide geographic area are other significant factors. For instance, application programs developed across several locations are much riskier than those developed at one site. Additional items in this category include the sophistication of project control techniques, and the extent to which the development team is homogeneous and accustomed to working together. This latter item should include the effects of subcontract programming, for example.

The fourth area of risk, project importance and commitment, can be measured by the aggressiveness of the schedule, the number of management agreements needed for implementation, and the extent of management commitment to the project. Each factor adds a measurable element to the risk assessment for the total project.

System hardware forms a fifth area of risk. If new or unfamiliar hardware is required for the system, or if the system has unusually stringent performance requirements, risk is introduced into the project. Programs developed to run within the capacity of existing hardware systems incur little or no risk in this area. Performance specifications for the system and capacity planning for the hardware system are important.

The last item of concern is the operating system and other software needed for application development and implementation. If the application uses routine system software and is developed using familiar languages, risk is minimized. Programs using new and unfamiliar languages are much riskier. Programs using prereleased or relatively untested software packages are even riskier.

There may be more risks. For example, if part of the application is purchased, there may be additional risks associated with the vendor. On the other hand, depending on the capability of the firm's programming staff, purchasing all or part of the application may reduce risks to the firm. Another example of an external source of risk is in the use of third-party telecommunications systems. This dependency brings with it additional potential risk that must be analyzed and quantified.

As shown in Table 9.12, a useful, but somewhat arbitrary, weighting or quantifying of these risk elements is shown to the right of the categories as a guide for the project manager. Managers may want to adjust the values to suit their individual situations. However, it is important that the methodology remain constant and unchanged during the life of the project. Because absolute values of risk cannot be measured accurately, the change in the value of risk is much more valid and useful. It is worthwhile to keep a history of risk measures to establish trends within the organization and across projects and project managers. Table 9.12 displays a more detailed breakdown of the weights for each category.

Minimally, the items must be scored and evaluated prior to each phase review and preferably more often. An item with no risk would be scored zero; an item having a great deal of risk would be given the highest score possible for that item. The weights should be consistent from review to review in order to determine risk trends. Precision is much preferred to accuracy in this analysis. The absolute level of risk is valuable in determining whether the project should continue, continue with modifications, or be terminated. For example, if at Phase 1 the risk totals 75, most prudent managers would seriously consider terminating the project. For the risk to equal 75, many of the individual items would have to be at high risk. Therefore, the entire project will be high risk at this time. On the other hand, if the total risk at Phase 1 is 20 or lower, this would be an indication of low or manageable risk. Numbers between these ranges require a detailed review and analysis at the phase review prior to decision making.

As the program proceeds toward implementation, the initial risk should decline. (A program successfully installed and operating has zero development risk!) Programs proceeding smoothly shed risk as they progress from Phase 1 to Phase 6. Increases in total risk over the life of the development project are danger flags. More detailed analysis is required. Yet, if the total risk is declining, the project may still be headed for trouble. An example of this would be a situation in which one of the items previously having a zero or low value suddenly increased in risk value. For example, if, during Phase 3, Item 3B, duration and complexity, increased from a value of 1 to a value of 5 because it was discovered the program was much more complex than previously predicted, that could be cause for serious concern. The concern will be present even though other items decline in value and the total risk declines.

TABLE 9.12 Detailed Risk Items

Risks	Relative Weight
1. Client activity	
a. Quality and quantity of requirements definition	6
b. Pre-requirements planning activity	3
c. Post-requirements activity	4
d. Knowledge and understanding of proposed system	4
e. Client training	3
2. Programming and management skill	
a. Implementor experience, skill, and ability	5
b. Management experience, skill, and ability	5
3. Application characteristics	
a. Size, scope, and complexity of the system	6
b. Duration and complexity of the project	5
c. Project deliverables	3
d. Project logistics	5
e. Project control techniques	7
f. Organization considerations	4
4. Project importance and commitment	
a. Aggressiveness of the schedule	4
b. Number of managers involved	8
c. Extent of management commitment	8
5. Hardware requirements	
a. New hardware	6
b. Stringent performance requirements	4
6. System software requirements	
a. Operating system and utilities software	5
b. Language requirements	5
Total	100

This analysis provides leading indicators of project success.[16] Management commits to deliver a product at some future time, at some cost, and with some stated function. This commitment is made with an understanding of the risks involved at the time the commitment is made. It is further understood that risks would be mitigated during the course of development in order to meet the commitments. If analysis over time indicates unfavorable risk trends, management must take corrective action to meet the previous commitments. Analysis of the sort developed above permits management to take corrective action before commitments are missed. The consequences of proceeding with the project in the face of high or rapidly rising risk can be very severe.

Seemingly easy tasks, such as switching to a new computer system, can be extremely troublesome even for firms experienced in computerization. For example, Sun Microsystems switched to a new management information system and, in doing so, lost control of customer orders and inventory, among other things. As a result, Sun failed to pay some of its bills on time and needed to perform some accounting work by hand. Because of the experience, Sun Microsystems suffered revenue and profit reductions. Other firms, including IBM, are displeasing customers because of uncoordinated or failing customer support systems. The difficulties of application management seem to spare no one.

RISK REDUCTION

Quantifying the risks inherent in the project is clearly beneficial. Prudent managers can take actions to reduce or mitigate the risks and take advantage of risk trends. It may not be possible to eliminate completely all the risk during the project, but it certainly is possible to focus attention on the areas of risk and to manage the risk proactively. Management action may consist of deploying special resources, instituting special control techniques, or using lower risk alternatives. Project managers have many resources at their disposal during the life of a project with which to manage problems. They need risk assessment tools to alert them to impending problems.

For instance, if training lags behind schedule, the project's risk will rise. Alert project managers will recognize this trend and search for the underlying causes. They know that training users to operate the application is very important and that poor or late training jeopardizes the successful implementation of the application. In the midst of development activities, it is easy to conclude that training will catch up later when there is more time. Aggressive managers will resist this temptation and take action to eliminate this risk item immediately, thereby mitigating further risk.

Problem management is a major part of a project manager's job. Risk analysis provides an analytical tool that yields early warning of impending difficulties. Alerted to future difficulties, IT managers can initiate action to cope with the difficulties when they are small and manageable. It is an indispensable part of the project management system for successful application development.

MORE ON THE LIFE CYCLE APPROACH

The life cycle methodology discussed in this chapter illustrates the management processes and procedures that must be used to achieve consistent results in application development. It is a popular method, but it is not the most sophisticated method in use. It has some disadvantages that other methods strive to overcome. The primary disadvantages to the phased approach are:

1. Tangible client results come late in the cycle.
2. It depends upon stable initial requirements.
3. It tends to be paper intensive and bureaucratic.
4. Parallel activities are permitted, but not encouraged.

During the first three phases in the life cycle methodology, developers and clients are dealing mostly in paper: analyzing current procedures, proposing new procedures, defining new system requirements, and developing new system designs. During this period, the results are words, symbols, diagrams, and resource statements like time, money, and people. Up to this point, no operational results are available for inspection.

In some cases, clients have difficulty in precisely stating system requirements while developers struggle to translate them into workable code. The clear but intense communication between developers and clients needed to attain a complete and workable statement of requirements is often difficult to achieve. For example, clients in marketing and IT analysts and programmers may use the same words but still not convey a common meaning. "Build what I mean, not what I say" is a frequent, implied sentiment during requirements definition.

The goal of alternatives to the traditional life cycle method is to circumvent these difficulties by introducing more parallelism into the development process, and by getting clients involved in different ways. In addition, the waterfall approach and its alternatives are improved significantly by using sophisticated development tools. Computer-aided software design (CASE) tools provide an automated formalism to assist both clients and developers in producing satisfactory systems.

Using sophisticated tools during development obtains observable results earlier for clients and for analysts to refine. Parts of the system become available to test before other parts are fully defined. This gives everyone the opportunity to develop ideas based on tangible results from earlier deliberations. The daunting challenge of defining all the requirements at the outset is removed.

There are now many variations or methodologies for software development, including prototyping, object development, incremental waterfalls, structured techniques, and information engineering. And there are variants of these. Some of these will be discussed in more detail in subsequent chapters.

Regardless whether the conservative life cycle approach or the free flowing prototyping method is used, managers need to understand their commitments and must have a method for evaluating the progress toward meeting them. To this end, managers need to have checkpoints to understand scope, content, resources, and schedules. Regardless of the method employed, managers need to quantify risk and be able to respond in a timely manner to contain or eliminate it. If managers cannot predict the risks or respond to them, then the project is operating out of control, generating unpredictable and, usually, disastrous results.

SUCCESSFUL APPLICATION MANAGEMENT

The ingredients of successful application management flow from a well-designed and smoothly functioning management system. The application management system is a controlled process capable of yielding predictable results and dealing with increased complexity. Its products meet all technical and functional specifications and are surprise-free. The resulting applications meet schedule and budget agreements and satisfy the conditions expressed in the business case regarding operating costs and realizable benefits. The purpose of the review process and the other efforts contained within it is to ensure that these goals are attained. Successful application management yields predictable products that contribute important assets to the organization.

The application management process defined in this chapter applies not only to programs, but to documentation, too. Programs, program documentation, and operating documentation are products of the development process and are managed in a similar manner. The documentation products must be critically scrutinized at the phase review, just as the emerging application is.

In most firms today, business managers are intensely interested in improving productivity and enhancing performance. Application project managers must concentrate on productivity too, and they must be able to demonstrate productivity improvements. The thoughtful, organized, businesslike approach to the development process detailed in this chapter is a necessary condition for productivity improvements, but by itself is not sufficient. It must be used in conjunction with other tools and techniques to be discussed in subsequent chapters.

SUMMARY

Managing application development projects is a difficult task. Because systems are becoming larger, more complex, and more strategic, the magnitude of the task is necessarily increasing. Application development involves more resources, occurs over a longer time period, and is inherently riskier than it was a decade ago. *Ad hoc* management techniques and routine project management methodologies need to be augmented by disciplined processes. These processes must be specifically designed to cope with the difficulties of application development and with the associated increased risk.

The phased development approach divides the project into manageable tasks. It requires a management review process to focus on fundamental project issues. The foremost issue is the validity of the project itself; this concern is addressed through sound business case analysis. The business case is reviewed periodically throughout the life cycle of the application to ensure continued project viability. Ultimately, without this essential foundation for the project, all efforts to develop a successful application will lead to failure, regardless of methodological subtleties.

Managers need control points during the life of the project to carefully reevaluate the business case and focus on issues of scope, content, resources, and schedules. The phase reviews offer decision points in the life of the project. They are designed to illuminate all important issues relating to the continuation of the project and to resolve them to management's satisfaction. Risk analysis is an important management tool because it yields leading indicators of project difficulties. Therefore, it provides management with an opportunity to resolve problems when they are small and easily handled. Failure to capitalize on these early warnings frequently leads to major problems later in the cycle. Major problems late in the life cycle of an application lead to extreme distress for project managers and for all who depend on the successful conclusion of the development effort.

Successful management of application development requires a management system tuned to the needs of programming projects. It demands a rigorous, thoughtful process and candid, open communication among all participants. The basis for much of this process lies in the antecedent activities of strategy development, strategic and tactical planning, technology assessment, and application portfolio asset management. Given a firm foundation in these activities and the rigorous employment of project management tools, techniques, and processes, the risks involved can be substantially mitigated, and chances for success greatly increased.

The management system for applications builds on previously developed management processes. This means that application development contributes to the business goals in a positive and well-understood manner. The firm and its managers know that IT's development activities are congruent with the firm's objectives. Strategy development and strategic planning, along with resource prioritization in tactical and operational plans, are effective processes. If projects are well managed, the firm's executives will be likely to hold realistic expectations of the IT function. The management system for applications is valuable because it focuses on the IT manager's critical success factors.

Review Questions

1. What is the connection between critical success factors and application development management?
2. What are the goals of successful application development? Why is this area a critical success factor for the IT manager?
3. Why is local development the only reasonable alternative for many applications?
4. Why does the life cycle approach divide the project into phases? What are the advantages of this phased approach?
5. What are the phases in a typical life cycle?
6. What activities take place in each phase? What information is required?
7. Why is Phase 6 important?
8. Do the system development life cycle and the phase review process apply to the maintenance of large applications? Under what conditions are these ideas most important?
9. What are the six essential elements of a sound project management system for application development?
10. What are the similarities and differences between managing a computer system application and managing the construction of an office building?

11. What are the ingredients of a computer system application business case?
12. Why is it becoming more difficult to develop and assess application business cases?
13. Why are phase reviews an indispensable part of the management system for application development?
14. What are the objectives of a phase review? How often should phase reviews be held?
15. How should the phase review process be modified for very large applications?
16. What role does documentation play in the review process?
17. Describe the risk analysis process. What are the elements that are reviewed in risk analysis?
18. Why is risk analysis considered to be a leading indicator?
19. What are the disadvantages to the waterfall methodology?

Discussion Questions

1. List all the factors that you think contribute to Perdue's success in application development. In your opinion, which factor is most important, and why?
2. Who should participate in systems analysis and design? How does the degree of involvement for analysts and programmers change from inception of the idea to implementation?
3. Discuss briefly what a systems analyst does in each of the six phases. What does the client manager do at the completion of each phase?
4. Discuss the phase review discipline as applied to very large and complex applications.
5. Describe the management system that must be in place in order to ensure project integrity from phase to phase.
6. Describe some of the intangible issues in the application business case and discuss why they are important considerations.

7. What are some of the intangible issues likely to be present in the development of a new on-line customer order entry system?
8. Consider a project to replace the current payroll program with a new program incorporating the latest tax changes to be implemented at year end. What would be the major risk factors? How would these risk factors tend to vary over the development cycle?
9. Where do the risks reside in the development of a totally new strategic information system?
10. Describe the management actions that form the core of the management system needed for successful application development.
11. What elements of the management system developed in this chapter would be most valuable to those firms discussed in the Business Vignette in Chapter 8?
12. Throughout the firm, describe the antecedent activities necessary for application development projects to proceed successfully.

Assignments

1. Outline the agenda for the Phase 3 review of an internal application program. Name the positions that should be represented and identify the order in which they present. If the program under development is going to be developed as a product of the firm, and Phase 3 just precedes product announcement, what additional considerations will be important?
2. Read an article on systems development in a scholarly journal, such as that in note 4, and summarize the management techniques in a report. Prepare your report for presentation to your class.

ENDNOTES

[1] Megan Santosus, "Perdue's New Pecking Orders," *CIO*, March 1993, 60. Reprinted through the courtesy of CIO. © 1993 CIO Communications Inc.

[2] Brenton Schlender, "How to Break the Software Logjam," *Fortune*, September 25, 1989, 100.

[3] The three most expensive known software errors cost the firms involved $1.6 billion, $900 million, and $245 million. Each was the result of changing just one line of code in an existing program. Peter G. W. Keen, *Every Manager's Guide to Information Technology* (Boston, MA: Harvard Business School Press, 1992), 46.

[4] Edward Yourdon, *Decline & Fall of the American Programmer* (Englewood Cliffs, NJ: Yourdon Press, 1993), 37.

[5] Charles R. Necco, Carl L. Gordon, and Nancy W. Tsai, "Systems Analysis and Design: Current Practices," *MIS Quarterly*, December 1987, 461. The authors believe the life cycle approach will benefit from structured development and increased use of tools, and it will be influenced by prototyping methodologies.

[6] There is little agreement on the number of phases into which the systems life cycle should be divided. However, the management principles apply regardless of the number of phases.

[7] Some things are harder to quantify than others and are called intangibles. If something is impossible to quantify, perhaps it is not real and should not be considered. If it can be quantified, then it should be considered tangible according to some who argue there are no intangible items.

[8] For example, see Paul M. Fischer and Werner G. Frank, *Cost Accounting* (Cincinnati, OH: South-Western Publishing Co., 1985), 219-226.

[9] C. James Bacon, "The Use of Decision Criteria in Selecting Information Systems/Technology Investments," *MIS Quarterly*, September 1992, 335.

[10] For more insight into this situation, see Eric K. Clemons, "Evaluation of Strategic Investments in Information Technology," *Communications of the ACM*, January 1991, Volume 34, Number 1.

[11] Joel D. Aron, *The Program Development Process: The Programming Team* (Reading, MA: Addison-Wesley, 1983), 340-343.

[12] A management axiom is that "you get what you inspect, not what you expect."

[13] System control and auditability features are critically important to applications programs. They are discussed extensively in Chapter 18.

[14] Stephen P. Keider, "Managing Systems Development Projects," *Journal of Information Systems Management*, Summer 1984, 33.

[15] The analysis of risk presented here is derived from the author's experience at IBM. Other organizations may use similar analyses.

[16] Aron, 343-348, presents risk assessment in a somewhat different manner.

10 Alternatives to Traditional Development

A Business Vignette

The American Programmer—An Endangered Species?

Edward Yourdan, a prominent authority on programming, is so concerned that he wrote a book based on this thesis.[1] There is considerable evidence to support his view from critics of the American craft and firms that have outsourced programming offshore. Unless U.S. firms adopt key software technologies, according to Yourdan, software development will continue to be outsourced overseas.

There are several reasons for this trend relating primarily to the availability of critical programming skills and the cost of program development.

Many firms are experiencing critical skills shortages in new technologies such as object-oriented design, C programming, and graphical user interfaces (GUI). According to Forrester Research, 86 percent of firms they surveyed reported that development plans presently are or will be limited by current programmer skills.[2] Key skills are being developed rapidly by programming teams from Eastern Europe, Russia, India, and mainland China. For example, the Shanghai Software Company advertises skills ranging from ADA to C++ to PASCAL on eight development environments and five operating systems.

In some countries, software development firms gain preferential access to college students trained in math or the sciences, some of whom were also schooled in the U.S. When jobs are scarce, labor is cheap by our standards and, when exports are desired, the software development industry often gets started with government assistance. That is the case in India, for example, where the government supports software technology parks in Bangalore, Bombay, Calcutta, and Delhi by giving the park residents tax breaks.

To facilitate international software development, some offshore developers have established links with U.S.-based marketing and management firms in a virtual organizational relationship. The U.S. firm manages the interface with the client and communicates with the developers via satellite data links, fax, E-mail, and Internet access. English-speaking managers at the Shanghai Software Company communicate with their U.S. counterparts and manage the documentation flow in English. In some countries such as India, communication is easy because English is the programming team's language of choice.

In addition to offering bright workers with key skills, offshore developers feature very attractive costs. For example, software writers in the former Soviet bloc countries are paid about one-fifth what their U.S. counterparts are paid.[3] Even after adding additional management and overhead costs, offshore development is a relative bargain. Citing statistics that coding is only about 10-15 percent of total project costs, some critics believe that software development is not a cost-based business. Others argue that the government should take a hard look at offshore programming before it siphons off corporations' money and diminishes the number of jobs in the U.S.

But firms like Motorola, Hewlett-Packard, Texas Instruments, Oracle, and Chase Manhattan Bank increased their skill base at reduced costs through overseas development. Lacking internal resources, Chase outsourced an application to Tata Unisys Ltd. in Bombay. Access Computers in Detroit found experienced C++ programmers in India for its multimedia software written in Visual C++.[4] State-of-the-art technologies are essential to the survival of the Indian software industry according to the Indian Trade association.

Whether offshore outsourcing is a permanent trend depends in part on the U.S. programming industry itself. Thousands of COBOL programmers need to be retrained in C++, object-oriented techniques, and software quality assurance methods. Not burdened with past habits and large mainframes, offshore programmers assume that the ISO9000 quality standards, graphical user interfaces, and client/server implementations are routine practices.

But some believe that the challenge from overseas developers will speed the adoption of technologies that increase quality and productivity and lower costs at home. Others seek comfort from the rapidly exploding need for programmers, arguing that there will always be enough work for everyone. For now, offshore development is relatively inconsequential. Only time will tell whether or not it remains a small factor.

INTRODUCTION

Chapters 8 and 9 developed a foundation for managing the application portfolio as one of the firm's major assets. Building on this foundation, Chapter 10 explores additional application-development tools and techniques and considers additional acquisition alternatives. This chapter discusses the topics of prototyping and object-oriented programming. It also considers the merits of purchased applications. It explores subcontract development and the employment of service-bureau organizations. Joint development activity through the formation of alliances will also be discussed.

This chapter concentrates on management systems issues important to managing the alternatives to local development. The approach in this chapter is to build on the traditional alternatives while exposing readers to additional alternatives. These additional alternatives are valuable in managing the complete portfolio asset. The final alternative, that of client/server and end-user computing, is covered in detail in Chapter 11.

Fourth-Generation Languages

During the past 40 years, computer programming has advanced through several stages or generations of technology, generally identified by the types of languages used in the programming task. The first computers were programmed in machine language, sometimes referred to as the first generation. Machine language used the language of the hardware, namely, the binary number system. This was followed closely by the second generation, assembly language. Assembly language replaced operation codes and addresses with easier-to-understand, natural, language-like terms or mnemonics. In each generation, one line of code resulted in one machine instruction.

Third-generation languages provided a major step forward in computer programming. Third-generation languages are more English-like; they generate several machine instructions from each language statement. Programs such as COBOL, FORTRAN, BASIC, and PL/1 are examples of third-generation languages. Running on computers today are a hundred or more billions of lines of code created in these languages, representing a $1 trillion investment. Every day, more code is being written with these popular programming tools.

Programmer productivity beyond that attainable with third-generation tools is desperately needed. In addition, the profession needs tools that are easier to learn and to use. These features will permit more people to create programs more effectively. Fourth-generation languages (4GLs), used with development tools supporting documentation, library functions, and other needs of the developers, promise to satisfy these needs in part. Indeed, James Martin defines fourth-generation languages by their ability to improve productivity. "A language should not be called fourth-generation unless its users obtain results in one-tenth of the time with COBOL, or less," states Martin.[5]

But firms are trapped by the past. They must maintain millions of lines of code that run valuable operational systems. Often, this code is written in earlier-generation languages by programmers no longer working for the firm. The task of converting these applications to improved languages and to different architectures such as client/server or other forms of distributed processing is formidable. Tools to assist in this difficult task are emerging.

There are many languages that qualify as fourth-generation languages. They can be classified by the categories shown in Table 10.1.

There are many products and languages that fall into the categories shown in Table 10.1. Some examples of fourth-generation languages are database query languages, such as SQL or QUERY-BY-EXAMPLE, information-retrieval and analysis languages, such as STAIRS or SAS, report generators like NOMAD or RPG, and application generators like MAPPER,

FOCUS, and ADF. There are many more language tools (perhaps a hundred or so) that are considered to have fourth-generation characteristics. Most of these languages or programming tools do not have the general capabilities offered by the third-generation tools we are accustomed to using. They are powerful, but more specialized. They are easier to use, but less flexible. Usually several 4GLs are required to satisfy the needs of most programming departments. In any case, if trained programmers use 4GLs properly, they improve productivity and quality. Table 10.2 summarizes the characteristics of fourth-generation languages.

TABLE 10.1 Types of Fourth-Generation Languages

Database query and update
Report generators
Screen and graphics design
Application generators
Application languages
General-purpose languages

TABLE 10.2 Characteristics of Fourth-Generation Languages

Advantages	Disadvantages
Can be learned easily	Low performance in large systems
Reduce programming time	Possible slow response
Improve productivity	Inefficient use of computer memory
Improve program quality	Restricted capabilities
Reduce maintenance effort	
Problem-oriented	

While operating, fourth-generation languages usually consume more machine resources than their predecessors. This means that their use must be reviewed and matched to the hardware and application tasks for which they are most suited. Very large programs supporting many simultaneous, on-line users are probably best written in conventional languages. Programmers use fourth-generation languages to optimize their programming capability. They are not designed to optimize computer performance. Therefore, transaction processing systems associated with very large databases may perform poorly if programmed in fourth-generation languages.

Fourth-generation languages reduce program development time at the expense of computer resources. Generally, this is a favorable tradeoff since the trend of programmer costs is upward, and the cost of increased computing power is declining. Some tools, called "cross-compilers," translate programs written in fourth-generation languages into lower level languages. The purpose of these tools is to combine the advantages of both language systems. This improves the performance of the development *and* the operational processes.

Programmer productivity improves considerably with 4GLs because they are easy to use and more powerful than their predecessors for many applications. Today, most applications in most firms can be developed using fourth-generation languages. Some firms have converted totally to one or more fourth-generation languages. They have improved productivity and program quality. Reliable reports of programmer-productivity gains of more than 20-to-1, and significant reductions in expenses are attributed to the use of fourth-generation languages. For example, Kawasaki stopped using COBOL entirely and is now writing all new functions and programs, including all new on-line applications, in Pro-IV. Arco Coal has eliminated COBOL programs from its portfolio and replaced them with programs written in Focus. These changes have been accomplished while reducing the programming staff.[6]

Many seasoned programmers resist these new languages because the new technology makes their skills, developed over time, obsolete. Using 4GLs requires retraining. Some programmers regard fourth-generation languages as technologically unsophisticated, suitable only for end-users or beginners. Managers frequently feel more comfortable with third-generation languages because they fear the introduction of another language. They are unsure whether or not the compatibility issues can be resolved, and they dread the transition period with its associated expense and frustration. Some programming departments and their managers remain firmly anchored in the past, seeking comfort in the lure of the familiar.

In some instances, fourth-generation languages are used as direct replacements for earlier languages. Often, tools to support the development process are considered a separate technology. In the future, however, the distinction between languages and tools will disappear. The developers will perform their tasks at programmer workbenches unable to distinguish one support mechanism from another. Truly, this will be the era of computer automated software engineering.

CASE Methodology

New and improved programming languages will be helpful only if the entire support environment, including the management system and people management practices, are tuned for success. Among other things, tools that take some of the drudgery and manual record keeping out of the programming task promise to improve morale and productivity if properly implemented, and if programmers are fully trained in using them. In a supportive environment, Computer-Aided Software Engineering (CASE) tools (individual workstations with extensive software supporting the programmer) are a valuable asset to development departments.

The application of fourth-generation languages through additional programmer-support tools adds further potential for enhancing productivity. These automated functions represent the results of trying to use computer technology for the benefit for computer professionals, programmers, analysts, programmer technicians, programmer librarians, and so on. The tools support the development process, from requirements definition through maintenance. Support functions include diagramming and modeling, code generation, test case development, and many forms of documentation. CASE tools are characterized by on-line, interconnected workstations containing functions and features to assist developers, clients, and managers produce quality applications.

Several hundred firms market CASE tools including CASE manufacturers, hardware and database vendors, consultants, and educators. Not all of these vendors provide full CASE function; many are biased by the type of business they serve. For example, firms supplying database software tend to provide tools that emphasize data, while consulting firms tend to stress methodology. Buyers of CASE tools need to understand the product thoroughly before committing to a vendor. Prices for CASE programs vary widely, depending on the function, ranging from $8,000 to $100,000 or more.

Sophisticated workbench technology uses a network of workstation devices. Networking assists in the complex task of communication between the members of the development team. These networks support timely and accurate information exchange. To be totally successful, CASE technology must support the entire project team, including the managers. CASE technology uses graphics extensively. It provides a controlled process for managing the versions and releases of the developing product, including the documentation. CASE provides an automated system for managing the test case library and for assisting with program validation. In some instances, the tools contain code-generation or code-rebuilding capability so that old programs can be rebuilt or refurbished rapidly and productively.

Some tools are designed to rebuild current programs. File descriptions, database configurations, and source code are placed in the systems design database where they are enhanced and updated, or migrated to new languages or data management systems. The process includes both reverse- and forward-engineering.

There are several types of CASE tools. Some support requirements-planning, analysis, and design. They create a model of requirements, check for consistency, and produce documentation of the results. Rapid production of design graphics and system documentation is a feature of these tools. These upper CASE, or front-end tools, assist in the early phases of the systems development life cycle. Lower CASE, or back-end tools, provide code-generation capability, test-case development support, and assistance in database development. They help produce documentation of the programmed application. Most tools support standards for specifications and implementation considerations. Some development tools assist in reusing parts of designs, coded modules, or test cases.

General-purpose products, or products that support the complete systems development life cycle are called Integrated, or I-CASE, tools. I-CASE tools are used in conjunction with a development methodology, usually the systems development life cycle. However, they may be employed with other methodologies. CASE systems support many of the common languages. They operate on individual workstations or receive support from workstations connected to mainframes or servers.

Modern CASE tools assist in the task of project management by collecting development statistics, displaying status reports, and communicating among and between members of the team and the project managers. These tools provide support to life cycle administration by developing and communicating project metrics and other project management information. Programming managers and program developers must be thoroughly familiar with the tools and firmly committed to their use in the organization. Half-hearted or partial commitment to CASE tools leads to confusion, frustration, and ineffective program development.

The U.S. Department of Defense recognizes the importance of CASE tools. In the DoD definition and analysis of software requirements, DoD standards direct developers to use systematic, well-documented methods. Many CASE tools help satisfy this requirement.

CASE is an important technology for several reasons. Users believe that CASE improves the quality of design and assists greatly in developing system documentation. CASE improves communication among developers. It has the ability to improve communication between developers and clients. Although some developers resist new tools or methodologies, improved tools will reduce this resistance and encourage programmers to adopt proven new methodologies. Management commitment to CASE involves significant expenditures for the tools themselves, and requires investments in programmer education and training.

Management control is vital in application program development. A controlled environment is a prerequisite to improved quality. High-quality development is also productive development. CASE tools used with proven project management systems, and with improved programming methodologies such as the object paradigm, have the potential to improve program quality and maintainability. The phase-review approach, valid business case analysis, and risk analysis focus on the management process. Metrics developed from these management processes and augmented by development metrics from the CASE tools permit the project to remain on schedule and within budget throughout its life cycle.

The growth and development of automated tools supporting highly productive languages will accelerate significantly in the future. The same is true for the deployment of these tools and techniques in application development. Businesses are demanding significant improvements in programming quality and productivity and are investing in CASE tools and fourth-generation languages to help attain these goals. At Con Edison, for example, programmer productivity improved from 65 to 400 lines of code per day in two years. CASE also increased quality, thus reducing future maintenance costs. Con Edison uses an application generator to turn design specifications into COBOL code. BDM International, a large systems integrator, reduced costs nearly $5 million on a fixed price contract for the Air Force. Error rates have declined by 75 percent. Souvran Financial Corp. saved $1.2 million on its first four projects using a CASE tool and has adopted the tool as a company standard. Souvran anticipates maintenance savings since programs developed with CASE contain fewer lines of code.

Improvements such as these can result from automating the programming task in conjunction with better management, improved training, and superior methodologies. The introduction of new languages, advanced tools, and new techniques creates some technical problems. However, it is mostly a people problem. Application development departments have little choice but to embrace new technology and new approaches. Astute IT managers will use their influence and skills to encourage early adoption and facilitate use of promising new developments.

The Object Paradigm

One new development, object-oriented programming, is gaining popularity in the academic world and in business and industry. The origins of the technology were in Norway in the 1960s, where scientists developed a language called Simula. A research team at Xerox continued this effort and developed an object-oriented language called Smalltalk. In 1981, the C++ language was developed at Bell Labs. C++ is widely used today, particularly in academic institutions. Object technology consists of object-oriented design, development, and databases. Object-oriented programming and object-oriented knowledge representation are also part of the technology.

Object programming approaches a problem from a different level of abstraction than does conventional programming. Conventional languages separate code and data, while object languages bring the two together in a self-contained entity called an *object*. Associated with each object is code appropriate for its use. These codes are called *methods*. Objects that share common methods and attributes are called *classes*. For example, a file may be defined as an object; and the methods appropriate to the object might be copy, display, edit, and delete. The object, and the methods common to it, form a single entity. For instance, another file may have the same methods as the previous file, and the two files would belong to the same class. There can be many members of any class.

One of the most powerful concepts of object technology is that of inheritance. New object classes can be defined as descendants of previously defined classes. The new classes inherit the methods of their ancestors, but these methods can be altered by adding new methods or redefining previous methods. For example, one could define a new object in the file class called *output*. Output would inherit methods of copy, display, edit, and delete. For use in an application system, programmers may remove the method called copy and the method called edit. The object programmer may add the method called print. Methods appropriate to output would then be display, print, and delete. Generally, object-oriented programming offers numerous possibilities.

Object technology is especially useful in forming user interfaces, screen applications such as menus, displays, and windows, and text, video, and voice databases. It isolates the effects of change and permits modular expansion of features. It greatly facilitates program reuse, thus providing a basis for increasing programmer productivity. Object technology provides a way of dealing with complexity through abstraction. It is useful in many situations, including operating systems, programming languages, databases, and user applications.

Already, the object paradigm is widely used in application development. Many developers are using C++, a version of C with object-oriented capabilities, available on a wide variety of platforms. Object-oriented COBOL is becoming popular, too.

There are many new developments in this rapidly emerging technology as computer manufacturers, application developers, government agencies, and others work to develop standards for object technology application. An international organization, the Object Management Group, has been formed to develop a common applications environment using object-oriented standards. This active group's mission is to establish worldwide standards, thus enabling the rapid and organized development of object programming. Some believe that object technology may be to the 1990s what structured techniques were to the 1980s.

Forward-looking program developers are using CASE tools, fourth-generation languages, and object-oriented techniques to improve productivity and product quality. Newly created development teams, such as those in overseas contract programming shops, are depending on these tools and techniques and on global networks to maximize competitive advantage. Resident programmers should embrace these tools and techniques and take advantage of close contacts with their clients to improve the development process.

Prototyping

Many systems incur problems from the very beginning of their life cycle caused by the extreme difficulty of the specification process. Problems originating in the early stages of system design are difficult to manage and expensive to correct later on. In fact, for many applications, the hardest part of system design may be specifying the problem. Languages, tools, or more training for developers or clients will not overcome these problems. For many people and in many situations, the preferred approach is to experiment a little with the problem, with the goal of obtaining a more realistic feel for the solution domain. This process is called *prototyping*.

In many cases, it is unrealistic for the development team, the analysts, and their clients to agree on the precise, final specifications early in the project's life cycle. On one hand, developers want to freeze the specifications because they know that changing specifications is a leading cause of schedule slippage and cost overruns. On the other hand, clients want to reserve some flexibility because they are unsure of their ultimate requirements, and they don't want to be locked into final specification prematurely. A compromise to these conflicting desires is needed.

This conflict arises naturally and has severe consequences if not resolved to everyone's satisfaction. According to a recent survey, 80 percent of development projects violate schedule and budget plans because requirements change after they are "frozen." Cost overruns or schedule slippage of 10 to 50 percent occur in 68 percent of these cases; more than 50 percent of all overruns occur in 9 percent of these cases. Poor initial requirements, unfamiliar applications, and prolonged project development are reasons for changing requirements.[7] The goal of avoiding creeping requirements is in everyone's best interest.

The ability to experiment with the design and to prototype the solution prior to final commitment is one promising compromise solution to the conflicting demands of developers and their clients. The notion of prototyping means that developers and clients must work together closely while avoiding communications problems (another source of schedule slippage), as incremental pieces of the system are built and evaluated. Successful prototyping requires rapid implementation of many small alterations with corresponding, rapid feedback on the merits of the changes.[8] Prototyping is best performed in a highly automated environment in which the tools support these needs. Third-generation languages operating in batch mode are not suitable for this task. Fourth-generation languages supported by CASE workbenches make prototyping feasible. Figure 10.1 illustrates the prototyping process.

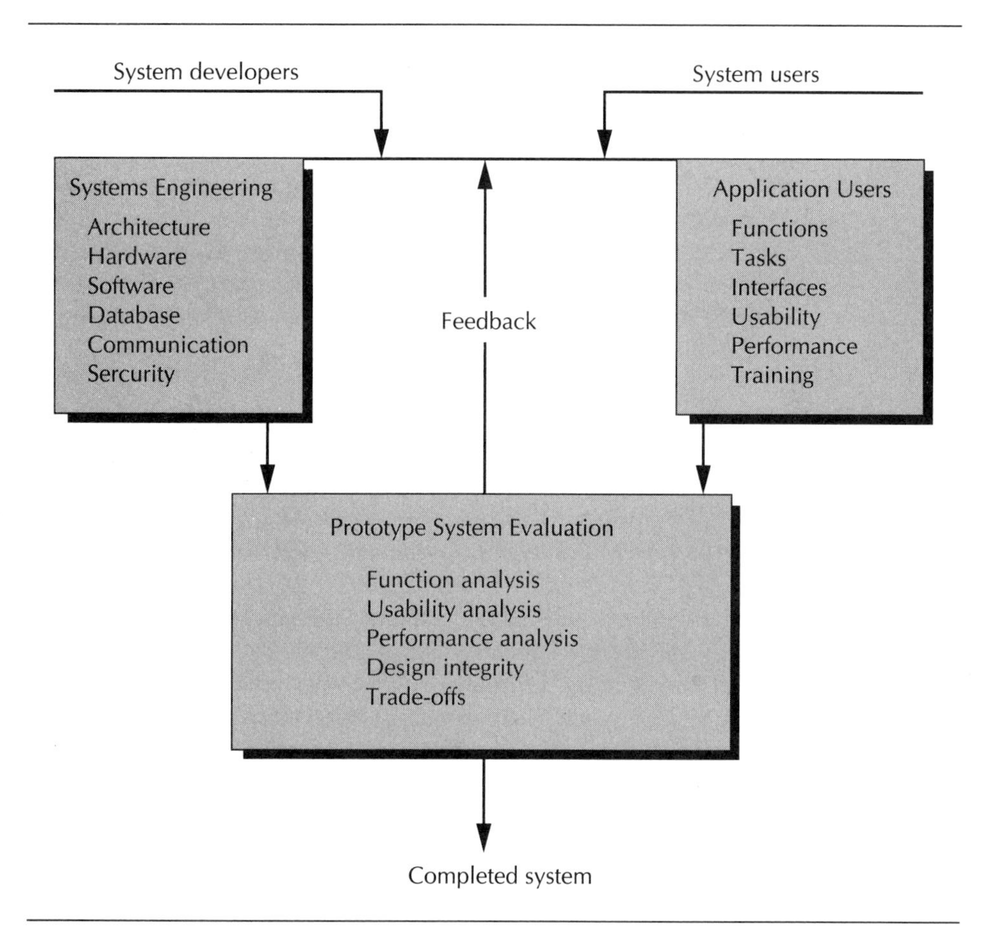

FIGURE 10.1 The Prototyping Process

During the prototyping process, system developers and system users generate ideas for the new system. An elementary model is devised for evaluation. Developers are concerned with hardware, software, architecture, and communication technology. Users are devising human/machine interfaces, function, and usability criteria. Their ideas are integrated, the model is evaluated, new ideas are generated; and the process is repeated until the problem and most solution parameters are specified. The experimentation continues until both parties are comfortable with the results.

The advantage of prototyping is that system specifications can be changed as a result of early experience with a running model. The expectation is that the final system will closely meet client needs and expectations. A disadvantage is that the prototyping effort is difficult to manage (when is the prototype finished?). Another disadvantage is that users may adopt the system prior to completion or integration with other systems or databases. It may be more difficult to obtain final documentation on a prototyped system, as well.

Prototyping activity can continue until the system is fully developed and operational. The prototype model is refined until it completely satisfies the user requirements. CASE technology generates most of the documentation and other supporting material. The system is then ready for production.

An alternative approach to prototyping employs a combination of the life cycle and prototyping methods. Prototyping is used to develop the specifications and remove some of the uncertainty during the initial stages of system development. Prototyping refines the user interfaces and solidifies the architecture. When these crucial tasks are finished, the life cycle methodology, beginning at Phase 2, is employed to complete the remaining stages. This is particularly attractive for large programs with complex human interfaces. Prototyping is very useful for developing critical specifications; the life cycle methodology disciplines the final development process. The advantages of each approach are combined to form a superior process.

One large computer printer is manufactured using a combination of prototyping and life cycle methods. Much of the user interface on these devices consists of back-lighted panels for operator messages and touch panels for operator response. The printer contains five microprocessors to manage the interface and control printer operation. All the interfaces were prototyped in software and hardware, and the specifications were developed and approved. Following this, both the hardware and the programming continued with formal development processes. The product was introduced on an accelerated schedule, at minimum development cost, and with high confidence that the human factors would be satisfactory.

Several other well-known techniques are designed to involve system users from the start and to provide early operational models. Rapid Application Development (RAD) and Joint Application Development (JAD) are techniques that require intensive developer/user interaction in the early design stages. Combined with CASE tools and prototyping, these interactions intend to resolve specification uncertainty prior to extensive coding and testing. Further, these techniques can be combined with traditional life cycle approaches to capture some of the benefits of each.

Managers have several alternatives for developing required additions to the application portfolio. The traditional life cycle can be used for applications that are easily and unambiguously specified. This is a valid and useful approach for applications that demand careful control because of risk factors in the business case or in the application itself. Prototyping, perhaps in combination with RAD or JAD, offers many advantages. It is becoming increasingly popular. When prototyping is appropriate, it solves several important and vexing problems. The usefulness of prototyping can be extended by combining it with others. In all cases, advanced languages and sophisticated supporting tools are mandatory.

The choice among the alternatives must account for development time and the cost of development, operation, and maintenance. Many of these considerations depend directly on programmer productivity, program quality, and user satisfaction. A careful analysis should accompany the decision to use one method in preference to, or in combination with, another. Increasingly, because of the steadily declining cost of computer hardware and the increasing cost of programming, the economics favor the prototyping methodology.

PROGRAMMING PROCESSES IMPROVEMENTS

Many organizations have difficulty meeting cost, schedule, or quality targets on programming projects because they lack basic techniques for understanding their programming processes. Classes for software engineers or application programmers teach analysis, design, and a variety of languages, but do not focus on programming processes or methods.[9] In many organizations, programmers work as individual artisans, having developed their craft (their individual methods) through experience on the job. Subjectively, "exceptional" programmers, sometimes called *eagles*, are known to be 10 times as productive as the average programmer. But, in most cases, the organization does not know why the disparity exists and cannot transfer the techniques of the high performer to the group as a whole. Although several factors are important, high performance programmers or programming groups study their processes, know their defect injection and removal rates, and utilize techniques that greatly improve their product quality.

Program quality is a serious problem. Programs delivered to customers typically contain three or more defects per thousand lines of code (KLOC). The standard for a high-quality product is three defects per million lines of code.[10] Poor quality causes product delays, raises development and service costs, and lowers customer satisfaction. Starting with a defect insertion rate of more than 100 per KLOC for the average programmer, defect removal is expensive in terms of labor and schedule costs. Techniques to reduce defect injection rates and to find and correct defects early in the development cycle when the removal costs are low, have a high payoff in schedule, cost, and quality of the end-product.

Formal concepts for process assessment and improvement have been developed by consulting firms, private research groups and, most notably, by Watts Humphrey at the Software Engineering Institute at Carnegie Mellon University.[11] Humphrey's maturity model describes five levels of programming process attainment from the Initial, or ad hoc level, to the Optimized level.

At the Initial level, little formalization exists as individuals ply their craftsmanship. (Most application development departments are at or near this level.) At level 2, the Repeatable level, statistical measures exist and statistical control is evident. At the third, or Defined level, quality and cost parameters are established and process databases are in place to gather and contain the statistics. At this level, the basis for sustained quality improvement is developed. At the Managed level, level 4, process databases are analyzed, programming processes are scrutinized, and process modifications are made. At the Optimized level, the fifth and final level, process measures are improved based on prior experience and processes are optimized further. This is a simplified description of the model; serious students of programming process improvements should refer to Humphrey's work referenced in the endnotes and other current writings.

Programming process improvement methods like Humphrey's maturity model are critically important to professional managers. First, managing application development successfully depends on having a repeatable and dependable process for code development. Making credible schedule, cost, quality, and function commitments implies a controllable design and development process. Without one, predicted schedules and costs are simply guesses—unworthy of professional managers. Second, advanced languages, CASE tools, and object technology become more effective in a disciplined environment. In an uncontrolled environment, sophisticated languages and tools produce more defects faster. This greatly increases the costs of defect removal, adds substantially to the schedule, negates the productivity improvements of the tools, and lowers the quality of the final product.

Individual and group quality improvement techniques are known: High performance organizations must adopt them. Organizations that fail to improve their development processes are candidates for application acquisition alternatives.

SUBCONTRACT DEVELOPMENT

Not all programs requiring development must be completed within the firm. Some application development can be off-loaded to a firm that specializes in this type of work. This activity is called *subcontract development* or application development *outsourcing*. For a fee, the firm may subcontract with another to develop some of the applications at their external facility. Several advantages of this form of program acquisition are listed in Table 10.3.

TABLE 10.3 Advantages of Subcontracting

Balances resources and skills
Trades money for people
Assists in managing staffing
Useful for obtaining experience

The advantage in taking this path to program acquisition is that it balances resources within the firm. Subcontract development trades money for application developers. It also reduces the firm's requirements for people, as compared with local development. It is a way to resolve skill imbalances within the firm. Subcontracting allows a firm to manage its long-term staffing more effectively because the subcontract tasks are a workload buffer for the firm's permanent staff.

Subcontracting may also be used to obtain skills on a temporary basis for specialized tasks, again reducing the need for permanent staffing. Because the subcontractor may have both skills and experience in the general problem area, the results may be superior to those of alternative methods. Taking all of these factors into account may also result in lower application acquisition costs to the firm.

Subcontracting also has some disadvantages, as shown in Table 10.4.

TABLE 10.4 Disadvantages of Subcontracting

Creates bureaucracy
Requires management attention
Requires rigorous requirements analysis
May expose confidential information

Subcontracting is bureaucratic because it involves contracts, reviews, reports, and more formal supervision than local development. Therefore, it is much more management intensive than in-house development. The task of analysis and requirements definition for subcontractors must be especially thorough and well documented. (The contract terms and conditions are based on this documentation.) After work commences, alterations to the work scope or content usually result in financial penalties and schedule changes. Unless the contract makes provisions for a prototyping approach, what is contracted for is what is delivered.

The act of subcontracting also results in diffusion or dissemination of information. If the system development involves confidential information, nondisclosure agreements may be required. The risk of exposure or breach of confidentially may be greater with a subcontractor than with the firm's in-house employees.

Conflicting objectives must be understood and worked out by both contracting parties. For example, the client firm wants the job completed on schedule, to specifications, and within budget. The contractor wants to maximize profits. A cost-plus contract could encourage the contractor to increase the job size or scope and lengthen the schedule. A fixed-fee contract may incline the contractor to perform as little work as possible to meet the contract terms. The contract terms and conditions must explicitly recognize this conflict. Also, the firm's management system must operate to enforce the contract in fairness to both parties.

An analysis of risk along the lines discussed in Chapter 9 is mandatory prior to entering into a development contract. Periodic risk analyses must continue during the life of the contract as well. Subcontract development places considerable burdens on the firm's managers and on the IT management system.

PURCHASED APPLICATIONS

Increasingly, businesses are turning to external sources for some of their application programs. The trend toward purchasing applications has been underway for several decades, but it has accelerated considerably since the introduction of personal computers. Indeed, the wide availability of highly useful applications has been the major driving force behind the explosive growth of personal computing. Although purchased applications are widely associated with PCs, they are becoming more common in the minicomputer arena and in mainframe installations, too. For instance, the IBM AS/400 has a portfolio of several thousand applications. Most of these applications have been developed in collaboration with IBM customers. They can be purchased with the hardware.

Software is big business. IBM is the largest supplier of software, but thousands of firms supply application programs as the mainstay of their business. Microsoft Corporation, Computer Associates International, and Oracle Corporation are the largest suppliers in the industry that, as a whole, generates nearly $20 billion in revenue and employes more than 100,000 people. These firms and others provide a smorgasbord of applications for organizations worldwide. They form the backbone of the professional software development industry.[12]

Advantages of Purchased Applications

As discussed earlier, in most installations there are many applications that must be developed and maintained by the resident programming staff. Some applications, however, can be performed very satisfactorily by purchased software. The decision to build or buy is many-faceted. Some portions of the decision rest heavily on the advantages and disadvantages of purchasing application programs. The advantages of purchased applications are numerous and are outlined in Table 10.5.

TABLE 10.5 Advantages of Purchased Software

Early availability
Well-known function
Known and verifiable quality
Inspectable documentation
Lower total cost
Availability of maintenance
Periodic updates
Education and training

Usually, in contrast with the extended in-house development cycles, purchased applications can be installed and utilized relatively promptly. Since the development time in the application life cycle may be as much as 50 percent of the total time, this time savings can be considerable. This usually translates into substantial financial benefits. The benefits from the application also occur earlier and improve the cost/benefits analysis. Thus, in many instances, purchased applications are more financially attractive.

Usually, purchased applications contain functions that are well defined or that can be determined easily. This functional certainty removes some of the risk, as compared with locally developed programs. Purchasers can be relatively certain of the functional capability of the program before investing their money. The functional capability of the purchased application can be objectively determined from the vendor's documentation, through test runs of the program, or subjectively from other purchasers. It also removes some of the biases associated with functional questions surrounding in-house development.

Through discussions with application users, industry experience, and pre-purchase application tests, the purchaser can obtain a good understanding of the quality of the code and the documentation. Again, this reduces risk. Documentation forms a major portion of the application product and is a source of concern with locally developed programs. This concern is twofold: Good programmers are not necessarily good writers and their documentation may be of low quality. Further, the documentation may not be available at the time when it is needed because if anything slips schedule, it's likely to be the documentation. With purchased applications, having both the code and the documentation available for inspection eliminates these risks.

Not only does early availability of the application improve the business case, but purchased applications usually have a lower total cost. This results from the vendor's ability to spread application development costs over multiple customer purchases. In contrast with those of hardware, per unit software manufacturing costs are very low. The primary manufacturing cost is that of documentation reproduction, packaging, and distribution. Therefore, there are large economies of scale in purchased applications.[13]

Many widely available applications have a continuing stream of enhancements available to users at reduced prices. These updates, built on previous versions, add functional improvements to the application. They can support new hardware features or functions, or provide integration with applications from the same or other vendors. These enhancements are designed to improve or broaden the product's function and to increase its value to customers. Usually these enhancements are available at relatively modest prices. Perhaps not all of the added function is useful to each customer, but usable enhancements typically offer a favorable cost/benefit ratio.

Lastly, purchased applications simplify user training. Many popular applications are supported by outstanding training materials: training manuals, reference manuals, user aids, help lines, user groups and, in some cases, vendor or third-party classes. Because of the popularity of some applications, most training is modestly priced. And, in contrast to most locally developed training material, vendor-supplied training aids that support popular applications tend to be of high quality.

Disadvantages of Purchased Applications

Given that purchased applications offer an impressive array of advantages, it is easy to understand why they have become so popular and the market has grown so rapidly. However, there are some important disadvantages to consider. Managers need to analyze and understand both the advantages and disadvantages so that intelligent decisions can be reached. Table 10.6 lists some disadvantages of purchased applications.

The principal disadvantage of purchased programs relates to functionality. In some instances, applications cannot be purchased to perform the functions that the firm requires. Examples of this can be found in the area of strategic systems. Because of the nature of the business and the need for confidentiality, or due to unique and proprietary information or processes, purchased applications may not offer a viable alternative to local development. Most strategic systems must be developed in house using the resident programming staff.

TABLE 10.6 Disadvantages of Purchased Applications

They may have functional deficiencies.
Program and database interactions may cause difficulties.
They are difficult to customize for the firm.
Management style elements may not be supported.

Some groups of application programs are so interrelated and so highly dependent on common databases that integrating a purchased application may be practically impossible. For instance, the array of financial application programs may depend so heavily on locally defined databases that the integration of a purchased general ledger program is difficult or impossible. The cost of integration will exceed the development savings in some cases. It may be infeasible to replace the entire set of financial and accounting applications for the same reasons. As the degree of interaction among and between application program sets increases, the insertion of a commercially available application becomes increasingly difficult.

Modifications to a purchased program may or may not be easily performed, depending on several factors. These include the availability of the source code, the nature of the source language itself, and the availability of program logic manuals or other forms of program documentation. There are other detracting factors such as the need for test-case development. If the source code is not available or if the programming documentation is incomplete or unavailable, program modification may be impossible or impractical. In any event, the firm must conduct a comprehensive cost/benefit analysis to determine the financial reasonableness of proposed modifications.

In some instances, it may be necessary or appropriate to surround the application with customized programs that bridge to the present application set. This may be preferable to modifying or customizing code within the purchased application. In addition, the need to preserve compatibility with future software releases is important. Any modifications to the application may need to be replicated in the next release. The vendor usually feels no responsibility to migrate your custom code to his next release. The financial considerations involved in this contingency must be factored into the business case for the proposed application.

An additional disadvantage can be that the function of the application may be deficient, incomplete, or implemented in a manner foreign to the firm's business operation. Application programs frequently implement the management system employed within the firm. Sometimes they include functions derived from management style considerations. To a large degree, customized application programs represent "how we do things around here." It is not likely that a popular, widely distributed application will include these nuances. A purchased application may require modification to either the application itself or to the management system. Resources required to perform these modifications detract from the business case for the application.

In most firms, the costs of modifying the management system to accommodate an application program are never evaluated. Many functional managers believe that programs are easily changed and that management systems are not. Some managers believe that programmers are employed to support management or the administration, and remain completely unwilling to consider the pros and cons of the issue. This type of myopic thinking has lead many organizations to spend large sums of money developing customized systems, the primary functions of which are common in the industry.[14] For many reasons, the issue of functionality looms large when considering purchased applications.

The quality, documentation, support, updates, and maintenance of purchased software is varied. Some purchased applications contain bugs or glitches that can be embarrassing or expensive for their purchasers.[15] And some purchased applications come with no training or educational support. There is no substitute for a thorough investigation of all aspects of the product prior to purchasing it. The reputation of the vendor is important, too. A reputable brand name may be well worth the additional cost, if any, included in the price. Good applications tend to be long-lived, and the association between your firm and your vendor will endure for years if it is mutually beneficial.

The many issues related to purchasing applications for the firm's portfolio are varied and complex. This is especially true if the application replaces or supplements a current application, and uses the firm's traditional databases as sources and sinks of information. The issues are somewhat simpler if the proposed application is a "stand-alone" product or if the application forms a new area of automation for the organization. Given all these considerations, purchased applications are becoming popular. This is because they tend to be financially attractive while offering a viable way to reduce the application backlog through contained financial investment.

From a business and financial perspective in the microcomputer world, the applications and the hardware on which they run are generally considered together. The purchase of applications in support of a microcomputer strategy is usually a given condition for the firm. As micros grow in capability and penetrate business processes more deeply, purchased applications will be an extremely important factor. Worldwide, new businesses, and many businesses not currently highly automated, will implement strategies centered on purchased applications. Some firms will opt for programmerless environments, and many that do so will succeed. For those that succeed, managing program development will be a non-issue.

ADDITIONAL ALTERNATIVES

Other alternatives to in-house application development are evolving. One of the most promising is the formation of alliances. For example, Kidder, Peabody Inc. discovered its systems to be outdated and faced the prospect of spending six years and $100 million, or acquiring technology and systems from its rival First Boston. Ultimately, Kidder worked out a deal with First Boston: By sharing resources, both firms benefited. As the costs of major systems increase, many firms are willing to share technology and resources to defray these costs even with competitors when that makes sense. Each firm separately develops the proprietary portions of the system, usually small in comparison with the total, for their own use. Some computer manufacturers form partnerships with customers to market customer-developed programs along with their hardware platforms. These marketing arrangements are beneficial to manufacturers and to their customers.

Alliances are also being formed between large and small businesses to advance their strategic interests. They are useful for many endeavors and are becoming increasingly popular in computer technology. These relatively new approaches to development can be very effective under favorable circumstances. When a satisfactory alliance partner can be found, the potential exists for each partner to benefit from reduced costs, improved schedules, and increased function. Successful product development alliances offer both partners the potential for increased revenue and profit.

Successful joint development and marketing efforts have been accomplished between Northwestern National Life and Infodata Systems Inc., Avon Beauty Group and IMI Systems Inc., and Hilton Canada and Control Key Corporation, among others. Some joint ventures are developed to solve mutual problems without the intent of selling the resulting product. For example, Security Pacific and five other banks are cooperating on an imaging technology development project. If successful, the project will eliminate millions of pieces of paper and will reduce data processing costs.

In another case, Baxter Healthcare and IBM formed a joint venture to sell computer hardware, software, and services to the health care industry. Baxter, with strong connections to the health care industry, and IBM, with solid image-processing, networking, and workstation technology, hope to change the way hospitals work. Patient records, X-rays, and CAT scan reports can be rapidly routed wherever they're needed. Electronic records can be stored at reduced cost and with improved efficiency. Considerations such as these and others are driving the trend toward alliances and joint ventures.

Yet another alternative for reducing expenses and off-loading program development work is presented by firms offering IS services. These service bureaus provide systems and operating environments capable of processing routine applications for their clients. Payroll processing is an example of an application that is frequently processed at a service bureau. The firm provides payroll input information to the service bureau. The service bureau processes the payroll, mails the checks or makes direct deposits, completes the payroll register, and returns it with other reports to the client firm.

The advantage of using a service bureau is that it off-loads work and responsibility while reducing in-house computer requirements. The service bureau makes program changes required by law or regulation and keeps the payroll functionally modern. For example, it will update the payroll program with new federal withholding tax changes or keep the program in compliance with changes in state regulations. The cost of this service is spread over many clients, reducing expenses for all.[16]

Service bureaus eliminate application development and maintenance costs for some applications. They also reduce hardware capacity requirements. The use of service bureaus is appropriate for many operational applications that have little strategic advantage. The discovery of these applications and the consideration of service bureau processing frequently arises during the prioritization process discussed in Chapter 8. They are a viable alternative through which to trade money for people and computer capacity and to reduce costs too. The notion of using service bureaus is the first step in considering outsourcing, an important current trend that will be discussed in detail in Chapter 11.

MANAGING THE ALTERNATIVES

Chapter 8 examined the resource prioritization problem inherent in maintaining the application portfolio. The alternatives available for portfolio acquisition are presented in this chapter and in Chapter 9. Given the range of alternatives, what methodology can be employed in the selection process? How can managers choose best among the alternatives?

In many firms, the prioritization methodology reveals that programming talent is the most constrained resource and that money is a less constraining factor. This realization frequently arises during the final discussion on prioritization, when it becomes obvious that the programming staff cannot respond to all requested work on the portfolio.

Five steps leading to optimum selection among the alternatives are summarized in the following questions:

1. Which applications can be processed at a service bureau, saving people resources and computer resources?

2. Which applications needing development or replacement can be purchased to save programming resources and time?
3. Are there ways in which the firm can enter into agreements with others, either on a contractual or a joint development basis, to optimize the use of the firm's resources?
4. Is it possible to improve development productivity by introducing improved development tools and techniques?
5. What alternatives are there to increase the human resources applied to application development? (End-user computing is discussed in Chapter 11.)

Given this array of alternatives, and the thought processes contained in earlier parts of this text, the firm is likely to achieve a balanced approach to the prioritization problem. Most likely, those programs with low strategic but high operational value will be candidates for a service bureau. Applications of high strategic value or potential strategic value are most appropriate for in-house development. These highly valuable assets must be managed carefully. They probably involve proprietary information, and they usually reside at or near the top of the prioritized list.

For the remaining applications, the choices must be made in a manner that applies available resources to optimize the firm's goals and objectives. The strategizing and planning processes discussed earlier are critically important foundations for the application program management system. IT management must ensure that the environment supports and encourages high productivity through the use of advanced tools, techniques, and management systems.

SUMMARY

Application acquisition offers IT and user managers a variety of opportunities for optimizing the firm's resources. It provides opportunities to develop the firm's important strengths. Managers must respond to these opportunities in a disciplined manner. This response begins with a clear vision of the application portfolio's contribution to the firm's success. This vision of the application portfolio resource is developed and enhanced through careful strategic planning. Tactical and operational planning fine-tunes the allocation of resources to strengthen the portfolio.

The portfolio of applications is augmented by using traditional life cycle methodologies or by employing alternative approaches. Although all the alternatives offer opportunities for gain, they are all accompanied by risk. These risks must be thoroughly understood and mitigated in some manner. Advanced tools and system development techniques are available, but managers must have technical and people management skills to implement and use them. Advanced tools shorten development cycles, offer great productivity improvements, improve product quality, and increase user satisfaction. Achieving these goals is a high priority for IT managers. Indeed, these objectives are critical success factors for IT managers and client managers alike.

The exploration of alternative approaches to system maintenance, enhancement, and acquisition is causing firms to reassess their programming development and computer operation functions. Purchased applications for large and small systems are rapidly gaining popularity. Developing commercial applications is a major industry. Firms recognize the economic value of the alternatives, and they are more willing to consider them. Many firms are very reluctant to take on the long-term commitments associated with a large, permanent programming staff. Many others are reconsidering their strategies of IT self-sufficiency and are actively pursuing alternatives. In many cases, CIOs are leading firms in these dramatic new directions.

Review Questions

1. Consider the Chase Manhattan effort. What unique management problems does offshore development create?
2. Where are the leverage points in using fourth-generation languages in conjunction with CASE tools?
3. Under what circumstances can prototyping, used in conjunction with normal life cycle development, be highly effective?
4. What are the advantages of prototyping in connection with the use of CASE tools?
5. What are the factors accelerating the trend toward purchased applications? Do you think these factors will increase or decrease in importance in the future, and why?
6. What are the disadvantages of purchasing application programs? Are these disadvantages more or less important for well-established information technology departments?
7. What are the risks in using purchased applications? How can these risks be quantified and minimized?
8. What is the role of purchased applications in connection with the explosive growth of personal computers?
9. In what ways does quality enter into the purchase decision?

10. What are the advantages and disadvantages of subcontracting applications development?
11. How does subcontracting application development assist in balancing resources?
12. What are the advantages and disadvantages of using a service bureau for some of the firm's applications?
13. What are the pros and cons of joint application development?

Discussion Questions

1. Discuss the advantages and disadvantages of using offshore programmers. For what kinds of systems, and under what circumstances, would this approach have the least risk?
2. Discuss the implications for American programmers, their managers, and their clients of increased global competition in the programming development business.
3. Using the techniques of risk analysis discussed earlier as a starting point, identify the elements of risk inherent in subcontract development. Prioritize your list of risk elements and discuss your rationale.
4. What trends in the information processing field are favorable to service bureau firms? What trends are unfavorable?
5. Draw a flow chart of the questioning process discussed in the section on managing the alternatives.
6. Discuss the relationship of alternatives to traditional development to the section on critical success factors in Chapter 1.
7. Why is programmer productivity such an important issue today?
8. Discuss the significance of the object paradigm.
9. How does the analysis leading up to Figure 8.4, Strategic vs. Operational Value, assist in the task of managing the alternatives?

Assignments

1. Using library resources, analyze two firms in the industry that produce commercial application programs. Compare and contrast these firms on the basis of their products and services.
2. Obtain descriptive material on two fourth-generation languages. Compare and contrast their capabilities and limitations. For what class of problems is each language most suitable?
3. Read the first chapter in Yourdon's book, *The Decline and Fall of the American Programmer* and summarize its main points. Present an argument that supports or contradicts Yourdon's thesis.

ENDNOTES

[1] Edward Yourdan, *The Decline and Fall of the American Programmer* (Englewood Cliffs, NJ: Prentice-Hall, Inc., 1993).

[2] As reported in Martin LaMonica and Elizabeth Heichler, "Operation Offshore," *Computerworld*, August 8, 1994, 73.

[3] G. Pascal Zachery, "U. S. Software: Now It May Be Made in Bulgaria," *The Wall Street Journal*, February 21, 1995, B1.

[4] See Note 2 above.

[5] James Martin, *Application Development Without Programmers* (Englewood Cliffs, NJ: Prentice-Hall, Inc., 1982), 28. There is no common definition of fourth-generation languages, but there are hundreds of vendors claiming to have superior offerings.

[6] Ben Chao, personal communication.

[7] Computerworld/First Market Research Corp. survey of 160 development professionals as reported by Gary H. Anthes, "No More Creeps!" *Computerworld*, May 2, 1994, 107.

[8] Edward Yourdon, *Modern Structured Analysis* (Englewood Cliffs, NJ: Prentice-Hall, Inc., 1989), 97–100. This text presents another discussion of the prototyping life cycle.

[9] Watts Humphrey summarizes the problem as follows. "Currently, software engineers learn software development by practicing on toy problems. They develop their own processes for these toy problems. These toy processes are typically not a suitable foundation for large scale software development processes."

[10] Quality improvement processes leading to the "6 sigma" norm are equivalent to 3 defects per million lines of delivered code.

[11] Watts S. Humphrey, *Managing the Software Process* (Reading, MA: Addison-Wesley, 1989), is the definitive work on this subject. See also *A Discipline for Software Engineering* by Humphrey, (Reading, MA: Addison-Wesley, 1995).

[12] "Top 50 Independent Software Vendors," *Informationweek*, May 31, 1993, 38.

[13] Lotus 1-2-3, Version 3, cost $7 million to develop and Lotus Development Corporation spent an additional $15 million on testing and quality control. This $22 million product can be purchased for less than one-one hundredth of the development cost alone. This is an extreme case, but it illustrates how economies of scale act to the purchaser's advantage.

[14] Examples of this phenomenon are widespread. Tens of thousands of programmers are developing and maintaining unique ledger systems, payroll programs, manufacturing applications, and inventory control programs. These unique programs offer no competitive advantage, but they do maintain the culture. They are terribly expensive. In some cases, these development activities are late, over budget, and sources of embarrassment for their organizations.

[15] Joan E. Rigdon, "Buggy PC Software Is Botching Tax Returns," *Wall Street Journal*, March 3, 1995, B1. Some popular tax-preparation programs contain errors causing the IRS to assess back taxes and interest charges later according to this article.

[16] Typical costs for this service are about $0.50 per person per payroll period. Most firms cannot process payroll for anything near this cost, yet many consider payroll processing part of their culture.

11 *Managing Client/Server Implementation*

A Business Vignette

Merrill Lynch Develops Client/Server Applications

Even before Merrill Lynch's CMA program was fully developed, the firm embarked on a massive project, the Professional Information System (PRISM), to support retail brokers with a completely automated and fully networked information system. PRISM is an advanced workstation tool that moves brokers from manual order-entry to a completely automated platform.[1] It was designed to support future growth in business volume and to capitalize on advancing workstation and network technology.

The primary objective of Merrill Lynch's PRISM was to give brokers the capability of scanning client information and stock market data simultaneously using multiple windows. Through PRISM, brokers retrieve client data, stock market information, and research opinions from the company's mainframe systems in New York.

The PRISM workstations (IBM PS/2s) in each retail office are connected to a LAN linked to Merrill's backbone network. All 500 domestic branch offices and international branches in Tokyo, London, Singapore, Sydney, Bern, and Toronto are linked to its New York and New Jersey headquarter's facilities with a $200 million backbone network supporting PRISM and other applications.

More than 17,000 workstations were installed jointly by IBM and Automatic Data Processing supervised by Merrill Lynch personnel over a period of 21 months ending in 1990. But, before the installations were complete, plans for more extensive capabilities were already being made.

The latest version of PRISM allows Merrill's financial consultants to view customer securities portfolios and trading positions, scan news services such as Dow Jones, Knight-Ritter, or Reuters, and watch internally generated video programs.[2] Consultants can analyze a customer's portfolio with current prices and get an instantaneous profit-and-loss statement of both realized and unrealized gains with just three keystrokes. And, with a few more keystrokes, they can generate pie charts of the portfolio mix and compare them with the desired mix. Hours of manual calculations are completed in seconds. The capabilities are stunning. Customers are getting much better service and consultants' performance is improving, too.

Merrill is considering how to link its customers directly to account and market data so that routine service requests and transactions can be handled without Merrill's intervention.[3] For knowledgeable customers, Merrill's consultants will not be involved in routine transactions.

But Merrill looks beyond systems excellence to internal efficiency, as well. Howard Sorgen, Merrill Lynch senior VP and managing director of global systems and technology reports that Merrill has standardized software across operations and combined 11 data centers into two in order to increase IS productivity.[4] Because of large business volumes, it's impossible for the company to eliminate its mainframes. Seven million accounts and related statement processing will continue to be updated on mainframes in batch mode, for example. The system consolidation into two mainframe sites reduced infrastructure costs by $100 million, and its host-based processing workforce by 40 percent.

All programmers responsible for host-based production systems have been consolidated in one central group. Sorgen states that Merrill Lynch has achieved benefits by other centralization moves. For example, Merrill combined into one group all personnel developing applications for business units. The firm has reduced its mainframe computing costs by 45 percent and reinvested half their savings into client/server applications.

Merrill's technology budget for hardware, software, personnel, services, and telecommunications expenses was estimated at nearly $1.5 billion in 1987. The 1994 budget is $950 million, up from $900 million in 1993, following large expense reductions in the late 1980s.[5] Stock exchange volumes doubled from 1987 to 1994.

Merrill Lynch ranked 29th on *Computerworld's* Premier 100 in 1994. Merrill employs 42,650 people of whom more than 2500 are in Information Systems. About 60 percent of its software budget is spent on new development. Seventy percent of their software investment is dedicated to client/server applications. Edward L. Goldberg, Executive VP, reports to Daniel P. Tully, Merrill's chairman, president, and CEO.

INTRODUCTION

Earlier chapters discussed the challenges associated with managing the firm's application portfolio and related databases. Several promising methods for prioritizing the development backlog and for managing the portfolio were presented. The text focused on several alternative approaches to development and explored the option of purchasing commercially developed applications. The text discussed the advantages and disadvantages of these alternatives; it reviewed situations in which one or more of these options would be attractive to the firm. This chapter turns to one popular alternative having great potential for coping with some of the difficulties mentioned earlier.

The situations described in the previous three chapters are frequently found in organizations that have installed some form of distributed computing. Capitalizing on advances in telecommunications technology and workstation products, firms are introducing capability at the employee's workplace that permits and encourages end users to participate in the application development and operation process. This participation increases the effort applied to application development. Also, it tends to reduce clients' frustrations with the application development group. The potential benefits of distributed computing are encouraging firms to invest in the technology.

Firms are stepping up their funding for distributed computing and devoting a larger share of IT resources to end-user support. Studies show that top IT organizations are spending approximately 20 percent of their budgets on end users, while client organizations are spending a significant amount of their own money on information systems and services. In some firms, more than half of the firm's expenditures for information technology is spent by or for end users.[6] Resource deployment of this magnitude significantly influences organizations and their people.

Distributed computing places special demands on employees, managers, and organizations, and its implementation raises many important issues. If firms are to attain considerable benefits through distributed processing, critical organizational and political barriers must be removed. Firms must understand the role of distributed processing, the form it takes, the situations in which it is advisable, and the management of its introduction and operation.

DISTRIBUTED COMPUTING

Distributed computing is a decentralized alternative to centralized or mainframe computing in which user workstations are linked to each other and to a controlling computer through a network. In the simplest form, terminals at the client's workplace are connected to a central processor in a star pattern. This was the earliest and least distributed form of computing since most of the processing capability resided at the host computer. This is a satisfactory arrangement for many applications not requiring much desktop processing power. Travel agency terminals accessing an airline reservation system are an example of this type of operation.

Two other forms of distributed processing are more popular now and are growing in popularity. These are cooperative processing, or peer-to-peer network processing, and client/server operations. In peer-to-peer and client/server processing, individual workstations are connected to a server (a CPU) that controls some operations and manages data stores via a local area net. In cooperative processing, any node workstation can initiate an application while other workstations may supply data or computing power in a manner transparent to the initiator. This mode of operation requires sophisticated network operating systems and is sometimes called *network computing*.

In client/server operations, the application is divided into two parts; one part resides on the server and the other part on the client workstation. Clients initiate transactions and rely on the server for some processing and for data management services. The server may seek information from databases located on other servers or from processors at higher levels in the architecture in order to complete the client's tasks. This is the most popular form of distributed computing. *End-user computing* is a more general term referring to stand alone or networked PCs programmed and operated by end users. Stand-alone operations are giving way to client/server operations as the PCs are networked to each other and to a server (sometimes a mainframe).

The networking of client/server LANs may be extended through internetworking to many other larger networks. For example, if the server is connected to the firm's central computer, that is in turn connected to the Internet or to the phone system. Through internetworking, enterprise and even global E-mail is possible. Figure 11.1 shows a hierarchical, client/server implementation. In the future, client/server implementations will be ubiquitous, providing E-mail services, multimedia applications, program development support, and group or collaborative work platforms. Getting to this state of information technology infusion and diffusion requires large investments and careful planning and implementation.

WHY ADOPT DISTRIBUTED COMPUTING?

Distributed, or end-user computing, is being widely implemented in business, industry, and government organizations because it is attractive for several compelling reasons. Table 11.1 itemizes the important factors encouraging the trend toward end-user computing.

TABLE 11.1 Factors Favoring End-User Computing

Empowers employees with new tools
Ability to optimize business processes
Declining workstation costs
Advances in telecommunications
Growing base of skilled users
Lengthy development backlogs
Availability of workstation applications

The major driving factor for distributed computing is the desire of organizations to optimize their operations by empowering their employees with tools, information, and responsibility. This blending of technologies creates entirely new types of business solutions built on communication, information generation and sharing, and new forms of collaboration. Decentralization of information and of the capability of employees to interact with it productively are augmenting and sometimes replacing older processes and centralization.

Very significant reductions in personal workstation hardware costs in conjunction with the growing availability of application programs for these workstations have encouraged the trend toward distributed or end-user computing. As the Business Vignette illustrated, Merrill Lynch capitalized on personal workstation hardware and highly functional networks to bring powerful new applications to the desktop. Merrill retains central processing for customer account and statement processing, but the trend toward decentralization is clear. Hardware and software that are inexpensive, reliable, and available at the individual workplace provide the tools to make Merrill's client/server architecture feasible. Economic considerations permit and foster widespread adoption of personal computing.

Advances in telecommunications, such as local area networks and sophisticated communications support to centralized systems and servers, have fostered program and data sharing. Workstation users have benefited from these advances. Networking is increasing the utility of workstations and is bringing the power of servers and mainframes to the employee's workplace. All the firm's employees working together through networked individual workstations is parallel processing at the firm level. It is rapidly becoming the norm.

Other driving factors for end-user computing are the growing backlog of work facing application development groups and the increasing costs of programmers and program development. Abundant evidence demonstrates that users of IT services can participate actively and productively in application maintenance, enhancement, and development. And user participation is advantageous to everyone. New development tools and techniques provide important advantages enabling technology users to be successful program developers.

The number of skilled individual users of personal computing applications is large and growing. Skilled users can be found within the ranks of professional, technical, and office workers and within the managerial population. The growth of this skilled population within the firm is accelerating rapidly as information technology penetrates the firm's operations. Numerous precedents for end-user computing have been established by firms that have implemented the technology. Collectively, these factors provide additional incentive to those who may still be contemplating the technology.

The adoption of distributed computing has very significant consequences for the firm, for the organizations within the firm, and for many of the firm's employees and managers. This technology introduction is a microcosm of the larger phenomenon of electronic data processing itself. Stages of growth, for example, have been identified as an important concept to distributed computing.[7] Distributed or end-user computing is worth studying, not only because it is highly relevant to today's world, but because it causes us to reflect upon the application of many principles that have been discussed thus far.

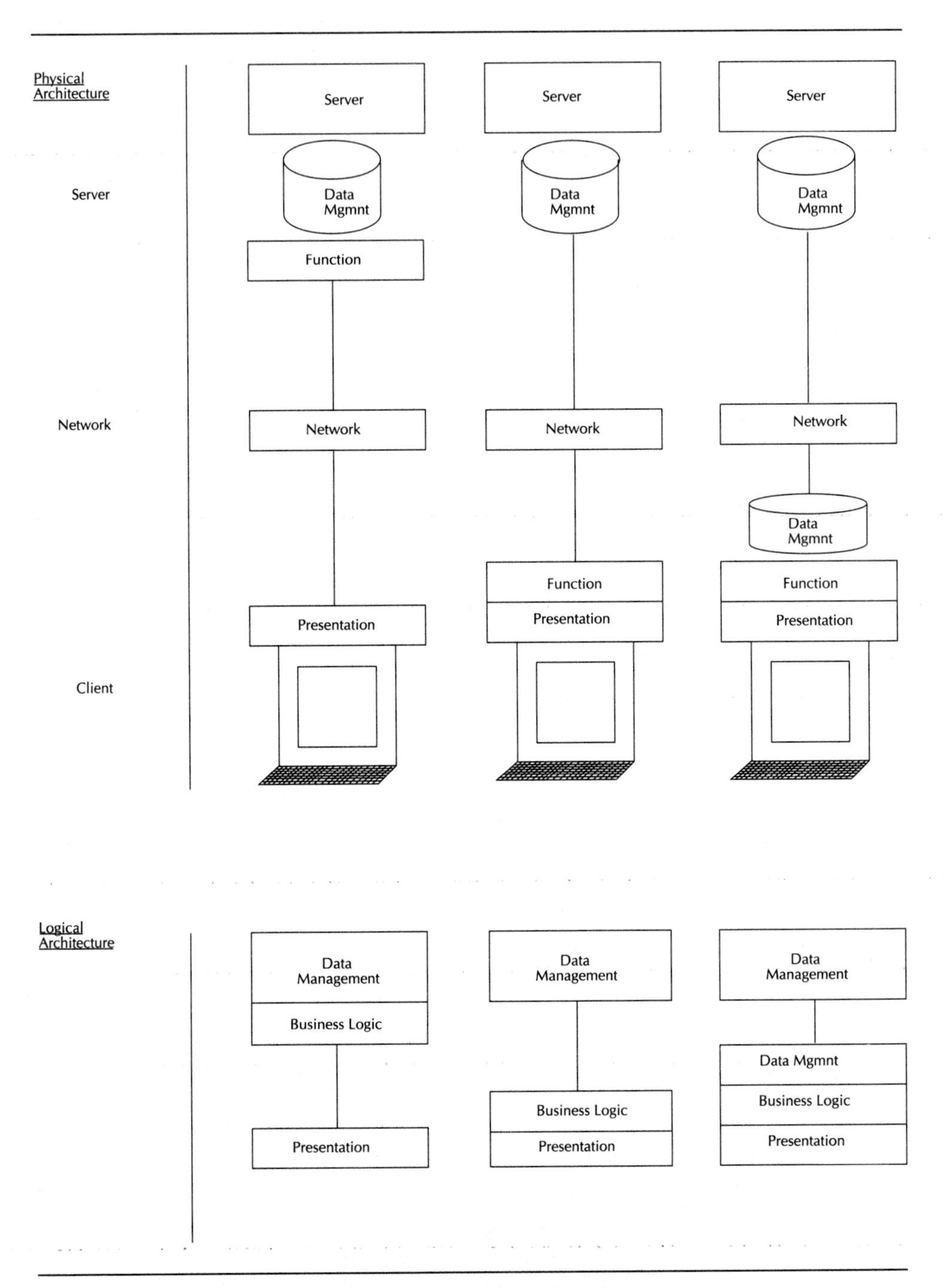

FIGURE 11.1 Client/Server Physical and Logical Architecture

REENGINEERING BUSINESS PROCESSES

Among all possible topics of interest to IT managers, client/server issues seem to top the list. Established publications are devoting considerable space to this topic and some new publications have developed around it. But is it a high-tech fad, or does it offer real advantages for organizations today? Do businesses benefit from client/server computing or does the hype exceed reality? Is it the right approach for most organizations and, if so, when is the correct time to implement it? These and other questions will be addressed in the remainder of this chapter.

In order to understand whether to adopt distributed computing and client/server architecture in particular, the firm must ask and answer some fundamental questions about its business and its structure. Are the business practices and business operations of the firm optimally tuned to the firm's business environment and, if not, how can they be changed and improved? In order to answer these questions, operations people develop process models of the current business system. These models analyze work and information flow from inbound processes, through operations, to outbound processes and sales and services activities. The analysts review the firm's value chain and all supporting activities as they strive to improve effectiveness by restructuring and applying information technology. Although this activity is not new, the most popular name for it is *reengineering*. Business process improvement or business process innovation more accurately reflect the nature of this activity.

"Information and information technology are powerful tools for enabling and implementing process innovation. Although it is theoretically possible to bring about widespread process innovation without the use of computers or communications, we know of no such examples."[8] In this statement, Tom Davenport minces no words in describing the relationship between information technology and improved business performance. Client/server implementations are the most popular means of linking technology and process improvements in the U.S. today.

In firms today, most work is performed sequentially, not because it needs to be, although it may in some cases, but because the means for parallel operations are not available, or are thought to be unavailable. Information technology, particularly distributed processing, allows individuals to perform parallel processing at the firm level. Powerful workstations, robust networks, and large, well-structured databases permit many employees to attack problems at the same time, in parallel. And, with current networking technologies, one firm's processes can be linked in parallel to those of supplier or customer firms. The tremendous potential of these concepts has driven much of the reengineering effort to date.

In performing reengineering, more accurately called *business process innovation*, the correct sequence of events is to outline current processes and workflow, search for superior process models, then find resources, including information technology, to implement the innovative new processes. Deoptimized business processes create the opportunity; information technology is usually part of the solution. By themselves, client/server architectures are a solution looking for a problem. Thus, ill considered client/server implementations may not constitute problem solutions.

For example, about $1 trillion worth of proprietary systems developed over decades are running on mainframe computers worldwide. These legacy systems have been expensed, the computers running them are fully depreciated, yet some firms are spending large sums of money rushing to convert them to distributed applications. In some cases, the rationale is that PC MIPS cost less than mainframe MIPS. But what's important is not the cost of MIPS, but the cost effectiveness of the organizations using the applications to run the business. This is an issue sometimes overlooked in the rush to embrace new technology. Legacy systems converted to client/server applications may be more effective but, in most cases, only if business process improvements are accomplished, too.

Outsourcing

In analyzing current business processes, one goal should be to identify things the firm should not be doing at all, and then terminate them. Another goal should be to find essential processes that can be accomplished more efficiently by someone outside the firm. Essential, routine processes outside the firm's core business should be considered for consignment to a firm specializing in this type of work. This activity is called *outsourcing*. For sound business reasons, outsourcing non-core business processes, including IT processes, is common practice today.

For example, the General Services Administration is assessing the benefits of outsourcing many of its IT facilities. These activities include data networking and local telecommunications facilities, computing centers, the Federal Information Center, and the Federal Procurement Data Center. The commissioner of GSA's IT Services said that data services were a good starting point to determine the value of privatizing these activities. The consulting firm of Arthur Andersen & Co. is doing much of the analysis for the GSA.[9] Outsourcing will be more fully discussed in Chapter 19.

When the new work processes within the firm have been identified and outlined, analysts and employees affected by them must work together to determine how the processes will be implemented. Usually, but not always, this means that technology will be used to facilitate the processes and empower employees to carry them out. In many cases, computing power will be placed at the workplace so that important information is available to the employee, and data on the process can be collected. Frequently, applications running on centralized facilities are restructured, enhanced, and rebuilt to operate in the client/server environment. This process is usually called *downsizing mainframe applications.*

DOWNSIZING

In many organizations today, executives are looking for ways to reduce costs, increase responsiveness, and improve flexibility. Information technology enables executives to achieve these objectives using decentralized operations while maintaining centralized control. Distributed data processing, linked to centralized control processors through advanced networks, is the enabling technology.

The trend toward distributed data processing is not new; it started in the 1960s with the development of minicomputers and departmental computing. More recently, the trend has accelerated as increasingly powerful microcomputers appeared in the workplace. Today's personal workstations contain eight or more megabytes of primary memory, they have several hundred megabytes of secondary storage, and they operate at 60 MIPS, or faster. More powerful devices will be available in the near future. Local area networks (LANs) connect these powerful workstations together and to high-speed printers, very large data stores, and huge central processors. The LANs are connected through gateways to other networks, making additional capabilities available to the workstations. Technology supporting distributed computing is very attractive.

Competitive pressures demand rapid responses from business executives. Many business managers prefer local control of their vital and important application systems because they relish the flexibility this control provides. In addition, they want to be responsible for their costs as they respond to changing market conditions. They prefer to manage their own information system. Accomplishing local control involves moving applications from centralized systems to individual or departmental processors. This process, called *downsizing*, may be accompanied by shifts in personnel.

The Attributes of Downsizing

Downsizing has advantages and disadvantages, and not all applications are appropriate candidates for consideration. Table 11.2 portrays the advantages and disadvantages of downsizing.

TABLE 11.2 Downsizing

Pros	Cons
Greater user control	Weakens central control
Empowers employees	Usually increases costs
Increases flexibility	Increases user skill demands
Decentralizes costs	User management distraction
Improves responsiveness	Database disintegration
Fosters purchased systems	Discourages common systems
Reduces IT workload	Increases user workload
Encourages innovation	Encourages parochialism

Downsized applications provide local control and flexibility to using organizations and improve their internal responsiveness. Electronic databases and computing power at the workstation empower employees and increase organizational flexibility. Some organizations report significantly lower costs and shorter schedules when users develop applications.[10] Costs incurred in operating departments are experienced directly, but IT costs are allocated or charged via a billing mechanism. Clients generally prefer to control their costs directly.

Downsized applications have hidden costs such as department attention and overhead, user training and operations costs, and costs derived from possible reverse economies of scale. For example, problem, change, and recovery management is much less efficient and costs more using decentralized systems. Indeed, user departments sometimes skip these essentials. Implicitly, they are trading higher risks for lower costs.

Downsizing encourages user innovation because there is little historical or cultural background to overcome. Traditions in IT may hinder innovative ideas unnecessarily. For example, IT organizations are prone to develop applications locally because that is what programmers do, but user organizations may opt for commercial applications. Due to the fact that they purchase so many other necessary items, they are inclined to purchase software, too.

Downsizing increases user workload and raises the demand for specialized skills. User managers must cope with these demands and must increase their management skill levels, too. Managing an application portfolio and running a small client/server network are non-trivial tasks. These activities require talent not usually found in operating departments. Discharging these responsibilities detracts from the main function of the department and diverts managers' attention from their primary tasks. Some managers refuse to take on these added functions and resist downsizing.

Data stores and large integrated databases are a deterrent to downsizing. Some databases are highly integrated and are difficult to separate into distributable units. Downsizing increases the data management problem and may lead to redundant data elements, increased storage costs, and, possibly, asynchronous conditions. Both application portfolio management and downsizing are critically sensitive to database conditions and architecture. Firms with highly developed information architectures suffer less from these difficulties than others. But, information architecture is an issue for most firms; many require major improvements in this area.

There are some additional disadvantages for the firm. Decentralized control of applications may lead to inefficiencies, particularly if corporate planning is poorly executed. This may mean, for example, that duplicate development occurs, or that corporate goals are deemphasized in favor of unit goals. Parochialism tends to favor unit performance at the expense of firm performance. Firms with highly disciplined strategic and planning processes are much better prepared to capitalize on downsizing opportunities. Disciplined management processes pay dividends in many areas.

What to Downsize

Not all organizations should downsize mainframe applications to client/servers, and not all centralized applications are candidates for downsizing. In addition to business reasons, such as reengineering to downsize applications, characteristics of the applications are important, too. Table 11.3 portrays applications suitable for downsizing and those that are not.

TABLE 11.3 Downsizing Applications

Candidate Applications	Unsuitable Applications
Small systems	Large systems
Single department applications	Enterprise applications
Isolated databases	Integrated databases
Stable applications	Evolving applications
Purchased applications	Customized applications
Technically simple systems	Technically advanced systems
Low-risk applications	High-risk systems applications

Small, stable, relatively simple applications with isolated databases are ideal downsizing candidates. Applications that are evolving, or growing in size and importance, are questionable candidates. Technically simple applications are preferred to technically sophisticated systems. Suitable applications are characterized by low risk; they can be managed easily by the using department. The number of people interacting with these applications is usually small, and downsizing is very likely to be successful. This type of system is an ideal candidate for downsizing.

Large, enterprise-wide applications involving many users interacting with integrated databases are generally best centralized. Evolving applications, customized to the firm's requirements, should not be disturbed. If the application relies on specialized or advanced technology, it is best left in the custody of IT professionals. High-risk systems or applications are poor downsizing candidates.

The management system developed in Chapter 8 to handle the firm's applications portfolio is very valuable for establishing the rationale for downsizing. Business goals and objectives are more important than political or emotional reasons when making downsizing decisions. The portfolio management system focuses on the firm's business objectives and on the objectives for individual functions and departments. It puts downsizing decision making at the proper level in the firm while reducing the emotionalism and parochialism that can surround these decisions.

Client/server computing distributes information technology activities, but it is also expands the scope of IT. IT must provide training in system development and operation, develop information architectures for distributed systems, and provide users with system development tools and other support. Although many applications are distributed to using department in client/server environments, frequently the program maintenance and enhancement activity remains the responsibility of IT.

Downsized applications operating within client/server environments are the ultimate expression of end-user computing. Downsized applications running on powerful workstations operated and maintained by departmental personnel that serve critical business functions represent distributed data processing at its best. Timely exchange of vital control information with corporate central systems through advanced telecommunication networks completes the architecture. Executives obtain reduced costs, improved responsiveness, and increased flexibility while retaining essential controls. For those firms prepared to capitalize on downsizing opportunities and deal with the issues it presents, the rewards can be significant.

THE ISSUES OF DISTRIBUTED COMPUTING

The issues of distributed computing can be found by considering seven potential problem areas and by introducing management practices and organizational changes fashioned to cope with these difficulties. Each problem area requires strategic, tactical, and operational reasoning and analysis for optimum resolution. Corporate culture, political considerations, and company policy also enter into the analysis of these issues.[11] Table 11.4 identifies the issues of distributed computing.

TABLE 11.4 The Issues of Distributed Computing

Software and applications
Hardware compatibility and maintenance
Telecommunications
Data and databases
Business controls
Finances
Staffing and personnel
Politics, culture, and policy

Compatibility among commonly used applications significantly aids the effective installation and application of distributed computing. Application software compatibility will simplify networking, training, and hardware installation. Installation and utilization tasks are simplified and benefits are recognized sooner if software compatibility is attained within the installation. Standards and guidelines for user-developed application programs assist in achieving program compatibility. User training programs for standard applications are mandatory. Training programs are simplified and are more effective in organizations that concern themselves with application compatibility.

Hardware implementation is also more effective if compatibility exists among the workstation devices. Hardware compatibility will simplify networking, software installation, and application portability. Plans to provide hardware maintenance for client equipment with minimum disruption to activities are essential. Also, it is reasonable to expect that hardware upgrades can be obtained conveniently and at minimum cost. Finally, a policy for workstation ownership must be established.

Communications issues include the physical network architecture, network software, and network policy questions. Important policies regarding the utilization of outside databases, linking through Electronic Data Interchange (EDI), and dial-in capability must be established. Network management and maintenance must be transparent to clients. IT must take the lead in managing these technical and policy items so that clients perceive a seamless environment. The firm's senior executives will provide input to answer or resolve policy questions or issues.

Data and databases raise many questions for end-user computing. Questions such as where the data resides, who owns the data, and how are databases controlled must be answered unambiguously.[12] The use of personal computers, and their access to large databases, greatly increases risk for the organization. Serious damage or loss can result from improper or nonexistant backup procedures. Also, data integrity issues must be resolved with well-conceived controls when uploading and downloading in networked environments. Proper resolution of these concerns is vital because client/server computing thrives on availability and security of information assets in a distributed environment.

In addition to the protection of information assets, security for application programs must also be provided. Some programs that maintain important records, or control valuable assets, require special attention. The firm's managers and the application owners must carefully consider the security risks and decide whether to distribute these programs to individual workstations. Programming processes and procedures must ensure that acceptable levels of programming quality are maintained.[13] Program documentation must be managed through enforcing appropriate standards. End users are more likely to omit documentation than professional programmers; they must be trained and counseled in disciplined documentation because their programs will need maintenance in the future. Providing network security is an IT responsibility.

As technology dispersion brings micros and minis to user departments, former IT responsibilities also migrate to user managers. Disaster recovery planning and management is one of these responsibilities. With end-user computing, department managers own and operate major information processing resources; they must take precautions against a variety of potential problems. Frequently, user managers do not understand what recovery planning is, or why it is needed. The central IT organization must provide training and assistance on this important subject and on other matters as part of its staff responsibilities.

As end-user computing develops, the issue of maintaining positive financial returns demands special attention. The conditions most likely to achieve satisfactory results include an aggressive attitude toward benefits accounting and specific attention to cost containment. In many cases, intangible benefits are prominent in the justification. Both tangible and intangible benefits should be evaluated and recorded. Excessive expenditures for hardware and software upgrades, failure to account for all the people-related expenses, and incomplete benefits recording will lead later to serious difficulties. Most studies show that the annual cost per user rises significantly when migrating from mainframe systems to client/server implementations. Thus, it is critically important to ensure value received for client/server investments.[14]

The costs of decentralizing are significant, and the business case partly depends on whether reductions in centralized computing can be achieved. In Merrill Lynch's situation, for example, 11 mainframe centers were closed, systems were consolidated, and some were distributed. Costs were shifted from centralized processing to distributed systems that closely support consultants. Investments in client support systems paid off in superior customer service and increased revenue and profit. These benefits resulted from increased organizational effectiveness.

Although all the conditions above are mandatory for success, the most important and demanding topic involves solving the people-related issues. Substantial training is required to convert non-programmers, or novice programmers, into productive and skillful end users. Skill migration takes place among the employees as they add information technology to their repertoire of professional job skills. Not all individuals will make this transition smoothly. Some employees will take the lead quickly and easily and will set the example for others.[15]

Managers must skillfully handle employee transitions. In some cases, employees experience traumatic job dislocations. This situation is so serious that one writer proposed that system changes be accompanied by an "organizational impact statement" that relates system changes with organizational changes to understand the effects.[16] In all cases, managers must be especially skillful in handling the disruptions. Management training should be stressed; it should not be confined to first line managers.

Political issues are important because end-user computing shifts the power structure in the firm. Individual managers are concerned about what the power shifts mean to them, and emotionalism frequently intrudes on rational decision making. Strategy development, long- and short-range planning, and application portfolio management are effective tools for dealing with political issues. These management systems put decision making on a business basis and help diffuse political concerns.

The issues are not all equally important at all times during implementation. In the early installation phases, user support is highly important. Good support encourages early adoption by hesitant users. It smooths the transition for early adopters. When adoption is progressing smoothly, financial considerations, data management, and business controls become increasingly important. During implementation, the organization is undergoing substantial and permanent change. At all levels, the firm's leadership must manage and shape the transition in accordance with long-term goals and objectives.

Given the magnitude of the issues and their pervasiveness, how will the management team cope? What actions can they take to deal with the potential problems? What must happen for the firm to capitalize on the potential benefits? The following sections will clarify the answers to these and other key questions.

ORGANIZATIONAL CHANGES

Structural changes within the IT organization will provide many opportunities to deal with the tasks noted above. IT must define support levels, preferably through service-level agreements, and must assign support tasks. The implementation of user support includes defining management processes and allocating resources in support services.[17] This can be performed most effectively by establishing two new entities within the firm. These organizational units must receive assistance from IT, but their staffing generally includes non-IT members, as well. These new units are the workstation store and the information center.

The Workstation Store

The workstation store usually reports within the IT organization. Its mission is to perform many vital support functions for end users.

One important support function that the store performs is purchasing workstations and software. In order to save time and money for the firm, the store purchases workstations in volume quantities and at discount prices for resale to client managers. The discounts may amount to 30 or 35 percent of the unit price. The store also purchases and distributes licensed applications with the hardware. In addition to reducing costs, this approach provides a control mechanism to ensure hardware compatibility for the firm. In some cases, it may be advantageous to secure compatible hardware from more than one vendor in order to satisfy the firm's requirements. The store can also ensure favorable prices for software by securing site licenses. Software compatibility can be achieved using this approach, as well.

The store is also responsible for obtaining and providing central maintenance for all workstations and for making substitute equipment available while maintenance is being performed. Maintenance support for clients is vital. They want to get their jobs accomplished and prefer not to be responsible for the maintenance of their equipment. The workstation should be handled in the same manner as the telephone—if it fails, someone must correct the problem promptly.

The store should manage upgrades and migration of hardware and software. The store obtains approved hardware and software upgrades and makes them available to clients with justified needs. These functions are important to clients and valuable to the firm. They must be performed efficiently and effectively.

The store also performs other important functions. For example, it gathers information from sales and inquiries regarding client requirements. IT can keep track of the pace of end-user adoption and can obtain trend information through sales analysis. This information is valuable as IT develops further plans for distributed computing. The workstation store also obtains benefits information from clients at the time of purchase. Clients are requested to submit this information when taking delivery of equipment or software. IT analyzes the benefits data, with other cost and benefits information, and uses it to provide financial justification for distributed computing on a continuing basis.

As the use of workstations progresses, there will be demands for both hardware and software improvements. The store's ability to satisfy this demand is accomplished in much the same way as the initial products were obtained and distributed. As a condition for obtaining upgrades, customers should provide an approved benefits statement that can be reconciled with the cost of the additional equipment or programs.

In addition to being a control point in the implementation of procurement policies, the store can serve very effectively as an internal control mechanism for physical assets. The store's record-keeping function is an important business control. When units are distributed, records are maintained that show the equipment location, the requesting manager's name, and the equipment configuration. This data is required for effective inventory control.

Information routinely gathered from clients while providing services is valuable to the organization for understanding trends and establishing client preferences. This information becomes part of the IT planning data and serves as a leading indicator of future demand.

The workstation store is not an optional entity. Effective implementation of distributed computing requires this unit to function efficiently and with an eye toward providing outstanding customer service.

The Information Center

Several additional activities demand careful attention to ensure that the implementation of distributed computing proceeds smoothly and reaches a successful conclusion. An organization called the information center performs these functions. This unit usually reports within the IT structure but requires close interaction with client groups. The information center is designed to facilitate distributed computing and performs many important functions, as Table 11.5 displays.

The information center provides training for employees on workstation hardware, software, and procedures. Most firms have formal training programs for end users, and most utilize a mix of internal trainers, contract trainers, and outside classes. The information center is responsible for coordinating this training activity. It also provides assistance to clients in the program development process and trains them in business controls and recovery management techniques.

TABLE 11.5 Functions of the Information Center

Conduct or provide user training
Provide development assistance
Evaluate new applications
Distribute customer information
Collect trend information
Determine problems
Gather planning information

The center evaluates new application programs and other software; it distributes information on client-developed programs. Data collection for determining trends in software requirements is an important function of the information center. The information center serves as a first-level clearing house for meeting the software requirements of distributed computing. Trend information gathered by the center is valuable to IT and to the firm for establishing future computing requirements and for shaping strategies and plans.

The center provides clients with guidance and assistance on software and hardware migration and upgrades and, if properly staffed and managed, helps prevent duplicate development activity. When the center is operating effectively, clients will seek advice and direction prior to taking significant action. The center provides coordination among the client departments as part of its staff responsibility.

The center also supports clients by maintaining a telephone help-line service and by performing initial problem determination. The center is responsible for customer support. This includes answering questions, even those that seem trivial to the experts. The center is a place to turn to when things go astray, as they occasionally will. When other sources of help are not available, the information center is the client's safety net. Adopting a helpful attitude toward clients is this group's key to success.

The information center plays a central role in distributed computing. It complements the workstation store. It solves or helps avoid problems that the firm is likely to encounter during the transition. In addition to the important task of employee training, the center must train managers on development processes and control issues. Training is a continuing activity for employees and managers. Management skills and employee proficiency with new tools are learned traits: The importance of training looms large.

Successful introduction and implementation of distributed computing requires the skillful employment of management techniques and the adoption of structural changes. These management actions are intended to reduce risk and to pave the way for capturing the benefits.[18] The firm will benefit substantially from this technology and will experience significant change in the process. Not only will the organizations within the firm be flexible and more responsive, but the individuals within the organizations will be better equipped, too, to assume more responsibility and to become more productive. Therein lies the payoff.

POLICY CONSIDERATIONS

The firm must make a number of policy decisions during the adoption and implementation of distributed computing. Some policies smooth the way for technology adoption and encourage client computing; others affect computing costs or bear on fundamental concerns such as business controls, security, or employees. These policy matters are usually important to the firm in the long term, and they warrant the attention of senior IT executives and others.

One such issue requiring attention at the policy level is hardware and software compatibility. The firm must decide whether to adopt one or more of the many popular application solutions available for employees. Adopting one application might reduce costs and make training easier, but it may reduce functionality overall. For example, should the firm adopt one word processing application, or should they permit employees to choose among more than one? Most firms limit the choice to one to ensure easy document interchange and to reduce training and cross-training. Since there are many applications and many choices within each application, the decisions on application policy are often complex. But allowing the issue to resolve itself by default is not a responsible alternative and will lead to problems later.

Selecting an application policy paves the way for a policy on hardware compatibility. Again, restricting the hardware options available to clients reduces costs, improves maintenance, and makes migration to future systems easier. Most firms limit the hardware options to one or two popular brands, but they may purchase compatible models or clones from several manufacturers. Some firms maintain a small advanced technology group for the purpose of testing new hardware and software to ensure that new technology developments are known and understood, when planning for the future.

The firm must develop policies regarding ownership and control of distributed hardware, applications, and data. Individual or local ownership of these items improves security and reduces the risk of loss or damage. Local or individual control may retard sharing and use of these assets and may result in negative consequences to the organization. Executive policy must choose among conflicting factors.

PLANNING CLIENT/SERVER IMPLEMENTATION

Planning for client/server operations is a nearly continuous process because the introduction and implementation of new tools for knowledge workers seems to be a never-ending task. Many planning questions address issues such as what technology should be introduced next, where should the introduction begin, and when and at what speed should changes occur. Since the primary purpose of downsizing is to enable improved business processes, client/server implementations must be part of the firm's strategy and planning process just an restructuring and reengineering are. The firm's routine strategic planning process serves to answer these questions.

Managers responsible for implementing client/server architectures must use several planning methodologies if they are to be successful. They should recall the basic concepts of planning, such as Nolan's stages of growth, critical success factors, and business system planning methodologies. It is likely that information technology already deeply penetrates many firms now implementing client/server architectures, so eclectic planning methods are appropriate. Sullivan states that firms in this position are in a complex environment. Essentially, they have entered the information age.[19] In general, this means that planning must be overt, systematic, and well developed. It also means that distributed system planners must be proactive and have strong strategic perspectives.

IMPLEMENTATION CONSIDERATIONS

The implementation of client/server architectures poses many challenges for IT and client organization managers. Some concerns are familiar to managers of centralized operations and some are new to the firm. Generally the considerations can be grouped into the following categories:

1. Organizational factors
2. Information infrastructure
3. Systems management
4. Management issues

Organizational Factors

The most critical factor in client/server implementation is to have a clear perspective on why the firm is making the change. Senior executives and line managers, including IT managers, should share this united perspective. It should be developed after careful examination of the firm's business practices and after the firm has developed a strategy to change and improve its practices. This process is thoughtful, introspective, and driven by the desire to improve operations by reengineering them to be more effective and efficient. The results of the business process improvement effort should be incorporated into business plans.

After new operational processes have been identified and developed in detail, the application of information technology, or other technology, should be developed to streamline their effectiveness. This is an iterative task because new business processes are being developed with a view of the available IT capabilities. IT managers and business managers must cooperate to develop the optimum plan based on technology and business conditions. This collaboration results in new ways of doing business and new technology applications.

Most new business processes lead to new, flatter organizational structures as employees are empowered with more responsibility and more tools. Thus, the plan to introduce client/server architecture must include a transition plan to the new structure. This rebalanced responsibility requires the attention of the firm's senior executives. Neglecting to consider these factors, or casual consideration of them, usually leads to severe problems later. Reengineering, restructuring, and advanced technology is a powerful mixture: It must be handled carefully to achieve the best results.

Information Infrastructure

Implementing client/server operations implies alterations in the firm's information infrastructure involving hardware, software, and databases. As shown in Figure 11.1, the hardware architecture includes individual workstations networked to a server that, in turn, may be linked to larger systems and/or to external networks. Selection of these components is a complex systems engineering task requiring detailed knowledge of the proposed applications, estimates of possible future applications, load factors, response times, scalability or expansion capability, and other technical details.[20]

Since the firm already owns or leases hardware and network components, the added equipment and services must fit with those already in the firm. Many vendors provide equipment, software, and services to support client/servers so firms must carefully select from among their options because they will live with the consequences for a long time. The claim of open systems must be fully investigated because interoperability must be a reality, not only for the present, but for the future, too.

Obviously, the operating system managing the client and the server hardware must support the firm's applications. The hardware and the system software must be selected to support database management systems needed by the applications. And the architecture of the database depends on the application, the defined infrastructure, and the firm's future needs. As a general rule, the tendency to underestimate system capacity requirements is high. This may be due to the fact that as users gain power, they find advantages in implementing tasks beyond those they presently perform. Planning sufficient capacity for now and allowing for future expansion are mandatory.

Systems Management

Many systems management tasks are similar to those associated with centralized systems and networks. Managing problems, changes, capacity, and performance and developing emergency plans and recovery actions apply to client/servers as they do to centralized operations. These topics, including network management, will be discussed in later chapters. But some tasks are unique to client/server operations. Most of these involve actions required of system users. The introduction and control of software, the security of individual workstations, password management, license management, and software distribution are some of these important tasks.

Some of these tasks require client training, others require client manager supervision, and most should be supported through the information center and the workstation store. This is a new and unfamiliar environment for clients whose primary responsibilities lie elsewhere. Successful client/server operation requires strong support in these areas from the IT organization. Regardless of which organization owns the hardware and the network, successfully operating them requires shared responsibility. Managers in IT and in the operating departments should be jointly measured on the results.

MANAGEMENT ISSUES

One of the most difficult management issues surrounding client/server computing is understanding the investment required and evaluating the return on that investment.[21] The value of client/server computing lies mostly in increased organizational effectiveness and is difficult to quantify for reasons cited earlier. Thus, the return is highly judgmental, intangible, and not easily measured with the usual financial tools. Unfortunately, quantifying the investment itself is equally difficult for most organizations, and the tendency to underestimate the costs is high.

Survey results show some of the difficulties of tying client/server implementation to cost savings. The question posed in the surveys to 500 or more large companies was "Will migrating to client/server save you money?" In 1992, 53 percent responded no; in 1993, 76 percent said no; and in 1994, 81 percent said no. Some of those surveyed admitted that the technology had been oversold and that their firms had unreasonably high expectations.[22] Most firms believe the technology ultimately will deliver benefits, but the migration to it will be painful.

Total costs of client/server technology over five years can exceed $48,000 per client, according to a recent Gartner Group study. In a large complex network, the cost per end-user can range from $50–60,000, a large increase from the $2–4,000 usually associated with desktop systems. Labor costs constitute more than 70 percent of client/server expenses, the study reveals.[23] Table 11.6 displays the cost ingredients that the study uncovered.

The report advised that costs could be reduced somewhat by reducing system complexity, installing development tools such as CASE for client/server, and developing and implementing architectural guidelines.

TABLE 11.6 Client/Server Costs as a Percentage of the Total

Cost Element	Percent
End-user labor	41
End-user support labor	15
Application development labor	8
Enterprise server operation and other labor	8
Education, training, and professional services	7
Purchased applications software	3
Wiring and communications	3
Client and server hardware and software	15

End-user spending on information technology is large and growing in the U.S. According to BIS Strategic Decision, a Norwell, Massachusetts, firm, end-user spending in the U.S. totaled $37.32 billion in 1994 and is expected to rise to $53.2 billion by 1999.[24] Much of this is in addition to the spending in the central IT organization, increasing amounts of which are spent on client-department systems. For many reasons, financial considerations are an important factor in client/server computing.

People Considerations

Client/server implementation may alter people's behavior and their attitudes toward their jobs. Roles and the organizational reporting structure may also change. For instance, in some firms that have installed distributed computing systems, IT analysts and programmers are reassigned to the client manager so their IT skills can directly assist the implementation in that business unit. In Merrill Lynch's case, the opposite occurred. When IT people in the business units continued to work there, they reported to the centralized organization so that the implementation would be uniform across various business units.

Client/server computing has a direct effect on traditional reporting relationships because empowered employees are more like the symphony orchestra musicians Drucker mentions in Chapter 1. Loyalties shift somewhat from the immediate manager to the function and to peer workers. For example, in the procurement function, loyalty to the function may become stronger while loyalty to the procurement manager declines somewhat. This is a direct consequence of empowering employees and flattening the organizational structure.

Today's integrated organizations are far more complex than those of 20 years ago. They are more difficult to manage because, like the symphony conductor, the manager must be coach, counselor, and teacher of skilled and empowered employees. The new organizational environment is greatly impacted by changes in technology, administrative reorganizations, and the nature of the work itself. Therefore, the firm must plan to address the human concerns of distributed computing. The plan should be centered around employee communication: Keep employees informed about system changes, obtain and use employee input on planned changes, explain changes in policies and procedures, and respond to employee grievances promptly. It is almost impossible to overcommunicate in a rapidly changing work environment.

Managing Expectations

When the firm first decides to install client/server computing, consideration should be given to selecting an operation that can serve as a prototype for future implementations. In a first effort, detailed results are sometimes difficult to predict and alternative conclusions can result. A feasibility study can provide insight for future installations. For instance, if client/servers are planned to support shipping and receiving at the firm's manufacturing plants, it would be wise to test the plan at one location, thus improving the plans at subsequent locations. The prototypical installation serves as a testbed for the technology, reinforces planning details, and is useful as a training site for subsequent installations. The installation of distributed computing usually results in difficult-to-foresee consequences, and the feasibility study or prototyping approach provides confidence.

The planning and implementation teams must carefully manage expectations.[25] Prototyping efforts must proceed with the understanding that senior managers support the effort because they have confidence that the work will potentially improve the firm's operations. Individuals involved in the installation must be prepared and trained to accept change and to cope with disruptions in routine work flow. Managers must tolerate the infrequent difficulties and permit the technology to be implemented according to plan without undue pressure or displays of overconfidence.

The prototype installation or feasibility study helps managers solidify the plan to introduce the technology throughout the selected functional areas. The study also focuses on current applications and assists in developing cost and benefit information about the new technology. Other results from the prototype include: confirmation of software and hardware choices, physical installation preparation, and finalized practices and procedures surrounding the new technology.

Successful introduction of distributed computing is highly dependent on human factors. It is very important that the software is easy to use. Easy-to-use software is characterized by smooth interactive capability and well-constructed option menus. Well-designed systems usually use a keyboard, a mouse, and a graphics screen that allow users to alternate easily between products or functions. Some current windowing systems are splendidly designed to make them easy to use. One window can be used for tutorials and help sessions, thus enabling the user to learn the software by using it.

Other environmental factors are also important. The physical environment, including lighting, seating, and dimensioning, must be carefully reviewed in order to reduce physical stress. Facilities planning must include sufficient space. But planning must also address items such as noise levels, temperature and humidity, and asthetics, such as color and office appearance. Careful attention to these human factors will reduce resistance to change, speed implementation, and improve morale.

There are many risks in placing sophisticated information technology in the workplace. Many of these risks can be averted by using the information center and the workstation store in the manner described earlier. Issues such as business controls and data security and control must be resolved. The information technology organization must provide guidance on these issues; IT must ensure consistent treatment of these topics across the firm. On many of these matters, the IT manager has staff responsibility for client/server operations just as for other distributed computing activities. The mutual interaction of the client/server implementors and the IT staff will assist considerably in ensuring systems success.

Change Management

Change management during client/server installation can be divided into three phases. During the first phase, when the prototype installation is underway and the early, more innovative employees are being trained, progress bulletins should be released, and managers should hold information meetings for latter adopters. The systems operation and impact should be explained so that expectations of the system can be established. Information center personnel should be available to answer employee questions; managers must respond to concerns. In particular, issues that may impact morale must be addressed promptly.

In the second phase, when employees are beginning to use the new tools, experienced employees should be assigned to assist beginners. Additional staff must be available to offset the reduced output of employees just learning new operations. Training activities must explain how each employee fits in with the system and how the system fits into the plan for the restructured environment. Some users will experience frustrations with the system. Managers must be especially sensitive to these employees and must take steps to reduce or eliminate their problems. Managers should provide frequent, positive feedback to employees as they make progress.

During the final phase, employees should be competent in using the new technology; they should be sufficiently skilled to recommend or implement improvements to applications on their own initiative. Managers should recognize and try to adopt innovative ideas.

Some employees have a difficult time adapting to changing work patterns and to the structural changes that usually accompany them. For these employees, success can be achieved by clearly explaining the reasons for change, by planning for change as it affects them personally, and by allowing them to appreciate the benefits of change. Both managers and employees must be flexible in order to respond to new opportunities. Successful change management strives to overcome personal anxieties and to reduce apprehension. Ideally, resolving these initial problems can lead to genuine enthusiasm about the changing environment.

Managers should recognize that changes can be implemented in the areas of technology, organizational structure, management style or technique, and organizational culture. Almost all changes, technological or organizational, result in shifts in employee's attitudes or behavior patterns. When technology creates jobs with higher skill requirements, traditional occupations are upgraded, modified, or eliminated.

When managers make changes that affect employee attitudes or behavior, the success or failure of the change may well depend on whether workers perceive personal rewards or benefits. It may also depend on whether employees have the ability to acquire new skills and whether they in fact want to learn them. The scope of communication and the quality of orientation and training play an important role in managing change successfully.

Several points are worth emphasizing. The impact of change will not be enthusiastically welcomed by all individuals—for some people change is threatening. The challenge comes from needing to start over again learning new skills. Attentive managers will cope with this difficulty through effective people management practices such as one-on-one communication, coaching, counseling, and training. The most important task for the manager is to recognize that each individual struggles with change in his or her own way. Successful managers deal with the problems on an individual basis.

SUMMARY

Client/server computing is one of the most important developments in the history of information technology. For the first time since the invention of the digital computer, immediate and convenient access to large computers is available to most professional employees in modern firms. There are many reasons to take advantage of these developments.

To capitalize on the technology of distributed computing, the firm's managers must select and implement software and hardware systems. Telecommunication systems must be implemented to support the new environment. In addition, managers must deal successfully with financial issues; they must resolve business controls problems; data management and data ownership issues must be settled; and, most importantly, managers must deal with people concerns skillfully.

The implementation of the information center and the workstation store will provide organizational resources to deal with many of these issues. These resources allow the IT organization to engage with end users in a meaningful and productive manner. These organizations form the IT manager's support staff to manage the change to distributed computing.

The firm's adoption of distributed computing causes the IT organization itself to undergo significant change. The information technology manager's job will encompass greater staff responsibility. The role of IT in the firm will expand; this translates into more responsibility for the people in IT, as well. Several new and exciting jobs will be created in the workstation store and the information center. Employees in the client organizations and in IT should be given the opportunity to fill these jobs. With a plan for job rotation, many people can benefit from these opportunities.

Distributed computing has high potential payoff but includes many significant risks. If managed properly, it can be a win-win situation for all involved. The firm gains substantial benefits, the management team has greatly improved resources for dealing with the problems and opportunities facing it, and the employees have new, more valuable skills and become more productive. Although the effort is great, the benefits can far outweigh the costs.

Review Questions

1. What changes in the computing environment are taking place at Merrill Lynch, according to the Business Vignette?
2. What did Merrill Lynch do for its consultants to make the new system easy and desirable to use?
3. What factors are encouraging the trend toward distributed computing?
4. What does distributed computing mean, and what role do advances in telecommunications play in it?
5. Define the terms reengineering, outsourcing, and downsizing.
6. What factors favor downsizing and what are its disadvantages?
7. What issues or problems must the firm solve in order for distributed computing to be successful?
8. Why is compatibility such an important consideration in the implementation of distributed computing?
9. What are some of the questions raised about databases as client/server computing is installed?
10. Distributed computing causes some problems faced by owners and operators of mainframes to migrate to client departments. What are some of these new problems for client managers?
11. Describe how the issues of end-user computing vary in importance over time.

12. What is the most important issue of end-user computing? Why?
13. What are the functions of the workstation store? How does the workstation store assist in maintaining the firm's sound business controls?
14. What are the responsibilities of the information center?
15. Where do the workstation store and the information center report in the firm? What is the source of staffing for these organizations?
16. What is the relationship between the store, the information center, and the firm's planning process?
17. What are some of the policy issues that accompany the thrust toward distributed computing?
18. What are the connections between client/server computing and information infrastructure?
19. What percentage of client/server computing costs are labor related? Why does this make cost control difficult?
20. How is prototyping related to managing expectations? How is it related to business case development?
21. Why is it important that managers deal with employees on an individual basis when implementing changes in the workplace?

Discussion Questions

1. The implementation of distributed computing raises the issue of expectations. What steps can be taken to ensure realistic expectations?
2. The text presents a number of factors motivating the trend toward distributed computing. Which of these factors relates to the firm's competitive posture in the industry, and why?
3. Earlier chapters discussed advances in hardware and telecommunications. How do these advances increase or decrease the risks associated with distributed computing?
4. Many new developments are occurring in telecommunications and networks. What are some of the consequences of these developments? Does this increase or decrease the probability of success? Why?

5. The firm may need to make some policy statements concerning the use of both purchased applications and user-developed applications. What policy questions may need to be resolved regarding these programs?
6. Client/server operations increase some of the management responsibilities in those departments using them, especially regarding data and business controls issues. What are some additional considerations that management needs to deal with in these areas?
7. What alternatives are available to the firm regarding the question of who should own the workstation hardware and purchased software?
8. What are the pros and cons of funding the workstation hardware and purchased software centrally, at the corporate level, versus the individual department level?
9. What financial advantages can an effective workstation store capture? Can you make some estimates and quantify these financial benefits?
10. Elaborate on the notion that the workstation store and the information center provide a "leading indicator" function for the IT organization. How does this relate to planning for the IT organization?
11. What benefits can be achieved for the organization and for some of its people by the staffing opportunities presented in the workstation store and information center?
12. Discuss the strategic implications of downsizing centralized applications to client departments.
13. Discuss the advantages and disadvantages of downsizing the general ledger application.
14. What is the importance of securing the involvement of senior management during reengineering activities?
15. Discuss the organizational implications of reengineering, outsourcing, and introducing client/server operations.
16. Discuss the difficulties in establishing investments and returns on investments in client/server implementations. How can executive managers help in this task?

17. How can managers ease the transition for employees into the client/server environment? What approaches might be most successful with the laggards?

18. Upon successful implementation of client/server computing, the responsibilities of the IT manager will include an expanded staff role. What skills are useful in discharging these new responsibilities? What measures of success can the manager employ in these endeavors?

19. Effective client/server computing empowers employees and increases the management span of control. Discuss the implications of these changes for both managers and employees.

20. Wireless networks and portable workstations enable the "logical office" concept, i.e., the office is where the workstation is. What opportunities does this concept present?

21. Discuss the potential opportunities for a firm that can install client/server workstations in a supplier's office. What new considerations does this scenario imply for the firm's application development department?

Assignments

1. CASE tools and object-oriented development are being applied to client/server implementations today. Research the issue in order to develop some of your own thoughts on what this means for both employees in operational departments and those in IT organizations. How might your conclusions relate to the structure of the IT organization.

2. Visit a firm in your city that uses distributed computing. Interview one of the operating department managers to develop an opinion on the suitability of the technology for colleges and universities. Discuss how your school could further employ client/server technology?

ENDNOTES

[1] Jennifer L. Janson, "PRISM Produces Productivity Spurt for Merrill Lynch," *PC Week*, October 1, 1990.

[2] Alice LaPlante, "Merrill's Wired Stampede," *Forbes ASAP*, June 6, 1994, 76.

[3] See Note 2 above.

[4] Craig Stedman, "Bullish on Technology," *Computerworld 100*, September 19, 1994, 33.

[5] See Note 2 above.

[6] *Computerworld's Premier 100*, September 19, 1994, 46-53. Client/server development expense varied widely from zero to 100 percent of software development expense among these top 100 firms.

[7] Sid L. Huff, Malcolm C. Munro, and Barbara H. Martin, "Growth Stages of End-User Computing," *Communications of the ACM*, May 1988, 542. The authors developed a five-stage growth model from the six-stage model of Nolan et al.

[8] Tom Davenport, *Process Innovation: Reengineering Work Through Information Technology* (Boston, MA: Harvard Business School Press, 1993), 300.

[9] Brad Bass, "GSA Mulls Outsourcing of Business Activities," *Federal Computer Week*, January 23, 1995, 3.

[10] Kathleen Milymuka, "Honey, I Shrunk the Mainframe!" *CIO* September 1989, 36.

[11] Peter G. W. Keen and Lynda A. Woodman, "What To Do With All Those Micros," *Harvard Business Review*, September–October 1984, 142.

[12] Lawrence S. Corman, "Data Integrity and Security of the Corporate Data Base: The Dilemma of End-user Computing," *Data Base*, Fall–Winter 1988, 1.

[13] CASE development tools are available and are essential to sound client/server application development. "Client/Server CASE:Oxymoron or Essential?" *Datamation*, September 1, 1994, 67.

[14] Estimated annual cost per user rise from $5600 to $9640 when moving to client/server from mainframe processing according to an article in *Fortune*. "Why Mainframes Aren't Dead Yet," *Fortune*, April 17, 1995, 19.

[15] James C. Brancheau and James C. Wetherbe, "Understanding Innovation Diffusion Helps Boost Acceptance Rates of New Technology," *Chief Information Officer Journal*, Fall 1989, 23.

[16] Richard Walton, *Up and Running: Integrating Information Technology and the Organization* (Boston: Harvard Business School Press, 1989).

[17] Robert L. Leitheiser and James C. Wetherbe, "Service Support Levels: An Organized Approach to End-User Computing," *MIS Quarterly*, December 1986, 337.

[18] William J. Doll and Gholamreza Torkzadeh, "The Measurement of End-User Computing Satisfaction," *MIS Quarterly*, June 1988, 259. This article describes a valid methodology for measuring end-user satisfaction.

[19] Cornelius H. Sullivan, "Systems Planning in the Information Age," *Sloan Management Review*, Winter 1985, 4.

[20] Atre Shaku and Peter M. Storer, "Client/Server Tell-All," *Computerworld*, January 18, 1993, 73.

[21] J. William Semich, "Can You Orchestrate Client/Server Computing?" *Datamation*, August 15, 1994, 36.

[22] Source: Sentry Market Research, Westboro, Mass., *Computerworld*, November 7, 1994, 71.

[23] Rick Whiting, "C/S Labor Costs Exceed Technology Costs," *Client/Server Today*, July 1994, 20.

[24] Source: BIS Strategic Decisions, Norwell, Mass., *Computerworld*, August 8, 1994, 90.

[25] Julia King, "Keeping a Grip on User Expectations," *Client/Server Journal*, November 1993, 46–50.

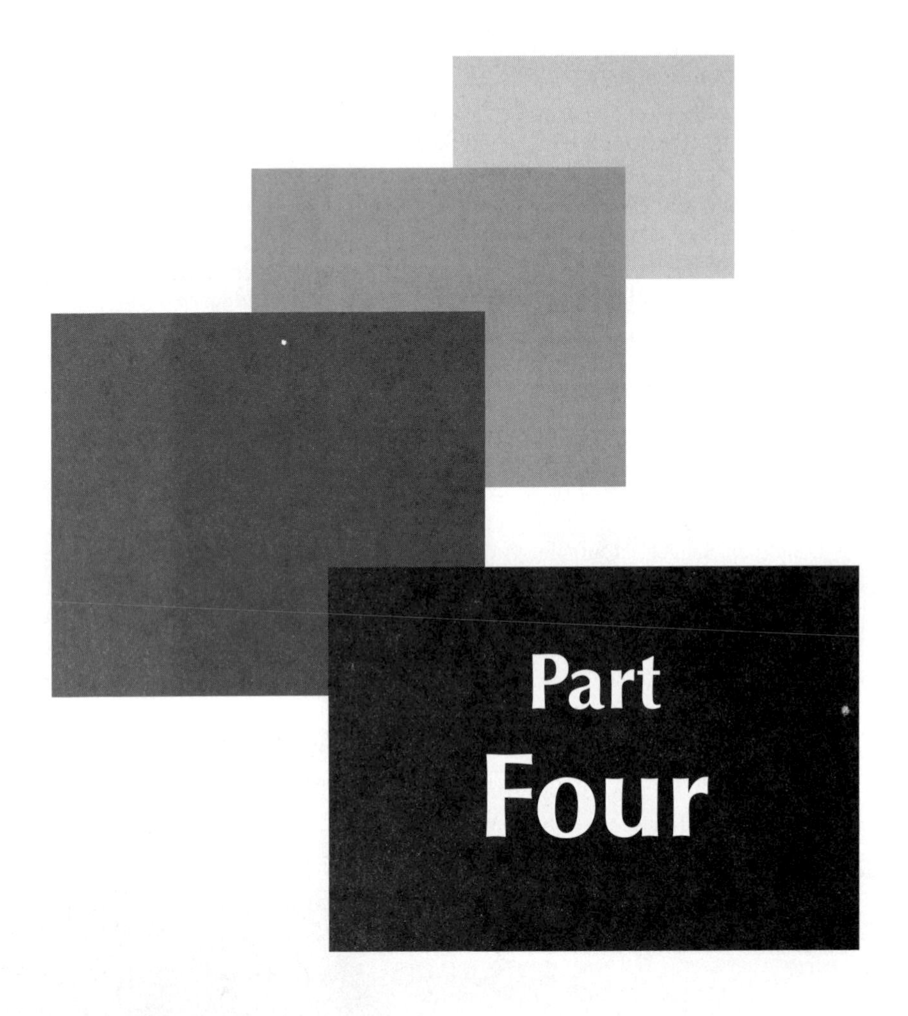

Part Four

Tactical and Operational Considerations

Business operations depend critically on the effective and efficient operation of information and telecommunications systems. A disciplined management approach to routine business systems operation is key to IT managers' success. A systematic, effective process for resolving numerous operational issues and potential problems is required in this complex arena. Part Four begins with the chapter Developing and Managing Customer Expectations and continues with Problem, Change, and Recovery Management; Managing Centralized and Distributed Operations; and Network Management.

A systematic, disciplined approach for handling operational issues enables IT managers to attain high levels of customer responsiveness from their information systems.

12 Developing and Managing Customer Expectations

A Business Vignette

An IS Management Triumph[1]

John Singleton began experimenting with service-level agreements and defining his Management by Results program when he was vice president for operations and data processing at Maryland National Bank in Baltimore. When he moved to Security Pacific as IS Chief, he refined and implemented these concepts with outstanding results. Under Singleton's leadership, Security Pacific Corporation reduced IS expenses, expanded its business, and improved IS performance. His accomplishments paved the way for his promotion to vice chairman of Security Pacific and chairman and CEO of Security Pacific Automation Company.[2]

Upon joining Security Pacific, Singleton was given the assignment of controlling IS spending which was growing at an annual rate of nearly 30 percent. Within a year, he had reduced spending growth about 10 percent, bringing it more in line with average rates for IS organizations of this size. He accomplished this reduction by implementing his Management by Results program and by centralizing the IS organization, contrary to the current trend toward downsizing and dispersing IS activities.

Singleton believes that centralized IS organizations have significant cost advantages over downsized operations and are better able to maintain control over IS activities. In addition, centralized IS operations offer bigger jobs to IS professionals and allow the firm to retain critical IS management talent. Decentralized operations fragment the job and create isolated pools of talent. This encourages aggressive managers to look outside the firm for advancement opportunities. But to be successful in large companies, centralized IS operations must be responsive to the needs of diverse and remote business units. This is where the Management by Results program comes in.

The program has four related components: IS strategic planning, service-level agreements between IS and users, IS performance tracking and reporting, and personal performance appraisal and salary programs. The goals of Management by Results are

- To link the IS business plan to the user's plan.
- To understand exactly what the customer needs are.
- To contract with users to fulfill their needs.
- To report accomplishments and results to users and IS people.
- To tie individual accomplishments to salary compensation.

The program ensures that IS people and the IS organization are committed to achieving company and user objectives.

Security Pacific Automation prepares an annual strategic plan tuned to user goals. The company uses a steering committee composed of board members from the parent organization to oversee IS activities. For example, Singleton reviews budget requests with the steering committee. To do this, he uses IS achievements obtained during the measurement process prior to going to the full board. This approach ensures that top level management agrees with IS strategies and plans, and virtually guarantees a strong link between IS strategies and overall firm strategies.

Service-level agreements (contracts between IS service providers and users) are the heart of the Management by Results system. IS works with each user to establish exactly what the user needs and how the user measures delivered performance. IS and users agree on performance standards, performance calculations, and methods for reporting unsatisfactory results. The agreement also links IS costs and performance because Singleton believes cost effectiveness is a common goal. He believes that a balance exists between performance, quality, and customer relations. The agreement must reflect this balanced approach among these critical performance areas. When finalized, the agreement is signed by IS managers and users to attest to their mutual agreement. There are more than 500 service-level agreements in place at Security Pacific.

Security Pacific Automation achieves high credibility with users by measuring and reporting all IS activities. The program sets standards and permits users to evaluate IS performance regularly and frequently—sometimes daily. User evaluations of performance help improve objectivity. Singleton attained excellent or above average ratings from 94 percent of his users during the first three quarters of 1988, a tribute to the program's effectiveness.

Individual performance appraisal and salary advancement are tied to IS performance, thus ensuring personal commitment to achieving results. Measurement techniques similar to those in the service agreements allow managers to track individual performance. Employees are measured on managing service levels, budgets, personnel, and new business. Singleton believes this feature helps employees take responsibility for directly managing their careers.

IS employees generate proposals that Security Pacific Automation uses to bid for customer business. This in-house IS division competes with outside vendors for client services or equipment. Largely because of its cost competitiveness and commitment to results through service-level agreements, the Automation Corp. is taking business away from outside vendors. Additionally, when IS underspends its budget or makes a profit on outside sales, it returns the excess to its internal customers. This strategy reinforces and cements relations between IS and its customers.

The Management by Results system has yielded high returns at Security Pacific. In-house staff was substantially reduced while expenses dropped five percent over two years, even though the company itself was growing by acquisition. During the past several years, Security Pacific purchased The Oregon Bank and The Arizona Bank. Cost containment at Security Pacific Automation partially financed these acquisitions. Electronic transactions grew by 20 percent in three years. Later, Security Pacific capitalized on its IS effectiveness by absorbing the data processing departments of three out-of-state banks with combined budgets of $100 million.

When Singleton joined Security Pacific, he found that unhappy users were his most serious concern. At that time, the company believed that IS costs were unreasonably high and that quality performance was lacking. IS also had a reputation for not finishing projects on time. In short, IS credibility with users was low, and they wanted improvements quickly. If Singleton had not been able to correct the situation within six months to a year, users were prepared to dismantle the IS operation, carve up the pieces among themselves, and take charge of their own destiny.

Singleton moved rapidly, but his program encountered resistance from IS managers and users alike. Some managers believed their responsibilities were not measurable, and users resisted centralization efforts until the benefits were apparent. IS people who could not adapt left the company voluntarily or were terminated. But Singleton demonstrated that centralized IS organizations can be flexible, responsive, and effective. He believes that IS must lead the way in reengineering the business and in restructuring the old ways of doing things.

For his accomplishments, Singleton received the 1988 Information Systems Award for Executive Leadership from the John E. Anderson Graduate School of Management at UCLA. The accounting firm of Peat Marwick Mitchell & Company found Security Pacific's IS expenses to be about 10 percent lower than comparable operations. IBM declared Security Pacific to be one of only two firms that measures and manages IS strategic resources effectively. Indeed, under Singleton's leadership, Security Pacific Automation Corp. is truly an IS management triumph.

INTRODUCTION

The acquisition, development, and enhancement of application programs and the management of associated databases involve many long-term, strategic issues. Strategic concerns are critical, but the applications raise many important tactical issues, as well. In many instances, since the operation and use of the applications and databases define the firm's modus operandi, they demand serious attention from managers in the tactical and operational timeframes.

Tactical and operational concerns arising in the production operation portion of the IT organization are the focus of this chapter. Production operation concerns itself with the routine implementation and execution of the application programs supporting the firm's mission. Some of these applications are on-line nearly continuously. They operate in a highly interactive manner as they support hundreds, perhaps thousands, of users. Other systems operate daily, weekly, or at month-end according to a predetermined schedule. The nature of this scheduled production work is operational and tactical. Because the firm's essential business activities depend on the production operation, the applications must operate effectively. This activity is one of the critical success factors for the IT organization.

The beginning of the disciplined approach to production operations management will be developed in this chapter by considering the notion of service levels and service-level agreements. This chapter formulates the process of reaching agreement with IT customers on service levels. It describes the components of a complete service-level agreement and outlines methods for acquiring measures of customer satisfaction. The central themes in this chapter are the management and satisfaction of a customer's operational expectations.

TACTICAL AND OPERATIONAL CONCERNS

The management of a computer operation, whether owned and operated by IT or by a client organization, lends itself to a disciplined approach. Careful management of an array of disciplines permits the organization to achieve success in this critical area. A large number of the firm's employees can quickly observe difficulties in computer operations. Server failure in a distributed processing environment usually leads to failures throughout the client network. Failure of the central processing unit supporting the office system will be noticed promptly from the executive suite to the remote warehouse. Production operations is a vital and highly visible function.

In contrast to routine computer operations, an error or an omission in the strategic planning process, while perhaps more devastating to the firm in the long term, may not become apparent for months or even years. Production operations managers' lives are dominated by what is happening today and what will happen next week or next month. It is mandatory that they have a management system that provides the tools essential to their success. More than other managers, they rely upon procedures and disciplines to cope with the wide variety of challenges that may come their way.

EXPECTATIONS

The first chapters in this text stressed the need for IT managers to meet expectations held by senior executives in the firm. The fulfillment of executives' expectations, whether or not they are realistic, is the principal standard against which IT managers are measured. We also learned that expectations are established through a variety of means. Some expectations are developed from sources external to the firm: from meetings with vendors, articles in the trade press, and meetings of business associations. These externally developed expectations are reinforced in the minds of executives by a number of internal factors, such as the amount of resources devoted to information technology. Expectations may not always be soundly based; in some cases, the IT organization itself forms unreasonable expectations. It is essential that the organization use a management system designed to cope with the process of setting and meeting realistic expectations.

Tools, techniques, and processes designed to deal with expectations related to the application portfolio were discussed in the previous part. These processes considered the tactical period, but they had strategic implications. This chapter concerns itself with tools, techniques, and processes for developing and managing customer expectations in the operation of these applications, processes that apply principally to the short term. The concepts are formalized processes for achieving service-level agreements between the IT organization and all its clients. These agreements establish acceptable levels of service and include mechanisms for demonstrating the degree to which service levels are attained.

One objective of the service-level agreement process is to ensure that the entire firm has a clear knowledge of what is expected from the computer center operation. The process must also provide an obvious means for recording and publicizing service levels delivered. Objectives always change over time, and some objectives are not always completely fulfilled. Given this reality, the goal of this chapter is to focus on how well the organization is working to achieve its objectives. A clear understanding of this issue will overcome debate and confusion within the firm concerning what the measurements really are. The process being developed establishes service-level agreements and reports service levels achieved.[3]

THE DISCIPLINED APPROACH

Service-level agreements (SLAs) are the foundation of a series of management processes collectively called the *disciplines*.[4] A discipline is a management process consisting of procedures, tools, and people organized to deal with an important facet of the production operation, whether located within the IT organization or in a client organization. Figure 12.1 depicts the relationships among the processes comprising the disciplines of production operations.

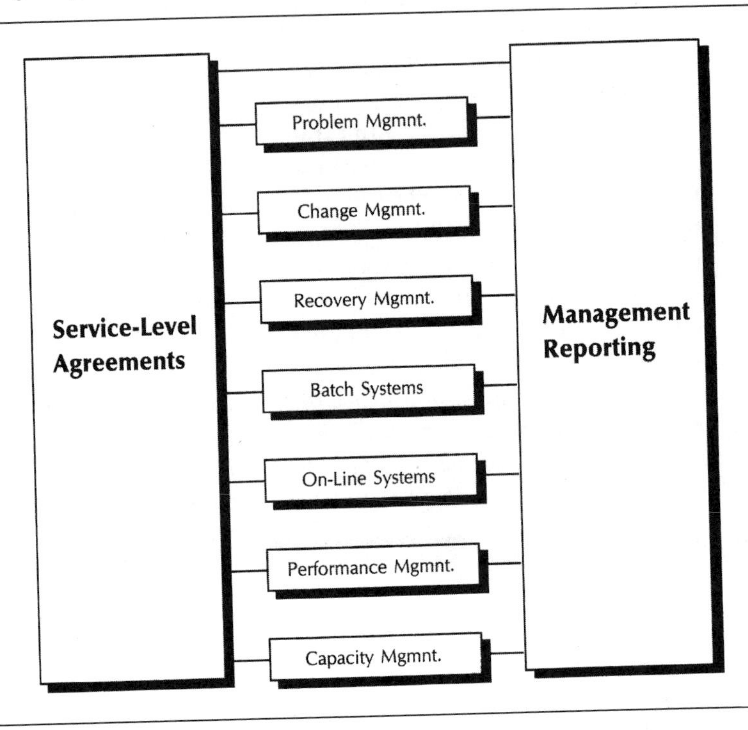

FIGURE 12.1 The Processes Continued

Computer center performance can best be judged against specific, quantifiable service criteria. Negotiated service agreements establish client-oriented performance criteria so that performance assessments can be made. Service control and service management depend critically on these criteria.

Computer center performance is made visible to management through specific reports. Service-level agreements involve all client organizations directly; but management reports are focused on specific organizations. Management reports are an essential tool for production operations managers because they organize and present the results of the intermediate disciplines. Managers in client organizations depend on them, also. The reports provide clients with information regarding the status of the service they are receiving, the actions being taken to correct any problems they may be having, and progress being made on improving system performance or capacity.

In order to provide superior service, operations managers must control all other aspects of the computer center. The items that must be considered are:

1. Control problems. Problems, defects, or faults in operation lead to missed service levels and expenditure of additional resources.
2. Control changes system-wide. Change control is vital because mismanaged change leads to problems and reduced service.
3. Recovery from service disruptions. Managers must plan to recover from the faults or defects in system performance that inevitably occur.
4. Batch and on-line operations.[5] Current batch and on-line workload must be scheduled, monitored, and the results delivered to client organizations.
5. Performance analysis. System throughput must be maintained at the planned level in order to meet service agreements.
6. Capacity planning. Capacity can be planned correctly only when the preceding disciplines are operating effectively.

All of these topics are essential to the successful operation of a computer center. They are considered in more detail in subsequent chapters. Collectively, these processes or disciplines provide the tools and techniques required for successful computer operations management.

SERVICE-LEVEL AGREEMENTS

The service-level discipline is a management process for establishing and defining levels of service provided by IT or other operators of computer centers. The objective of the service provider is to deliver service to its client organizations at an established level. Service planning is the basis for tactical planning in the operations arm of IT. The process that establishes service levels involves a significant degree of negotiation between organizations. Clients, their managers, and the firm's senior managers must be involved.[6] In most cases, it is an iterative process because of the dynamic nature of the business, changing requirements, and new technology. Additionally, the process evolves as expectations are developed and leveled.

The process culminates in a mutually acceptable agreement; it documents the service level each IT client will receive.[7] The purpose of the agreement is to ensure that the client organization using the service and the organization providing the service mutually consent to the terms set forth in the SLA. The agreement discusses all aspects of service expected by the client. During the negotiation phase, the balance between service levels and the costs of providing service needs to be explored in detail. Both parties must strike an effective balance between these issues for the agreement to be mutually satisfactory. A properly constructed service-level agreement establishes a cost-effective means for client organizations to use IT services.

The agreement must outline the objective methods that will be used to measure and report service levels. The cost of providing service to the client organization must be justified by the responsibilities discharged by that organization on the firm's behalf. For example, the costs of providing improved on-line response to computer-aided design system users must be less than the value of the improved productivity resulting from the response improvements. Much of the cost/benefits resolution should have occurred during the tactical planning process.

Although the IT organization must initiate the service-level agreement, client line managers and IT managers must negotiate the agreement. In especially complicated and difficult cases, the negotiation may occur between the CIO and other executive managers. A higher level of attention is required when disagreements cannot be resolved at lower levels due to valid differences of opinion or insufficient breadth of vision. Frequently, senior executives are better positioned to define required service levels and justify additional expenses. This is especially true in cases where costs and/or benefits are intangible or difficult to quantify.

The services that IT will provide must be expressed in terms meaningful to client managers. For instance, response times measured at the user's terminal are more meaningful to clients than CPU seconds measured by the service provider. Turnaround time (the time from job submission to job completion) must be measured at the user's workstation, not at the central processor.

All client organizations must be included in the SLA process, and nearly all computer operation services should be considered. Applications that are infrequently executed or run on an as-required basis, or that place low demands on the resources of the organization, need not be specifically included in the agreement. Applications that the firm widely uses require special attention. An earlier chapter stated that each application requires an owner-manager who takes charge of the application asset and discharges ownership responsibilities. One of the owner's responsibilities is to negotiate service levels with service providers on behalf of all application users. It is the owner's responsibility to use cost-effective trade-offs to achieve a balanced agreement with service providers. The most effective service-level agreements are achieved when the negotiating managers exercise corporate statesmanship, keeping the interests of the firm foremost in their minds during the negotiation process.

WHAT THE SLA INCLUDES

The standard service-level agreement begins with administrative information.[8] This includes the date the agreement was established, the agreement duration, and the expected renegotiation date. It may specify unusual conditions, such as significant workload changes, that will mandate renegotiation. The agreement includes key measures of service required by the client organization and service levels to be delivered and measured by the service provider. Resources and associated costs required by the provider to deliver the service are also included.[9] The agreement also describes a mechanism to report the actual service levels delivered. Table 12.1 lists the elements of a Service-Level Agreement.

TABLE 12.1 Service-Level Agreement Contents

Effective date of agreement
Duration of agreement
Type of service provided
Measures of service
Availability
Amount of service
Performance
Reliability
Resources needed or costs charged
Reporting mechanism
Signatures

Negotiation of service-level agreements occurs when the operational plan is being prepared. At this time, resources for the coming year are being allocated, and near term requirements for IT services are becoming clear. The effective date of the agreement usually coincides with this portion of the planning cycle. For services that are relatively stable in operation, the duration of the agreement will probably be one year. Payroll or general ledger applications are examples of such systems. For applications of this type, the agreement will be renegotiated during the following planning cycle one year later.

Applications that have volatile demands or growing processing volumes may require more frequent SLA renegotiation. For example, a new strategic system experiencing great implementation success may have rapidly growing demand for IT services. It may require a renegotiated service-level agreement every six months. The agreement should spell out the circumstances that trigger renegotiation. Figure 12.2 illustrates these concepts in a partially completed SLA.

Figure 12.2 illustrates the administrative information included in an SLA between a firm's IT organization and the Personnel function. The agreement notes that personnel will be charged for the actual service used according to rates established by the firm's controller. (This is not a fixed price contract.) The manner in which firms recover IT costs vary widely. This illustrates one of the possibilities. The subject of charging methods is covered in detail in Chapter 16.

The agreement is negotiated in mid-October, but does not take effect until the beginning of the next year. In this case, the negotiation took place during the planning cycle. It becomes effective with the new calendar year.

FIGURE 12.2 SLA Adminsitrative Information

SERVICE-LEVEL AGREEMENT

The purpose of this agreement is to document our understanding of service levels provided by IT Computer Operations to Personnel Function. This agreement also indicates the charges to be expected for the service described herein and the source and amounts of funds reserved for these services. This is not a fixed price agreement. Charges will be made for services delivered at rates established by the controller.

The date of this agreement is October 15, 1995.

Unless renegotiated, this agreement is in effect for 12 months beginning January 1, 1996.

SLAs must specify the type of service required. The service may refer to batch operations on a regularly scheduled basis, batch runs on demand, or complex on-line transaction processing applications. The service provider should negotiate the agreements by class of service since capacity, performance, and reliability depend on the total demand levels in the overall organization. Figure 12.3 is an example of a service specification.

FIGURE 12.3 SLA Service Specification

Specific type of service	Application service	Hours of use
Batch processing	Payroll application	2 hr/week
On-line data entry	Payroll application	daily*

System availability

Monday – Friday Payroll input data entry will be available during normal working hours Monday through Thursday and Friday 8:00 through 24:00.

Saturday Payroll will run around 03:00 Saturday morning and will be available to personnel 08:00 Saturday.

Sunday Emergency service only.

System performance

*Four special workstations in the personnel area will be used for this activity, and the general agreement to subsecond response time is part of our understanding.

Figure 12.3 illustrates the service to be delivered to the Personnel function for payroll processing. It specifies data entry during the week with extra hours available on the last work day of the pay period. This occurs from workstations physically located in the Personnel area for security reasons. Subsecond response times at the workstations and availability of the payroll output are documented.

Schedule and Availability

The schedule describes the period during which the system and the application are required to operate. For on-line applications, the schedule should describe the availability throughout the day. It must also outline the conditions for weekend and holiday availability. The IT manager must allow for scheduled maintenance or make provision for alternate service during maintenance. The client manager should focus on peak demand periods, which may include weekends at the close of an accounting period and other significant events. Both parties should consider reduced availability or reduced performance during off hours, if these considerations yield cost savings. Likewise, during especially critical periods for the client, it may be attractive to negotiate exceptionally high service levels. A clear mutual understanding of system requirements and capabilities will lead to the most effective agreement.

Timing

For batch production runs, the job turnaround time is usually a key measure. (Turnaround time is the elapsed time between job initiation and delivery of customer output.) Many batch production runs are scheduled through a batch management system. The batch management system accounts for the many relationships between the applications and the databases. The data used as input on one job may have been created as output from several other jobs. Hence, the timing of the jobs becomes crucial. Clients should understand these dependencies. In most cases, the negotiating process will help them develop a good understanding of scheduling constraints. Frequently, large production runs occur overnight. In many instances, output availability from all overnight jobs at the beginning of the work day is the critical measurement. The output may consist of reports delivered to the client offices, or, more likely, on-line data sets available for the user's review through personal workstations.

The most critical parameter for on-line activities is response time. (Response time is the elapsed time from the depression of the "program function" key or the "enter" key to an indication on the display screen that the function has been performed.) Response time is highly dependent on the type of function being performed. Therefore it should be specified by type of service. For example, trivial transactions should have subsecond response; i.e., the operator should not be constrained by the system when performing simple transactions. Trivial transactions require so little processing that the response appears to be instantaneous, on the order of several milliseconds or less. If subsecond response time cannot be achieved for trivial transactions, productivity drops substantially and the system becomes a source of annoyance to the user.

Many transactions are distinctly nontrivial. For instance, an on-line system for solving differential equations may provide subsecond response time when users enter parameters, but the numerical methods used to obtain the solution may require the execution of many millions of computer instructions. The client usually understands this type of condition and appreciates the response time required to solve the problem. Many other situations, particularly applications that query several databases to search for specific data, are decidedly nontrivial even though the customer's query seems simple enough. Everyone concerned needs to understand these situations thoroughly prior to entering into agreements. Education through the information center can achieve the necessary level of understanding.

During the negotiation process, the IT management team must provide service availability and reliability information. An example of an availability statement is: "The system will be available 98 percent of the time from 7 A.M. to 7 P.M. five days per week." An example of a reliability statement is: "The mean time between failures will be no less than 30 hours, and the mean time to repair will be no more than 15 minutes." The availability and reliability measures must be unambiguous: Users and IT personnel must be able to measure them easily. The remaining items in the sample SLA are shown in Figure 12.4.

FIGURE 12.4 SLA Final Items

Reporting process: Reports will be made to personnel on a quarterly basis documenting the service levels delivered on the payroll application.

Financial considerations: Workstations will be charged at the rate of $5 per hour of use; batch processing will be charged at the rate of $35 per CPU minute. We estimate this will cost approximately $90,000 during this 12 month period. Corporate overhead will absorb these charges.

Additional items: none

Signatures

by:_____________________	by:_____________________
IT organization	Application owner

Workload Forecasts

Workload forecasts and required IT resources are essential to service-level agreements. The operations department must install production capacity sufficient to process the load generated by all applications in total. The capacity must be sufficient to deliver the service promised to all the clients. IT workload should be defined for the period of the service agreement in terms that client managers can understand. IT managers must translate these workload statements into meaningful and measurable units for capacity planning. The forecast should cover batch workload volumes, volumes of on-line transactions, amount of printed output, and other resource demands that the client requires of the service provider. In many cases, average workloads will suffice. But the negotiations must cover changes in workload if significant daily, weekly, or monthly departures from the average are expected.

On-line transaction volumes can vary by an order of magnitude during the day. They frequently display a morning peak, a lull in activity around noon, and a peak again during the afternoon. Some client/server applications display this pattern. In some cases, the peak loads generated by one application occur during the valley in another application's load. Firms operating nationally experience workload waves throughout the day. Departments on the east coast open for business three hours before departments on the west coast. The air traffic control system in the United States illustrates this phenomenon, as flights begin to depart from Boston and New York airports around 7:00 A.M. local time, several hours before airport activity picks up in Seattle or Los Angeles. Workload follows the sun in international operations.

The coincidence of workload peaks from many applications is a more critical issue for IT managers. Month-end closing of the financial statements is an example of activity that usually generates this form of peak activity. In some cases, a number of activities conspire to form a huge bubble of workload for the organization. Year end is a prime example of this phenomenon. Not only are the December closings in process, but other year-end activities are taking place as well. Year end is also a convenient time to implement new applications to supply the customer with new and different functions for the coming year. Many times, these new applications are mandated by annual changes in the tax laws or by other government regulations. If the fiscal year differs from the calendar year, the workload pattern will take on yet another form. Satisfactory service-level agreements anticipate all these events and keep them in clear focus.

Unanticipated or unusual increases in workload are a frequent cause of missed service levels. Generally, increases in workload do not come without warning. An alert organization will reopen the discussion on service levels when such increases are first anticipated. Everyone benefits from well-planned workload management. Generally, all organizations suffer from one organization's poor planning. Sometimes the workload is difficult to predict because of changes in the organization's needs, or because of changes in the organization itself. Acquisition or consolidation are examples of global organizational change. In these cases, a reanalysis of the load and a new forecast should be prepared for the client organization.

Reasonably accurate forecasts are essential for providing satisfactory service, although workload forecasting is often tedious and time consuming. History is a good guide for most forecasts. IT should provide clients with current workload volumes and trend information; a particularly easy task if a cost accounting or charging mechanism is installed. IT charging or cost recovery mechanisms are valuable in the SLA process because they help focus on cost-effective service levels. (IT accounting processes will be discussed in Chapter 16.) It is mandatory to IT and clients that the load analysis and load forecasting proceed successfully.

Measurements of Satisfaction

Key service parameters contained in the service-level agreement must be routinely and continuously measured and reported. Specific, quantifiable parameters are preferable to more ambiguous measures. In general, if the IT organization reports these critical measurements ambiguously, or if the reported data contains errors of fact, the client organizations will develop their own tracking mechanisms. Uncoordinated measurements and reports usually lead to conflict, accusations, and finger-pointing, and needlessly consume energy. The most effective approach is to provide unambiguous, credible reporting techniques at the outset.

One additional consideration must be addressed. Although it appears obvious, service providers must measure and report service from the user's perspective. It doesn't help to report job completion time if the output isn't available to the client organization until later! It only confuses matters to measure on-line transaction response times at the server. What the user sees at the workstation is all that counts.

Response times are easy to obtain if the computer operations department has a personal computer programmed to execute a representative sample of interactions with the CPU or server. The PC can log transaction data and report statistics on response times. It can be switched to various ports on the I/O channels, and will provide valuable data on user response time. This device can be the benchmark for establishing and monitoring on-line response times.[10]

It is essential that IT's measurement system have high credibility with client organizations. This credibility leads to mutual trust and confidence and results in open and objective discussions when events don't transpire as planned. For any number of reasons, small excursions from the norm may occur. These deviations are corrected most expeditiously in an atmosphere of open communication and mutual respect. Precision measurements of delivered service build trust and confidence between the service provider and its clients. Objective, credible reporting on service-level achievements paves the way for improvements in other activities.

THE ROLE OF USER SATISFACTION SURVEYS

Periodically, the IT organization should solicit informal customer opinions on service-level satisfaction. Such surveys provide valuable insight into the customer's perceptions of IT performance.[11] Measured performance against the service-level targets should be tempered with user perceptions of service. It is not enough for service to meet stated criteria; the perception of service must also be satisfactory. IT managers must know the degree of user satisfaction, and the most useful approach is to ask the users directly.

The IT organization can choose one of several avenues to obtain this information.[12] One method is to conduct a periodic opinion poll of on-line users through a brief questionnaire on their workstation screens. The questionnaire appears at the conclusion of the session and asks for the user's perception of service delivered during the preceding period. To be effective, the questionnaire must be optional, anonymous, and well designed. A second approach is to survey a wider audience over a longer period and include a broader scope of questions. The results of these surveys are very useful in detecting problem applications. The results can uncover areas where service is poor or where perceptions of service are poor. Surveys are useful for detecting incorrectly established service agreements, and for discovering clients who are not sufficiently educated in the service-level process.

Client surveys should be used to improve internal service measures and to relate internal measures to perceived and actual customer service. The goal is to achieve a high correlation between internal and external service measures. Attaining this goal will result in resource optimization, efficiency improvements, increased customer satisfaction, and reduced costs. IT organizations have an abundance of tools suited to performing this task.

Collectively, client managers and IT managers should take action to correct unsatisfactory situations that the surveys or measurements detect. Managers can use the survey and measurement data to establish more effective relationships between users and providers of service. Their actions may take the form of better communication, better service, or more education. Both organizations benefit from such actions.

ADDITIONAL CONSIDERATIONS

In a complex environment, there are usually some additional operational items that the service agreement must include. During infrequent critical periods, some departments may need dedicated processors, large amounts of auxiliary storage, or other unusual services. Other departments may require special handling for data entry, or specific, one-time analyses. Prudent IT managers will recognize the importance of responding enthusiastically to these vital client requirements, even though the request is beyond the scope of the SLA. Satisfied customers must be a priority goal for every manager.

Unfortunately, not every manager will endorse this process of establishing service-level agreements. In user-owned client/server operations, managers tend to regard these processes as unnecessary formalities. Problems can be solved, they believe, by installing larger workstations, faster LANs, or more powerful servers. But eventually, as the applications become more complex and the operation grows, some formal processes are necessary to maintain order and sustain service. Operating by instinct and reacting to difficulties will not prove to be satisfactory in large or complex situations.

In some firms, political considerations may frustrate the process. Establishing SLAs may be contrary to the corporate culture. In some organizations, strong-willed managers prefer to demand service regardless of cost. These managers believe their function requires high service levels purely because they perform important functions for the firm. They may believe that other organizations in the firm exist to serve their function, and that IT must do so also, no questions asked. Some managers abhor the scrutiny that the service-level process brings to their activities. They prefer to conduct their business in an atmosphere of near secrecy. This attitude can seriously jeopardize the successful functioning of computer operations.

In some firms, the corporate culture is not conducive to the detailed planning and commitment process inherent in reaching service-level agreements. Discipline is required; some firms simply don't have it. These firms tend to behave in an *ad hoc* manner, reacting to situations as they occur. IT managers and other managers in these firms are continuously subjected to myriad conflicting forces that make a smooth operation nearly impossible. Establishing service-level agreements requires cooperation among many of the firm's managers. In this sense, the responsibility for successful computer operations transcends the service provider.

CONGRUENCE OF EXPECTATIONS AND PERFORMANCE

The goals of the SLA management process are to attain mutually acceptable levels of expectation, to develop an atmosphere of joint commitment, and to foster a spirit of trust and confidence between organizations. In this atmosphere, essential business information flows freely and openly.

It is not always easy to achieve these goals. For example, how can the firm handle the dilemma that occurs when clients demand better service than IT can deliver? Should the SLA describe service that can't be delivered, or should the SLA describe achievable service that is unsatisfactory to the client? One way to resolve this problem is to consider affordable costs. Through management processes described earlier in this text such as strategizing, planning, budgeting, plan review processes, and steering committee actions, affordable IT costs should have been developed. If the firm can afford more capacity, it should procure more, and develop the SLA accordingly. If costs limit capacity, then client's expectations and the SLA must reflect this consideration.

In some cases, congruence of expectations and performance is achieved only after one or two plan cycles have elapsed. For firms just starting to systematize operational activities, the first iteration may not satisfy everyone perfectly. In most firms, however, the service-planning process is not new.[13] It is an integral part of their planning cycle. For these firms, leveling expectations and performance is a structured and ongoing process.

Successful implementation of the service-level process is required for effective and efficient functioning of computer operations. SLAs are the cornerstone of the disciplined process leading to success in computer operations. Success in this area is vital for the firm. It is a critical success factor for IT managers.

SUMMARY

Service-level agreements are the foundation on which the management systems of production operations rely. Service-level agreements are essential and valuable tools in the IT management system. But they are important for other reasons as well. The process of negotiating the agreements develops and enhances understanding and mutual respect between the service providers and their clients. The IT organization will develop a much clearer understanding of the business and their clients' requirements for accomplishing their missions for the firm. The client organizations will develop a better understanding of the opportunities and constraints inherent in the IT organization. A high level of respect, trust, and confidence mutually benefits all organizations within the firm and, of course, the firm as a whole.

Review Questions

1. What are the elements of the disciplined approach to managing production operations?
2. How do SLAs deal with the issue of expectations in production operations?
3. What are the elements of the Management by Results system at Security Pacific?
4. What is the purpose of the SLA? Who participates in establishing it?
5. What are the ingredients of a complete service-level agreement?
6. What is the distinction between turnaround time and response time?
7. Can the concept of a service-level agreement be considered standard business practice, and why?
8. What considerations should be used to break ties in the event of a deadlock during the SLA negotiation process?
9. Why are workload forecasts necessary for a satisfactory SLA process?
10. What difficulties arise in selecting a unit of measure for workload forecasts?

11. Why is it essential for the IT organization to measure and publish its service-level performance in a highly credible manner?
12. In what ways do user-satisfaction surveys benefit the IT organization?
13. What are some of the social or political benefits that accrue to the firm from the SLA process?
14. What modifications to the SLA process may be valuable for strategic information systems management?

Discussion Questions

1. Describe the elements of Singleton's management system used at Security Pacific.
2. Discuss the advantages of centralized IS organizations as perceived by Singleton at Security Pacific.
3. Many production operations activities are operational or tactical. Some are strategic or long range. Discuss some examples of strategic activity?
4. Discuss the financial considerations that must be addressed in establishing the SLA.
5. One step in gaining line management involvement in information technology is institutionalizing the process of IT/line management cooperation. How do the disciplines, particularly the SLA, help accomplish this goal?
6. Customers of information technology organizations sometimes state: "I can't predict my demands on IT more than three months in advance." How would you overcome this belief?
7. Discuss the relationship of the service-level process to the concepts of tactical and operational planning as discussed in Chapter 4.
8. In addition to those mentioned in the text, what other means can you propose for obtaining customer feedback on IT service. Discuss the advantages and disadvantages of these methods.

Assignments

1. Design a screen that appears at log-off time to obtain customer feedback regarding the quality of the on-line service provided during the session.
2. Contact a service business firm in your area and investigate its service-level process. Does it have the equivalent of the SLA? Does it measure and report service levels? If customer satisfaction surveys are conducted, how are they handled?

ENDNOTES

[1] Byron Belitsos, "A Measure of Success," *Computer Decisions*, January 1989, 48. See also John P. Singleton, Ephraim R. McLean, and Edward N. Altman, "Measuring Information Systems Performance: Experience With the Management by Results System at Security Pacific Bank," *MIS Quarterly*, June 1988, 325.

[2] Security Pacific employed nearly 41,000 and earned $740 million in 1989. It was merged into Bank of America in March 1992.

[3] Additional details can be found in Edward A. Van Schaik, *A Management System for the Information Business: Organizational Analysis* (Englewood Cliffs, NJ: Prentice-Hall, Inc., 1985), 162.

[4] This nomenclature stems from the author's experiences with these management processes.

[5] Batch processing (or scheduled production processing) refers to batching or accumulating transactions for processing. Accumulating hours worked and rates of pay for processing payroll checks is an example of batch processing.

[6] David R. Vincent, "Service Level Management," *EDP Performance Review*, April 1988, 3. Well-executed service-level agreements improve technology integration and assist users in finding more IT opportunities according to this article.

[7] See the appendix to this chapter for a sample service-level agreement.

[8] C. N. Witzel, "Service-Level Agreements: A Management Tool for Technical Staff, *Journal of Capacity Management*, June 1983, 344.

[9] John P. Singleton, Ephraim R. McLean, and Edward N. Altman, "Measuring Information Systems Performance: Experience With the Management by Results System at Security Pacific Bank," *MIS Quarterly*, June 1988, 325.

[10] Response times for trivial transactions at the user's terminal or workstation can vary significantly from those measured at the CPU or server. Response time is mostly a function of application complexity, system loading, or contention for secondary storage devices. Response time at the terminal includes delays due to processing within the terminal, contention at the terminal control unit or server, loading of the telecommunication links, and delays within the communications controller or processor.

[11] Sophisticated firms conduct statistically significant customer satisfaction surveys daily and compare the results to internal service measures. The internal service measures are refined and improved based on survey data. These firms are able to improve service and to reduce costs significantly through improved resource utilization.

[12] Robert E. Marsh, "Measuring and Reporting User Impact of Substandard Service," *EDP Performance Review*, March 1989, 1. The author develops a service-level indicator that measures responsiveness, availability, and reliability. The article describes its construction and use.

[13] Kathryn Hayley, "CIO Challenges in the Changing MIS Environment," *Journal of Information Systems Management*, Summer 1989, 8. This article states that departments having service-level standards range from about 88 percent in the insurance industry to about 54 percent in the energy industry.

APPENDIX Sample Service-Level Agreement

SERVICE-LEVEL AGREEMENT

The purpose of this agreement is to document our understanding of service levels provided by ______________________ to ______________________.

This agreement also indicates the charges to be expected for the service described herein and the source and amounts of funds reserved for these services. This is not a fixed price agreement. Charges will be made for services delivered at rates established by the controller.

The date of this agreement is ______________________.

Unless renegotiated, this agreement is in effect for 12 months beginning ______________.

Specific type of service	*Application service*	*Hours of use*
______________	______________	______________

Describe agreements reached on:

System reliability ______________________

Reporting process ______________________

Financial considerations ______________________

Additional items ______________________

Signatures of parties to this agreement

by: ______________________ by: ______________________

Service provider *Application owner*

13 *Problem, Change, and Recovery Management*

A Business Vignette

Aberrant Circumstances Cause Massive Outage[1]

On September 17, 1991, a power outage at an AT&T switching center in lower Manhattan caused a seven-hour shut down that was labeled *the* worst disruption of air traffic in U.S. history. Air traffic controllers in the Northeast U.S. were unable to operate, leaving 85,000 passengers stranded. The Federal Aviation Administration (FAA) vowed never to let this happen again.

The trouble began early that morning when Consolidated Edison asked AT&T to reduce commercial power consumption in anticipation of a high electrical load later, due to exceptionally warm weather. At around 10:00 A.M., AT&T cut over to its own diesel generators at its Thomas Street switching center in lower Manhattan. What followed, according to Kenneth Garrett, senior vice-president of network services at AT&T, "was largely caused by a combination of aberrant circumstances. The problem was not a result of any systemic or fundamental failure of AT&T's technological and human resources," he said.

The switching and other equipment at Thomas Street is operated on direct current (DC) converted with rectifiers from alternating current (AC) supplied by Con Edison. Upon switching to the generator, an electromechanical overload relay that was set too low shut off power to the rectifiers, forcing the Digital Access and Cross-Connect system (DACS) driving three large phone switches to operate on battery power alone. The battery life with this load is six hours.

Moving to battery power normally triggers audible and visual alarms on several floors of the building housing the switches. Visual alarms on the 15th and 20th floors operated normally, but were not seen because of their location in unoccupied areas. Earlier, employees had disabled the audible alarms on these floors. On the 14th floor of the switching center, the bulb in the visible alarm was burned out, and the wires to the audible alarm had been accidently cut during remodeling and had not been replaced. In short, all the alarms in the building signaling the shift to battery power were either not working or not observed.

Meanwhile, at AT&T's remote monitoring site in Georgia, things appeared normal; battery operation was not monitored there. Eventually, operating personnel at the Thomas Street station noticed one of the visual alarms signaling operation on battery power but, by this time, only a few minutes of battery power remained. The system shut down at 4:50 P.M..

Efforts to recharge the system and get the phone network operating again began immediately and, by 11:00 P.M., about 90 percent of domestic and international lines were restored to service. For some, phone service was unavailable for more than 6 hours.

This incident provoked numerous federal inquiries: the House Telecommunications and Finance Subcommittee, the Federal Communications Commission, and the House Government Operations Committee. In addition, the FAA began to take corrective action to improve its unreliable network.

During the previous 12 months, the U.S. air traffic control system around the country experienced a total of 114 network outages, about 8.7 per month, averaging 6.1 hours each. In July 1992, 10 months after the Manhattan phone outage, the FAA announced a nearly fail-safe national network for the FAA's National Airspace System. MCI developed the network and began phasing it in over three years.

The network provides digital links to 20 regional centers, 80 major airports, and 5,000 other FAA facilities. The net is designed to offer 99.999 percent availability, meaning no more than 5.3 minutes of annual downtime. The mean time to repair the net is less than 30 seconds. The system is built on redundancy so that at least two paths lead to any site, and all sites can be monitored by the network management center in Reston, VA, which itself is duplicated in Sacramento, CA.

The 10-year outsourcing deal with MCI, worth $856 million, is expected to reduce operating costs in addition to improving reliability, according to the FAA.

INTRODUCTION

The preceding chapter laid the foundation for a disciplined, systematic approach to managing production operations, a critical success factor for service providers. This chapter develops the tools, techniques, and processes for managing three more essential disciplines: problem management, change management, and recovery management. The relationship of these processes to the complete production operations management system is displayed in Figure 13.1.

The processes of problem, change, and recovery management are grounded in the discipline of service-level agreements. In turn, they form the basis for the discipline of management reporting. Service-level agreements are the foundation, and management reporting is the capstone, of the disciplined approach. Problem, change, and recovery management are some of the processes forming the management system to ensure the attainment of service levels. The remaining processes are described in subsequent chapters.

The focus of problem, change, and recovery management is on departures and potential departures from acceptable operation. The purpose of these management processes is to correct deviations from the norm and to prevent future deviations. When a problem occurs, management takes action to correct the problem and to remove its source. The problem-management discipline guides managers in taking corrective action when problems arise, and ensuring that the same or similar problems do not recur. Frequently, managers introduce system changes to increase system capability or to take corrective action to overcome problems. Change, however, is a rich source of potential problems. Change management concentrates on minimizing the risk of system alterations and providing management control for change activities. However, uncontrollable events may occur that lead to major computer operations problems regardless of management diligence. Recovery management deals with these eventualities.

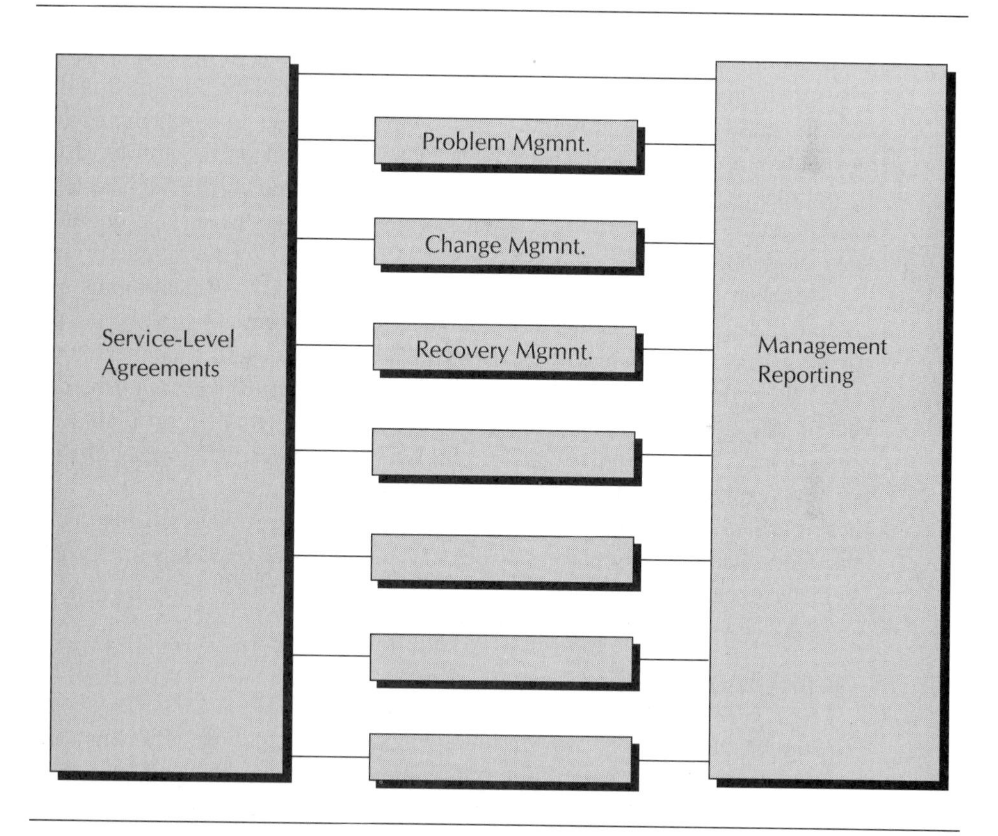

FIGURE 13.1 The Disciplines of Problem, Change, and Recovery Management

These disciplines apply not only to IT and other service providers but must be part of the user's management system. System users depend on system operations; they must be intimately involved in the operational processes. In addition, as user departments acquire computing hardware and application software, they begin to assume responsibility for managing problems, for controlling system changes, and for developing recovery and contingency plans. Users who own and operate equipment and applications must implement management disciplines or suffer degraded performance from poorly controlled systems. For user-owned systems, IT assumes a consultancy role. The disciplined approach to operations described in this chapter applies equally to IT systems and to user owned and operated systems.

PROBLEM DEFINITION

The word *problem* means different things to different people. In the business environment, problems are in the mind of the beholder. All observers do not necessarily perceive problems in the same way. For the purpose of this discussion, and to prevent ambiguity, *problems* are defined as incidents, events, or failures, however small, that have a negative impact on the ability of the operations department to deliver service as committed in the service-level agreements. Problems include actual failures that result in service degradation, or potential failure mechanisms that could lead to degraded service.

Necessarily, this definition excludes many other kinds of problems that computer operations managers might face. For example, the salary budget for next year may be less than the requested amount and managers must revise their financial plans accordingly. Parking for employees may be disrupted due to temporary construction activity thus creating problems for some, especially during periods of inclement weather. Management may need to resolve these issues but, unless they lead to missed service levels, these problems are excluded from the discussion of problem management. Managers face many important issues, but the thrust of the disciplined management process is to provide committed customer service.

PROBLEM MANAGEMENT

Problem management is a disciplined process for detecting, reporting, and correcting problems impacting the attainment of service-level objectives by the service-providing department. The problem sources include hardware, software, network, human, procedural, and environmental failure that disrupt or potentially disrupt the service level delivered. Problems are mostly generated within the firm, but may originate from external sources, as well. Problems may originate in the client organization or in the IT organization. The problem management system must deal with problems from all sources that affect the delivery of satisfactory service.

The service provider uses the problem management system to achieve service objectives. These objectives are listed in Table 13.1.

TABLE 13.1 Problem Management Objectives

Reduce defects to acceptable levels
Achieve committed service levels
Reduce the cost of defects
Reduce the total number of incidents

Implementation of an effective problem management system will reduce the number of incidents or events to an acceptable level while reducing their associated cost. Successful implementation of the problem management system will enable the service provider to deliver committed service levels to user groups. The process will minimize the total number of problems requiring attention. Dealing with fewer problems conserves organizational energy and reduces costs. In addition, the problem management system provides managers with tools for understanding the root causes of the relatively infrequent failures that ultimately occur.

Problem management is a required process for successfully managing production operations. It is one of the first steps in the overall management system. Problem management is a necessary condition for success, but is not necessarily sufficient for success.

The Scope of Problem Management

Problem management broadly encompasses many sources of potential and actual impacts to service levels. The scope of problem management is outlined in Table 13.2.

TABLE 13.2 Scope of Problem Management

Hardware systems
Software or operating systems
Network components or systems
Human or procedural activities
Application programs
Environmental conditions
Other systems or activities

Failures in hardware systems are indicated by unscheduled system restarts, unanticipated or intermittent operations, or other abnormal operating conditions. These difficulties may arise from failures in I/O channels or in peripheral devices such as tape drives, disk storage devices, or workstations. Failures can originate within the CPU, as well. System software is also a problem source. Parts of the operating system may malfunction or some important function may not be available to the applications. Intermittent or unanticipated results may also originate within the system software.

Most major systems are connected to networks that are an occasional source of difficulty. Networks may function intermittently or may fail totally. The failure may originate in network hardware, network software, or operating system communication programs. Failures may result from incorrect human interactions with the network or improper procedures governing the human interaction.

System failures may occur at the person-machine boundary or may result from improper manual procedures. The procedures themselves may lead to failure if they are incomplete, incorrect, or subject to misinterpretation, or if they do not cover all situations. Operator failures may also result from not following established procedures or from poorly executed procedures.

The application portfolio is a frequent source of failure particularly if it is undergoing major enhancement or maintenance activity. Chapter 8 pointed out that failures are especially likely if the applications are old or have had frequent, past maintenance. The applications may yield unintended or unexpected results, or they may display unusual or abnormal program termination. The expected function may be missing or incorrect. These situations usually result in unsatisfactory performance and often lead to missed service levels.

The environment is another source of difficulty. Power disruptions, air conditioning difficulties, or failures in the heating system may all lead to service-level disruptions. On a larger scale, earthquakes, hurricanes, or other natural disasters may be an infrequent source of serious difficulty. In addition to the causes outlined above, if service levels are missed for any other reasons, these defects are also included in the scope of problem management. In general, any situation that impacts service or has the potential to impact service must be addressed within the scope of problem management.

The Problem Management Process

The process of problem management includes tools and management techniques designed to detect, report, correct, and communicate the particulars of problems and their resolution. It embraces an informal organization and defines responsibilities for its members. It includes protocols and regimens for solving problems and for analyzing and reporting the results. Individuals and groups committed to minimizing incidents rely on the process to maintain service at the highest attainable level.

The Tools of Problem Management

The essential tools and processes of the problem management discipline are displayed in Table 13.3. Complete and effective implementation of these steps comprises the primary element of the problem management system.

TABLE 13.3 Tools and Processes of Problem Management

Problem reports
Problem logs
Problem determination
Resolution procedures
Status review meetings
Status reporting

Problem reports record the status of all incidents. Each report begins with problem detection and records corrective actions until the issue is satisfactorily resolved. The person who discovers the problem initiates the report, then files it with the individual responsible for the problem management process. The problem report includes the items listed in Table 13.4.[2]

TABLE 13.4 Contents of the Problem Report

Problem control number
Name of problem reporter
Time and duration of the incident
Description of problem or symptom
Problem category (hardware, network, etc.)
Problem severity code
Additional supporting documentation
Individual responsible for solution
Estimated repair date
Action taken to recover
Actual repair date
Final resolution action

The problem report receives updated information during the time the problem is open or unresolved. When the incident is closed, the report is posted with the problem resolution data, and it is filed. A record of all problems is maintained in a problem log. The problem log is a method for recording incidents, assigning actions, and tracking reported problems. The log contains essential information from the problem reports for all resolved and unresolved problems during some fixed period.[3] A convenient and useful period is the rolling last twelve months. The problem log and associated reports provide management with data for analysis. Careful analysis of prior problems is very useful for anticipating and reducing future problems.

Problem Management Implementation

Problem status meetings are an essential part of the management process. Status meetings should be held regularly to discuss and document unresolved problems, to assign priorities and responsibilities for resolution, and to establish target dates for corrective action. A regularly scheduled, daily problem meeting is a reasonable approach. Major problems require additional reviews. Some questions for these extraordinary reviews are: "How could this problem have been prevented?" and "How could the impact of this problem been minimized?" Status meetings and reviews should focus on recurrent problems and careful analysis of unfavorable trends. Areas experiencing high problem levels should be given special attention. Trend analysis will provide clues to weak or vulnerable areas. Special consideration should be given to rediscovered problems because this may indicate incomplete or ineffective problem resolution.

Representatives of each business function having open problems and key individuals representing IT should participate in the status meetings. The IT representatives are members of systems programming, computer operations, and the applications programming groups. The hardware service organization liaison person should also be present. The problem management team leader should chair the meetings. When the process is operating effectively, it is not necessary for managers to be present. Managers are responsible for supporting the effort and for ensuring participation and results, but generally the meeting will be more effective if participation is limited to nonmanagers. Managers tend to assess people, but what is required at the problem meeting is the assessment of problems. People assessments should be reserved for another forum.

Problem determination and resolution are aided by classifying the incidents by categories such as hardware, software, and networks, and by establishing problem severity levels. The team members should determine the categories and severity levels. Individual problem resolution assignments should be based on documented individual responsibilities. The documentation will help clarify each individual's role and ensure that proper skill levels are applied to problems. For example, if the problem is found to reside in the operating system utility programs, the systems programmer with responsibility for these utilities should be assigned this problem.

Problem resolution procedures include action plans and estimated resolution dates from the individual to whom the problem was assigned. Resolution status is reported to the team leader and to the affected areas or users. Implementation of corrective action is scheduled after review with all affected parties and with concurrence of the team leader. If the problem involves system changes, the change management process may be invoked. In any event, when it can be demonstrated that the problem has been resolved to the team's satisfaction, the problem report is closed and posted to the log. Figure 13.2 illustrates the flow of the problem management process.

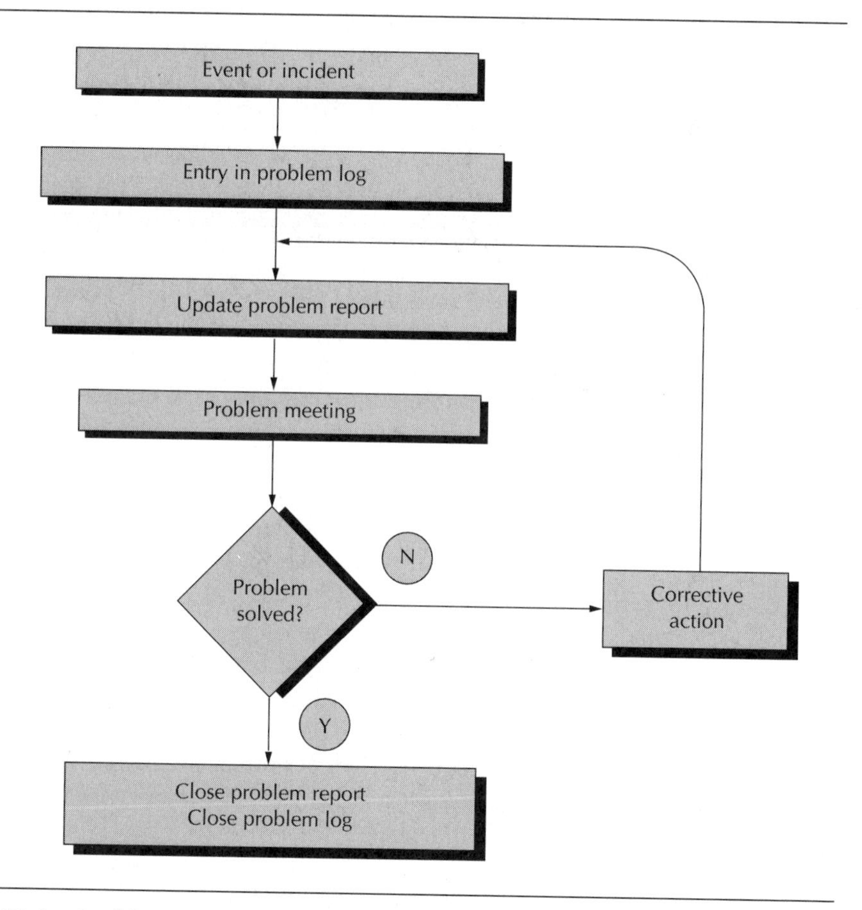

FIGURE 13.2 Problem Management Process Flow

Automated tools can help create problem reports and maintain the problem log. Computer-based tools support interactive databases and make the task of generating reports for the team members and for management easy and efficient.[4]

Management's attitude toward the people involved in problem management is vital for the discipline's success. The process is designed to solve problems and to help prevent incidents. Managers must refrain from using it as a vehicle for individual performance assessment. Instead, managers must establish and foster an atmosphere of free and open communication among all the participants. The environment should exclude finger-pointing and assigning blame. A productive environment encourages a spirit of mutual cooperation between service providers and system users. Jointly and separately the team members should feel responsible for resolving incidents promptly and for preventing incident recurrence. Taking responsibility for resolving an incident is not equivalent to accepting responsibility for causing the incident. Some of the most effective status meetings occur in a supportive environment. When the environment is favorable, the presence of managers is not necessary.

PROBLEM MANAGEMENT REPORTS

IT managers must receive summary reports documenting the effectiveness of the problem management process. The summary data should reveal trends and indicate problem areas impacting the delivery of satisfactory service levels. These summary reports are useful in establishing future service levels. IT plans should account for problem activity when establishing resource and expense levels.

Problem reports may be produced weekly, monthly, or on request. The reports highlight numbers and causes of incidents by category and severity. The duration of service outages, length of time for problem determination, and time for resolution must also be reported. An aged problem report is very useful in determining the organization's responsiveness to incidents. An important result derived from the report is an analysis of duplicate or rediscovered problems. The rediscovery rate is important because it measures problem resolution effectiveness. Another important result stems from an analysis of problems caused by changes applied to the system. Problems caused by system changes can measure the effectiveness of the change management process.

Well-conceived reports enhance the process by providing concrete evidence of its effectiveness. Problem reports demonstrate the service provider's responsiveness to customer needs. They lend credibility to service-level commitments and reinforce the alliance between the service recipient and the service provider. Problem summary information is especially useful when renegotiating SLAs. The entire problem management process is essential to operational success.

CHANGE MANAGEMENT

Change management is a management system for planning, coordinating, and reporting system changes that have the potential to impact negatively the service that the operations department provides. It is a disciplined approach to planning, coordinating, and reporting system changes. Changes are the focus because they have the potential to influence service delivered by the service provider. Changes can involve modifications to hardware, software, or networks; they can originate from human processes, manual procedures, and the environment. The objective of the change management process is to ensure that system changes are implemented with minimum or acceptable levels of risk and that service levels are not jeopardized by planned change. The process foundations are service-level agreements. Service providers deal with many changes, but only those with the potential to impact service delivery are included in the change management discipline.

Modifying a complex, human-machine system raises significant, often unrecognized risks. These include disrupting system processing and reducing service levels. Changes tend to generate problems. Frequently, well-intentioned individuals are motivated to introduce small changes in areas that may appear remote from the operation. Their intent is to improve system operation in some way. Ill-considered changes to sophisticated environments often lead to difficulties. For these reasons, a disciplined approach is required.

Rapid technological advances, coupled with information technology growth within the firm, make change a way of life for IT managers. IT and client organizations must cope with many issues when introducing change. A disciplined management approach to change implementation will reduce problem incidents and yield significant benefits to the firm.

Problem management is tightly linked to managing changes since problem resolution frequently requires change. Change management and problem management must act in concert to ensure that the solution to one problem does not result in another. In an undisciplined environment, it is common for the solutions to two problems to yield one additional problem requiring action. Some organizations report that 80 percent of corrective actions result in additional problems. An effective, disciplined approach to problem and change management has the potential to reduce the ratio from 2:1 to 200:1 or more.[5] The resulting benefits to the firm are greatly improved effectiveness and significantly reduced costs. There are other less obvious but important advantages, as well. For example, the image of the IT organization is enhanced and employee satisfaction in the client and IT organizations is improved when problems are minimized and changes are controlled.

The Scope of Change Management

Change management addresses many activities that could negatively impact the delivery of committed service levels. The major sources of change that bear on service levels are listed in Table 13.5.

TABLE 13.5 Change Management Scope

Hardware changes
Software changes
Application program changes
Changes in the environment
Procedural changes
Equipment relocation
Problem management-induced changes
Other changes affecting service

Most changes to computer-based information systems or to their environment involve the risk of service disruption. All changes that induce risk must be controlled using the change management process. Hardware changes of all types should be included in the change management plan. Any alterations to operating systems, network software, utility programs, or other support programs must be considered. Application programs undergoing maintenance or enhancement are prime candidates for the change-management discipline. Earlier chapters noted that the application-enhancement process is error prone; it frequently leads to operational failures. With this in mind, changes to application programs must be monitored carefully, too.

The Change Management Process

Change management includes tools, processes, and management procedures used to analyze and implement system changes. The ingredients of the change management process are itemized in Table 13.6.

TABLE 13.6 Change Management Ingredients

Change request
Change analysis
Prioritization and risk assessment
Planning for the change
Management authorization

Production environment changes enter the change management process via a change request.[6] The change request document is a permanent historical record detailing the change actions from initiation through implementation. It records the type and description of the requested change and identifies prerequisite actions. This document is a record of the risk assessment, the test and recovery procedures, and the implementation plan. It is also an approval vehicle that documents the individual responsible for introducing the change, and management authorization. Table 13.7 identifies the major contents of the change request.

The change request document is developed as the change is analyzed, planned, and implemented. Upon completion of the change, the change request is filed for future reference and analysis. The change request must also be recorded in the change log.

TABLE 13.7 Change Request Document

Description of the change and log number
Problem log number if resulting from an incident
Prerequisite changes
Type of change
Priority of change and risk assessment
Test and recovery procedures
Project plan for major changes
Requested implementation date
Individual responsible for managing the change
Individual requesting the change, if different from above
Management authorization

Changes are approved for implementation by the change management team. The team is composed of individuals representing computer operators, systems engineers, application programmers, and hardware and software vendor liaisons. The facility manager must be represented because changes in the physical environment (heating, power, space, etc.) are critical to the operation. One of the team members representing the operations group leads the change management team. Often, there are advantages to having some representatives serve on both the problem and change management teams. The change team meets regularly and frequently for short periods to analyze, review, and approve the plans for change.

The change management team focuses on risk analysis. It is essential that changes are categorized by potential risk to the operation. Change analysis should focus on factors that quantify the magnitude and significance of alteration to ongoing operations. For example, if the change significantly impacts system performance or capacity, or if it has a dramatic impact on the customers, it probably carries major risk. If this change interacts with other pending changes, or if it mandates future change, it must be reviewed very critically. Changes that affect more than one functional area or that require individual training are riskier than those without these conditions. A sound test and implementation plan assists greatly in reducing risk.[7] Successful teams ensure low or acceptable risk to the ongoing stream of changes.

Major or extraordinary changes require a thorough project plan that identifies people, responsibilities, target dates, and implementation reviews. Change team attention to numerous details is required for major changes. The change team technically audits the process and, in special circumstances, may request additional reviews by nonpartisan technical experts. The change will be approved only when the change team is confident that it can be implemented with low risk of failure. The team must also be confident that, if the implementation is not successful, the recovery plan will be effective.

The goal of the change management process is to minimize the risk associated with system alterations. This goal applies to all changes, whether they are major and substantial or relatively minor in scope. The replacement of a large mainframe may require change planning and management over the period of a year or more. Installing the latest version of the operating system on network servers may take months of planning, but adding another direct access storage device may be completed in a week or so. Altering the format of a batch program may require only a day or two. In any case, all changes, large and small, should be implemented only after all risks to the firm have been considered and evaluated.

Changes must be capable of being tested, and the team must evaluate the test plan for thoroughness and completeness. The test cases and expected test results are documented in advance and compared with actual results upon implementation. Variances between expected and actual results require analysis and rationalization. Variances may reveal incomplete preparation or ineffective implementation. Significant variances may indicate more problems or additional changes in the future.

REPORTING CHANGE MANAGEMENT RESULTS

Successful change implementation must be communicated to all affected parties, and the details must be entered into the change log. The log should be reviewed periodically to ascertain trends and assess process effectiveness. The assessment focuses on problem areas, expectations versus results, required emergency actions, and areas requesting or requiring numerous changes. One measure of process effectiveness is the percentage of changes implemented in a problem-free manner. This number should exceed 99 percent in a well-disciplined environment.

Periodically, the service provider must analyze trend data and summary information obtained from the change log review. Managers providing information services are responsible for ensuring effective operation of the change management process. They need information from the change log to make their assessments. Like problem management, change management does not need continuous, detailed management involvement. The process needs management direction and support but not continuous intervention.

RECOVERY MANAGEMENT

Information technology managers recognize that the resources they manage are critical to the firm's success. They are keenly aware that unavailability of major on-line applications will have immediate and serious consequences and that degraded operation of customer-support applications jeopardizes customer relations. The consequences may be loss or deferral of sales and revenue, or possible damage to the firm's reputation.

A dramatic example of system failure occurred on May 12, 1989, when the American Airlines Sabre reservation system crashed. Downtime lasted approximately 13 hours. The crash was caused by the failure of a utility program formatting large, direct-access storage devices. The program went astray, destroying the volume serial numbers on 1,080 volumes of data and making access to the data impossible. The entire system, based on eight IBM

3090-200E mainframes, was inoperable from 23:38 on May 11 until 10:29 on May 12, after which time it gradually returned to full operation within 90 minutes. This system failure disabled American's load management process and permitted travel agencies and corporations to book deep-discount tickets in an unlimited fashion and to rebook higher fare tickets at discounted rates.[8] The extent of American's losses from these activities has not been published.

In other situations, the firm may incur degraded internal performance and loss of efficiency. Vital information assets can be damaged or rendered inoperable by numerous natural or man-made actions. In the face of this reality, prudent IT managers will develop plans designed to cope with the potential loss or unavailability of information resources. Recovery management deals with the possibility that IT resources may be lost or damaged, or that they may become unavailable or unusable for any reason.

In recent years, major disasters caused loss or damage to computer and communications facilities and surely, more are in the offing. In 1991 Florida experienced a major hurricane, the Chicago business district was flooded in 1992, in 1993 the World Trade Center and the London financial district experienced bombings and the Mississippi river flooded, parts of California were damaged in 1994 by earthquakes followed by floods and fires later, and a federal building in Oklahoma City was bombed in 1995. These events underscore the importance of recovery planning.

Many managers experience difficulties dealing with the issue of recovery planning. They view the risk of losing the computer center from tornado damage, for example, as extremely low. Or they consider the solution to this situation to be extremely complicated and difficult to manage. For these reasons and others, some IT managers rationalize that this type of planning has a low priority—low enough that they devote little or no effort to recovery concerns.

On the other hand, events such as an emergency power outage or a head crash on a direct-access storage device are more likely to occur. Management of these contingencies is easier to visualize and implement, and managers usually appreciate the need to prepare contingency plans for these events.

IT managers in financial institutions are required to have a recovery plan. Federal regulations require that banks and other financial institutions maintain disaster recovery plans on file for inspection by federal auditors.[9] The Office of Management and Budget has directed federal agency attention to this subject since 1978 through Circular A-71, "Security of Federal Automated Information Systems." Updates to this circular have directed the Commerce Department to develop standards and guidelines that regulate recovery planning and assist other federal agencies in developing their plans.

IT managers are dealing with a continuum of possibilities involving the likelihood of disaster versus the seriousness of the event's consequences. The recovery management process deals with these uncertainties and articulates actions in cases where management believes the associated risk is worth the planning effort. Figure 13.3 shows the relationship between risk of loss, probability of occurrence, and management action required.

Recovery plans deal with potential situations in which the risk of loss can vary from low to high. Planning is appropriate because of the high probability that the event may occur. Virus invasion, server CPU failure, and telecommunications link severance are examples of events for which recovery plans should be prepared. Contingency planning is the term reserved for high risk events with low probability of occurrence. Low risk events with low probability of occurrence can be handled as they arise through the problem management system.

		Risk of Loss	
		High	Low
Probability of Occurrence	High	Recovery Planning	
	Low	Contingency Management	Problem Management

FIGURE 13.3 Recovery Planning and Contingency Management

Each organization must analyze its situation and place possible events into the appropriate category. For example, a virus infection may not be related to geographic location but floods or earthquakes probably are. Firms also differ in degree of loss from identical events. For example, a firm that has EDI links to suppliers and customers is probably more at risk from telecommunications failures than one which uses WANs exclusively for internal communications. Sound recovery management demands a thorough analysis of the firm's risks and probabilities.

Recovery plans provide backup resources that help restore service in the wake of likely events. In most cases, these plans are like insurance policies: they're important to have, though the hope is not to need them. Prudent managers implement recovery plans because they know that complex and vital resources can suffer from many types of loss. Recovery from minor emergencies is more than a once-in-a-lifetime experience. Recovery from major disasters is something no manager wants to experience, but an event for which every manager must plan.

Recovery management is one of the disciplines of production operations. It involves people, processes, tools, and techniques. Recovery management originates from the conviction that the IT organization has a commitment to defined service levels. Recovery management is based on the realization that contingency planning can effectively ensure service levels. Like the previous two processes, recovery management involves collaboration between managers and nonmanagers representing the service provider and the client organizations. Recovery management requires that managers cooperate with user groups to achieve mutual goals.

Because recovery management is a critical part of every firm's plans, IT managers must be especially attentive to recovery planning. The importance of recovery management is illustrated in the following Business Vignette.

CONTINGENCY PLANS

Contingency planning addresses high risk events that have a low probability of occurrence. It ensures successful performance of critical jobs during periods when service or resources are lost or unavailable. The application owner assumes responsibility for contingency planning that ensures the operation of critical applications. For example, the production control manager who owns the materials requirements planning system is responsible for managing production in the event of a major system hardware failure. However, the responsibility for correcting the system hardware failure resides with the provider of service. In this case as in most others, the owner of the application has more options and more flexibility for dealing with the problem than the service provider.

The split in responsibilities outlined above calls for more, rather than less, interaction between application owners and service providers. The application owner should obtain critical planning information such as probable frequency of outages, likely duration of outage, and anticipated extent of service loss from the IT organization. Additionally, IT may recommend helpful action plans to the application owner. As with the disciplines discussed earlier, all affected parties must work together in a disciplined manner to serve the firm's best interests.

Crucial Applications

In consultation with application owners, IT managers should identify the most critical applications in the firm's portfolio. This is not an easy task, but it is a very important one. It is not easy because the critical nature of an application is subjective and can only be seen in context with other applications.

A Business Vignette

When Disaster Threatens[10]

In 1989, Hurricane Hugo and the California earthquake reminded IS professionals of an important principle: Good contingency planning can counteract the effects of natural disasters. However, providing protection against disasters can be very costly. IS can survive almost any contingency, providing the plans, the equipment, the backups, and the money are in place before the disaster strikes. Firms that elect not to prepare for disastrous events may need to rethink their positions.

On August 15, 1989, a medium-sized credit union in Charleston, SC, decided to sign on a hot-site disaster recovery contract. A month later, hurricane Hugo blew Charleston apart, along with its power supplies. The credit union had a mainframe computer upon which all its operations, including automatic teller machines, depended. Victims of the hurricane needed cash.

The disaster recovery contract with Sun Data, of Norcross, GA, came to the rescue. Under the terms of this contract, the credit union transferred its data processing to a hot site in Atlanta and, with the help of local communication links, it was able to provide service to its customers. Credit union depositors needed money to get back on their feet in Hugo's wake, and federal officials wanted to know when funds would be available. By using the backup facility, installing some manual processes, and cooperating with other financial institutions in the area, the credit union provided critical service to their 52,000 account holders.

Reciprocal arrangements with other similar institutions do not always work. This credit union might have set up a disaster plan with another Charleston Bank, for instance—and Hugo would have blown them both out of business. One of the most common traits of major disasters is prolonged, widespread electrical power outages. While effective for local difficulties such as burst water pipes, facility sharing with other local firms is ineffective for dealing with broad regional disasters. Power problems usually accompany other problems and, increasingly, firms are considering uninterruptible power sources that can bring down the mainframe and large data stores gracefully, prior to switching over to the hot site. These UPS devices, though expensive, also protect against power surges.

Sound disaster recovery planning and effective plan implementation protects the firm and its reputation, provides continuous customer service (sometimes when it's needed most), and indicates responsible IT management. No firm can afford to be without it.

That is, an application may or may not be critical depending on whether or not other applications are available. For example, order processing for the manufacturing plant is an important function, but on-line order entry may be more critical. Failure in order entry means loss of business, but failure in order processing may only delay the start of production. Interactions among applications are important in determining which applications are critical.

Timing of the emergency is another factor to consider. The hardware and software supporting on-line applications such as air traffic control, oil refineries, and reservation systems, for example, are always critical. The payroll application is generally critical on a weekly basis, but the ledger may be critical more often. Finally, the expected duration of the outage is another complicating factor. Most applications become critical if the service outage is prolonged.

There are two reasons for performing the analysis in spite of the obvious difficulties and ambiguities. The first is to prioritize the applications that must be recovered first. In the case of widespread failure, a predetermined priority sequence organizes the recovery work and helps prevent chaos. The second reason is to understand the possible trade-offs inherent in the prioritized sequence of application recovery. If hardware or telecommunications capacity is degraded, then it is very useful to know what load can be shed in favor of higher priority applications. Definition and analysis of these trade-offs must involve system owners and users. A well-organized approach directs user's attention to the most critical tasks.

Environment

Each critical application requires a specific environment for successful operation. The important environmental factors are identified in Table 13.8.

TABLE 13.8 Application Environmental Factors

Hardware system components
Operating system and utility software
Communications resources
Databases
Sources of new input data
Knowledgeable users
Critical support personnel

Plans that include all the critical environmental parameters must be developed for each important application. The interaction between the critical applications and all other related applications or systems must be developed as part of the plan. This planning effort may be quite large. Prior to undertaking this work, managers must understand the trade-offs between planning costs and the cost of loss or operational failure.

Emergency Planning

The emergency planning process addresses situations resulting from natural disasters such as floods or wind storms, and from events such as riots, fires, or explosions. Emergency plans must contain the steps needed to limit the damage and deal with the problems. Typically, these events are clouded with uncertainty and have a low probability of occurrence.

The most effective mechanisms for dealing with these emergencies involve early detection and containment procedures to limit the damage. Detection mechanisms include fire alarms and detectors for smoke, heat, and motion. In some cases, liaison with the civil defense authorities is appropriate. Additionally, the plans for handling emergencies usually include procedures for evacuation, shelter, containment, and suppression.

The plan must identify evacuation conditions, establish means to communicate evacuation plans, and contain procedures to ensure that evacuation has been successfully completed. Shelter plans include protection from rain, leaking pipes, and water from other sources in addition to protection from wind damage and flying debris. The plans for containment usually involve storage rooms for fuel and other hazardous materials and for vital information such as tape volumes and critical documents. These storage facilities are constructed with fire- and water-retardent walls, floors, and ceilings. Suppression plans describe the procedures to extinguish fires, stop water flow, clear the facility of smoke, and maintain security for the event's duration.

Emergency plans must be clearly and succinctly documented. They must be communicated to everyone potentially involved with emergency situations. Responsibilities must be established and directed, and individuals must be trained to respond effectively. Emergency plans must be tested periodically to ensure that employees and managers are completely familiar with their responsibilities and know how to respond effectively. IT managers are responsible for developing effective emergency plans for their facilities.

Strategies

Many options are available to contingency planners. Among others, these options include manual operations, backup systems, or data servicers. Backup systems may be located at the same firm or may exist at cooperating firms. The firm may also use the services of hot site providers. These enterprises maintain operating hardware and software configurations for use in emergencies. The credit union described in the previous business vignette survived Hugo by using a remote hot site.

Disaster recovery service is a large and growing industry. With growth rates of 20 percent per year, total industry revenue exceeded $1 billion in 1995. Services consist of hot-site providers, contingency planning and consulting services, and firms that sell software systems for disaster recovery planning. There are many vendors willing to work with IT managers to orchestrate disaster planning and recovery.

In most cases, resorting to manual procedures is usually the least effective disaster response. If the system is relatively simple and the outage of short duration, manual processing may be effective. But in many instances, manual processes are impossible to implement and completely ineffective. For example, manual processing will not sustain an airline in the event of a reservation system failure nor will it help engineers when their computer-aided design system fails. However, with relatively uncomplicated systems and for short outages, some form of manual fallback is usually appropriate. For example, with a low-volume, on-line order-entry system, clerks can manually complete order forms while the system is recovering from a brief outage. When the system returns to full operation, the data on the forms is entered into the on-line system. Short-term manual methods may suffice for simple, noncritical systems or applications.

For severe outages, some additional form of backup is required for all systems. A multiple system with distributed architecture may be effective if the firm has geographically separated processing centers linked with broadband communications lines. The distributed architecture can be particularly effective if there is a reasonable degree of hardware and operating system compatibility. Compatibility is very important for backup or recovery purposes and may be desirable or necessary for other reasons, also.

Telecommunication networks are potentially effective in transmitting programs and data between sites for backup purposes and for communicating results during recovery. IT managers, sensitive to recovery management issues, strive for this architecture.

Sometimes, mutually beneficial arrangements can be negotiated with cooperating firms. Each firm needs to provide for contingencies. If technical details receive sufficient attention, cooperation may be mutually advantageous.

Documents outlining the terms and conditions of the cooperative agreement must be prepared and signed by representatives of each firm. Effective agreements outline procedures for testing the backup and recovery procedures and for implementing the recovery plan.

Some firms provide data processing centers for other firms to use in the event of a severe outage. These "insurance installations" write a policy which provides fee-based system capacity for emergency use. Technical issues of hardware and software compatibility, and program and data logistics must be addressed. Network availability and capacity must also be factored into the decision to use these services. In addition, some major firms provide disaster planning tools or total disaster planning services for organizations unprepared to undertake this activity independently.

Service bureau organizations are another alternative in the recovery-planning process. Frequently, these data processing organizations are well equipped to handle additional processing loads. Your firm may already employ data servicers for peak-load processing or for special applications. It would be natural to develop an emergency backup arrangement with them. In these cases, the logistics of backup and recovery may have been partially developed, thus simplifying emergency contingency plans.

Usually, telecommunication systems require special contingency planning because they are highly critical to the firm and because they offer unique planning opportunities for critical applications.[11] Firms with multiple locations linked together via telecommunication networks must consider two issues. The first issue is how will the firm maintain intersite communications in the event of network disruption; the second is how the firm will use the intersite network to assist in backup and recovery for its individual locations.

There are several important considerations in maintaining intersite network systems. Usually, network redundancy, alternative routing, and alternative termination facilities are considered first. Modern networks usually provide ample opportunity to exploit these alternatives. In an emergency, it may be possible to exchange voice and data facilities, or to use the traditional dial network for data. Some firms employ value-added network firms for backup purposes. As an absolute last resort, manual processes may be considered. However, manual processes for network operations are very unattractive, even for emergency purposes.

Using the network for major application hardware and software systems backup and recovery must be addressed as part of the firm's system architecture. If recovery management is an architectural consideration in network design, considerable advantage can be obtained in recovery planning, and significant side benefits are realized, as well. For example, networks that exhibit recovery advantages usually permit efficient load sharing with attendant cost reductions. Networks pose special problems and offer special opportunities for the recovery management planner. Because of their critical role, networks require special consideration in the recovery planning process.

RECOVERY PLANS

The goal of the recovery management process is to develop, document, and test action plans covering the contingencies facing the IT organization and the firm. The actions must include all portions of the organization that engage in information technology activities.[12] The firm must be prepared to cope with adversity in a logical, reasonable, and rational manner. The firm must protect its major information technology investments located throughout its constituent parts.

Individuals in the IT organization are a critical and perhaps indispensable resource. Because of their unique knowledge, skills, and abilities, IT personnel are especially vital in times of emergency. Most recovery plans assume that the people will survive the disaster and be able to carry out recovery activities. But sources of outside help should be considered in the planning, since this assumption may not always be true. The personnel strategy must address the availability of skilled, in-house staff and adjunct personnel from external sources. Recovery plans must include provisions for notifying key people and for managing work assignments during the recovery process.

Disaster planning must consider the equipment and space necessary to conduct essential operations. Since some data processing equipment requires power, air-conditioning, and a raised floor, alternative space within the firm's facility should be earmarked for emergency use. Additional space, not currently owned or leased by the firm, should be considered for planning purposes. Sources and availability of this space will change constantly, thus the options should be reviewed periodically to ensure appropriateness.

The recovery plan must include documented emergency processes outlining actions to be taken by the recovery teams. Departures from standard operating procedures are expected during emergencies. For example, in the event of an emergency, systems programmers may be assigned to all local and remote data centers to implement network recovery procedures. Since networks are critical, the first priority of key technical people will be to ensure their operation.

The emergency personnel roster and recovery team assignments and responsibilities are an important part of the documented recovery processes. This information must be distributed to all employees and managers involved in the recovery operation. To avoid destruction, key individuals must file the plans off site rather than storing them at the firm's facility.

Routine recovery plan testing is necessary to ensure successful implementation. It is as important to test recovery plans periodically as it is to perform occasional fire drills. Testing will maintain employee awareness of their roles and responsibilities and will focus attention on this easy-to-defer activity. Telecommunication links must be exercised and data retrieved and restored to test the completeness and validity of the recovery process. Not all processes can be tested, but the firm should schedule sufficient testing to ensure recovery readiness. IT managers must be confident that recovery plans are thorough, effective, and ready to be implemented when needed.

The importance of recovery management and the extent to which it applies in the daily operation of the firm is illustrated in the following Business Vignette.

A Business Vignette

Risk Management Succeeds

The 1989 San Francisco earthquake brought death and destruction to the Bay area. Roads, bridges, buildings, and utilities were damaged or destroyed, but many computer installations survived virtually intact—in many cases with less damage than the buildings in which they were housed. Computer system survival in an area known for its computer design and development activities was the direct result of disaster planning and advance preparation.

Many Silicon Valley computer technology firms are critically dependent on computer systems for their operation—indeed their survival—and they employ extensive risk management techniques.

One of the predictable risks of operating in the Valley is the loss of electrical power. Protection against a momentary or prolonged power outage requires these companies to install uninterruptable power supplies which take over when routine power fails. These UPS systems offer great protection against a common threat, but are quite expensive to install.

At one firm in the Valley, mainframe computers continued to operate during the quake though less well protected personal computer equipment sustained damage. Mainframe operations were supported by a two-tiered backup power supply—a huge battery system which kicked in immediately, followed later by diesel-powered electrical generators. The company closed one facility to check for damage and to make minor repairs but reopened the following day.

Another large company in Santa Clara County kept most of its facilities open immediately after the quake. In anticipation of earthquakes, the company reinforced its buildings at a cost of about $1 million. This precautionary investment reduced the extent of the damage and saved money in the long run.

One Silicon Valley firm considers these elaborate and expensive plans so essential that the firm has appointed a director of corporate disaster recovery to oversee these operations.

SUMMARY

The disciplines of problem, change, and recovery management are central to the effective operation of a computing facility. These disciplines are built on service-level agreements; they support service commitments and provide some of the necessary conditions for success. Effective management of these disciplined processes and others to be discussed in the following chapters helps ensure successful computer operations. Successful computer operations is a critical success factor for IT managers.

The disciplined approach lends itself well to daily operations in a computing center. This approach also meshes well with the tactical and operational planning essential to production operations. The disciplines discussed thus far are based on the premise that thoughtful, conscientious individuals can work together to minimize the effects of the inevitable difficulties that arise in a complex environment. Both service providers and service users benefit from this structured approach because it maximizes the results to everyone's advantage.

Problem detection, correction, and prevention are vital steps in protecting committed service levels. Changes resulting from problem correction, or required for other reasons, must be carefully controlled and implemented. Occasionally, activities go astray for reasons beyond a manager's control; plans must be formulated to contend with these contingencies. IT managers and other service providers must handle these issues confidently, competently, and effectively in order to succeed.

As distributed computing evolves and grows, and with decentralization on the rise, the disciplines of production operations become more widely applicable. Service providers and IT organizations in operational departments must contend with problems and changes. The disciplines of problem, change, and recovery management are vital to owners of data-handling facilities regardless of their organizational affiliation.

Review Questions

1. Define "problem" in the context of problem management. What is problem management?
2. Why are human and procedural issues included in the scope of problem management? In what ways should these issues be treated differently from the others?
3. What sociological considerations surround problem management implementation?
4. What actions can management take to sponsor and promote effective problem management meetings?
5. Discuss the relationship between the disciplines of problem management, change management, and service-level agreements.
6. What are the tools and processes of problem management?
7. What items are contained in the problem management report?
8. What are the ingredients of an effective problem resolution procedure?
9. Change management deals with understanding and controlling risk. What change management actions accomplish these goals?
10. How can managers ensure that they are sufficiently involved in the change management process?
11. What is recovery management? How is it related to contingency planning?
12. What interactions take place between service providers and service users during the development of contingency plans?
13. Why is it difficult to develop a list of critical applications?
14. What special problems and opportunities does a telecommunications network present in developing recovery plans?
15. Discuss the reasons why recovery strategies should be developed as part of the usual IT strategy and planning process.
16. Why is it difficult to test recovery plans? What are some ways to overcome these difficulties?

Discussion Questions

1. Discuss the recovery management issues that the Manhattan phone outage raised for IT managers.
2. What special considerations enter into the recovery management process when third-party service providers are used?
3. Because of the Manhatten incident and others, the FAA took action to correct its network defects. What similar options are available to firms in the same area? What actions can they take to reduce their vulnerability?
4. Relate this chapter specifically to the notion of critical success factors introduced in Chapter 1.
5. For the topics presented in this chapter, discuss the balance between bureaucracy, effectiveness, and efficiency.
6. How might one attempt to quantify the economic benefits derived from effective operation of problem and change management?
7. Discuss some approaches a firm might use to evaluate contingency and emergency plans and test recovery plans.
8. If you were the Chief Information Officer of a firm that relies on client/server computing extensively, how would you organize people to implement the disciplines discussed in this chapter?
9. What are the advantages and disadvantages of relying on informal organizations to accomplish the goals of these disciplines? If you were the firm's CIO, would you prefer using formal or informal organizations for these tasks? Explain your rationale.
10. Discuss some ways in which you could quantify the effectiveness of the processes discussed in this chapter.
11. Assume you are the manager of a department in which each of your employees has a workstation interconnected to all others in the department through a LAN. The LAN itself is connected to the firm's central processing facility. Discuss the important factors involved in the recovery management process for your department.
12. During which stage of growth is the subject of recovery management likely to arise? Why?

Assignments

1. Itemize the elements of information from the problem management process that you think are important to report. Design the format for this management report. How frequently should it be published? To whom should it be sent? If you were the IT manager, discuss the follow-up action you would expect to take after issuing this report.
2. Visit your university computer center or the computer center of a local firm and review its problem management system. Prepare a critique of your findings and report them to the class.
3. Design a change log patterned after the problem log in Appendix B. Be sure the log clarifies the relationships between changes (e.g., the log must identify prerequisite changes, corequisite changes, and mandatory future changes).

ENDNOTES

[1] Anita Taff, "AT&T Relates Events Leading to Massive N.Y. Net Outage," *Network World*, October 14, 1991, 15; Ellen Messmer, "AT&T Outage Spurs Federal Action on Reliability Issues," *Network World*, September 30, 1991, 9; and Gary H. Anthes, "Phone Outage Drives FAA to Backup Strategy," *Computerworld*, August 3, 1992, 63.

[2] Appendix A to this chapter contains a sample problem report.

[3] See Appendix B to this chapter for a sample problem log.

[4] Most major computer vendors supply automated, on-line tools to assist in problem management.

[5] These figures are based on the author's experience with problem and change management.

[6] A sample change request is contained in Appendix C to this chapter.

[7] John Kador, "Change Control and Configuration Management," *System Development*, May 1989, 1. This article discusses how to test and implement changes with low risk.

[8] David Coursey, "Sabre Rattles and Hums, Crashes," *MIS Week*, May 22, 1989, 6.

[9] Ron Levine, "Disaster Recovery in Banking Environments," *DEC Professional*, January 1988, 104.

[10] John Mahnke, "Planning Saves MIS From Disasters' Wrath," *MIS Week*, October 30, 1989, 1.

[11] Common carriers must have disaster recovery plans also or risk loss of customers to competitors. Some now stress their ability to provide service in the face of loss of facilities.

[12] Phillip J. Rothstein, "Up and Running: How to Ensure Disaster Recovery," *Datamation*, October 15, 1988, 86.

APPENDIX A Sample Problem Report

Problem control number ______________________________

Individual reporting the incident ______________________________

Time and duration of incident ______________________________

Description of the problem ______________________________

____

____

____

____

Problem category ______________________________

Problem severity code (*circle one*) 1 2 3 4 5

Individual assigned to correct the problem ______________________________

Estimated repair date ______________________________

Actions taken to recover (*append additional pages if required*)

____

____

____

____

____

____

____

____

____

____

____

____

____

Actual repair date (*problem closed*) ______________________________

Final resolution actions, if any ______________________________

____

____

____

APPENDIX B Sample Problem Log

Problem sequence number ______________________________

Date of incident ______________________________

Problem category ______________________________

Brief problem description ______________________________

____

____

____

____

Severity code (*circle one*) 1 2 3 4 5

Duration of repair action ______________________________

APPENDIX C Change Request Document

Change log number ______________________________

Problem log number (*if applicable*) ______________________________

Description of requested change ______________________________

____

____

____

____

Prerequisite changes (*indicate change numbers*) ______________________________

____

Corequisite changes (*indicate change numbers*) ______________________________

____

Category of change ______________________________

Priority (*circle one*) low medium high

Risk assessment (*circle one*) low medium high

Requested implementation date ______________________________

Individual requesting change ______________________________

Change manager ______________________________

Management authorization ______________________________

Attach test and recover plan.

Attach project plan if applicable.

14 *Managing Centralized and Distributed Operations*

A Business Vignette

Lenders Maintains Information Flow[†]

Profile: Jerry Lenders
Position: Director of Technology, Infomart
Mission: "To learn enough from mistakes to prevent them from happening again."

Journalists are a demanding bunch, and nobody knows that better than Jerry Lenders, director of technology at Infomart in Toronto. He oversees what the company calls an "electronic printing press," an on-line database system that delivers news retrieval and library services to journalists at some of the top newspapers in North America.

"It has been said that we are the heart of the organization, and I think that is true," he notes. "If the system is not up, it causes aggravation for everyone, not just the people on-line at the time. Obviously, there is no revenue coming in, so it is crucial that we stay up because that revenue does not get replaced."

The database service, called Infomart On-line, also serves as a gateway to Southam News newswire, Dow Jones News/Retrieval Service, and Datatimes, three on-line systems offering full-text retrieval of newspapers and other services. Lenders also manages an on-line system called Private File Service, a customized on-line database service used by 45 clients who need to store and manage text and numeric data.

The success of an electronic information business hinges on being able to offer subscribers ready, easy access to the information stored in its "electronic warehouses." "Newspapers don't want anything to get in their way," Lenders says. For Lenders, that means he must make certain that his system can provide more than 1,000 subscribers with instant access to information published in dozens of daily newspapers and from a variety of other sources round the clock: virtually 24 hours a day, seven days a week

At the core of the Infomart system is a cluster of four DEC VAX minicomputers. Four other VAXs are available for overflow or as backup in the event that the primary cluster crashes. A triple redundancy in communications links helps ensure that lines stay open. Some 20G bytes of information, about half of it newspaper text, is stored on the system's disk farm of 15 drives. Managing the information that must be constantly available at a subscriber's fingertips can be daunting.

"This business has a lack of historical data as to what users really want: archival information or current information," Lenders says. "My gut feeling is that they only want to access current information, but we're monitoring it to find out." Lenders estimates that the database is growing at the rate of about 11M bytes per day. "We recently bought four DEC RA90 hard disk drives capable of storing 1.4G bytes each," Lenders says. "Those will last us until May or June of next year."

Lenders says that keeping this growth under control is a challenge. "We need to decide what information must be put on-line, what is put on high-speed storage drives and what will not be kept at all," he says. The newspaper files are backed up every night. "We operate the system for newspaper customers seven days on 22 hours; for others it is seven days on 20 hours," Lenders explains.

Lenders, who majored in computer science at the University of Waterloo in Toronto, began working at Infomart as a systems engineering representative (a programmer) in 1981 and two years later moved up to become a systems engineering manager. In June of 1986, he become the director of technology with responsibility for the entire computer operation and its 16 employees.

He says that he constantly wrestles with providing adequate service to the company and its subscribers at the lowest possible cost. "The bigger challenge is that the technology is changing so rapidly," Lenders says. "That makes it difficult to stay on top of what is cost effective. I am very conscious of the bottom line," he adds. "How do we maximize the bottom line and service availability? That is a fine, tricky line."

Since computer failure can be disastrous if its cause cannot be found and resolved quickly, guarding against that prospect is one of Lender's primary responsibilities. "We have an all-in-it-together attitude here, and we methodically resolve problems," he says.

"I don't mind when we make mistakes, but I always make sure that we learn enough from them to prevent them from happening again," Lenders reports. "Once is fine as long as we learn from our mistakes and plug the holes so that we can go on."

INTRODUCTION

The preceding chapters discussed service-level agreements, problem management, change management, and recovery management as the first elements in the disciplined approach toward managing computer facilities. Successful computer operations are a critical success factor for IT managers; a well-organized process is essential for successfully managing this activity. Jerry Lenders understands this very well. His mission is to learn from problems and to prevent their repetition. He manages change in a rapidly growing environment.

This chapter builds on the foundation elements of service-level agreements and the processes of problem, change, and recovery management. It develops tools, techniques, and processes with which the remaining disciplines can be implemented. Discussion of the disciplined approach to computer operations continues with the essentials of batch systems, on-line systems, performance management, and capacity planning.

Figure 14.1 displays the relationship among these processes and relates them to management reporting.

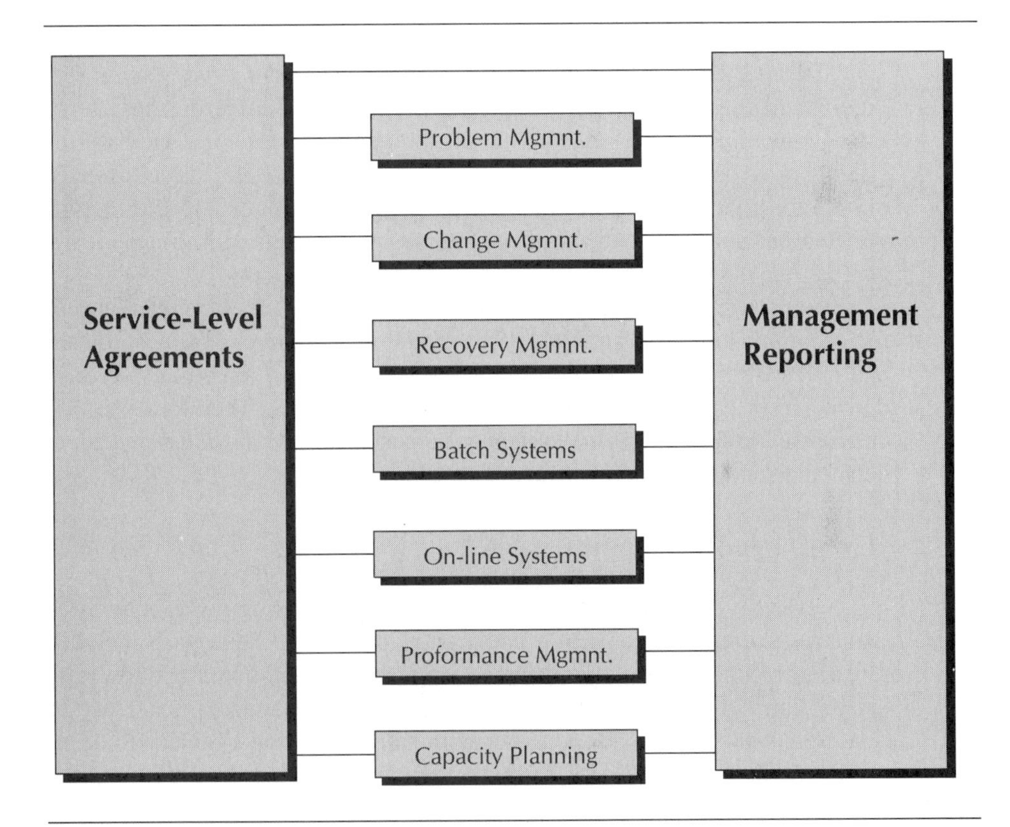

FIGURE 14.1 The Processes Continued

Problem, change, and recovery management processes, which are grounded in the discipline of service-level agreements, provide information for management reporting, the capstone of the disciplined approach. The underlying reason for developing and implementing batch management, online management, performance management, and capacity management is to ensure attainment of service-level objectives. Attaining these objectives is the key to achieving customer satisfaction.

MANAGING SYSTEMS

Careful system performance management helps ensure that service-level objectives are achieved through the effective and efficient use of systems resources. Users of system services and the service providers have expectations regarding service delivery. These expectations should be quantified through formal service-level agreements as discussed in Chapter 12. But whether or not they are specifically quantified, it is a basic management responsibility to use system assets effectively and efficiently.

And, as the Business Vignette suggests, management must provide service according to customer expectations in spite of uncertainties. Jerry Lenders must provide continuous access to a large and growing database for customers who are uncertain about their specific needs. His success hinges on extraordinary system availability and adequate capacity in the face of growing but uncertain demand. Consistently high system performance is his number one objective.

The performance management process defines the management system for accomplishing measurably satisfactory levels of output from hardware and software systems. System performance delivered to the customer is derived from both centralized and distributed programs. Therefore, the discussion of batch and on-line system management processes must precede a detailed discussion of performance management.

Batch Systems Management

Batch systems management is the process for controlling the execution of regularly scheduled production application programs. This work includes receiving incoming transactions, processing batch data and transactions, storing or distributing output data, and scheduling the resources necessary to accomplish these tasks. Scheduled work must be planned and defined in service-level agreements. Reports of measured activities must be prepared for management review and analysis.

Major computing centers have many examples of batch systems. Some manufacturing plant examples are inventory reconciliation processing and production requirements generation. Daily accounts-payable processing or routine updating of the general ledger are examples from the firm's financial department. Many product development laboratories have large simulation programs that are processed as batch systems. Usually, production jobs from personnel, sales, marketing, or service are regularly scheduled batch work also. As client/server computing develops, some of these batch systems are being converted to on-line applications.

Typically, the scheduling and controlling of these production applications is handled through an automated workload scheduling system — itself an application program. The scheduling program initiates jobs upon successful completion of prerequisite jobs. Thus, the system manages application dependencies. The scheduling system maintains an orderly flow of work through the computer center. The job scheduling process ensures that operational systems are used efficiently, that batch work proceeds in an orderly and organized manner, and that service-level agreements for batch operations are achieved. In complex, large-system environments, a computerized scheduling system is required to manage and control the hundreds of batch operations in a disciplined manner.

Routine computer-center operating procedures include complete and unambiguous instructions for processing the scheduled work in the absence of unusual occurrences. This means that job initiation information is complete, and that the operations staff knows and can meet the conditions necessary to complete the job successfully. These instructions and this information are part of the input data for the scheduling program. In addition, instructions for handling potential difficulties, such as unsuccessful completion of a prerequisite job or job failure resulting from insufficient secondary storage space, must be part of these procedures. These instructions are developed as part of the recovery management process.

Developing the schedule for the computer center is part of operational planning. Operational planning consists of a long-range plan, a daily plan, and a current plan.[2] The long-range plan describes the applications scheduled for the next several weeks or so. This plan includes daily, weekly, and month-end applications, and jobs scheduled on demand. A plan for each day's production is developed and is entered in the daily scheduling system. The scheduling program provides the system operators and managers with the current status of the batch production work

Effective problem management requires the manager to record deviations from established procedures that impact service-level agreements, or that have the potential to do so. Unusual events such as unsuccessful job completion, unanticipated output, or atypical operator intervention must be logged and entered into the problem management system so that corrective action can occur. Changes to production schedules, resource levels, or types of resources required for production work must be reviewed by the change management team. Batch operations are closely linked to the problem and change management processes.

In a multi-shift environment, production operations managers must ensure that the transfer of responsibility between shifts proceeds effectively throughout the day. An operations transition meeting helps communicate the system's current status to the incoming operations crew. An operations log, maintained by each shift of operators, should be reviewed with the senior people from the oncoming shift so that an orderly transition occurs.

Periodically, the production managers and their senior staff members must assess the batch management process. The review focuses on process effectiveness. Thus, it must evaluate process scope and methodology, job scheduling and resource management activities, and the effectiveness of the relationship between batch, problem, and change management. The review should develop recommendations and directives for process improvements. The reported findings are very important to those individuals who are engaged in service-level agreement negotiation and preparation.

There must also be a strong link between individuals responsible for hardware performance and those controlling the batch management process. Alterations to the work schedule, variations in the volume of input transactions, or changing requirements for on-line storage frequently affect system performance. They may also precipitate changes in the hardware or software system configuration. Configuration changes to obtain optimum throughput may be required to satisfy customer needs or to achieve service-level commitments. Prompt, effective communication of changing conditions will reduce the number of future problems and service-level disruptions.

IT managers and their key staff members are vitally concerned with production operations since it is a critical element contributing to their success. The process of batch management and the reports of process effectiveness highlight trends in operational activity. Successful IT managers scrutinize the reviews and reports carefully, looking for variances between planned, scheduled, and actual workload. Variances from normal operations that have been reported to the problem management system, and potential or developing bottlenecks in the work flow, are high-interest items for IT managers. Successful batch management achieves service levels and maintains satisfied customers.

On-line Systems Management

On-line systems management is a disciplined process consisting of the information, tools, and procedures required to coordinate and manage on-line application activities. On-line applications support users, both internal and external to the firm, in their use of remotely connected devices. System software and hardware required for on-line applications are included as part of the application activity. The objective of on-line management is to ensure that on-line application service levels are achieved through the efficient and effective use of computer resources.

There are many examples of on-line systems that operate as client/server implementations. Some well-known programs, such as the order-entry systems and airline reservation programs discussed earlier, are on-line systems. Other examples include electronic mail systems, inventory update programs, computer-aided design systems, and service dispatch and reporting programs. Many application systems formerly operated as scheduled batch programs have been converted to client/server implementations and are now on-line programs. Typical examples of these applications include language compilers for application development, many database update and query applications, manufacturing floor control systems, and most decision support systems.

The scope of on-line management includes coordinating the people, tools, techniques, and facilities to monitor and control on-line application systems services. This coordinating activity is best accomplished through common, well-defined interface points for users, vendors, and support groups. Specific groups to be addressed include operations personnel, technical support individuals, and applications development and maintenance programmers, among others. In routine practice, a single interface point or control point should be established for each group. For example, the operating-system support department might be the designated contact point for operations personnel and for hardware vendors. Complex, on-line systems such as multiple processors in various locations may demand several coordinated interface points.

Several functional responsibilities must be assigned to specific persons at the assigned control point(s). These responsibilities consist of user-interface functions and network-system functions. The user-interface function relates to on-line application users and to application development and maintenance programmers. On-line application programming groups may reside in the IT organization or in the application-using organization. For example, an on-line application specialist from the system support department may be responsible for addressing technical application issues with application developers. Or an experienced application programmer may provide end users with technical guidance regarding on-line application development. In each case, application developers have a contact point through which they channel their concerns.

The information center is a useful hub for coordinating these responsibilities. People seeking help should always have a fall-back contact point that can either resolve their concern or direct them to someone else more qualified to help. This contact point is frequently termed the *help desk* and, in many instances, is located in the information center. User interface or control point responsibilities are shown in Table 14.1.

TABLE 14.1 User Interface Functions

Receive user communication
Respond to user queries
Perform problem determination
Obtain technical assistance when required
Initiate problem management action
Provide guidance to users and support groups

The user-interface control point receives, records, and reports all inquiries from on-line system users. Through the help desk, callers may describe current problems or may seek assistance or advice regarding on-line application usage. System users also frequently inquire about the status of problems and changes. The person at the control point performs elementary problem determination and may obtain assistance from appropriate support functions for complex problem determination. Unresolved problems are recorded and reported to the problem management system for resolution. In addition, the user interface may provide guidance to users and to technical support personnel for using on-line as well as batch applications.

Figure 14.2 lists the activities of the on-line and network control point.

TABLE 14.2 On-Line and Network System Functions

Initiate, monitor, and terminate applications
Coordinate system terminal operator's activities
Monitor volume and performance indicators
Implement predefined network procedures
Develop problem determination procedures
Train personnel to use procedures
Coordinate application problem determination

The on-line and network control point can be an individual or function that initiates key on-line applications, monitors their operation, and terminates the applications according to procedures. The control individual coordinates the activities of remote master terminal operators and system console operators. Volume and performance indicators for on-line applications are also maintained at the control point. Examples of these indicators include the number of individuals using the application, current response times, and system resources devoted to the application. This data is required for service-level management and is an important element of the management reporting process.

The on-line application control point individual or function must implement predefined system and network backup and recovery procedures. These procedures are usually developed by technical support specialists in application development or network management departments. Control point persons develop and test problem determination procedures and ensure that these procedures are customized for system users and their applications. They coordinate the resolution of problems associated with network devices such as servers, bridges, and communication controllers. Their activity requires close cooperation with the problem and change management systems and with network management facilities.

In order to improve business processes, the on-line management process must be reviewed periodically to assess its effectiveness. Findings and recommendations from each review must be documented for management's attention and action. Summary data documenting the effectiveness of the on-line management process should be directed to IT management and, when appropriate, to owner-managers of on-line systems. Managers in client departments are very interested in service levels and appreciate information on process effectiveness. Effectiveness reports help IT management recognize trends in on-line services and service objective attainment. When this information is shared, trust and confidence between managers in both organizations is improved. Good relations are important to both groups.

PERFORMANCE MANAGEMENT

Performance management is a management system for defining, planning, measuring, analyzing, reporting, and improving the performance of hardware and operating systems, application programs, and system services. The objective of performance management is to ensure that performance targets and performance service levels are achieved through the effective and efficient use of systems resources. Six processes form the performance management discipline. They are outlined in Table 14.3.

TABLE 14.3 The Processes of Performance Management

1. Defining performance	4. Analyzing measurements
2. Performance planning	5. Reporting results
3. Measuring performance	6. System tuning

Performance management processes are applied to systems hardware, systems software and operating systems, and application program resources. Services provided by centralized or distributed computer operations such as application scheduling are included in the management process, as well. The coordination of resources used in combination with each other determines the overall performance of computer operations. Therefore, the performance management discipline embraces these resources in total.

Defining Performance

System performance, including hardware, software, networks, and people, is defined as the volume of work accomplished per unit time. Performance of the central processing unit is measured by the rate at which it executes instructions. Comparison of the actual instruction rate to the maximum rate determines the effectiveness of the CPU. For example, if the CPU is capable of 80 MIPs but is in the wait state half of the time (the hardware is not executing instructions but is waiting for work to do), the effective performance is 40 and the efficiency is 50 percent.

System performance measurements encompass the CPU performance and the associated hardware and software performance. Planning, measuring, analyzing, and reporting hardware performance of large centralized processors are centered around the concept of system performance. Tuning activity balances the CPU hardware, I/O channels, I/O devices including networks, and the software and application programs. The objective of performance management is to keep all hardware components relatively busy while maintaining continuous, high-quality service to system users.

From the clients' viewpoint, performance is defined by the number of jobs completed per unit time, the number of transactions processed per unit time, or the job turnaround time for batch jobs. Turnaround time is the elapsed time from job submission to job completion. For all these activities, the unit of work is not uniform. Therefore, the planning and measuring processes are statistical in nature. The reporting and tuning activity must account for the statistical nature of the work.

For distributed systems and client/server systems in particular, response time to on-line transactions is critical. This is true because the hardware, software, and network components are specifically organized to improve the user's productivity. In other words, the overall system (hardware, networks, and software) should optimize the operator's productivity and not some hardware or software metric. Thus, for critical on-line systems, response time is *the* key performance measure. Planning, measuring, analyzing, and tuning hardware and software focus on the concept of improving response to system users.

The goal of performance management for centralized and distributed systems is to ensure that performance objectives are achieved through the efficient and effective use of system resources. Performance objectives are established through performance planning.

Performance Planning

Performance planning establishes objectives for the throughput of human/computer systems. Performance planning also develops processes, techniques, and procedures to ensure that the system delivers the desired throughput. Performance planning is an integral part of the disciplines discussed previously; in particular, it is an essential step in service-level attainment. The performance plan must account for the amount and type of work to be performed as well as its distribution over time.[3] The plan is based on known or assumed capabilities of the combined hardware and software system and must include the performance characteristics of the applications themselves. For example, if rapid response is required for on-line transactions, system loading must be carefully controlled. Responsiveness declines rapidly as system loading increases.

There are other examples. If program enhancement makes an application more efficient, the total system will perform better. If the database is reorganized and tuned for the application, response to data requests will improve. If the workload is more evenly distributed throughout the day, system throughput will improve. In other words, system performance and system capacity are tightly linked. Improving system performance is equivalent to increasing system capacity. For this reason, performance management and all activities affecting system performance must be analyzed and clearly understood prior to capacity planning. Capacity planning prior to or in the absence of performance management is a wasted effort.

There are no generally established parameters that define system performance, but some measures are usually found to be valuable. These common measures are system response time, transaction processing rate, CPU time in the wait state versus the system state, CPU time in the system state versus the problem program state, and system component overlap measures. For instance, if the CPU spends an unusually high percentage of time in the problem program state, this may indicate increased application throughput. Throughput of large systems also increases when input/output and processing operations are highly overlapped. System programmers in the technical support department can identify many additional performance factors.

However, detailed performance measures depend on the type of processing that the system performs. For example, system performance measures for an airline reservation system are different from those used to evaluate a computer-aided design system. Response time to travel agents' requests is highly important, but instruction processing rates are more important for the lengthy, complex mathematical calculations that design systems perform.

Installation performance planning is usually accomplished by the technical support staff using both general and installation-specific performance indicators. These indicators also are the basis of performance measurement.

Measuring Performance

Response time and throughput measurements under a variety of workload conditions are the basis of performance management. Key performance data and measurements include the time interval or elapsed time for processing key categories of user work. Examples of these measurements are the service time for a transaction, transaction rates or number of transactions per unit time, and the response time for transaction initiation. Additionally, the amount of work or the number of transactions the system performs in total and for each user is an important performance measure.

The process for collecting measurements, establishing the range of acceptable performance levels, and tuning applications and services must be well documented. Many performance measurements relate directly to service-level agreements and are referenced in them. For example, the service-level agreement may specify subsecond response times for trivial transactions or may indicate some number of transactions per minute for the on-line order-entry system. The primary responsibilities for performance management, such as collecting and analyzing performance data and enhancing systems performance via corrective actions, must be assigned to the unit's technical support manager.

Performance management in large systems is concerned with the use of key hardware components, such as direct-access storage space, input/output units, main memory utilization, and paging/swapping subsystems. Performance management also includes the control and management of system software components such as supervisor modules, program modules, and system input/output buffers. System programmers in the technical support department have sufficient knowledge and skill to adjust system parameters and configurations using performance monitoring tools to improve system performance.

For large systems, there are several ways to measure or monitor computer system performance.[4] Hardware devices or monitors that can be attached to the computer system to accumulate statistics on system component utilization are widely available. For example, a hardware monitor can be attached to a computer input/output channel to measure the number of transactions processed and report the percentage of channel busy time. This information is useful to system programmers for balancing input/output activity among the channels on the system.

Large computer system performance can also be measured with software monitors. Software monitors can either be a part of the operating system or a part of the applications themselves. Software monitors measure CPU activity such as CPU busy versus wait state, CPU in the system state versus the problem-program state, channel busy time, channel overlap time, and many other measurements.

Application program monitors can be customized to provide specific application data. This data may include utilization of storage devices and frequency of module execution. From the application monitor, the programmers may determine transaction processing times in the application. They may generate frequency distributions of workstation usage by workstation type or location. The amount and type of performance data available is limited mostly by the application programmer's imagination.

Hardware monitoring devices are independent of the computer system operation and provide precise, detailed measurements for later analysis. Their use requires specific hardware system knowledge. Hardware monitors are relatively expensive. Software monitors degrade system performance somewhat, and they only measure information available to the computer instruction set. But software monitors are easy to install, easy to use, and relatively inexpensive. For these reasons, software monitors are much more popular than hardware monitors.

Analyzing Measurements

Performance measurements must be obtained for appropriate workload periods including peak and off-peak times. Measurements are most useful if they provide information that helps identify trends in system performance. Performance data analysis, especially from peak workload periods, may reveal system bottlenecks. Identification of bottlenecks and subsequent corrective action improve performance and yield additional capacity at no additional cost. The analysis of performance parameters must include comparisons with the performance plan. The reasons for departures from the plan must be understood so that service-level integrity can be maintained and the planning process improved. Intelligent use of performance measurements and well-planned system tuning efforts will greatly improve system performance. They are also valuable in the capacity planning process.

Performance measurements, and comparisons to historical trends and to the plan, must be reported. System owners and/or IT managers require these results to perform service-level risk assessments and to assess the integrity of the performance management discipline.

System Tuning

The role of user satisfaction surveys was introduced in Chapter 12. You will recall that user surveys should be developed and used to validate the credibility of performance measurements. Measurements must reflect actual user experience. Surveys are valuable in establishing user satisfaction and relating customer opinion to current levels of key performance measures. Continual correlation of customer perceptions with internal measurements is a key element in ensuring customer satisfaction.

When surveys or measurements detect that established performance levels are not being achieved, managers must develop plans to correct the unsatisfactory situation. Corrective actions for alleviating low performance levels include hardware configuration changes (capacity increases, if warranted), input/output balancing, operating system performance improvements, system memory tuning, and direct access storage space reorganization. Additional activities such as tuning application programs that consume large amounts of critical resources and limiting the maximum number of concurrent users may also be required. Restricting the number of concurrent system tasks or other limiting actions may also be required to improve system performance.

Periodically, the performance management process itself should be reviewed to determine its effectiveness. The analysis of findings and the recommendations should be presented to the IT management team for review and possible action. Reports documenting performance levels achieved for all major application programs must be produced regularly for IT management, for application owners, and for users of IT service, as well. Performance level reports for each application and service are essential for capacity management and planning activity.

CAPACITY MANAGEMENT

Capacity management is the process of planning and controlling the quantity of each system resource required to satisfy users' current and future needs. Capacity management also includes forecasting the physical facilities (electrical power, air conditioning, cable trays, and raised floor) needed to install additional system resources. The disciplined processes discussed earlier in the text are prerequisites for effective capacity management and planning. Computer capacity can be determined only when managers deal effectively with problems and changes and when they control the performance of batch and on-line systems. Capacity management is an essential step toward meeting service levels.

The objective of the capacity management process is to match system resource levels with those required to achieve service levels, while system loads and service demands change. Additional resources may result from workload changes, additional applications, or service level improvements.[5] Capacity management processes are also useful for identifying surplus or excess capacity, or obsolete capacity that can be removed from the installation. Capacity management relates directly to IT's financial performance and indirectly to the firm's financial performance. Effective capacity management is absolutely essential to successful IT managers.

Capacity Analysis

The capacity management process includes an analysis of current system resource requirements. Based on present-day applications and services, this analysis establishes the benchmark for comparing proposed system configurations. Current system resource utilization is obtained from the system itself. The data is obtained from performance measurements. It can be analyzed to determine average workload and peak periods such as peak hour, peak workload day, and monthly financial closing days, for example. Additional analyses should be conducted by service elements such as transactions processed, applications serviced, user-group workload, and department activity. These analyses are the basis for capacity planning.

As an example of capacity analysis, consider on-line secondary-storage capacity. Secondary storage is critical to computer operations and is relatively easy to measure. If the measurements reveal that dataset storage extensions during peak processing take most of the available storage capacity, then it is predictable that jobs will fail for lack of storage as transaction volumes increase. In this instance, storage capacity measurements indicate that adjustments in system configuration or workload parameters are required to avert failure. Capacity analysis of storage devices should also detect datasets that are no longer used, or so infrequently used that they should be moved to off-line storage. Similar measurements and analyses can focus on CPU instruction rates, channel-busy percentages, and input/output activity. This type of capacity analysis is required for all system resources on a continuous basis.

The results of capacity analysis permit systems managers to know the extent to which system components are being used throughout the day. The analysis reveals processing bottlenecks or potential bottlenecks and also indicates unused or underused system resources. Capacity analysis is the basis for capacity planning.

Capacity Planning

Capacity to satisfy future requirements is based on business volume growth, new application services, and improved service levels. This load information must be obtained through careful consultation with system users and is usually obtained at the time service levels are established. The fundamental basis for establishing workload information is the business plan. Workload data must be analyzed for consistency with the business plan and with current service levels. Load forecasting begins with SLA negotiation; it forms the basis for system resource capacity forecasting.

The task of capacity planning involves the input parameters of volume changes in current applications, load increases stemming from new applications or new functions, and planned changes in service levels. Current capacity excesses or deficits from the previous analysis are the source of additional input. The data are used to forecast future capacity requirements. For example, if present direct access storage is 90 percent utilized, and volume changes and new applications are expected to increase storage requirements by 20 percent, then additional storage of at least 20 percent will be required. Conservative planning may indicate the need for a 30 percent storage increase.

All other system resources must be analyzed in a similar manner. For instance, if response times to on-line transaction processing is deteriorating, capacity planners have several choices. They can tune the on-line system to improve its performance, or they can adjust the operating system configuration or priorities to improve response. They can adjust or reschedule other work running in the multiprogramming environment to improve response, or they can increase system resources. System resources can be increased and response improved by installing higher speed storage devices for the on-line program or its data. As a last resort, a faster CPU or server, or an additional server, can be acquired and the network reconfigured. System planners have many options for maximizing performance and capacity.

The final result of capacity management is an optimized configuration of hardware, system software, and application programs. This optimum future configuration effectively satisfies the needs of the business and achieves the performance committed in the service-level agreements. Information derived from the capacity management process is vitally important in the IT planning process.

The capacity management process must be documented for IT managers and for managers in client departments. The responsibility for collecting system requirements and forecasting systems capacity must be assigned to an IT manager. There are many techniques ranging from simple to complex that IT managers can use to forecast systems capacity. For example, if CPU capacity is critical to the operation, then capacity management may hinge almost completely on CPU parameters. For most systems, however, capacity analysis involves a composite of many measurements. In these cases, system simulations may be required for complete analysis. The key to capacity management is selecting the simplest technique that provides a satisfactory degree of accuracy.

Most successful computer installations maintain a database containing previous workload projections and capacity requirements. These historical databases are essential for studying trends in workload and capacity growth and for improving forecasting techniques. The combination of service levels, system tuning, and capacity planning tends to permit heavier loading of CPUs, less unused capacity, and lower costs, as new capacity purchases are deferred.[6] Capacity management is a cost-effective process.

Capacity planning and management is the conclusion of all previous disciplines. This is true because all the disciplines impact capacity requirements. For example, the manner in which problems are handled directly relates to workload: Good problem management reduces problems, lowers workload, and reduces capacity requirements. The same is true for change management. Efficiencies in the operation of batch and on-line systems directly translate into capacity reductions. In other words, poorly managed batch and on-line operations require capacity increases in comparison with well-managed operations. Finally, in the absence of performance analysis and planning, capacity planning will be ineffective. Capacity cannot be well planned if significant performance uncertainties exist.

Client/Server Capacity Considerations

Capacity analysis and planning in distributed systems, and client/server systems in particular, hinge on system performance as measured by system response time from the user's perspective. In many implementations, the system involves hierarchical servers, databases, and network components. In these systems, it is especially critical to understand the system's operation in order to head off potential bottlenecks. The system's operation is subtle: Sound measurements are required.

But, with hardware prices declining rapidly, managers tend to believe that additional capacity can be obtained cheaply and easily, and that measurements are difficult and cost more than they're worth. For example, if response at the workstation declines, some managers would opt to replace the 386 workstation with a faster, 486 device hoping the faster CPU will improve response. It may or it may not, depending on the initial cause of the bottleneck. If the bottleneck was workstation speed, then a faster CPU may improve performance. But the bottleneck may be at the server or in database access. If so, performance remains largely unchanged by a faster CPU. The organization may have expended resources for little or no gain.[7] The belief that one can buy performance off the shelf is correct, but there is no substitute for knowing exactly where to spend the money.

As networks get more complex and as firms become more dependent on them, it becomes more important to have a sound basis for capacity planning.

Bottlenecks are much more difficult to identify in a complex distributed environment and up-front planning becomes more urgent.[8] Capacity planning for client/server systems is more difficult than for large, centralized systems, but tools that streamline the process are slowly becoming available.

Capacity planning in complex, user-oriented distributed systems cannot be achieved satisfactorily using "seat-of-the-pants" methods. Managers pay the price of systems failure when they substitute guesswork for solid analysis in complex systems. Network planners must start developing historical databases and modeling and simulation tools to assist them in the planning process.

Additional Planning Factors

In order to be effective, capacity analysis and planning, which are essential to any successful computer operation, require antecedent processes. Some IT organizations try to determine capacity without completing the prerequisite steps. And some publications address capacity without recognizing the system's inherent dependency on changing business conditions and performance analysis and planning.

Senior IT managers must remain alert to conditions that potentially affect computer system capacity. They must carefully scrutinize additional business information that bears on capacity forecasting. Some factors that IT managers must consider in the capacity planning process are:

1. Changes or alterations in strategic direction destined to improve or increase IT services
2. Business volume changes in either direction
3. Organizational changes (always a potential impact on IT resources)
4. Changes in the number of people who use IT services[9]
5. Changing financial conditions within the firm
6. Changes in service-level agreements or service-level objectives that bear on system performance requirements
7. Portfolio management actions such as new applications or changes to current applications that impact system throughput
8. System resources required for testing new applications or modifications to current applications
9. Applications schedule changes initiated by operations or user managers
10. Schedule alterations for system backup and vital records processing
11. System outage data and job rerun times from the problem management system

The first five items result from changing business conditions within the firm and are usually reflected in business plans. Some important factors such as volume, financial, or budget changes can occur between plans, usually as the result of external conditions. IT managers must be particularly sensitive to these factors. For example, if the firm is experiencing unplanned growth in sales revenue, some client organizations may receive an unplanned budget increase. Typically, the client organization may spend 10 percent of its budget on IT services, but its marginal spending for information technology may be 50 percent. This occurs because as revenues rise and functions receive additional discretionary resources, they increase their spending on automation to handle the increased volumes. The reverse is also true. The fluctuations in demand for IT services are likely to be wider than the fluctuations in the firm's business activity. This phenomenon is critical to senior IT managers.

Information about the remaining six items on the list is available from the disciplines of production management or from individuals within the IT organization. Individuals responsible for problem and change management, batch and on-line application specialists, information center personnel, and IT customer service representatives are responsible for gathering information that assists in formulating workload projections. Clients also must be coached to alert service providers when anticipated changes in requirements are first detected. Effective capacity planning depends on a continuous stream of critical information. Information gathering is a continuous process, but it must be accompanied by validation procedures. As part of the validation process, the IT steering committee should examine and concur with unusual or unplanned workload projections.[10]

THE LINK TO SERVICE LEVELS

Equipment plans developed from the capacity management process specify the hardware components required to satisfy client service levels. The equipment plans translate into equipment, installation, and setup costs, and they specify additional supporting facilities. After developing the optimal system configuration with the additional required components, the computer and network facilities must be evaluated to ensure that the proposed configuration can be satisfactorily housed. Additional equipment and space may be required. The capacity management discipline is highly important because it drives the budgeting and planning process and specifies resources needed for vital applications.

Periodically the capacity management process should be reviewed to assess its effectiveness. Close agreement between previous capacity forecasts and actual capacity requirements is one effectiveness criterion. The predicted versus the actual results of workload, capacity, and service levels should be compared for each client application. These comparisons should be made even if hardware or software changes have not been made. The results of the capacity review, including findings and recommendations, should be documented for further scrutiny and analysis, and should be retained for future reference. Reports documenting the capacity required for each application should be produced for IT and for client managers. Subsequent service-level negotiations require this information.

MANAGEMENT INFORMATION REPORTING

Service-level agreements are the foundation of the management system for computer operations, and management reporting is the capstone. Throughout the discussion of the disciplined processes within this management system, reporting has played an essential role. Each process informs managers of its results so that sound decisions and process improvements can be made. This management system for computer operations is effective. It enables managers to succeed in a vital and critical area. The free flow of important and introspective information is essential to process success. Each management process must be examined for effectiveness. This evaluation must be used to improve the process incrementally. Thus, for several reasons, management reporting is essential.

Management reports are not used exclusively by the IT organization: They are essential to IT customers, as well. Mature, successful IT managers take every opportunity to share information because they know that sharing important data improves performance. As Jerry Lenders pointed out, "We have an all-in-it-together attitude here, and we methodically resolve problems." Lenders' attitude is extremely important to all Infomart managers and to managers in other successful firms. In most firms, managers are problem solvers and, given the chance, prove reasonably adept at it. The system must give them a chance.

Management reporting and all informal communications are highly important to efficient operations, even though reporting reveals flaws and paves the way for criticism. Reporting problems, failures, and successes exposes individuals and the organization to criticism and to praise. But more importantly, open communication leads to increased trust and confidence. Building trust and confidence is important to managers, even if they remain vulnerable in the short run. Good managers understand the value of good communication. Effective managers insist that reporting processes be honest and complete.

SUMMARY

At Infomart, Jerry Lenders balances requirements for customer service, system availability, and recovery protection by providing double redundancy in CPUs and triple redundancy in communications links. Part of the redundant capacity stems from uncertain customer demands and part from the need for high availability. All capacity costs money, and his essential trade-off is between system costs and system benefits to Infomart. In addition, Lenders must understand technology advances and relate these to the needs of the business. Lenders exemplifies the role that many successful IT managers play in firms today.

This chapter presented tools and techniques for managing production systems—batch systems and on-line systems. In large centralized systems, departures from normal operations lead to problem-management action. Defined control points and responsibilities are required for effective on-line operations. The control points ensure that user questions are answered, that problems and changes are properly resolved or coordinated, and that on-line or distributed systems are coordinated and controlled. Because distributed systems users are often physically dispersed, the control point functions as a communications hub among system developers, operations personnel, and system users.

Centralized and distributed applications critically depend on computer capacity and on the performance of system resources. There is a strong link between system performance and system capacity because improvements in performance, achieved through system tuning, effectively increase system capacity. In fact, this may be the lowest cost capacity available. Capacity planning can be effective only when all other disciplines are operating well.

Reports from all the disciplined processes are presented to service provider and service user managers to inform them about the performance of the operation and to help them independently assess the effectiveness of each process.

The management processes described in this and preceding chapters form a roadmap leading to successful computer operations: a critical success factor for service providers in IT or operational departments. This management system ensures success when implemented completely. Implementation may seem arduous and time consuming, but shortcuts or circumventions lead to inefficiencies and, ultimately, to system failure. Failures in centralized or distributed operations lead to failures elsewhere. Ad hoc management of complex operations such as client/server implementations always causes extreme distress for managers. Successful managers embrace the management system described here; they thrive on its benefits and enjoy its rewards.

Review Questions

1. Describe the relationship between service-level agreements and management reporting, and the disciplines discussed in this chapter.
2. What are the essential reasons for developing and implementing the management processes described in this chapter?
3. Describe the relationship between system performance and management expectations at Infomart.
4. Define batch management.
5. What are some examples of batch systems? Do you think on-line systems are more difficult to manage? Why or why not?
6. Why are batch systems difficult to schedule? What functions can a computer program perform to streamline this task?
7. Describe the input that computer operators may provide to the problem management system.
8. Some computer systems operate around the clock. What special problems does this pose, and how are they solved?
9. Describe the relationship between operations personnel and the processes of performance management, capacity management, and problem management.
10. Define on-line management. What kinds of communications difficulties do on-line systems create? How can they be overcome?
11. What is the scope of on-line management?
12. Compare and contrast the user interface function and the network system function. How do their responsibilities differ?
13. The on-line and network system function initiates, monitors, and terminates applications. Where in the organization should this function reside? Explain your rationale.
14. With whom do the network control point personnel interface?
15. What is performance management? What are the processes of performance management?
16. How is CPU performance usually measured?

17. How do system users measure performance?
18. What does performance planning accomplish?
19. What tools are available to measure system performance?
20. What is the connection between system bottlenecks and system tuning?
21. Explain capacity management. What are its objectives?
22. How are capacity analysis and capacity planning related?
23. In what ways does capacity management depend on the previous disciplines?
24. What is the link between capacity planning and service levels?
25. Why is management reporting called the capstone of the management processes?

Discussion Questions

1. Discuss the essential difficulties that Jerry Lenders faced in his job at Infomart.
2. Discuss the reasons why the disciplined processes are essential in attaining committed service levels.
3. Discuss the connections between batch and on-line systems and the disciplines of problem management and change management.
4. Many batch programs are being converted to on-line systems. Describe the implications of this conversion to performance management and capacity planning.
5. You have read that the management processes must be examined periodically for effectiveness. In your opinion, why is this important? How might you go about doing this?
6. Reporting the results of the performance management process is important to several groups. Discuss the importance of these reports to system operators, system programmers, application programmers, system users, and managers.

7. System performance is subjective. It may mean something different to system programmers than it does to application users. What accounts for these differences? Why must a common ground be found?
8. Discuss the advantages and disadvantages of hardware monitor devices and software monitoring tools. Why might a combination of devices and tools be most effective?
9. Analyzing performance measurements, reporting results, and tuning systems are iterative processes. Discuss the IT manager's role in these processes.
10. Discuss the reasons why capacity management must be preceded by the processes of problem management through performance management. What difficulties might arise if capacity analysis is not preceded by the recovery management discipline?
11. Discuss the trade-offs among the processes of IT planning, service-level agreements, recovery management, and capacity planning. What risks are always present regardless of the manner in which trade-offs are made?
12. Discuss the reasons why organizational changes should always stimulate a review of system capacity factors.
13. Information technology managers are change agents. How does this statement relate to your answer to the previous question?
14. What issues discussed in Chapter 1 are overcome using the process of management reporting?

Assignments

1. List the reports generated by the management system outlined in this and the two previous chapters. In your opinion, who should prepare these reports? Who should review them? Prepare a schedule for these reporting and reviewing processes and describe the resulting management process.

2. The processes described in the disciplined approach to computer operations are more easily applied to mainframe installations. How can these principles be applied to minicomputer installations? What ideas are relevant to client/server implementations? How would you implement them?

ENDNOTES

[1] Michael Alexander, "Lenders Maintains Information Flow," *Computerworld*, December 12, 1988, 87.

[2] Israel Borovits, *Management of Computer Operations* (Englewood Cliffs, NJ: Prentice-Hall, 1984), 191.

[3] H. Pat Artis and Alan Sherkow, "Boosting Performance With Capacity Planning," *Business Software Review*, May 1988, 67. This article states that workload characterization is the cornerstone of all performance and capacity planning programs.

[4] Various types of performance measurement techniques are discussed in Borovits, pages 153–179.

[5] A quantitative approach to hardware selection based on capacity measures is given by Borovits, pages 55–64.

[6] Craig Stedman, "Capacity Planners Press CPU Limits," *Computerworld*, December 19, 1994, 57.

[7] Dennis Hamilton, "Stop Throwing Hardware at Performance," *Datamation*, October 1, 1994, 43.

[8] Craig Stedman, "Capacity Planning Takes Client/Server Steps," *Computerworld*, November 28, 1994, 75. This article lists some tools available for capacity planning in client/server environments.

[9] Douglas J. Howe, "A Basic Approach to a Capacity Planning Methodology: Second Thoughts," *EDP Performance Review*, October 1988, 1. End-user computing increases the complexity of capacity planning, and it extends the planning horizon out to 10 or more years according to this article.

[10] The steering committee may not always agree with the projected additional load. The committee can help rationalize predicted workloads and additional capacity with business financial objectives.

15 Network Management

A Business Vignette

Boston Edison Integrates Computers and Telephones[†1]

Forced by competitive pressures and downsizing to do more with less, Boston Edison did just that, thanks in part to a computer-telephone integration (CTI) application it rolled out last year. The company provides electricity to nearly 700,000 residential and business customers in 40 cities and towns.

"We had been looking at CTI for years," says John Dubiel, manager of planning and technology at the Massachusetts utility. Boston Edison began using CTI in its billing department, where 40 agents fielded 570,000 toll-free calls in 1993. The move to CTI was one of several responses to Boston Edison's reduction of 700 jobs, a move to hold down costs.

Three AT&T Definity G3r PBXs that Boston Edison bought in 1992 helped prepare for the move to CTI. The same year, the company committed to the Novell NetWare platform.

In early 1994, Dubiel says, Boston Edison agreed to become a controlled introduction site for a CTI application. The pilot went so smoothly that the utility met its goals in half the time. The connectivity expected for Tuesday was achieved by Monday; instead of three agents "live" by Thursday, there were three live on Tuesday, six on Thursday.

The system prompts billing department callers to enter their 12-digit account numbers. The account number triggers a database lookup and, as soon as the match is made, four screensful of information are delivered to the answering agent's Compaq DeskPro 486 PC in under two seconds.

Forcing callers to enter a dozen digits may sound like an invitation to confusion, but not for the capable customers of Boston Edison, Dubiel says. "We did a study and found that 95 percent of people get all 12 digits of their account numbers without needing a second chance."

Customers may not need to do that touchtone work much longer, however. At the end of 1994, Boston Edison put in a primary rate ISDN line to its emergency calls department. The new line delivers automatic number identification (ANI) information. ANI will speed the database lookups and instantly deliver customer data to agents. ANI could be extended beyond emergency calls if this test implementation succeeds.

CTI has improved productivity in several ways, Dubiel says. "We save about 14 seconds on a 150-second call," he says. "Also, agents can handle many calls without touching the keyboard." Agents used to average 24 keystrokes per call.

One other saving is the time involved in mandatory, after-call report logging. With CTI, the time spent logging reports matches almost identically the time spent on the call, Dubiel says, so shorter calls mean shorter wrap-ups.

Boston Edison employs 4100 people and earned $109.8 million on sales of $1.49 billion spent in 1994.[2]

INTRODUCTION

The previous three chapters emphasized the management of information systems operations, discussed the role of customer expectations, and developed the disciplines of problem, change, and recovery management. Performance analysis, capacity planning, and management reporting completed the management processes for production operations. Both centralized and distributed computer operations in today's firm are tightly linked and highly dependent on data communications networks because they transport most of the data coming into or going out of computer systems. Many of the firm's important systems are on-line; they process the data as it arrives. Because networks play an extremely important role in modern information systems, their management deserves very special attention.

Networks are a central part of the information processing infrastructure in firms of all sizes. LANs connecting personal workstations are the most rapidly growing segment of the telecommunications market, indicating that small firms and departments in large firms are rapidly adopting the technology. Major telecommunications projects designed to link operations electronically have been undertaken at Ciba-Geigy, H. J. Heinz Company, G. D. Searle, and many other firms. Target Stores and Van Heusen are implementing T1 nets within and between their facilities. And Merrill Lynch links its New York and New Jersey headquarter's facilities with 500 domestic branch offices and international branches in Tokyo, London, Singapore, Sydney, Bern, and Toronto. By now, most firms have implemented LANs. Thirty percent of the Premier 100 firms have networked 95 percent or more of their employee workstations.[3] Network management is an important topic for these firms and for aspiring IT managers.

Networks greatly expand the firm's ability to process information. They add value but they also add complexity to the firm's information processing infrastructure. However, many of the management practices and management systems described in preceding chapters for traditional information systems apply to network management. For example, strategy and plan development, the management of expectations, and operational disciplines are all especially important to network managers. As we learned in the Business Vignette, carefully planned and implemented new telecommunications technology reduces costs and improves customer service.

This chapter addresses network performance and many network operational and management issues. Its purpose is to develop a foundation of management systems and principles upon which network managers can rely for success.

THE IMPORTANCE OF NETWORK MANAGEMENT

In this decade, well-managed networks are at the heart of many important information technology developments. They are the basis for many well-known and emerging strategic systems and are stimulating the growth and development of global information systems and international businesses. Within the firm, information technology is a potent force for organizational restructuring and business process improvement. Between firms, networks enable modern alliances and joint ventures, couple suppliers and customers, and facilitate cooperative endeavors of all kinds. Therefore, networks and their management must be accorded the utmost attention.

Networks Are Strategic Systems

Network management is becoming increasingly important to businesses and to their IT managers for several reasons. Intense interest in networks stems largely from their obvious and growing operational and strategic value. As we learned earlier, most strategic systems rely heavily on telecommunications technology. Reservation systems, brokerage products, and many other information systems are telecommunication based. And new, rapidly emerging telecommunications technology is known to have great tactical and strategic importance to modern organizations. Examples like E-mail, VSAT, EDI, groupware, video conferencing, mobile computing, and wireless LANs are currently important and have great potential.

The Boston Edison example points out that the integration of computers and telephones provides superior customer service. Thousand of firms, from trash recyclers to catalog merchandisers to insurance companies, use automatic number identification linked to workstations and customer databases to extend real-time personal service.[4] Most insurance companies are active users of telecommunications services. Security-General, for example, expects independent agents to sell its policies, rather than competitors' because of superior agent service. Their system provides E-mail communications to headquarters, a library of policy forms, and the ability to generate policies on-line. Security-General expects to recover its costs by charging its agents a flat fee for using the on-line service. Agents will recover their fees by writing more policies.

Well-planned networks integrated into important business systems generate advantage by reducing costs, saving time, establishing service excellence, and upstaging competition. Business processes improved through telecommunications technology can deliver leverage to the firm through the benefits of increased organizational effectiveness.

Networks Are International

International trade and commerce and the globalization of business enterprises are facilitated by telecommunication technology. For example, significant capacity expansion is being planned for the Pacific basin, not only to carry additional traffic, but to provide backup in case of network failure. Fiber-optic cables connect the mainland U.S. with Hawaii and Japan. Additional connections will link the U.S. with Hong Kong, Korea, China, and the Philippines.[5] Additional cables are under construction in the Atlantic, the Caribbean, and the Mediterranean by firms from several countries. Connections through the U.S. enable communication between the Far East and Europe. In the near future, high-speed, broad-bandwidth telecommunication lines will encircle the globe.

The European Economic Community agreements and current political alterations in the Eastern European block of nations have generated intense interest among Western companies. Their interest spans a broad front but is especially strong in the field of information systems and telecommunications. Companies that already have a foothold on the continent, like IBM, are likely to compete head to head with well-established firms in West Germany and other EEC countries for a share of the Eastern European information-processing market. Within Europe, the Netherlands hopes to be a center of electronic communication and a hub for package and mail delivery, building on a 500-year-old tradition.

Leading-edge firms around the world recognize the opportunities and are capitalizing on them. The Volvo Corporation uses international networks to link manufacturing and sales operations in Europe with those in the U.S.; it cannot operate its business without these linkages. Kodak links tens of thousands of employees with its voice-messaging system and connects computers internationally through its communications facility. DEC manages its worldwide business operations through extensive use of international communications facilities. Messaging through global networks of PROFS systems enables IBM employees to communicate internationally on product development, manufacturing, or sales issues. High-speed data pathways permit firms to develop hardware and software products on an international basis.

The global infrastructure for international communications is rapidly falling into place, with U.S. firms such as AT&T and many of the regional Bells playing a major role. Large segments of our economy already take advantage of the system; the travel industry is one example and financial institutions are another. More than $1 trillion move through international telecommunications systems daily, it's estimated. Executives in many industries in all countries are intensely interested in capitalizing on the potential of international networks. Telecommunications promises to be one of the great business opportunities of the 1990s.

Networks Facilitate Restructuring

"Communications networking is an extremely important—some have argued the most important—part of a company's process information infrastructure. Once a network is established, a company often discovers many possible process innovations that might result from its use."[6] In this statement, Tom Davenport forcefully underscores why decentralization of business activities and of IT operations are important current trends shaping corporate strategies. Today, decentralized business activities include using information technology to facilitate alliances, partnerships, and joint ventures, and information coupling between firms and their suppliers, and customers. Networks are important because they enable decentralization and corporate restructuring: They support the information flow that dispersed or remote units demand. Networks are critical assets of decentralized organizations because they provide efficient and effective communication of knowledge; they facilitate the flow of vital control information between dispersed organizations.

Information technology, particularly telecommunications systems, enable business processes to transcend the firm's physical boundaries. Technical and business information flows via electronic data interchange (EDI) from one firm to another, linking similar processes with interorganizational systems. Sales and service data streams through wireless LANs, and volumes of business and management information move on the Internet or through commercial services such as America Online. It's hard to overestimate the value of this type of communication.

As an example, five companies in five countries are collaborating to build a new generation jet engine. The companies located in the U.S., Great Britain, Japan, Germany, and Italy will jointly build jet engines, worth approximately $4.3 million each for 2000 new aircraft over the next 20 years. This cooperative effort is possible only because a 24-hour communications network supplies data on design control, bills of material, parts catalogs, tool design, and many other items among the firms.[7] Connectivity of systems, data, applications, and processes is the critical component of decentralized businesses. Organizational restructuring and trends toward decentralization are facilitated by network technology. Thus, network management is critically important to corporate strategies.

There are other reasons for the growth in the size and the importance of telecommunications systems. Other advancing technologies, principally the rapid growth of personal computers, are making networks and network management more important. It is estimated that about 30 PCs per 100 people are operational in the U.S., about four times as many as in any other country. Many home PCs and business workstations are connected to nets. But PCs and workstations stimulate only part of the growth in networks. Much of the rest comes from electronic mail, facsimile, telex, telephone, voice mail, audio conferencing, computer conferencing, electronic document transfer, and video conferencing. An all-purpose workstation handling our personal audio, video, text, and image communication will become available in the near future.

Network management is also important because networks are becoming more complex and difficult to manage. Networks are becoming more complex because the number of nodes is increasing rapidly and communication paths are longer and more convoluted with the growth of internetworking. There are many more network products such as concentrators, multiplexers, modems, and routers. Customer equipment is proliferating, too. There are many suppliers of network and customer equipment frequently offering incompatible products. In spite of significant progress in standardization, incorporating multi-supplier products into networks poses a management challenge.

IT managers and their firms desire network integration, but this objective strains present network management systems. Most firms are planning significant increases in expenditures for network management systems, but the availability of technical and management skills limits some firms. Due to the large installed base of networks and network devices, and because of their rapid growth, there is a large pent-up demand for skilled network managers and effective network management systems.

Business and government executives are intensely interested in networks, their capabilities, and the opportunities they afford their organizations. The telecommunications revolution embracing divestiture in the U.S., privatization overseas, and dynamic worldwide company repositioning has drawn critical executive attention to telecommunications.

THE SCOPE OF NETWORK MANAGEMENT

Network management consists of systems, processes, and procedures that help achieve efficiency, effectiveness, and customer satisfaction in network operations. These management practices are applied to hardware, system software, and application software which transport voice, data, or image information throughout the firm. Network management focuses communication assets. Some communication assets will be found in the firm's portfolio of application programs. But not all assets themselves are necessarily owned by the firm; some, like bandwith, may be leased. Thus, the scope of network management is broad and diffuse, rather than narrow and concentrated.

Physical assets subject to network management include customer terminals, local cabling, concentrators, modems, multiplexers, and lines or links. Private branch exchanges and computer processors frequently form part of the physical system, as well. Communication software, some application programs, and databases supporting the hardware and software are also included within the scope of network management. The traditional boundary between telecommunication management and information system management is rapidly disappearing because the technologies themselves are indistinguishable. The separate management subjects have been replaced by one: information technology management.

Network management and the discipline of on-line management are also merging. Firms with extensive traditional data processing operations use the management systems discussed earlier to manage these processes. Organizations with extensive telecommunication systems have implemented the management systems found in this chapter. Firms with both must blend network management throughout their management systems.

IT managers must not make arbitrary distinctions that may cause discomfort to system users. For example, if application users experience problems, they want solutions. They don't care whether the problem originates in RBOC equipment, in local cabling, or in their application. Effective customer support can be achieved only when IT integrates and coordinates portfolio management, operations management, and network management. The IT management system must provide users with seamless, transparent telecommunication services. Users expect high-quality service; their expectations are entirely reasonable. Successful IT managers meet this challenge.

Critical Success Factors

As with other facets of information technology, network managers should understand the handful of factors that are critical for success. Critical success factors developed in Chapter 1 for the IT business apply to all its many parts, including telecommunications systems. Sound business management, strategy development, and planning pertain equally to networks as they do to other areas of information technology. Responding to customers' operational needs is always critically important.

However, some additional factors are critical for network managers according to Terplan.[8] He considers the following items to be critical success factors for network managers.

Methodology: How to proceed under certain circumstances.

Tools: What are the most meaningful instruments for facilitating human work?

Human Resources: The ultimate responsibility remains with people in any system managing communication networks.

Managing networks is easier if there are some general principles or broad guidelines that apply to the job. Critical success factors help in this regard. Methods, tools, and human resources are valuable considerations for network managers. This chapter expands on these concepts and explains their application.

MANAGEMENT EXPECTATIONS OF NETWORKS

Users' requirements for applications incorporating telecommunication capability are driving the integration of voice, data, and image communications. Software developers are formulating new kinds of applications as they tie computer screens into the telephone system and anticipate full motion video. The firm's PBX is playing an expanded communication role in the organization as customers expect access to databases through the telephone system for services ranging from health care to metal reclamation to loan application. The integration is occurring at the application level and at the data transport level.

The growth of network applications is accompanied by increased requirements for more and better network management systems. Appropriate tools, processes, and organizations must be available so that managers can maintain functionality, security, and availability. Corporations are willing to make sizable investments in large and sophisticated networks, but they insist that the implementation be manageable when complete. Users expect that vital network capability will be reliable, available, and cost effective. In short, investments in network hardware, software, and applications demand management tools, techniques, and processes that ensure successful network operation.

Executives expect to use telecommunication systems to link employees and managers throughout the firm thus expanding everyone's span of communication. They expect to tie customers and suppliers to the firm through sophisticated electronic data interchange (EDI) applications. Corporate executives expect information technology to provide convenient business linkages to others, as needed. They have high expectations for achieving competitive advantage from telecommunications technology.

As we learned in previous chapters, unfulfilled expectations are an important source of difficulty for IT managers. The emergence of valuable telecommunications capability and the integration of this capability into the firm's mainstream operations increase further the attention that managers must devote to managing expectations. Successful IT managers are well served by management systems designed to cope with the expectations of executives and users throughout the firm.

High expectations and an abundant supply of impressive technology are fueling the demand for new products. There are many network component vendors and a variety of options for managers to consider. Even with the growing attention to standards, many system components are vendor specific or proprietary, complicating network management. Multi-vendor network systems increase the difficulties of network management and foster the demand for open systems and better network management systems.

THE DISCIPLINES REVISITED

Customer expectations must be carefully managed. Expectations of networks include availability, reliability, and responsiveness at affordable and reasonable costs. Network managers require specific management tools, techniques, processes, and organizational structures to maintain network components that are highly responsive and highly available. Communication network management is a system of people using management information and tools to maintain physical and logical control over network operations. Operational control involves solving physical and logical network problems and monitoring performance to ensure attainment of contracted service levels. Like production operations, network management benefits substantially from a disciplined approach.

The disciplines of network management are applied to large and important networks, even those that are global in scope. For example, Merrill Lynch's worldwide telecom network, jointly designed and operated by MCI and IBM, gives network design, capacity planning, and disaster-recovery planning to MCI and the operational disciplines of performance, problem, change, and configuration management to IBM. Merrill will spend $200 million over a ten-year period on its global network.[9]

MCI and IBM use the disciplined approach to manage the Merrill Lynch network, just as firms throughout the world use these disciplines for computer-center management. Service-level agreements incorporating customer expectations regarding reliability, responsiveness, and availability are the basis of the network management system. Management reporting is the capstone of the system. Network managers must concern themselves with problem, change, and recovery management. Performance planning and analysis, capacity planning, and configuration management are crucial to network managers in running the management system.

Some references define the activities or disciplines of network management somewhat differently than this text. For example, the International Standards Organization (ISO) defines the management-system functions as:

- Fault management (problem management)
- Configuration management
- Accounting management
- Security management
- Performance management
- Adjacent areas such as planning[10]

This text considers change management and recovery management as essential disciplines; accounting and security are also important but are discussed in a broader context in subsequent chapters. As Terplan noted, computer-based network management systems incorporate most of the activities noted above, however, the ultimate responsibility lies with humans; a critical success factor. The disciplines of network management related to service-level agreements and to management reporting are shown in Figure 15.1.

Network Service Levels

Network service levels are established in much the same way as service-level agreements for computer-center operations and they serve the same purpose. The agreement establishes mutually defined levels of service that the network manager will deliver to client organizations. Usually, network service agreements are negotiated at fairly high levels in the firm; frequently, individual user managers are not involved in these negotiations. For example, the service agreement for the network supporting the firm's office system will probably be established between the network manager and the administrative manager responsible for office systems. Likewise, service levels for order-entry systems will be resolved between the order-processing manager and the network manager. A well-functioning network management system should be transparent to most individual managers and to all network users.

Network service-level agreements describe the type of service provided and contain measures for monitoring network service. In general, this means that network and user managers must agree upon and document network availability, reliability, and responsiveness. The later usually involves both response times and workload or volumes and is frequently referred to as network performance. Network managers must ensure that user managers understand the inverse relationship between volume and response time for their applications.

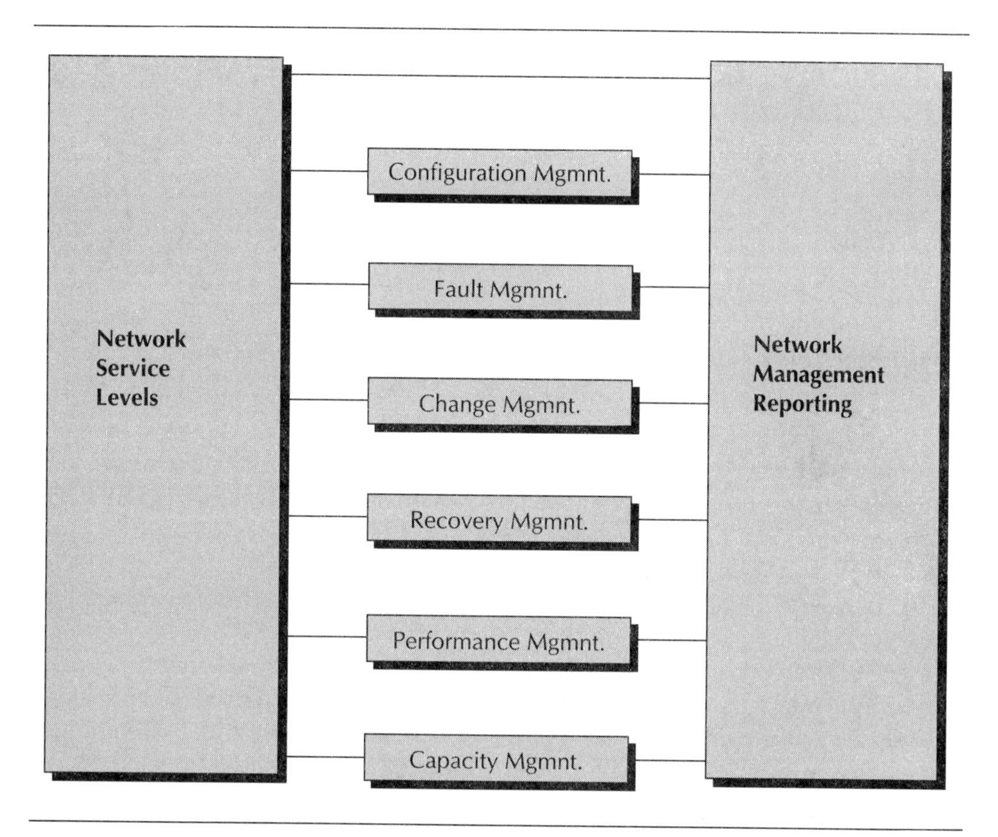

FIGURE 15.1 The Disciplines of Network Management

For most on-line applications involving simple transactions, subsecond response time is a reasonable expectation. Response time for on-line transactions is highly visible to users: Poor response degrades both productivity and morale. Research demonstrates the existence of a direct, positive correlation between system responsiveness and improved productivity. When responsiveness can be improved at low cost by network tuning or reconfiguration, for example, unit effectiveness and productivity are substantially enhanced. Thus, for several reasons, network managers have considerable leverage to improve business operations.

Network service-level agreements are documented in much the same manner as the SLA for production operations. The main ingredients of the network SLA are shown in Table 15.1.

TABLE 15.1 Network Service-Level Agreement

Date of agreement
Duration of agreement
Network service to be provided
Availability of service
Service reliability
Response times committed
Network cost/benefits statement
Service reporting
Signatures

The network SLA describes the types of network services that the client organization will receive. The service could be as simple as a local area net for the immediate office area, or as complex as a broadband international net, employing satellites for transoceanic communication. The services transcend the physical implementation. They must include types of transmission (voice, data, or image) and transaction volumes. This part of the agreement intends to capture the users' expectations regarding the types and volumes of communication and the scope of geographic coverage.

Users have expectations concerning the availability, reliability, serviceability, and cost of network service. The network SLA describes and documents these expectations. For some networks, availability is limited to normal working hours. For others, the network must be available continuously. For example, the office system LAN is operational during usual working hours, but the service may extend beyond these hours upon request. International operations may require almost continuous long-haul network capability. Response time requirements will also vary widely. LAN users expect subsecond or near instantaneous response for simple transactions; international applications may be less demanding. The service agreement describes the level of network service committed to the users.

In many cases, firms purchase or lease all or part of their network from vendors, such as local phone companies or independent suppliers. For long-haul services, AT&T, MCI, Sprint, and many others can meet most demands. Most of these suppliers are anxious to describe their high-quality products. Network managers should obtain the information they need and factor it into their customer agreements.

It's important that the network SLA account for the cost/benefits factors of network service. Network managers and applications managers must establish cost-effective service levels so that the operation, frequently new or reengineered, attains its financial and operational objectives. As Boston Edison through its use of CTI and Merrill Lynch with client/servers demonstrates, advances in telecommunication and computer technology are valuable in improving overall business, customer service, and response time.

Service levels attained versus service objectives must be carefully reported. The reported results must be highly credible easy to verify for all interested parties. As with production operations, user satisfaction surveys and opinion polls can be very effective in understanding and maintaining the client satisfaction that is so important to service providers.

Configuration Management

Configuration management includes the facilities and processes necessary to plan, develop, operate, and maintain an inventory of information-system resources, attributes, and relationships.[11] It is an organized approach for controlling network topology, physical connectivity, network equipment, and for maintaining supporting data. It also includes allocation of transmission bandwidth to various applications and consolidation of low-speed traffic onto higher speed circuits for more economical transmission. It is an essential step in the disciplined process of network management. Table 15.2 indicates the scope of configuration management.

TABLE 15.2 The Scope of Configuration Management

Physical connectivity
Logical network topology
Bandwidth allocation
Equipment inventory
Equipment specifications
User information
Vendor data

Configuration management is concerned with the physical network elements such as customer terminals, controllers, bridges, routers, modems, multiplexers, and links, and their physical and logical connections and topology. The discipline requires an accurate inventory of equipment, including physical location, technical capability, and intended customer application. The firm does not necessarily own all the equipment; some may be leased from equipment suppliers, and some communication capabilities may consist of leased bandwidth. Whether owned or leased, network system managers must know all capabilities.

Configuration management controls the physical connections and interrelations of the telecommunications equipment. It manages the logical topology of the network through routing tables. It assigns and controls bandwidth allocation between applications. Configuration management is a critical discipline because it allocates and controls network assets. The configuration databases must describe the network configuration at any given moment. They must be maintained for use by subsequent disciplines.

In addition to inventory data, other databases are required for efficient network management. Physical device addresses, application requirements, and other technical information must be readily available to reconfigure the network, if required. Reconfiguration may be necessary because of defective component operation or changing bandwidth requirements. Information about vendors must also be available to track service calls, install additional equipment, and manage equipment enhancements and upgrades. Configuration management databases are essential for managing problems and changes.

Network Problems and Changes

Network problem management (fault management) is similar in many respects to problem management for computer-based application systems. Faults or problems are indicated by network users, network operators, or by automatic alarming devices. The fault is documented in a problem report and tracked in a problem log.

Problem resolution may involve network technicians, vendor equipment service personnel, network users, and application specialists. During the problem resolution process, information contained in configuration databases is available for analysis and diagnosis. The process of problem management consists of identifying or locating the fault, isolating the faulty component through reconfiguration so the network can continue operation, replacing the faulty component to restore the network to its original status, and repairing the faulty component and placing it in inventory.

When the problem has been satisfactorily resolved, the configuration databases are updated with pertinent new information. For example, if the problem originated in a router, and the device was replaced with a new unit, the field inventory records would be updated and vendor maintenance activity would be initiated. If a link became inoperative, routing table information would be changed after proper analysis, traffic would take alternate paths, and link repair would begin.

The trouble report is closed when the problem has been corrected. Then the problem log is updated to reflect the corrective action.

Not surprisingly, network problems arise from a variety of sources, however, physical problems are most common. Table 15.3 displays the interesting relationship between problem frequency and OSI protocol layers.[12]

TABLE 15.3 Network Problem Sources

Protocol Layer	Percentage of Problems
Application	3
Session	7
Presentation	8
Transport	10
Network	12
Data link	25
Physical	35

There are many tools and pieces of test equipment to help network technicians isolate faults. Some rather simple testers can discover faults in the physical cabling, connectors, and switches. These devices cause many of the problems with networks. Incorrect addressing or unplanned configuration changes also lead to problems. User intervention with network components should not be permitted because it usually causes more problems than it solves.

Frequently, networks are designed to be highly flexible systems. Their physical and logical configurations change constantly. The changes result from fault correction, bandwidth allocation adjustment, application or user mobility, and network growth. Changes to the network's physical or logical configuration are a prolific source of faults if not managed carefully.

In some network systems, bandwidth and some configuration changes occur routinely and automatically. The network system may include dynamic bandwidth allocation, for example, or it may perform automatic routing changes. Some changes result from the problem correction process and some from the need to improve performance. Other changes are designed to expand the network to serve more users; others to add equipment, expand services, or reduce costs. Such changes originate from thoughtful planning and occur slowly over time. The change management process described in Chapter 13 is a good model for managing these nondynamic network changes. Yet all network change involves risk. The disciplined process of change management manages and controls risk.

Network Recovery Management

It is not possible to eliminate all risk of failure in network operations because some events are beyond the network manager's control. Indeed, in many firms, third parties own and operate vital network resources. Natural disasters such as hurricanes or earthquakes, or unpredictable events such as the World Trade Center bombing, underscore the need for network managers to have disaster recovery plans.

Network managers must have plans to recover from local disasters that affect LANs and their connections to local mainframes. The firm's phone system and PBX are vital communication assets. Recovery plans for them and for the connections to common carrier facilities or to centralized computer systems must be in place. Most, but not all firms have recovery plans for the phone network.

According to the Communications Managers Association, disaster plans are in place for networks in most firms. Disaster plans exist in 82 percent of firms for data networks, in 68 percent for voice networks, and in 6 percent of the firms for image networks. The data center, itself, is covered with a disaster plan in 78 percent of the cases, according to the report.[13] About one-fifth of these firms have no disaster recovery plans for their network assets!

Networks that rely on common carrier services face some specific risks. The rapid deployment of fiber-optic links is being concentrated along rights-of-way owned by communications companies, railroads, pipeline companies, and utility companies. Because rights-of-way are expensive, common carriers often install multiple fibers in one trench. To economize further, several carriers use the same trench, in some cases. These networks are dangerously vulnerable because an accidental breach of the trench may sever all the links of several carriers. Network managers cannot be sure that leasing links from several companies offers backup: They must examine the routing, as well.[14]

Recovery management is an important and critical task for network managers. But the many alternative network configurations available and the need to provide network solutions for computer system recovery planning make the task manageable.

Network Performance Assessment

Most network applications require that the network perform at or above certain minimum performance levels. For example, voice transmission requires a minimum bandwidth of 4000 Hertz in order to meet typical user expectations. Likewise, high fidelity sound transmission requires significantly higher bandwidth, 20,000 Hertz or more. Network performance must be maintained at levels that satisfy application and user expectations, as recorded in service agreements.

Network performance must be monitored and controlled. Monitoring allows managers to understand network performance, while controlling permits managers to adjust capacity and change performance. Performance tools measure traffic and enable managers to compare throughput to capacity. Managers must understand changes in throughput, they must discover bottlenecks, and they must measure user factors, such as response time. In addition, network managers must measure mean time between failures and mean time to repair. These measures are required to comply with availability, reliability, and serviceability commitments in the service agreement.

In general, network availability expressed in percentage is calculated by dividing the mean time between failure (MTBF) by the sum of the mean time between failure and the mean time to repair (MTTR). The following formula expresses this relationship:

$$\text{Availability (\%)} = \frac{\text{MTBF}}{\text{MTBF} + \text{MTTR}} \times 100$$

For example, if a network has a mean time between failures of 200 hours and has a mean time to repair of one hour, the availability is about 99.5 percent.

Availability increases as the mean time between failures increases, so it's important to have reliable network components. However, most network managers are strongly motivated to achieve availability by reducing mean time to repair to the lowest possible number, or by eliminating it entirely. This can be accomplished by installing network redundancy, as the Infomart example suggests (Chapter 14 Business Vignette), providing alternative capacity, or by automatic rerouting. This strategy provides network reliability to the customer, as well.[15] Reducing MTTR is a favorite way to achieve service levels. However, redundant capacity is not a substitute for service agreements, as some managers believe.

Transmission accuracy is usually not a concern with today's networks because of built-in error-detection and error-correction mechanisms. However, error rates must be monitored because retransmissions add to network loading. They may also be a leading indicator of future failures or faults.

Capacity Assessment and Planning

Network capacity assessment and planning were relatively simple when terminals were directly connected to a mainframe in a star or hierarchical network and standard applications were being used. Today's decentralized, client/server networks are used in far less predictable ways; new uses for creative applications can generate large, unplanned loads. Client/server networks permit users to be innovative and more productive thus diminishing network planners' ability to estimate loads and plan capacity.

Because of the dynamicism of today's networks, capacity must be assessed regularly and capacity planning must be a nearly continuous process. The rapid pace of technology introduction and adoption and the rapid growth of applications mean that network managers must constantly monitor throughput, user-experienced response times, and network utilization. This data helps determine and detail the current load. When analyzed over time, it also gives managers the trend information needed to project future network bottlenecks and constraints so that they can consider design alternatives.

Load projections of current applications and load forecasts from planned future applications are the basis for accurately projecting future load. The growth in current application volumes may result from increased number of users (end-user computing may be expanding its customer set), or from increasing transaction volumes with current systems (E-mail usage may be increasing). New applications, such as linking the distribution system to the firm's customers through EDI or physically locating client/server workstations with suppliers, may place new, increased demands on the network. Business plans, functional plans, and service agreements may reveal additional demands on new network uses.

Educated guesses about future loads and equipment requirements lead to excessive expenditures, poor performance, and inefficient network configurations. In large networks, such as Digital Equipment Corp.'s internal Easynet, with more than 95,000 nodes, computerized modeling and simulation tools are needed to maintain satisfactory service levels.[16] Except for small, simple nets, fine-tuning and capacity planning cannot be accomplished manually.

A new capacity plan must be developed to handle the incremental load. It may specify additional bandwidth, more user-oriented devices, additional controllers or concentrators, different, more effective links, or a new, more optimum topology. The objective of the capacity plan is to install cost-effective network solutions that achieve user expectations specified in service-level agreements. Because of the dynamic nature of telecommunication systems and applications, network managers should update the capacity plan quarterly. It should be reviewed periodically through monthly status reports.

Demand for network facilities always seems to rise to fill expanding supply. Additional bandwidth, more devices, and improved capacity usually lead rather promptly to increased load. Network planners must focus on service-level agreements and cost-benefits analysis to optimize business under these changing conditions.

Network planners should use the strategic business plan as the starting point for their capacity plan; they should develop tactical plans to implement it. Network architectures must be flexible enough to accommodate changing business conditions. The additional planning factors discussed in Chapter 14 are valid for network planners, too. Network planners should not overlook changing financial conditions, organizational restructuring, and the other items noted in Chapter 14.

If possible, the network should be kept simple and easy to understand. Even simple networks require skilled people and take significant amounts of time for testing and reconfiguration.[17] But, unnecessarily complex configurations of links and components increase installation and maintenance effort significantly. (Network complexity is thought to grow as the square of the number of nodes.) When capacity enhancements are required, the change-management discipline is very helpful in implementing them.

Management Reporting

Many of the performance assessment tools provide management data that can be shared with network users. Users must receive reports relating to their service-level agreements: They should be factually informed of service availability, reliability, cost, and system responsiveness. In turn, they should respond to satisfaction surveys. As with most management situations, there is no substitute for high-bandwidth communication between service providers and service users.

Network managers and senior IT managers need additional information. Throughput, or the transaction-processing rate, network utilization, problem analysis, change-management results, and capacity plans are essential. The results of performance analysis and capacity planning are important for tactical planning purposes.

The disciplines discussed above, from service-level agreements to management reporting, form the methodology. This disciplined methodology is the first of Terplan's critical success factors for network managers.

NETWORK MANAGEMENT SYSTEMS

Network management systems consist of computer hardware, network specific hardware such as performance monitors, software, and a variety of application programs. They are designed to provide operational control, to collect data about network operations, and to monitor and report network usage and performance. Data collection, storage, and retrieval capabilities are useful for managing problems, monitoring and managing performance, and planning network capability and growth. These are some of the tools Terplan speaks about, and they comprise his second critical success factor.

Systems or tools that perform these tasks are becoming more sophisticated and more integrated. Managers of networks require systems that support multi-vendor hardware configurations and provide automated operational controls. Suppliers of network management systems are driving toward these objectives. But according to a recent *Computerworld* article, "never have there been so many competent management products available, yet still the prospects are dim for getting a single one of these products to manage the total network on an end-to-end basis."[18]

Until recently, network management systems were designed and operated to support specific network elements. For example, a LAN would receive management support from software installed in the server or the master client on the net. A large mainframe system would obtain support from system software in the network control program, a part of the operating system. Support for each network element was separate from the others, and it was vendor unique. The concepts of open systems and integrated network management systems are designed to coordinate and integrate network management support across network elements and vendors.

The goals for an integrated network-management facility have not yet been attained, but progress is evident. Major vendors like IBM, AT&T, DEC, and many others are assembling the essential elements. However, these systems are still largely vendor-unique. In a recent publication, more than 500 network management products were listed and described.[19] Sales of network management products are exploding, as managers search for tools to control and operate rapidly growing networks.

Most network management tools are workstation based, consisting of specific hardware and software. Most contain alarm monitoring, traffic monitoring, network status reporting, and remote facility testing capability. Some support capacity planning, resource utilization analysis, and configuration management. Others provide chargeback and billing functions. The most capable systems cost upwards of $250,000. The choices are many, and the selection depends upon the firm's current network configuration and its future network plans.

The Open System Interconnect (OSI) standards offer hope for resolving some of the network management difficulties. International standards are being proposed for network management systems based on the OSI model. The implication is that as network product vendors move toward OSI network standards, system-management vendors will provide tools to operate in the standard environment. Future network management systems will go beyond providing tools; they will contain expert systems to automate further some of the tasks previously reserved for network specialists.

Expert systems will be valuable for problem determination, recovery operations, and some operational control tasks. They may help design new network configurations and advise network managers during change management. Sophisticated expert systems may also analyze performance data in near real time and advise management about impending system overloads and capacity constraints. Because network monitoring systems gather enormous amounts of data, and because much of the data has high short-term value, expert systems have the potential to act quickly on the information to affect system changes. They have the potential to optimize system resources and to reduce system costs. Phone companies are their early adopters.

INTERNATIONAL CONSIDERATIONS

As the enabling technology, telecommunication enables firms to capture strategic advantage through voice, image, and data communications. Using these forms of information exchange, software applications enable the firm to relate to its customers and suppliers more effectively and to attain internal efficiencies. The firm increases the effectiveness of its value chain through improvements in time to market, for example. Telecommunications improves effectiveness, increases efficiency, and improves old relationships or creates new ones.

Global networks enable the firm to attain these advantages on a worldwide scale. Global networks shrink time and space on an international scale permitting organizations to achieve global advantage. "The globalization of commerce and communications is changing the way we work, the way we do business, the way we learn, and the way our nation competes in the global marketplace. Why global? We might as well ask: Why breathe?" stated William G. McGowan, former chairman of MCI Communications Corporation.[20]

International links based on ISDN are being established with both Europe and the Far East. Nipon Telegraph and Telephone and Illinois Bell Telephone Company joined forces to offer an ISDN link between Tokyo and Chicago. The link utilizes AT&T's Switched Digital International services and the Japanese carrier KDD's international ISDN service to join the countries. Within Europe, network technology is a hot topic as firms jockey for cross-border capabilities. More than 60 percent of firms are using low speed, leased lines or packet switching, and 44 percent are leasing T1/E1 lines. More than one-third of these will adopt ATM or basic rate ISDN in the near future. In addition, satellite services are being rapidly deregulated, creating a whole new dimension in European cross-border data flows.

France Telecom interconnects its ISDN system, called Numeris, with ISDN in the U.S. This connection will give its customers access to U.S. applications and provide U.S. customers with connections to applications in France. Numeris serves the entire country of France and is expecting to grow rapidly as ISDN gains popularity in the U.S. France Telecom widely advertises its service to and within Europe, but it faces stiff competition from U.S. firms and from alliances between U.S. and European firms.

Recognizing these trends, the Netherlands privatized its PTT in January 1989, not only to meet its domestic needs, but to capture a share of the world's telecommunication business. The Dutch are positioning themselves at the center of the European open market. In their first year as a private business, they invested $1.8 billion installing ISDN and fiber-optic cables.[21] The new Dutch company gives the Netherlands a strong competitive edge in the global telecommunications market.

The dynamic nature of worldwide telecommunications places many responsibilities on international network managers. They must be knowledgeable across a broad range of topics such as the technical characteristics of long-haul links including satellites, services provided by international carriers including value-added networks, tariff schedules from many suppliers in many countries, and many facets of national and international rules and regulations. Transnational data flow is subject to numerous guidelines, regulations, and laws. International network managers be aware of these rules and must have a means for staying current on changing policy and regulations in this country and abroad. The telecommunications manager in a company with international operations has a challenging position.

THE NETWORK MANAGER

Networks and network management are rapidly growing businesses fueled by technological, economic, and policy considerations. Rapid advances in telecommunication technology are coupled with an explosion of products, services, and bandwidth driven by increasingly favorable economics. Policy and political changes are cooperating to accelerate system adoption and implementation on a global scale. Adoption of the technology both enables and mandates structural changes in organizations. Structural changes usually require advanced systems, so the cycle reinforces itself.

Management tools, techniques, and processes lag technology adoption by a considerable margin in many organizations. Thus, IT managers are not completely prepared to orchestrate the communication technology revolution. But orchestrate, they must. Because network assets cross organizational, political, and geographic boundaries, their management demands generalist skills. Corporate networks probably place higher demands on the IT manager's general management skills than traditional information systems.

In fulfilling their functions and responsibilities, corporate network managers must develop the network strategy for the firm in support of the firm's business strategy. They must evaluate technical and regulatory developments in the industry on a worldwide basis and provide input to the firm's strategic plan in order for the firm to capitalize on emerging trends. They must shape the firm's strategic direction with their technical and business insights. In short, the corporate network manager must be a superior business strategist.

Corporate network managers must also be outstanding tacticians. They must clearly understand the costs and benefits of the current network system and must evaluate emerging technologies and new products such as FDDI, Frame Relay, ATM, and others. They must understand changes in the legal, regulatory, and business environment and must use this knowledge to develop and implement near-term plans. The near-term plans must move the IT organization and the firm toward broader objectives set forth in strategies and strategic plans.

Corporate network managers must also attend to operational considerations. They must review the reports from the management disciplines and ensure that the disciplines are effective. Operational control is one of their important responsibilities.

Corporate network managers have both line and staff responsibilities; they must consult with the firm's executives, and advise and counsel functional managers throughout the organization. They have large, important jobs with considerable responsibility.

Senior network managers have a primary responsibility to cooperate with the firm's executives in establishing telecommunications policy for the organization. Rules of governance in the acquisition, application, and operational use of network assets establish the framework for profitable network exploitation. One policy that should be established is that the disciplines and tools of network management outlined above must be followed, regardless of who owns and operates the net. Policy must unify network implementation and make it seamless and cost effective. Ad hoc rules concocted by rogue departments when implementing departmental LANs lead to trouble later, when the networks must be integrated. However, with networks, as Strassmann says of information systems in general, "it is good information policy to have a set of rules that are unusually permissive to innovation."[22]

And, as in all endeavors, people are the enabling resource. Skilled and talented employees and managers are the sustaining force in network applications. Terplan's third critical success factor is Human Resources: The ultimate responsibility remains with people in any system managing communication networks. Much more will be said about this in later chapters.

Largely, network management has been learned on the job by individuals whose training and experience has been in computer science, information systems, or telecommunications. But many telecommunication managers were trained prior to deregulation. They worked with the local telephone company to arrange for local and long-distance service; corporate executives considered their service to be a utility similar to water or electrical power. Today, the job is much more complex because there are many more products, services, and vendors; telecommunications is much more than a utility. It generates revenue and creates strategic advantage while shaping and molding the firm, and perhaps the industry itself. Today, the firm's telecommunications function has merged with the information systems function as the technologies themselves have merged.

Until recently, there were scant opportunities for students to learn network management during their college training. About 35 colleges offer degrees in telecommunications, but only 3 grant Ph.D. degrees. In contrast, information systems is offered by about 360 colleges and universities in the U.S., and 60 of these offer the Ph.D. Some universities allow students to combine telecommunications, information systems, and business courses while pursuing a Masters degree. Academic interest in networks and network management is on the rise.

SUMMARY

Networks form an important part of the information processing infrastructure of most firms. They are important because they are an integral part of many operational and strategic systems and help provide many advantages for the firm. Telecommunication systems shrink time and distance. They provide leverage to firms through improvements in operational efficiency and effectiveness. And they can make these improvements on a global scale. Networks are also important because they facilitate process innovation and provide communication and control mechanisms to firms as they restructure, or as they move toward decentralized operations.

Network management includes methodology; processes and procedures to ensure efficiency, effectiveness, and customer satisfaction. Network managers have tools; instruments for facilitating the work of managing networks. And network management hinges critically on human resources. Ultimately, people are responsible for achieving the potential of telecommunications.

Corporate network managers need many skills to carry out the tasks of strategizing, strategic planning, tactical management, and operational control. If they manage or use global networks, international considerations must be a part of their thinking, as well. The network manager has an important and challenging job full of excitement and potential rewards.

Review Questions

1. What motivated Boston Edison to roll out the CTI application, according to the Business Vignette? What management issues does the Vignette raise?
2. Why is network management important to most firms today?
3. What is the connection between the disciplined approach to production management discussed in the previous chapters and network management?
4. Describe some of the network capabilities and explain their strategic importance to firm.
5. What technologies are spurring the growth of networks? How are some of these technologies working together to hasten the growth of networks?
6. Review the critical success factors listed in Chapter 1 and discuss those applicable to telecommunication systems and opportunities.
7. For what reasons are networks important to corporate organization and to restructuring?
8. What are network managers' critical success factors, according to Terplan?
9. Describe the scope of network management.
10. What do users expect of networks? What do executive managers expect?
11. What are the disciplines of network management?
12. What are the essential ingredients of the configuration management discipline?
13. What databases are maintained in the configuration management system?
14. What processes are involved in network problem management?
15. According to the text, most faults originate in the physical protocol layer. Why do you think this is so?
16. What network changes occur automatically in some networks? What types of changes are planned through change management?

17. What risks are associated with using common-carrier networks? How can network managers minimize these risks?
18. How can network managers maintain high system availability?
19. What information is needed to develop a new capacity plan? From what sources does this information come?
20. Why is it difficult to develop automated management tools for today's networks?
21. Describe the specialized knowledge that managers of international telecommunications systems must have.
22. What responsibilities do corporate network managers have? What responsibilities do they have regarding the disciplines?
23. What are the differences between network service-level agreements and applications service-level agreements?

Discussion Questions

1. Discuss the importance of telecommunication systems with respect to each of the strategic systems discussed in Chapter 2.
2. Discuss the reasons why international networks are expanding very rapidly. Consider the technical, economic, regulatory, sociological, and political factors that impact this expansion.
3. Discuss how a firm might use a telecommunication system as part of its implementation strategy for decentralization. What role might telecommunication systems play in the strategy of a firm that wants to grow by acquisition?
4. Discuss the reasons why network management is a difficult task. Which reason is most important in your opinion? Why?
5. Define the boundaries of network management. Where is the boundary with respect to other internal operations of the firm? How far does the boundary extend outside the firm?
6. What are the sources of the expectations that executives may develop regarding telecommunication systems? What role should IT managers play in developing executive expectations?

7. How do you think information infrastructure and organizational learning are related to telecommunication systems in an organization?
8. Compare and contrast the service-level agreement for a traditional information system with a network system SLA.
9. What are the reasons that the configuration management discipline is so important to network managers?
10. Why must configuration management be a real-time operation for managers of sophisticated networks? There may also be a modest configuration management system for the mainframe computer center. Discuss the similarities and differences between configuration management for the mainframe and configuration management for networks.
11. Discuss the evolution of the disciplined management systems as a firm migrates from traditional, centralized data processing to client/server computing. What changes might take place as the firm adopts EDI?
12. Referring to the formula for availability in this chapter, discuss the relationship between availability, reliability, and serviceability. What actions can managers take to improve availability?
13. Discuss the capabilities that are available with today's network management systems. What additional functions would you like to have in such a system?
14. Discuss the potential applications of expert systems to network management. Why do expert systems have more potential in network management than in managing large centralized systems?
15. Discuss the responsibilities of a network manager in an international firm by writing his or her job description. What characteristics would you like to see in an applicant for the top network management position at FedEx?
16. Consider a firm's goal of linking its dispersed operations to its headquarters through networks. Discuss the implications of application portfolio management on this strategic direction.

Assignments

1. Using Chapter 13 as a guide, design a fault or problem report and a problem logging system appropriate for network managers. Develop a rationale for combining network and data processing problem meetings or for holding them as separate events. What are the most important factors in this decision?
2. Study One of the major network management systems such as Hewlett-Packard's Open View Network Node Manager, NetView/6000 from IBM, or Cabletron System's Spectrum. Develop a list of its functional characteristics and discuss its strengths and weaknesses.
3. If a continuously operating network must be available 99.998 percent of the time and has an average mean time to repair (MTTR) of 10 minutes, what must be its mean time between failures (MTBF)? Discuss the implications of your answer.

ENDNOTES

[1] Kevin Tanzillo, "CTI Shaves Off 14 Seconds Per Call at Boston Edison," *Communication News*, January 1995, 26.

[2] Boston Edison serves 650,000 customers in and around Boston. BE is expected to earn $120 million on sales of $1.6 billion in 1995 according to *Value Line*, December 6, 1994, 162.

[3] Premier 100, *Computerworld*, September 19, 1994.

[4] Computer technology integration (CTI) refers to applying computer control and functionality to telephones and includes many applications. Connecting servers and PBXs is the first step in this linkage. Computer telephony is predicted to grow rapidly. See "Anticipating Telephony," *Lan Times*, March 13, 1995, 53–63.

[5] Dennis W. Elliott, "North Pacific Cable Meets Asian Connectivity Needs," *Telecommunications*, February 1990, 39.

[6] Tom Davenport, *Process Innovation: Reengineering Work Through Information Technology* (Boston, MA: Harvard Business School Press, 1993), 254.

[7] Wayne Ryerson and John Pitts, "A Five-Nation Network for Aircraft Manufacturing," *Telecommunications*, October 1989, 45.

[8] Kornel Terplan, *Communication Networks Management* (Englewood Cliffs, NJ: Prentice-Hall, Inc., 1987), 2.

[9] *Trends in Communication Policy*, September 1989, Economics and Technology, Inc., 101 Tremont St., Boston, MA 02108.

[10] See, for example, William Stallings, *Business Data Communications* (New York: Macmillan Publishing Company, 1990), 722.

[11] Stallings, 716.

[12] Rolf Lang, "Diagnostic Tools Decipher LAN Ills, Reduce Downtime," *Federal Computer Week*, June 26, 1989, 48.

[13] *Communications News*, October 1994, 6.

[14] James B. Pruitt and David Kull, "Tenuous Connections: Public Network Flunks Key Test," *Computer Decisions*, April 1989, 28.

[15] David A. Stamper, *Business Data Communications* (Redwood City, CA: The Benjamin/Cummings Publishing Company, Inc., Second Edition, 1989), 462–5. Stamper shows that redundancy in network components increases availability nearly but not quite to 100 percent.

[16] Sheila Osmundsen, "Expanded Networks Need Expanded Planning," *Digital News*, August 3, 1992, 1.

[17] Michael Puttre, "How to Avoid the Classic Blunders of Networking," *MIS Week*, October 2, 1989, 21.

[18] "Fruit Salad Anyone?" *Computerworld*, March 8, 1993, 101.

[19] "Product Guide," *Data Communications*, October 21, 1994, 113-46.

[20] Willam G. McGowan, *Telecommunications*, April 1989, 5–6.

[21] Peter Lablans, "Privatization of the Dutch PTT: New Telecommunications Opportunities," *Telecommunications*, November 1989, 61.

[22] Paul A. Strassmann, *The Politics of Information Management* (New Canaan, CT: The Information Economics Press, 1995), 33.

Part
Five

Controlling Information Resources

Business control is a basic management responsibility. Controls are required by law in many instances and are increasingly important with advanced IT systems. Managers must understand financial and operational control issues surrounding information technology and must maintain satisfactory control as technology penetrates the organization and business practices are reengineered. This part's chapters include Measuring IT Investments and Their Returns, and IT Controls and Asset Protection.

Managers must be increasingly aware of business controls as technology penetrates the firm. Managers must develop and use refined tools, techniques, and processes to discharge their control responsibilities effectively.

16 *Measuring IT Investments and Their Returns*

A Business Vignette

The Real Language of Business†1

Compared to the heavy accent now being placed on return on investment (ROI), competitive advantage was just small talk. Inquiring CEOs want to know: "How much bang is the corporate technology buck really buying?" If money talks, then ROI is one of its most common languages. Yet CIOs seem to have a hard time achieving fluency.

Despite all the money their companies have spent on information technology, CIOs still struggle to explain the benefits of an IT investment in hard dollars. They are more likely to justify IT expenditures by citing improvements in productivity, response time, customer complaints, error rates, and nonfiscal measures than by measuring the return on investment in information technology (ROIT).

Nonfinancial measures are useful and valid, but CEOs and CFOs who command the power of the purse rarely consider them sufficient. These ultimate arbiters frequently insist on proof that money spent on IT results in earnings and savings that surpass the investment. CIOs often feel caught between an irresistible force and an immovable object as they are compelled to provide financial justification for an investment that resists such hard measurements.

ROIT analysis can do more than keep CIOs in their boss's good graces. It can also direct them to the most lucrative opportunities for using technology. It can be a tool for evaluating existing and potential systems and a mental exercise to sharpen strategic thinking. And it can serve as a common language for explaining information systems to other executives.

For all these reasons, CIOs should learn to speak ROI more effectively. They need to find more techniques for measuring and demonstrating hard-dollar benefits and for persuasively articulating soft benefits—where possible, in hard-dollar terms. But CIOs should never become slaves to old, even obsolete, accounting formulas. After all, not every IS investment decision should be made strictly on the basis of ROI. An unbending insistence upon quantifiable returns can lead an organization to pass up investing in systems or applications that are ROI-resistant but still significant and worthwhile. Some of the most important advantages of IT, such as winning customer loyalty by improving quality and service, are hard to translate into dollars. An organization has to have the wisdom to know when to waive ROI for good reasons.

For the Perrier Group of America, a Greenwich, Connecticut–based company, ROIT combined restructuring and a supporting IT plan to yield $20.4 million annual hard-dollar return on an initial investment of $9 million, and subsequent annual costs of $5 million. Although difficult to evaluate financially, customer retention rates also are on the rise, yielding additional soft-dollar returns.

Perrier chose to overhaul its route operations, which distribute bottled water to customer businesses and homes, in order to bolster its standing in those markets. To accomplish the goals of reducing costs and increasing its customer-retention rate through improved service, the company consolidated redundant functions and departments at 40 stand-alone branch offices into regional centers.

Efforts to convert to a regional model of operation began in 1989, with completion scheduled for the end of 1993. The benefits to date have been impressive, with a $5 million annual investment netting a $15.4 million annual return. Over the five-year life of the plan, Perrier expects to pocket a cool $77 million in savings, primarily resulting from reduced headcount and improved productivity.

Before 1989, each of the 40 branches operated by five Perrier companies (Arrowhead, Great Bear, Ozarka/Oasis, Poland Spring, and Zephyrhills) located throughout the country maintained its own distribution, data-entry, teleservice, and credit and collection personnel. In addition, each branch had its own decentralized minicomputer and phone system. "Essentially, it was like operating 40 individual businesses all operating their own departments," said James Waldeck, Perrier's senior vice president of the route division.

The technical agenda consisted of overhauling and enhancing the existing route-management software used by distribution personnel, creating common databases, providing handheld computers for drivers, implementing advanced telecommunications, and consolidating into regional data centers in Los Angeles and Greenwich. The consolidations standardized equipment, training, and maintenance.

The new technology has directly affected the bottom line. The route-management software overhaul alone has reduced headcount by 32 for an annual savings of $800,000. The handheld computers have eliminated 59 full-time positions, saving $2.3 million annually, while also reducing the cost of printing and storing invoices by $565,000. Consolidating the data centers also reduced headcount: All told, the company was able to reduce personnel by 275, mostly in data-entry departments.

The improvement in customer service corresponds directly with an increased rate of customer retention and improved overall market share. Putting a dollar value on retention rate is somewhat tricky. "The key," says Rowan Snyder, corporate director of MIS, "is to recognize that it is very definitely costly to lose customers." Market share has grown 2 percent and revenues have increased significantly, according to company reports.

Waldeck also credits regional operations with helping Perrier stay ahead of the competition. "The water industry is in some tough times right now," he said. "While everyone is struggling to stay even, we're doing much better this year."

Perrier proved that regionalization and consolidation, supported by a comprehensive IT plan, could generate both hard- and soft-dollar returns on investments.

INTRODUCTION

Information technology resources comprise a large and, in some cases, a growing portion of most firms' budgets. Because of cost pressures and increasing competition, many organizations today are focusing on ROI and intensely scrutinizing IT budget items. Many firms are spending 2 to 5 percent of revenue on IT activities; IT managers are expected to explain these expenditures and account for them to the firm's officers.[2] As corporations become more information intensive, these expenses grow rapidly and become increasingly visible. The consolidation of telecommunication departments with information processing departments, the growing importance of telecommunications to the firm, the widespread adoption of distributed computing, office and factory automation, and the increasing complexity of the application environment all heighten executives' interest in IT expenses.

Interest in IT expenses is also intensified because of the manifold order-of-magnitude declines in the cost of computers and computing components, and the steadily increasing cost of labor. Thus the IT payoff stems from the benefits of using technology to reduce the total labor costs, or to greatly increase the effectiveness of money spent on labor. Reducing the cost of computing by employing lower cost technology is much less important in the overall scheme of things.[3] Managers need to know the cost of using human and computer resources in business processes in order to measure the benefits.

Accounting for IT resources is a complex and difficult task. The task is complex because the resources are allocated in a variety of ways. Some of the expense mechanisms are tied directly to the internal workings of complex hardware and software systems, some are incurred in the operation of sophisticated networks, and some are related to labor. For example, accounting for the operations of a client/server networked system requires a detailed accounting of operating and support labor, training costs, and maintenance costs of the hardware and the network. Few firms have cost accounting systems designed to handle these details.

Accounting for IT resources is difficult because the resources themselves have many forms and are generally widely scattered throughout the organization. In addition, knowledge workers and their equipment are inseparable in terms of performance. But the firm's accounting system treats people as an expense and machines as assets and is incapable of tracking the performance of people/machine systems. When people, networks, and machines are entwined, the situation becomes more difficult to understand. These systems increase organizational effectiveness, but accounting systems are incapable of displaying these effects. In most firms, there are no accurate ways to measure the total investment in technology and quantify its returns.

Measuring information systems investments and returns is a critical task for corporate executives; they cannot ignore the problem, nor can they expect it to go away. Increasingly, CEOs and CFOs are restraining IT expenditures, not so much because they believe the returns are unsatisfactory, but because IT executives cannot quantify the return. In the absence of a sound measurement system, reluctance to increase IT funding is warranted. In the firms making the most effective use of information technology, budgets are tightest and measurements of IT cost effectiveness are most pronounced.[4]

Measuring information technology investments and expenses and making these measurements available to beneficiaries of these expenditures is a critical first step in solving the ROIT problem. Accounting for IT resources and charging for their use alters incentives and impacts organizational behavior. High costs discourage use. Low costs or free services encourage high and perhaps unwarranted consumption. Consequently, accounting and charging processes alter cost and expense flows, influence the economics of organizations within the firm, and change the behavior of individuals and organizations. In addition, the administration of the IT cost-accounting system may be a time-consuming task, but it is an important task.

This chapter introduces IT resource accounting and discusses some common chargeback methods. Several alternatives for handling application development, program maintenance, and centralized and distributed operations are discussed. These variations illustrate the subtleties of charging mechanisms and provide insight into evaluating the return on IT investments.

ACCOUNTING FOR IT RESOURCES

Skilled people, sophisticated networks, complex and strategically important systems, and modern hardware and software systems are very valuable. People with skills in object programming, or those trained in client/server implementation, are in critically short supply. In some cases, IT resources take a long time to acquire or develop and, when developed, represent a large investment. IT investment decisions and the allocation of future IT resources are critical tasks for the firm's managers. To make these investment decisions intelligently, managers need a reasonably accurate knowledge of past expenditures and plausible projections of future expenditures. A relatively accurate cost-accounting system is essential for beginning to quantify the costs and benefits of IT activities. The accounting system also quantifies and calibrates the costs of delivering IT services to user organizations.

There is no meaningful alternative to accounting for IT expenses. Some firms treat IT expenses as overhead and pool similar expenses together. For example, all labor expenses for IT are collected and summarized, hardware expenses are pooled, and supplies and other costs are also summarized. In some firms, these pooled expenses are treated as corporate overhead affecting the firm's profit at a very high level. In other firms, expense categories for IT are granular: Expenses are passed through to the ultimate beneficiaries of the service. While there are many alternatives between these extremes, there is no alternative that avoids accounting for expenditures. But the choice of accounting methodology has many important consequences.

Accounting is a process for collecting, analyzing, and reporting financial information about an organization. Accounting can have two forms: financial accounting and managerial accounting. Financial accounting provides information about the organization to outside individuals or institutions such as banks, stockholders, government agencies, or the public. Because the information is used by the public, financial accounting is subject to carefully crafted accounting rules. Financial accounting is oriented toward providing a balance sheet and an income statement for the firm.

On the other hand, managerial accounting is concerned with information that is useful in the internal management of the organization. The organization can establish its own rules and tailor them to meet its specific needs. The goal of managerial accounting is to provide the firm's managers with information that will enable them to optimize the firm's performance. Accounting within the IT function provides the IT management team with information to optimize and control IT activities. IT accounting information is a valuable and critical resource.

Traditional financial accounting methods do not reflect the importance of intangible assets in the service and information industries. Estimates indicate there are more than $1 trillion invested in software and databases that have been expensed and not capitalized. "Companies spend enormous amounts of money on internal computer systems for production, purchasing decisions, sales analysis," says Arthur Siegel, vice chairman for accounting and auditing at Price Waterhouse, "but general practice is to expense those costs as incurred. We need to rethink this."[5] Firms can change their internal accounting systems to reflect their assets more accurately, even though current conventions require reporting differently to the public.

Innovative managerial accounting can overcome some of the deficiencies of traditional accounting systems. For example, application development costs can be capitalized and amortized over the anticipated life of the application. Within the firm, purchased system software and applications can be treated like balance sheet assets and depreciation can be taken to reflect technological and business obsolescence. Using this view, managers will reflect on the total IT investment, on its probable life, and on the costs of depreciation and obsolescence. This type of information is valuable and not difficult to obtain, but usually remains unknown in most firms. Accounting for human or social capital is a problem that also needs to be solved. These and many other aspects of accounting for IT resources are the main themes of this chapter.

Managers are concerned about the financial aspects of information technology. IT assets and their associated costs and expenses are being increasingly dispersed throughout the firm as distributed data processing and end-user computing blossom. And, as computer resources become widely dispersed, the budgetary responsibility for these expenses usually becomes dispersed too. In addition, the mix of costs and expenses is changing due to the declining cost of hardware, the increased use of personal workstations, the rapid growth of networks of all kinds, and the trend toward purchased software. For all of these reasons, it is becoming more difficult to understand the total cost of information processing in the firm.

As the dispersion of information processing continues, expense rebalancing is occurring between centralized and distributed operations. According to *Computerworld*, budgets of the companies in the premier 100 are rising more slowly than the rest. They are spending about 20 percent of the budget outside IS and more than 40 percent on client/server investments.[6] The trends in many central IS organizations are to reduce spending on mainframe computers while increasing the purchase of application programs, network hardware and software, and investing in client/server systems. In addition, the costs of personnel have been rising at a steady rate. As a result, in most firms, the total expenses for information processing have been rising.

As the technology becomes increasingly pervasive in the firm, costs and expenses are more difficult to identify and quantify. Training costs associated with the introduction of new information technology are a good example of this difficulty. When training in information systems activity was largely confined to the central IS organization, accounting for this activity was relatively straightforward. In most firms, as the technology became diffused, training expenses increased. As office systems and client/server computing flourish, training and other startup expense are increasing and less easily identified. The dispersion of the startup efforts and labor associated with operation and informal maintenance make accurate accounting for this activity difficult. For instance, the introduction of office automation may cost as much a $12,000 per workstation. About half of this cost is attributable to additional support staff, startup expenses, and individual training. In client/server operations, more than 75 percent of the long-term costs are people related. The usual accounting system fails to account for many of these expenses.[7]

In spite of the difficulties, accounting for IT expenditures is not optional. What is optional, however, is the methodology used in the accounting process. Current rules and regulations and commonly accepted accounting practices leave considerable room for customization. The degree of customization and the manner in which IT expenses are allocated to IT service users are important issues for the firm and for the IT organization. IT resource accountability is a critical issue.

THE OBJECTIVES OF RESOURCE ACCOUNTABILITY

Accounting for IT resources is a logical continuation of the information technology planning process described in Chapter 4. The management system for developing strategies and plans leads directly to the topic of resource accountability. The IT planning model discussed earlier includes applications, production operations, resources, and technology. In addition, the planning process includes feedback mechanisms based on measurement and control.

The objectives of resource accountability are to help measure the progress of the operational and tactical plans and to form a basis for management control. Control is a fundamental management responsibility. It is a critical success factor for managers. Specifically, IT managers must deliver services of all kinds, on schedule, and within planned costs. IT resource accountability deals directly with these measurement and control processes.

The IT planning process and the firm's planning process culminate in a budget for the firm and a budget for IT resources that support the firm's objectives. The firm's budget describes the resource expenditures required to meet the plan's objectives throughout the plan period. Likewise, the budget for the IT organization includes the resources necessary for IT to meet its organizational goals and objectives. A successful planning process results in an overall firm budget that incorporates the separate budgets for each organization within the firm. The budgets are tightly coupled and collectively support the firm's goals and objectives.

Control and planning activities are tightly linked. Control is the process through which management assures itself that members of the firm act in accordance with the policies and plans of the firm. In other words, policy setting and planning precede control. Generally, the same people are involved in both planning and controlling the firm's operations. IT accounting systems are one way for the firm's personnel to communicate with each other regarding IT plans and actions. IT accounting information can also motivate staff when used for appraising individual and unit performance.

For example, consider a firm that embarks on a strategy to improve its product development and manufacturing effectiveness through the introduction of computer-aided design and computer-aided manufacturing systems (CAD/CAM systems). The firm's strategy will express this long-term goal. The details will be found in the functional strategy statements for the product development, product manufacturing, and IT organizations. The firm's plans describe how each of these organizations will expend resources to accomplish this goal. Each organization's plans will describe how resources will be applied to accomplish its part of the overall task. The IT organization will develop strategies and plans to support the CAD/CAM installation, and the IT budget will also allocate resources to accomplish the CAD/CAM installation. To complete the process and to control the installation activity, the IT organization needs a management system to measure and control the expenditure of budgeted resources.

Budgetary control is the process of relating actual expenses to planned or budgeted expenses and resolving the variances, if any. All firms have a process for accomplishing this task. The IT organization will be included in this process; the IT manager will receive information periodically from the controller on its financial position relative to the budget. IT managers must monitor the important task of controlling budgetary spending. Generally, the firm's financial staff assists them.

The tasks of planning, controlling, communicating, and budgeting are all forms of management decision making. Accounting information from the IT organization is very useful in making decisions regarding IT activities. These activities are also valuable because they cement relationships between IT and other parts of the firm. The budget plays an especially important role in this process. IT resource accounting processes accomplish several important objectives. These are listed in Table 16.1.

TABLE 16.1 What IT Resource Accountability Accomplishes

Provides continuity between planning and implementation
Establishes a mechanism to measure implementation progress
Forms a basis for management control
Links IT actions to the goals of the firm
Communicates plans and accomplishments
Provides performance appraisal information

Is it sufficient to control IT activities with budgetary processes alone? Are the best interests of the firm served if client organizations are not financially involved in IT expenses? Since the IT organization provides valuable firm services, perhaps it is best if the costs of these services are recovered from the users in some way. Many firms believe so; chargeback systems are receiving increased attention.[8]

The tendency to link technology costs to operational departments is also accelerating as firms become more focused on costs and cost containment. IT managers want to demonstrate that operational requirements drive their expenses and want operational department heads to support these expenditures at the executive level. Thus, most firms charge users for IT services in one form or another.

Most firms believe that the advantages of an IT cost recovery process outweigh the disadvantages. The most important advantage of IT cost accounting and cost recovery is that they provide a basis for clarifying the costs and the benefits IT services. A well-structured and smooth running IT cost-accounting system attacks an important IT issue—the role and contribution of IT. An important secondary advantage is that the process strengthens communication between IT and user organizations. Table 16.2 lists some benefits of IT cost recovery.

TABLE 16.2 Benefits of IT Cost Recovery

Clarifies costs/benefits of IT services
Strengthens communication between IT and user organizations
Permits IT to operate as a business within a business
Increases employee's sensitivity to costs and benefits
Spotlights potential unnecessary expenses
Encourages effective use of resources
Improves the cost-effectiveness of IT
Enables IT benchmarking
Provides a financial basis for evaluating outsourcing

IT cost-accounting and cost-recovery methods focus managers' attention on services that have low or marginal value. For instance, if users are charged for the operation of a regularly scheduled application, the output of which has low value, they will analyze the costs and benefits and may terminate the application. If production operations are free, the processing of marginal application programs will continue, perhaps indefinitely. Cost accounting provides tools to detect unnecessary or cost-inefficient services. Inefficient services can be altered or eliminated, thus reducing expenses and improving overall effectiveness.

Users who receive a bill for IT services have information they can use to maximize their gain while minimizing costs. They may search for more effective processing schedules or increase their use of services having high marginal value. For example, the operation of an on-line application may be expanded when users have tangible evidence that the expansion in services is financially attractive. Likewise, if users are charged for application program enhancement, they will be motivated to request enhancements that are financially viable. The charging process encourages the effective use of scarce resources.

A well-designed charging mechanism also enhances the IT organization's effectiveness. IT costs are scrutinized more carefully when they are well known and have sufficient granularity. The need to explain IT charges to clients directs IT manager's attention to the cost elements. Attention to the cost elements motivates IT to improve the cost effectiveness of its customer services. Insufficient attention to the cost elements or poor granularity in the cost structure promotes ineffective resource utilization. For instance, if all telecommunication costs are grouped together, degrees of effectiveness among the various services may go undetected, and the resulting inefficiencies will negatively effect the IT organization.

Pricing IT services and billing users for them heightens awareness of IT costs and benefits among the firm's employees and managers. IT personnel become sensitized to the money the firm is spending to support their activities; they become more cost conscious. Likewise, clients billed for IT services become more value conscious. They appreciate what the firm is spending on their IT activities and tend to use them more wisely. Chargeback processes increase the cost and value consciousness of employees and managers alike.

The primary disadvantage of IT cost recovery is its administrative overhead. The cost recovery process is not free and must be cost effective itself. Cost recovery may not be justified for organizations that are not information intensive. Also, cost recovery may not be acceptable within the cultural norms of some organizations that prefer not to involve users in the costs of IT services.

Not all IT organizations favor user chargebacks. Some IT managers prefer to keep their finances clouded in secrecy rather than expose their operations to whatever criticism may result from charging. Some firms operate with relative insensitivity to costs once the budget has been approved. In some organizations, employees are relatively insensitive to expense. In these organizations, chargebacks are ineffective. Not all firms favor IT chargeout, and not all firms are prepared for the effective operation of a chargeout system.

One alternative to chargeouts is to distribute IT costs through a committee or to allocate them during the planning process. While these are simpler approaches, they suffer from many disadvantages. The allocation or distribution methods are frequently political, inaccurate, relatively inflexible to changing conditions, and they reduce line management accountability and responsibility. Another alternative is to bury IT costs and expenses in general corporate overhead, keeping them relatively invisible. This approach completely eliminates the advantages of chargeout systems and is not favored by cost-conscious firms.

Given the relative advantage of cost recovery processes to most firms, what goals and objectives should the firm strive for in the process? What can the process gain for the firm and how must it be established for maximum effectiveness?

The Goals of a Chargeback System

IT chargeback mechanisms should serve the firm and its constituent organizations by improving the firm's effectiveness and efficiency. The IT organization should benefit, but the firm as a whole should benefit as well. Table 16.3 shows some goals of an effective chargeback system.

TABLE 16.3 Goals of a Chargeback System

Easy to administer
Easy for customers to understand
Distributes costs effectively
Promotes effective use of IT resources
Provides incentives to change behavior

Budgetary controls are important, but they are subject to the same effectiveness criteria as other activities within the IT organization and the firm. The chargeback process that IT adopts must be easy to administer and must be easy for clients to understand. The administrative cost must be small in proportion to the resources being administered. For example, if the cost to account for and bill out an IT service is $5, then the total amount billed over the accounting period should be considerably more. In other words, the granularity of chargeable services and the chargeback process itself must be cost effective.

In order for IT customers and managers to accept the chargeback mechanism, the algorithm for computing the customer charge must be easy to explain and easy to justify. A complex billing system based on obscure and hard-to-understand parameters will alienate customers and create planning and budgeting difficulties. If the algorithm is constructed from simple parameters and is easily understood, then the charging process is easier to administer.

The chargeback mechanism must account for IT costs in a fair and equitable manner. For example, the charge for programmer's time may be based on hours worked using two rates, depending on the programmer's skill level. Since there are many more than two different levels of productivity among programmers, this algorithm is not completely accurate, but most client organizations will consider it fair. It is also relatively routine to administer and, therefore, cost effective.

IT chargeback mechanisms alter the financial incentives for organizations to use IT services. The choice of chargeback method and the manner in which it is used influence usage patterns and the relationship between the IT organization and the client. Therefore, another goal of the chargeback process must be to promote the cost-effective use of IT resources. For instance, if there is no charge for the use of on-line direct-access storage devices (DASD), then there is no financial incentive to use the storage effectively. The charging process doesn't encourage users to delete obsolete datasets, for example. On the other hand, if direct-access storage is priced unreasonably high, then users may resort to inefficient methods of data handling to reduce costs. For example, they may use tape storage when DASD is more effective for the organization.

If clients are sensitive to charges, then charging mechanisms can be used to steer the organization toward certain technologies and away from others.[9] New technologies can be attractively priced to encourage use, and older systems intended to be replaced by new technology can be priced unattractively. The price may not reflect the actual costs: The new equipment may be more expensive than the older, fully depreciated hardware. In this instance the pricing algorithm is designed to provide incentives toward more efficient operations, not to recover actual costs.

Chargeback systems that are easy to understand and administer, distribute costs effectively, and promote effective use of IT resources are valuable additions to the IT management system. Successful IT managers rely extensively on them. Also, successful IT managers carefully consider the advantages and disadvantages of the alternative chargeback methods before selecting and implementing one.

ALTERNATIVE METHODOLOGIES

There are two major accounting alternatives for handling IT cost recovery: the cost center method and the profit center approach. Both methods distribute IT costs to users, but they vary considerably in other respects. The profit center method is designed to generate revenue that exceeds the IT cost. The IT organization may use these profits for their own approved purposes. Of course, if costs and expenses exceed revenue, then the profit center loses money. Ultimately, IT must find some way to recoup their losses.

On the other hand, the cost center method seeks to break even financially. The amounts received for services are expected to match costs very closely. Methods to make costs and revenues match exactly are often used. Details of these financial arrangements are discussed in the following sections.

The Profit Center Method

IT organizations established as profit centers operate as a business within a business. The IT organization's costs and expenses must be recovered through charging customers for services rendered. The typical relationship between revenue and expenses prevails; the organization may operate at a profit or a loss. The profit center method is easy to understand and explain.[10] Other advantages of this method are listed in Table 16.4.

TABLE 16.4 Advantages of the IT Profit Center

Easy to understand and explain
Promotes business management
Provides for outside comparisons
Establishes financial rigor
Enables outside sale of services

IT managers who operate a profit center develop important skills as business managers. They use many of the disciplines that independent business people use. They learn business management in a way not usually possible in other environments. IT profit center managers appreciate the sometimes invisible expenses of corporate overhead, employee benefits, and equipment depreciation. And they learn the fundamentals of pricing strategy. Business management experience extends to IT customers as well who may gain exposure to the financial consequences of issues of which they were not previously aware. Firms favor the profit center approach for this reason among others.

Profit center prices reflect the real cost of doing business within the firm. They can be easily compared with the price of similar services from outside suppliers. These comparisons, or financial benchmarks, provide a mechanism for measuring IT effectiveness, and they promote IT efficiencies. If the firm has a policy permitting clients to purchase competitive services, then IT is usually highly motivated to remain competitive. If the IT organization cannot remain competitive with external suppliers, then the firm may decide to purchase services externally. This process may also highlight inefficiencies within the firm because it directs attention to general overhead expenses and corporate accounting policies. The profit center approach has benefits that extend beyond IT and its clients.

Another advantage of the profit center approach is that it provides a high degree of rigor and discipline to the financial relationship between IT and its client organizations. Costs are more carefully established, expenses more thoroughly evaluated, and prices and charges more insightfully determined. This disciplined method assists in other processes as well. For example, accurate costs to achieve service levels are available, thus negotiations on service-level agreements proceed more objectively. Application portfolio management is improved and make/buy decisions are made with heightened confidence. In general, financial management improves.

Establishing the IT organization as a profit center enables IT to sell services outside the firm with confidence that the activity is profitable for the firm. Selling IT services may be advantageous to the IT organization and to the firm. The profit center methodology is a prerequisite to becoming an external revenue-generating service bureau. The IT profit center may be established as an independent business with its own financial accounting system. In this case, the profit center accounting system will link to and closely support the financial accounting system for the organization. The accounting rigor of a profit center enables the firm to capitalize on its IT resources as a source of revenue and profit, if it chooses.

There are some obvious disadvantages to the profit center approach as well. Financial rigor comes at the expense of administrative overhead—rigorous financial treatment may not be worth the price. Detailed knowledge of all expense ingredients, however small, and soundly based prices may not yield improved performance overall. In addition, the firm is somewhat dependent on successful management of the interface between IT and client organizations. There is no guarantee that prices will be properly established, or that clients will anticipate fully the costs of their IT services during their budgeting process. The IT organization may not earn the anticipated profit necessitating high level adjustments. If IT is highly profitable, its customers may have paid more than required; perhaps sacrificing alternative investments. The degree of financial coupling between IT and its clients may be insufficient to avoid these difficulties. Skillful profit center managers achieve success in spite of these pitfalls.

The Cost Center Approach

An alternative for handling IT cost recovery is the cost center approach. The cost center methodology is based on a process in which each client organization budgets for its anticipated IT services during the planning and budgeting process. The sum of the client-budgeted amounts for IT is the cost center support from its customers. Managers expect that this amount will approximately equal the anticipated IT expenses for the planned period. Because there may be mismatches between planned IT expenses and client-budgeted support, this approach forces intense interaction between organizations during the planning stage. Planning and budgeting is intense and iterative. It has other characteristics as well, that are displayed in Table 16.5.

In order to complete the annual planning process, client organizations must know their anticipated requirements for the planning period. They must also know IT prices for the period so they can submit their budgets. Their budgeted support for IT is their anticipated usage multiplied by the price for the service. During the planning stage, IT also prepares a budget that anticipates all the customer demands and plans to satisfy them. The total of all the client budgets for IT must approximate the budget IT prepares.

TABLE 16.5 Cost Center Characteristics

Cost Center Characteristics
Promotes intensely interactive planning and budgeting
Establishes prices in advance of known support
Forces mangers to handle variances
Exposes the plan process to manipulation
May lead to conflict (this may be beneficial)
Forces decision making
Reinforces the SLA and capacity planning processes

If the budgeted support is not close to the planned IT expenses, then the process recycles in the IT organization. This recycling effect means that the IT budget is the last to be completed. Also, because of changes in the support level, prices may change. Price changes alter the budgets in the client organizations by a small amount. Usually this phenomena is handled by permitting some variance at plan time between the sum of all budgeted support for IT and the expense budget prepared by IT.

In the ideal situation, customer budgets for anticipated IT services and the IT budget for anticipated expenses are approximately equal. Ideally, actual customer demand for services equals planned customer demand, and prices from IT equal planned prices. In actual practice, many variances must be handled during the execution of the plan. Demand changes generate price changes during the year with financial consequences to the clients. For some services, such as centralized production operations, if the demand increases slightly, IT tends to make a profit; if demand decreases, IT starts incurring a loss. By year-end, the variances must be resolved.

There are several ways to resolve these variances. One approach to resolving the overage or underage is to carry the variance at a higher level in the firm, and generate a profit or loss at that level. Second, the profit or loss can be distributed to the customers as a retroactive rate adjustment. This results in an overage or underage in the customer budgets. Third, the rates or prices can be changed slightly during the year to keep the running variance close to zero throughout the life of the plan.

As one might expect, the cost center method is subject to customer manipulation. For example, if a client organization underbudgets for IT computer services and later increases its demand, the extra demand, if it can be satisfied, may result in lower rates for all customers. In this case the gamble pays off for that particular client. If many clients underbudget, IT may be unable to satisfy the true demand and customers will receive poor service. Those clients who budgeted accurately also suffer.

If the corporate policy is to cover variances at a higher level at year end, clients are likely to overbudget for IT. IT will acquire greater capacity than required, clients will obtain better service than planned or justified, and the higher level must cover the excess expenses. Clients may spend the excess in their IT budget on other items, justified or not. There are other variations of this process as well. Given the above, the cost center method works well when the firm has strong financial discipline and the organization's controller takes action to protect IT and the firm from potential abuse.

The cost center approach is contentious and leads to conflict, some of which may be healthy. If the capacity planning process is not very effective, the cost center methodology tends to flush out latent demand. If capacity planning is well orchestrated, cost center planning proceeds quite well and the results can be very satisfactory. In any case, cost center planning forces decision making. This may be advantageous if other management processes within the firm are not completely effective.

In addition to the complete profit center method and the total cost center approach, other alternatives combine aspects of each, or use other methodologies for certain IT services. For example, some services may be priced separately, partial recovery of certain costs may be allowed, or costs may be based on long-term contracts with the user. What are some of the alternatives and enhancements to the cost recovery methodologies discussed thus far? The following section will address this concern.

ADDITIONAL CONSIDERATIONS IN COST RECOVERY

IT services are a varied lot; no one single scheme for recovering costs is satisfactory to everyone. Consider, for example, the labor intensive process of application development as compared with the operation of on-line mainframe applications or the data communication network operation. Application development has no economies of scale. Generally, it has little excess or latent capacity, and suffers from a variety of people concerns and issues such as communication and motivation. The mainframe operation however, may have economies of scale, may have large amounts of latent capacity, and is less susceptible to people issues and concerns. Telecommunication networks are different from each of these. Because of these differences, cost recovery methods can be optimized to take advantage of the service characteristics.

Funding Application Development and Maintenance

Portfolio management and the management of application development lend themselves very well to customized or unique cost accounting and cost recovery approaches. For example, the costs of application program maintenance and minor enhancements can be recovered through the profit center or the cost center methods using rates established for application programmers. If a programmer works for 10 hours performing minor enhancements to an application, the owner of the program is charged for 10 hours of service at the established rate.

Since minor enhancements are a way of life for important applications, it is better to recognize this condition in advance, and prepare a long-term contract describing the on-going support for the application.[11] The contract for "period" support may state, for example, that the programming department will provide 20 hours per week of programming effort devoted to program enhancements that the client manager requests. Probably, the contract period will be for a year or more at a rate sufficient to recover the programmer's expenses. This type of contract has an inherent degree of flexibility. It avoids the problem of accounting for the programmer's hours except in a general way. It gives the client manager the opportunity to make minor enhancements without contending with other users for programming resources. The IT organization and the client department both benefit from this arrangement.

Application development projects may also benefit from alternative cost recovery methods. Consider, for example, a variation of the pay-as-you-go method for recovering costs on a major programming project. At the beginning of each phase, a cost is established to complete the next phase. At the end of the phase, the customer is billed for the contracted expense, and IT recovers its costs for the phase. This approach closely relates the level of effort to the objectives established during the phase review. In effect, IT commits to producing the function on schedule *and* within budget. Since IT usually will not agree to changes in function during the phase, client and IT managers are motivated to produce a high-quality plan for the next phase. This approach offers the advantage of reinforcing the discipline of the phase review process with financial incentives. Charging for application development by phases is generally preferred to recovering costs on a pay-as-you-go basis.

Charging for application development by phases exposes poor planning, reveals the financial impact of change and discloses the costs of poor implementation. In addition, this approach makes the decision to terminate the project, if that becomes a possibility, less difficult. The temptation to continue the project, making up for past excesses by trimming future planned expenses, is less likely.

Another method for handling application development is to recover the development cost from the application owner over some predetermined life of the application. This approach recognizes that benefits from the application are realized after installation. It relates development expense to benefits realized over time. The advantage of this method is that early risk is taken at some higher level in the firm thus making application development more attractive for users. The advantage for the firm is that applications of high potential value for the firm can be developed when they might not be financially feasible otherwise. To the client, this method appears to capitalize the development effort, though it may be expensed at the higher level.

Some projects should be funded at the corporate level. Applications that have great strategic value for the firm should be managed financially by the firm and not by any single client manager. Though the application may be used primarily by one class of customer, strategic decisions regarding the application should be made at higher levels in the organization. For example, an electronic design-automation program that gives advantage to the firm engaged in designing and building electronic products cannot be managed effectively by managers in the development laboratory. Product managers are likely to make short-range decisions; their view focuses on the product and not the firm and, as a group, their interests are diverse. The solution is to make decisions at a level where long-range considerations are preserved, the firm's interests are paramount, and divergent motivations can be reconciled.

Widely used computer systems must also be managed financially at some high level. Office automation systems are of this type. Office managers should drive the requirements process, but the funding for office systems is best performed in a central administrative function. This approach recognizes the firm's office automation costs, but it avoids the difficulty of allocating costs to individual workers or departments. Some telecommunication systems and large database applications fall into this service category, as well.

Cost Recovery in Production Operations

There are many algorithms for recovering costs in production operations. A common theme behind most of these algorithms is to charge for the resources used in the production of useful output. For instance, CPU cycles used, the amount of primary memory occupied and its duration, channel program utilization, pages of printed output, and other measures of production resources are common usage parameters. Generally, these various parameters are linked together in some fashion to form a charging algorithm. IT establishes a price for each resource so that the charging algorithm reflects the cost of work performed for the customer. In some cases, this type of charging algorithm can become quite complex; it may also be very accurate.

Although these measures can be quite precise, they are not very satisfactory to IT customers. They are quite difficult to develop because they require a detailed knowledge of the system and the economics of its parts. Because of these complexities, precise charging algorithms are difficult to explain and hard for most clients to understand. Clients have trouble relating some of these measures to the useful work the system performs for them. Simple methods, such as charging for the application's elapsed time, are probably more effective.

For applications or processes that occur over extended periods, elapsed time is a common denominator of the charging algorithm. For example, an on-line application operating continuously may be priced according to the fraction of the CPU resources it consumes. If the on-line order-entry system requires one-third of a major CPU, the service is priced to recover one-third of the cost of operating the machine. Frequently, large on-line data stores are priced at so much per track per day. The idea is to recover the cost of on-line storage from the users in proportion to their storage volume. If large on-line data stores are associated with continuously operating applications, the elapsed-time charge should include both CPU and DASD costs.

Many other variants are useful in charging for production operations. One such variant is the use of price differentials for classes of service. For instance, users of prime-shift capacity (8:00 AM–6:00 PM) may be charged more than off-prime shift users. The price differential can be justified since prime-shift time has more value to most users than off-shift time. The price differential encourages system use when there is usually surplus capacity. The shift of workload from prime to off-prime is equivalent to a capacity increase; it represents a real cost saving to the firm. Using the same rationale, weekend work may be processed at reduced rates while priority or emergency work may cost more. These are examples of price incentives designed to encourage more effective use of IT resources.

Services dedicated to one class of user should be charged directly to that user. For example, if one CPU and associated equipment and support personnel are devoted to one client department, that department should pay the full cost of that service. This charging methodology is equitable, easy to justify, and easy to explain. Dedicated telecommunication links fall into this expense category.

In production operations, it is important to plan ahead and to recognize the effects of technological obsolescence. This can be accomplished very effectively if costs are based on a multiyear plan. The multiyear plan prevents wide fluctuations in prices as equipment is replaced or additional capacity comes on stream. Advanced planning recovers costs early and matches long-term revenue and expenses. Successful IT managers use this approach to everyone's advantage.

Finally, some services should be sold outright. End-user computing and office automation benefit considerably from this approach. Rather than billing the customer for the monthly cost of the terminal or personal workstation, IT should sell the hardware to the client and bill the connect time or network time based on system use. This simplifies the accounting process, is easy to justify, and is easy to understand. Most client managers are quite comfortable with this approach. Upgrades or additions to the customer's personal hardware or software are best handled on a purchase basis, too.

Recovering Costs in Distributed Systems

As firms move toward customer-controlled systems from centralized systems and resources are rebalanced, methods to understand and account for the costs and expenses in distributed environments may also shift. The degree of change in financial treatment may vary widely, depending on the firm's policies regarding distributed processing. For example, if the IT organization continues to own the hardware and continues to develop or procure applications for the client department, little change occurs. On the other hand, if client departments own the hardware and develop or procure their software, financial rebalancing takes place and accounting is altered. Usually, the situation falls somewhere between these extremes.

As a general rule, if the distributed system serves one function, marketing, for example, that function should justify and pay for the system. If that function also develops or procures the software, it should pay the bill. The firm should be aware of these costs or expenses, however. This occurs easily for items purchased from the outside but is more difficult for client-developed programs. Again, a policy decision requiring the client department to record costs of program development is needed. Without such a policy, client program development adds to the other hidden costs of distributed processing, increasing the difficulty of related decision making.

Usually, IT supports distributed processing in various ways. Network and operating system support, server hardware and backbone nets, and programming and technical support are usually obtained from skilled IT professionals. The considerations discussed in the previous section apply to this kind of support.

In distributed processing environments, what matters most is that costs are identified and controlled in some manner; who does the work or who owns the assets is far less important. As IT line activities decline in magnitude and as staff activities increase, financial rebalancing accompanies responsibility shifts. One staff responsibility that increases is ensuring that corporate executives have the information needed to make decisions. Establishing policy guidelines that account for IT expenses in reasonable and effective ways, wherever they are incurred, is an increasingly important staff responsibility.

Network Accounting and Cost Recovery

Networks significantly complicate the IT cost accounting and cost recovery processes. Production computers connected to networks see the network as another source or sink of information for the application operating in memory. IT accounting routines can deal with application accounting, but when applications in several computers communicate with each other and place computing load on the CPUs, the accounting process becomes very complex. To account for all the costs accurately, and to do so in a way that clients will understand, is not possible.

The accounting problem becomes more complex when public carriers and value-added networks enter the picture. International operations add yet another level of complexity. Corporate network managers must have a solid understanding of the impact of expenses on service levels and must know the expenses to the component level. They should keep detailed trend information that can be related to trends in performance and utilization. Their knowledge of network costs should permit them to make trade-offs between various types of new technology. They must be able to relate new technology costs to capacity increases and performance and service improvements.

IT managers must be very careful about charging clients for telecommunication services. Some services, such as long-distance dial-up voice or data communication, should be charged to the user's account directly. Other network services, such as those supporting EDI applications, are probably best recovered at the corporate level. Other applications must be analyzed on a case-by-case basis. If the network expenses readily correlate with an application, the application user should be billed for them. If the expenses are small, or if they are co-mingled and not easy to separate, it may be best to recover them through corporate overhead.

Some network management systems contain accounting packages to help solve some of these problems. The need for an accounting solution may influence the selection of a network management system.

The process of accounting for and recovering IT expenses from IT customers is a creative one that benefits from a businesslike approach. The variations discussed above can be used with either the profit center or cost center structure. Ultimately, the methodology and variations depend upon the information technology maturity of the firm, the relative importance of IT, the corporate culture, and the objectives that the firm's executives hold for IT. As these factors change over time, the cost accounting and chargeback systems will also change. These processes will go through iterations as the organization matures.[12] In addition, the relationship between the IT organization and client organizations will develop as the accounting process matures.

RELATIONSHIP TO CLIENT BEHAVIOR

One important goal of user chargeback processes is to encourage and promote the cost-effective use of IT services. There are many financial motivations that result from cost recovery actions; other motivations can be developed as well.

Consider, for example, a firm that underutilizes information technology and wants to promote increased technology adoption. What strategy should this firm adopt? For how long should the strategy be employed? One strategy for this firm would be to increase IT funding and services, but accumulate the increased expenses at the corporate level. For example, if a firm wanted to encourage intersite communication, it could install a network linking several of its sites and cover these expenses at headquarters. This, in effect, makes the telecommunications network a free utility to the users. This approach encourages user managers to develop intersite communication applications under these circumstances.

This approach to funding is appropriate if the firm lags the industry in technology adoption or is at the initiation stage with a new technology. The methodology hastens the firm's migration into the contagion stage where, perhaps, financial incentives are no longer needed. When the contagion stage is reached, special incentives can be reduced and another approach used.

There are many incentives inherent in the chosen approach to funding IT applications, some that were evident earlier in the discussion. As the firm matures in its use of information technology and as the management team becomes proficient in using the IT management system, variations can be employed that maximize the chances for attaining corporate goals. For example, the firm may abandon the cost center methodology in favor of the profit center approach when the opportunity to sell IT services outside the firm is recognized.

If an appropriate degree of maturity and sophistication is reached, and if the firm owns attractive services, the IT organization may begin to generate revenue for the firm. Many firms make their production facility available as a service bureau. Some firms sell their applications and some sell the use of their applications—airline reservation systems are an example of this. And some firms engage in contract programming. As these examples illustrate, corporate goals for the IT function may require the firm to adopt alternative accounting methods.

IT accounting systems that use carefully constructed chargeback methods are fundamental to effective IT operation. They provide the basis for cost-effective use of IT services. They greatly enhance communication between IT and client organizations. The firm's performance improves through the motivations present in the accounting and chargeback mechanisms. Successful IT managers fully exploit these concepts.

SOME COMPROMISES TO CONSIDER

The cost recovery approaches discussed in the previous sections are subject to many variations that can be expanded for many purposes. For example, the firm may not require that the cost center operation fully recover its costs, or that overages be returned to the using organizations. The purpose of the managerial accounting system is to serve managers, not to be highly precise in the accounting processes. If the system controls, communicates, motivates, and helps measure and plan as management desires, additional accuracy and precision in the accounting processes may not be justified. Precision and accuracy can be sacrificed for reduced administrative overhead and improved effectiveness. Likewise, the accounting methodology can change as the firm matures in its use of information technology. Precise comparisons over time can be sacrificed for improved motivation or control as business conditions change. The purpose of the system is to serve management, not to be a pinnacle of accounting purity.

EXPECTATIONS

IT clients expect the cost recovery system to accomplish certain things. They expect to have a system that is easy to use and easy to understand. The cost recovery system must be designed for the clients and must help them advance their business relationship with the IT organization. Clients expect IT costs to be distributed fairly and effectively, and they demand consistency. Frequent or unusually large price changes upset their plans and diminish their confidence in IT's ability to manage its affairs properly. Cost recovery processes are valuable to the organization, but their use and administration requires sound management skills.

IT managers and the firm's executives have larger expectations for IT's accounting system. They expect to understand the manner in which IT is spending resources to advance the firm's goals: Well-designed accounting techniques help executives understand the IT organization as a business. They also expect to be able to measure IT's value to the firm. IT accounting and chargeback systems are invaluable in this regard.

However, as the Business Vignette noted, many IT expenditures can be measured only in terms of improved organizational effectiveness; Perrier increased its customer retention rate, for example, and used information technology to restructure. Generally, traditional accounting systems are inadequate for measuring organizational effectiveness. Innovation in IT accounting systems can overcome some of these difficulties, but many challenges remain for the IT manager or the CIO, and for the firm's senior executives.

CIOs can challenge the firm's accounting system. They can raise questions about the firm's ability to measure organizational effectiveness. And, they can take action to correct the deficiencies in the traditional accounting system by installing innovative managerial accounting systems for the IT organization. In the process, they will have forged stronger ties and linkages between IT and the rest of the firm. In addition, they will have strengthened their business relationships to the operating executives and to the firm's corporate staff.

MEASURING IT INVESTMENT RETURNS

Thoughtful CIOs must consider improved ways to value investments in machines, software, and people. Improved valuations of these resources will lead to better ways to measure productivity, performance, and organizational effectiveness. IT's role and contribution have been a leading management issue for a long time. Innovative accounting and measurement of costs and benefits have the potential to resolve this issue, in part.

Traditional ROI calculations, as discussed in the section on application program investments, are a good starting point in evaluating the worth of IT investments. But for many IT projects, such as investments in technology infrastructure or those leading to improved quality or customer service, strict ROI calculations frequently show a negative or low return because of uncertain results or the inability to fathom the business dynamics resulting from the investment. Typical ROI calculations are based on a static business environment, but frequently, IT investments are designed to alter the environment. This makes traditional evaluations difficult.

To help overcome these difficulties, John Henderson built an evaluation model adapting techniques from stock-trading analyses that permit multiple outcome scenarios from the initial investment.[13] Henderson's Option Model augments ROI calculations by permitting risk assessments of various investment decisions as business strategies and systems requirements change. The model recognizes risk explicitly and encourages making investments in stages to learn about the consequences before committing to major expenditures. As investments are put in place, future decisions are based on information gleaned from recent results and altered views of the future.

The Options Model does not remove all uncertainty about IT investments, but it does provide some additional insights. The model is complex to use; some managers may avoid it for this reason. In addition, people, systems, and business strategies are not subject to the same financial rigor as stocks and options, thus introducing additional uncertainty. Its use should supplement traditional evaluations and management judgment not substitute for them.

During the past decade, there has been much discussion about the value of information systems to business in general and much uncertainty regarding their worth to the service sector of our economy in particular. During the 1980s, service industries invested more than $1 trillion in information technology, yet productivity rose less than one percent annually, according to U.S. government statistics. Some questioned the value of IT, reasoning that IT investments offered intrinsically low return, but others claimed that the problem was measurement, or lack of it. The dilemma, called the *productivity paradox*, is a serious matter, concerning enough to lead to many studies and analyses.

Efforts to resolve the issue for individual companies led Paul Strassmann and others to search for new measures of business performance that go beyond earnings per share and revenue growth.[14] Strassmann believes that "there is no demonstrable correlation between the financial performance of a firm and the amount it spends on information technologies. What matters is not how much you spend but how well you spend it in supporting business missions and contributing to productivity." His thesis is that quality of management is more important in technology and service businesses than assets. Therefore, according to Strassmann, return on investment measurements must be replaced by measures that evaluate management's added value.

Strassmann developed an Information Productivity Index based on publicly available data that measures the effectiveness of corporate management. The index is a ratio of economic value-added (calculated by subtracting from operating profit after taxes the value of shareholder equity, multiplied by the cost of capital) divided by the cost of management (cost of sales, general overhead, and administrative expenses are a reasonable approximation of management costs). Measures such as Strassmann's Information Productivity Index are gaining popularity in determining the effectiveness of managers and their use of information to improve firms' productivity.[15] As competition increases and productivity becomes more important, measurements of returns on information technology investments will increase in importance.

SUMMARY

Information technology resources are a large and growing portion of the budget for many firms. The firm expects its managers to use these resources effectively and to account for them properly. In many cases, the IT resources are dispersed throughout the firm making the accounting process difficult. In addition, the complex and varied character of the resources themselves makes the accounting task more difficult. For all these reasons, accounting for IT resources is important to IT managers and to the firm.

Accounting for IT resources assists in planning, controlling, communicating, and assessing performance. It is essential to the effective operation of the IT organization and the firm. But if the IT organization elects to recover costs from customers, additional elements of motivation and effectiveness come into play.

Chargeback methods are very important to IT managers; their implementation must be skillfully handled. A blend of psychology and accounting is required, tempered with the practical considerations of implementation. Accounting systems for IT organizations that intend to sell services outside the firm must be more rigorous than for those that do not. Accounting systems for IT are desirable, but they are not an end in themselves. The purpose of managerial accounting systems is to permit business managers to operate more effectively. The rules for their construction and implementation must be designed to meet this goal.

Accounting for IT resources and measuring the return on IT investments raises some difficult but very important questions. IT investments impact on business operations at the organizational level, but their effect may not be directly financial. Dollars saved or costs avoided are relatively easy to understand, but most IT investments have a broader impact. Increasingly, for example, application developers are designing and installing network-based systems to improve market position or provide superior customer service. However, improved organizational effectiveness is often the only measure of their return on investment—a measurement that traditional accounting systems cannot provide. Clearly this area requires a creative approach from CIOs and other senior executives in the firm. Innovative managerial accounting systems are the first step in the process of measuring organizational effectiveness.

Review Questions

1. Why is IT resource accountability important to the IT organization and to the firm?
2. Define financial accounting.
3. How is managerial accounting different from financial accounting?
4. What trends in information handling are making IT accounting more difficult and more important?
5. What is the relationship between the IT plan and the IT budget?
6. The IT accounting system serves what purpose?
7. What are the benefits of an IT cost recovery system?
8. How does a well-designed charging mechanism enhance the IT organization's effectiveness?
9. What is the primary disadvantage of the IT cost recovery processes?
10. What are the goals of an IT chargeback system?
11. What are the advantages and disadvantages of the profit center method for recovering IT costs?
12. What are the characteristics of the cost center method of IT cost recovery?
13. The cost center method can be contentious. Under what circumstances is this advantageous?
14. Why might it be best to use several methods to recover various IT costs?
15. What is meant by period support for application programs?
16. What alternatives can IT managers use to charge for application development? What are the advantages of each?
17. The costs of production operations can be recovered in several different ways. What variations are useful in production operation cost recovery?

18. The accuracy of cost recovery systems is not necessarily important to IT managers. Explain why. When is it very important?
19. What do clients expect from IT cost recovery methodologies?
20. What is the CIO's role in IT accounting and chargeback systems?

Discussion Questions

1. Discuss the differences and similarities between financial accounting and managerial accounting for IT organizations.
2. Discuss the relationships between planning, budgeting, measuring, controlling, and accounting for IT activities.
3. How would the activities mentioned in Question 2 work together to successfully implement a new market analysis program?
4. How does a well-designed cost recovery system improve the effectiveness of the IT organization?
5. Discuss the reasons why the IT cost recovery mechanism need not be accurate, but must appear to be equitable to the clients.
6. The text discusses some instances in which charging methods alter user behavior toward the consumption of IT resources. Can you give some additional examples of this phenomenon?
7. Compare and contrast the profit center and the cost center methods for recovering IT costs.
8. Under what circumstances do you think the cost center method is a better choice than the profit center method for IT cost recovery?
9. The costs of application development and maintenance might be recovered in a different manner than the approach used to recover the costs of production operations. Explain why.
10. Identify the goals in the management processes of application development and maintenance that the alternative cost recovery approaches satisfy.

11. Discuss the advantages and disadvantages of price differentials in the cost recovery of production operation resources.
12. Describe how a firm might price computer services based on a multiyear plan that includes the replacement of a major CPU.
13. What type of cost recovery system would you recommend for a firm in the integration stage of growth discussed in Chapter 1?
14. What type of cost recovery system would you recommend for a firm that believes that IT's role and contribution are a major issue?
15. Describe the possible changes in IT cost recovery methods that might occur as the firm grows from a modest user of technology to a data servicing operation.
16. Discuss the expectations clients may have of the IT accounting system. What expectations might the firm's executives have of the system?
17. Accounting for networks is difficult and charging users for them is even more difficult. How can the databases of configuration management help account for some of the fixed costs of networks? Discuss the advantages and disadvantages of accounting for data networks as corporate overhead.

Assignments

1. Draw a flow chart of the budgeting process that takes place in a firm using the cost center method to recover IT costs. What prevents the iteration from continuing endlessly?
2. Visit a local firm and determine what methods the firm uses to recover IT costs. Why has the firm adopted the approach they are using? Why is it effective for them or what improvements would you suggest?
3. Analyze the accounting package for one network management system and prepare a report for the class. In your report, indicate what goals this accounting system accomplishes.

ENDNOTES

[1] Allan E. Alter, "The Real Language of Business," *CIO*, January 1993, 37, and Megan Santosus, "A Watertight Case," *CIO*, January 1993, 40. Reprinted through the courtesy of CIO. © 1993 CIO Communications Inc.

[2] "The Premier 100," *Computerworld*, September 13, 1993, 50. The 100 most effective users of information systems as measured by *Computerworld* spend 2.4 percent of revenue on IS; nine of these firms spend more than 5 percent.

[3] Paul A. Strassmann, *Information Payoff* (New York: The Free Press, 1985), 80. "Payoffs are realized by managing the benefits. Costs are important, but secondary," according to Strassmann.

[4] Michael L. Sullivan-Trainor, "Best of Breed," *Computerworld Premier 100*, September 19, 1994, 8.

[5] Richard Greene, "Inequitable Equity," *Forbes*, July 11, 1988, 83. Alfred Rappaport, an accounting professor at Northwestern University, commented, "As we become a more information-intensive society, shareholders' equity is getting further away from the way the market will value a company."

[6] Michael L. Sullivan-Trainor, "Best of Breed," 9.

[7] Paul A. Strassman, "The Real Cost of OA," *Datamation*, February, 1, 1985, 82.

[8] Jeanne Buse, "Chargeback Systems Come of Age," *Datamation*, November 1, 1988, 47.

[9] David A. Flower, "Chargeback Methodology for Systems," *Journal of Information Management*, Spring 1988, 17. This article discusses the chargeback system at Prudential Insurance.

[10] "Many, many years ago I coined the term profit center. I am thoroughly ashamed of it. Because inside a business there are no profit centers. There are only cost centers. Profit comes only from the outside. When a customer returns with a repeat order and his check doesn't bounce, then you have a profit center. Until then you have only cost centers." Notes from an informal talk by Peter Drucker as quoted in Charles Wang, *Techno Vision* (New York: McGraw-Hill, Inc., 1994), xvi.

[11] Flower, "Chargeback Methodology for Systems." Prudential Insurance uses this approach.

[12] Buse, "Chargeback Systems Come of Age."

[13] Jeff Moad, "Time for a Fresh Approach to ROI," *Datamation*, February 15, 1995, 57.

[14] Paul A. Strassmann, *The Business Value of Computers* (New Canaan, Conn: The Information Economics Press), 1990.

[15] In a departure from methods used in previous years, *Computerworld*'s Premier 100 in 1994 were determined using Strassmann's methods. See Paul Strassmann, "How We Evaluated Productivity," *Computerworld Premier 100*, September 19, 1994, 45.

17 IT Controls and Asset Protection

A Business Vignette

Robbery on the Information Superhighway[1]

At 2 A.M. February 15, 1995, FBI agents arrested Kevin D. Mitnick at his home in Raleigh, North Carolina, after weeks of sleuthing on the information superhighway. Mitnick, 31, described by prosecutors as the nation's most-wanted computer hacker, was accused of infiltrating numerous computer systems from New York to California and stealing information worth more than $1 million, including thousands of credit card numbers and software that controls the operations of cellular telephones.

The trail leading to Mitnick was picked up on Christmas Day 1994 when Tsutomu Shimomura, a computational physicist and outstanding cyber-sleuth, discovered that someone had broken into his computers near San Diego from some unknown remote location. The hacker, who used sophisticated techniques to steal several thousand files, angered Shimomura who promptly terminated his vacation and dedicated himself to apprehending the cyberthief. With the aid of California lawmen and the FBI, Shimomura established monitoring posts to catch the thief at work.

At his beach cottage north of San Diego, Shimomura discovered someone had systematically looted his powerful workstation of hundreds of files of information particularly useful in breaching computer networks and cellular phone systems. This was not a random act but a deliberate attempt to obtain important information, including advanced security software that could be exploited for illicit purposes. The intruder posed as a familiar node on the Internet; he commandeered a computer located at Loyola University in Chicago for his attack. He also left a computer-altered message on Shimomura's voice-message system heckling him. Shimomura's interest in this case intensified.

The attacker clearly infuriated the wrong person. Shimomura, 30, is a valued consultant on computer security to the Air Force, the National Security Agency, and the FBI. Over the years, he designed security tools for networked systems and developed a reputation as a computer security expert.

On January 28, Bruce Koball, a computer programmer and an organizer for the public policy group Computers, Freedom and Privacy, linked a newspaper account of the break-in with a puzzling message he received from an on-line service called Well in Sausilito, California. The group's file on Well's system had grown by millions of bytes, Well's officials told Koball. Upon investigation, Koball discovered his files now contained Shimomura's stolen information. Well notified Shimomura who recruited two colleagues and established an around-the-clock monitoring system on Well's property. With the help of another security expert, the team discovered another break-in and theft of 20,000 credit card numbers from Netcom Communications in San Jose, California.

On February 9, the team moved from Well to Netcom where they set up equipment to capture the hacker's every keystroke. Calls came from many locations leading the FBI to conclude the intruder was in Colorado, Minneapolis, or Raleigh. Meanwhile, the U.S. attorney in San Francisco, aided by subpoenaed phone records from GTE, Sprint, and others, determined the calls were being placed to Netcom's phone bank in Raleigh from a cellular phone modem. The intruder cleverly disguised his call path by altering software in the phone company switches, but after hours of tracking records, they determined the calls originated near the Raleigh-Durham airport. The action quickly shifted to North Carolina.

Shortly after midnight on Monday, February 13, Sprint technicians and Shimomura were cruising the area with a radio direction finder attempting to locate the signals from the attacker's cellular phone. Quickly they identified an apartment complex near the airport as the location of the intruder's cellular calls. On Monday, the FBI sent a surveillance crew to assist in pinpointing the calls. They needed a precise address in order to obtain a search warrant.

By Tuesday evening, February 14, the agents had determined the specific address. At 8:30 P.M. a federal judge issued a warrant from his home in Raleigh. FBI agents knocked on the door to apartment 202 at about 2:00 A.M. After five minutes, Mitnick, who had been living under the assumed name Glenn Case, opened the door, claiming he was on the phone to his lawyer.

Mitnick, now jailed without bond and with carefully controlled phone privileges, has been charged with computer fraud and illegal use of a telephone access device. In 1989, Kevin Mitnick had been convicted of stealing software from Digital Equipment Company. He had also been convicted of hacking MCI phone computers.

Six years earlier in November 1988, the first celebrated case of superhighway mayhem occurred when Robert T. Morris released a virus infecting more than 6000 computer systems from Massachusetts to California, causing millions of dollars in damages. Speaking of the perpetrator, a computer scientist at Argonne National Laboratory said, "He's somebody we would hire. The right to hack is held higher than the right of someone to tell you not to. It's an inalienable right."[2] Society clearly believes differently. In this case, Morris was caught, prosecuted, and convicted of felony acts. He was sentenced to a $10,000 fine, three years' probation, and community service.[3]

INTRODUCTION

Controlling the information technology business and protecting its assets from theft and damage are fundamental responsibilities of managers who own or operate information resources. In nearly all firms, controls are becoming more important as information technology penetrates the organization. In most instances, business controls are required by law. Operational systems in today's sophisticated and fast-paced business environment must be grounded on a solid base of operational and accounting controls. The applications, databases, networks, and hardware must be carefully protected against loss or damage. Successful IT managers understand these issues; they install control and protection mechanisms and operate them effectively. IT managers must be prepared to demonstrate effective controls to auditors and to the firm's senior executives.

Because information technology is widely used throughout the firm, IT managers' responsibilities extend beyond their own organizations. As part of their staff responsibility, they are expected to lead in establishing control policies for the entire firm. Their knowledge of technology and its accompanying risks and exposures is critical in developing vital security and control policies; they must guide their firms in this difficult and important activity.

In this chapter, business controls and asset protection issues associated with application systems and their supporting hardware and networks will be discussed in detail. The participants in this activity are identified, their responsibilities defined, and their activities explored in depth. Control and protection of application programs begins in the development phase and continues throughout production. This chapter develops control disciplines for the applications and presents mechanisms for auditing and reporting control status. The firm's auditing activities rely in part on the foundation of control and protection policies developed in this chapter.

THE MEANING OF CONTROL

Control requires managers to know the details of the significant activities taking place within the organization. The details consist of knowing the particulars of what, when, where, why, how, and who as they apply to all important computer functions. In order to operate under control, IT managers must first understand their mission—what they are supposed to be doing. They must know what is expected of them—what is acceptable and what is not.

Control also requires that IT managers have a routine means for comparing actual versus planned performance. The deviations between planned and actual performance are obvious in a well-controlled organization. Also, managers must have the means to respond promptly to plan deviations or to out-of-control conditions. They must have a way to acquire this information regularly and a process for responding to it quickly. They must also be able to detect improvements in performance resulting from variance corrections. They must be able to do this for all the activities within their scope of responsibility.

THE IMPORTANCE OF CONTROLS

Control is one of the primary management responsibilities. Successful IT managers establish and maintain effective business controls as a part of their routine activities.

Business controls are becoming more important in automated organizations because lack of control or out-of-control conditions are becoming more subtle, less obvious, and potentially more damaging to the operation. For instance, insufficient controls in a manual accounts payable function may lead to some unauthorized payments, but an uncontrolled accounts payable program can write unauthorized checks at the rate of thousands per hour. (In one well-known situation, this actually happened.) An error in an inventory program can add discrepancies to the database much faster than manual reconciliations can detect and correct them.

Control is especially important to IT managers because many of the organizations that the IT function supports rely on computer-generated reports and other automated tools to establish their controls. For example, inventory-control process auditing in a manufacturing plant quickly leads to computer-produced reports presenting transaction activity and describing inventory status. Control weaknesses in the IT function can directly affect inventory control and control throughout the firm. Table 17.1 summarizes the reasons why business controls are important to the IT function.

TABLE 17.1 Why Business Controls Are Important to IT Managers

Control is a primary management responsibility.
Uncontrolled events can be very subtle and damaging.
The firm relies on IT for many control processes.
Control is required by law in public corporations.
Controls assist the firm in protecting assets.
Environmental and executive pressures require controls.
Technology introduction requires controlled processes.

Control requirements and accurate recordkeeping are mandated by law for publicly held corporations. Specifically, organizations must provide proper authorization of transactions and perform recordkeeping in conformity with accounting principles established for the firm. The firm must provide and maintain asset protection and must physically verify and reconcile assets with inventory records. Managers must document the extent to which corporate accounting principles have been followed.

Managers must evaluate the sufficiency of controls and must appraise actions taken to correct control weaknesses. The firm's officers must certify that these actions have been taken. Management performance and judgment in controlling assets also require assessment. The following executive statement is typical of those found in the annual reports of publicly held companies:

> Management has established and maintains a system of internal accounting controls that provides reasonable assurance as to the integrity and reliability of the financial statements, the protection of assets from unauthorized use or disposition and the prevention and detection of fraudulent financial reporting. The concept of reasonable assurance recognizes that the costs of an internal accounting controls system should not exceed, in management's judgment, the benefits to be derived.[4]

Controls are also important because of environmental and legislative pressures and executive concerns about control activities. Because of the growing complexity of the business environment, executives expect both manual and automated control mechanisms to be maintained at peak efficiency. Large on-line applications consisting of sophisticated hardware and software systems cross functional boundaries and interrelate business activities throughout the firm and sometimes link the firm to suppliers and customers. Business executives demand automated and controlled processes for these complex activities. Most firms heavily depend on information systems; the systems must be carefully controlled.

Frequently, the adoption of new technology creates the potential for greatly improving employee productivity, and for improving the effectiveness and efficiency of the operation overall. Computerized automation also increases the firm's ability to control errors and omissions and to prevent fraud. However, new and increased control risks, with which management must contend, accompany these benefits. Effective controls must precede or at least accompany the introduction of the technology.

The introduction of new information technology to the firm usually coincides with the need for new control and protection measures. For instance, the deployment of personal computers and the use of new networking technology greatly increase the need for data security and physical inventory control processes. However, it is common for the introduction of new technologies to lead the adoption of the necessary control mechanisms by a large margin in many firms. For these firms, insufficient controls or out-of-control situations usually accompany new technology. Alert IT managers anticipate these events and plan control systems prior to introducing new technology.

PRINCIPLES OF BUSINESS CONTROL

The primary job of managers in an organization is to take charge of the firms assets entrusted to them. Successful managers must control and protect the assets, use them to capitalize on and advance their part of the business, and grow, develop, or add value to the assets. Managers who have custody of information assets, both tangible and intangible, must control and protect them in order to succeed. In order to control and protect the firm's property, managers must know what assets they control and understand their value to the firm.

Asset Identification and Classification

Tangible information assets include physical property, like CPUs, servers, routers, cabling systems, and personal workstations. Many information assets are intangible, intellectual assets such as operating systems, application programs, and databases; often they are more valuable than the physical assets. In almost all instances, program and data assets are worth much more than the devices on which they are stored. In most firms, the value of information assets generally exceeds the annual IT budget by a wide margin. In some firms, the value of the enterprise is considerably understated on the balance sheet because of the tremendous value of intangible information assets.[5] IT managers have a great deal of responsibility for protecting all of these assets from loss, damage, or improper use. Information asset protection is an important task for them.

A manager's first step in controlling and protecting assets is to conduct an asset inventory—to develop an organized list of the assets for which he or she is responsible. In confining the list to information technology items, some examples might be: computer hardware, system software, application programs, databases, documentation, and passwords. The list of possible assets is rather long because documentation can include many additional items such as strategies, plans, designs, algorithms, and others. Taking inventory identifies those assets that the manager must control and protect.

The second step is to establish a value for each of the inventoried items. This step generally reveals three types of assets: assets with intrinsic value such as money, stock certificates, or checks; assets with possible proprietary value such as new product designs; and assets that are valuable because they control other important assets. The payroll program is an example of a controlling asset. The appraisal step organizes the assets by value and provides a rational basis for establishing controls and protection for them. This step is called *asset classification*. Managers can develop and implement sound controls once this step is complete.

Most organizations have an asset classification scheme for proprietary information. One such classification structure has four categories of information: top secret, secret, confidential, and unclassified. Unclassified information is public and available to anyone in the firm. The remaining categories of information are available *only* on a need-to-know basis. This means that if an individual is cleared for secret information, he or she can access such information only if it is required for the job. Having a secret clearance does not mean that the individual has access to all information classified as secret.[6]

In addition to these classification categories, most firms have an additional category for personal information. Information of this type includes salary, performance, and medical data. Some firms classify this information under personnel, others identify it as personal and confidential. Access to this information is also restricted to those with a need to know.

The classification category must be indicated on the document itself or contained within the dataset. The classification must be obvious to anyone viewing the information. This is an important principle of information control.

Separation of Duties

One of the most effective control measures in business operations is the principle of separation of duties. Separation of duties means that several individuals are involved in transaction processing, and that no single individual processes the transactions from beginning to end. As an example of how this might work, let's consider payroll processing. In this process, one person prepares the time card information, another validates the totals and transmits the information for processing, another controls the blank checks and supervises the processing, another validates the processing through check register data, and yet another distributes the checks. In order for fraud to occur, several people must act together to make it happen. Separation of duties reduces the possibility of fraudulent acts.

Separation of duties is relatively easy to administer and control; managers can validate the control mechanisms at each interface. In addition, if managers periodically change the individuals responsible for the tasks, this control mechanism can be very effective.

It is also extremely important to validate the output with the input. In the case of payroll, this can be accomplished periodically by hand delivering payroll checks. The person delivering the checks must work outside the immediate organization, positively identify the recipient, and verify hours worked with the employee. There are many variations on this theme for accounts payable, customer shipments, incoming inventory, and other activities.

Efficiency and Effectiveness of Controls

Controls are most satisfactory when they are simple to operate and easy to understand. They are most effective when they are routinized and operate in a timely manner. To be totally effective, controls must cover all possible exposures. When action is required, control mechanisms must respond and produce action in a timely manner. For instance, when input data errors are detected in the operation of an application, management must receive prompt notification. They must act quickly to identify the cause of the error and to initiate corrective action. Managers must promptly invoke the problem management system to take final corrective and preventative action.

In all cases, managers must relate the cost of control and protection processes to the expected frequency of unfavorable events and to the anticipated loss resulting from these events. Over-control and excessive expenditures on controls are both possible. Controls must be cost justified. To establish the proper balance between these conflicting forces, managers must analyze the application and managerial judgment used in implementing controls.

CONTROL RESPONSIBILITIES

Business controls operate most effectively when clear responsibilities are assigned to specific individuals. For application systems, several individuals or groups are important in this process:

1. The application program owner (almost always a manager)
2. The users of the application (some applications have many)
3. The programming manager for the application
4. The individual providing the computing environment
5. The IT manager (either line or staff responsibility)

Each of these individuals has definite responsibilities which must be discharged correctly for application controls to be effective.

Owner and User Responsibilities

The application owner is the manager of the function or department that uses the application to conduct its business activity. For instance, the owner of the perpetual inventory program in the manufacturing plant is the materials manager, and the owner of the accounts payable application is the organization's accounting manager.

Application owners are responsible for providing business direction for their applications. The owner/manager is responsible for establishing the functional capability of the application program and for providing the justification or benefits analysis for any expenditures related to the application. The owner authorizes the program's use, classifies the data associated with the application, and stipulates proper program and data access controls.

The payroll manager, for example, owns the payroll program and controls access to it and to the related payroll data. The owner authorizes the processing of the payroll program. If the program requires modification due to changes in the withholding rate, or if the firm decides to obtain payroll processing from a vendor, the payroll manager provides the business case and establishes the business direction for the application. The payroll manager is completely responsible for the payroll program. Application owner/managers throughout the firm are responsible for the applications required to operate their part of the business.

Application users are the individuals or groups of individuals authorized by the owner to use the application and related data according to the owner's specifications. They are required to protect the data in keeping with the owner's classification. Users are responsible for advising the owner of operational difficulties and functional deficiencies. Individuals in the payroll department who update payroll records and initiate the payroll process are examples of application users.

Some applications have many users. The network of personal workstations comprising the firm's office system is one example. With networked PCs, one manager in the administrative function is named the owner of the application and all the secretaries and administrative personnel are considered users. Another example is the inventory control system in a large manufacturing plant. The production control manager owns the inventory application system, but workers throughout the plant, from production planners to shipping clerks, are the users. In these examples, the owners and the users all have responsibilities for the application and its data.

IT Managers' Responsibilities

The responsibility for organizing and managing the work of development, maintenance, or enhancement on the application resides with the programming manager. He or she is accountable to the owner for meeting programming objectives. These objectives include functional capability, schedule attainment, cost control, and quality performance. The programming manager is also the custodian of the application and associated data during the development, maintenance, or enhancement activity. The programming manager who modifies the payroll program under the direction of the payroll manager is an example of an individual in this custodial role.

The supplier of computing services is responsible for providing the computing environment within which the application is processed. The supplier of services is responsible for negotiating service-level agreements with the owner and for achieving the agreed-upon levels of service. Maintaining a secure environment for the application and associated data is a major responsibility of the supplier of service. For example, if the firm's payroll is processed in the central computer center, the operations manager is the supplier of service and is responsible for the computing environment. If an operational department manager owns a client/server computing system, he or she is responsible for the computing environment.

IT managers are responsible for ensuring that these individuals receive proper guidance regarding their responsibilities. IT managers must establish procedures for the development and use of applications. They must routinely obtain information allowing them to evaluate the control posture of the applications on a continual basis. In some cases, they may need to conduct audits to validate business controls.

These assignments and responsibilities are not optional; they are mandatory for the controlled operation of application programs. Failure to establish one or more of these positions or to ensure effectiveness in these positions will lead to control weakness and possible failure. IT managers must establish these positions and manage them properly.

APPLICATION CONTROLS

Application controls are most effective when they generate documentation that validates the appropriate functioning of the application on a continual basis. Application owners should use this information in conjunction with other control information to certify their processing. The controls themselves should be well documented to ensure prompt, accurate applications audits.

Automated and manual control mechanisms should be classified as confidential information and should be handled accordingly.

Application systems benefit from most of the principles associated with business controls, broadly applied. For example, the principle of separation of duties applies to applications as well as other, more common activities. But, in addition to general controls, some principles stem from the intangible nature of application programs as operating assets. However, application programs are best controlled using a combination of programmed and manual controls. In the case of application program assets, the characteristics of the assets are usefully employed in their control.

Application Processing Controls

The control and protection of applications consists of two parts: ensuring that application programs perform according to management-established specifications, and maintaining the integrity of programs and data. The first part deals directly with the operation of the programs. It is concerned with the correct functioning of the application and with handling input and output data properly. The second part deals with security and protection of the program and data assets. It is concerned with controlling access to the programs and data files and with the integrity of the information in the files.

Correct functioning of the application and proper handling of input/output data requires the application to have auditability features and control points. They are part of the original system specifications and are most effective when they are designed into the system at the outset. Chapter 9 pointed out that application control and auditability is a design issue. Operating within the phase review process, managers must establish requirements and specifications for control elements that later become part of the application design. They must audit the development process during the phase reviews to make sure that design specifications contain their requirements.

System Control Points

System control points are portrayed in Figure 17.1. Controls can be applied during transaction origination and when the data is input into the system. Once in the system, control can be established when the data is transmitted into or out of storage, when it moves over the teleprocessing system, or when it is processed within the CPU. Lastly, control should be maintained over the data when they exit the system.

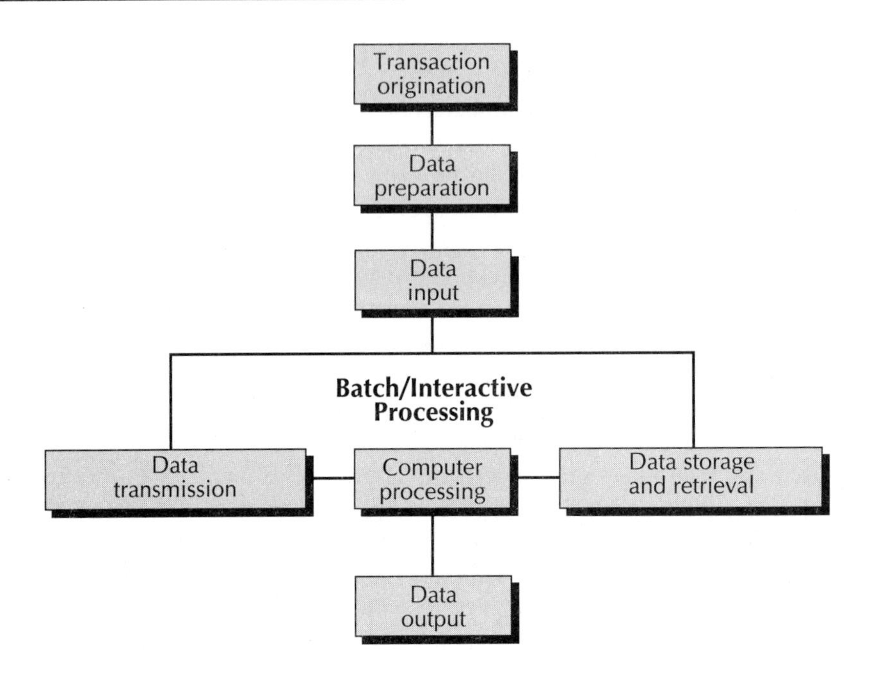

FIGURE 17.1 System Control Points

Figure 17.1 identifies points in the data processing system where control exposures exist and where controls must be exercised. Exposure points occur at transaction origination and at the time that data is electronically input to the system. Data can be stored for later retrieval and processing, or it can be transmitted to the computer system for immediate processing. In either case, the system processes the information and generates output for the application department to use. The results of computation may be stored for retrieval later or for further computation.

Some important control actions that must be considered at the point of transaction origination are displayed in Table 17.2. Transaction origination usually involves one or more input documents that must be controlled and monitored. Source documents are designed for this specific task. They usually have separate, printed squares for each character or number corresponding to the data requirements of the application and, usually, they are prenumbered for transaction control. Whenever possible, preprinted information and electronic scanning should be used to reduce clerical errors.

TABLE 17.2 Considerations at Transaction Origination

Document design
Manual review of source documents
Authorization
Separation of duties
Transaction numbering
User identification
Transmittal logs between organizations
Error detection and correction
Document retention and storage

Only specifically authorized users should handle the source documents; whenever possible, the handling process should be divided in order to separate individual duties. Since the transaction documents contain identification numbers, logs can be maintained to record document transmittal between units. Because the input documents are developed in the user department and transmitted to the computer operation, the transmission logs validate that the proper documents have been processed.

Error detection and correction during the input process require special attention. Resubmission of input data must be recorded and verified. Error handling must be performed according to carefully constructed procedures to prevent additional errors and fraudulent transactions. After the input process has been completed, the source documents must be stored and retained for some established retention period.

Batch-data input and interactive-data input controls are very important. They must be clearly established and carefully monitored. Table 17.3 indicates some of the common means for establishing controls over input processes.

TABLE 17.3 Input Data Controls

Batch Data Input	Interactive Data Input
Input processing schedules	Terminal access security
Source document cancellation	Terminal usage logs
Editing and validation	Data editing and validation
Control totals	Display and prompting formats
Batch control processes	Interactive control totals
Error handling procedures	Error detection and correction

Batch processing includes scheduling the batch runs and ensuring that the input for the run matches the source-document data. Usually the batch application edits the input data and validates the information against the source input through control totals. For instance, the input of financial data will contain a count of the number of transactions and totals of important data in an accounts summary. This guards against the addition of extra input or the loss of previously input data. When the input has been validated, batch processing begins and source documents are marked as processed.

Control over batch processing resides with the production operation department. This department is responsible for advising the user department when improper batch execution occurs. Error handling is especially crucial because error correction through rerun procedures is itself an error-prone process. Errors must be carefully examined. The presence of input or processing errors sometimes indicates fraudulent processing. The operations department must maintain error logs and the IT and user managers must review them regularly.

Interactive processing has many of the same control considerations as batch processing, but it is more complicated. The complications result from the use of remote devices for interactive operations. Proper physical control over the terminal devices is a mandatory first step. The devices must be secured from all but authorized users. Authorized users are identified to the system through passwords or personal identification numbers. The specifics of password construction and use will be discussed later.

Input data must be edited for reasonableness and tested for validity. Constructing the algorithms for effective editing and testing requires sound knowledge of the application and considerable ingenuity. The rules may be complex, and there may be many of them. The design of interactive displays and prompting formats can be very effective in this regard. For instance, if the field to be entered is numeric and can never exceed four digits, the screen design should allow for no more than four digits. The input editor should test for the presence of nonnumeric characters. If the dollar figure to be entered cannot exceed three figures to the left of the decimal point, screen design and prompting can control to this limit.

For some important applications, expert systems perform these front-end processes in part. Expert systems are especially valuable for interactive data entry where individuals can receive expert assistance and take immediate corrective action. These considerations are part of the system design process. They must be an important element in the user's functional requirements statement.

Tables 17.4, 17.5, and 17.6 contain the control and audit tasks that must be incorporated in processing, data storage and retrieval, and data output.

Processing, Storage, and Output Controls

Operating system software and the application itself enhance the validity of the program execution. The system software performs specialized functions that help control the use of data stores such as tapes and disk files (see Table 17.4). The system checks tape-label information to ensure that the correct tape file is being accessed. This helps prevent processing against the wrong file. It also verifies control totals to ensure that the correct amount of information has been processed. For all types of secondary storage, the system validates that the correct version of the data set has been used. This is especially valuable for applications that create today's output from yesterday's output and today's transactions, and in cases when previous datasets must be preserved.

The program execution process must be accompanied by programmed routines validating that the process is complete and that the results are correct. The processes or algorithms for accomplishing this are highly application dependent and must be included in the program specifications when the application is developed. In some cases these algorithms are confidential. The application owners should be involved in developing error detection and correction processes. They should review all error correction activity carefully for improper or fraudulent conduct.

TABLE 17.4 Computer Processing Controls

Validate the input dataset
Validate the dataset version
Verify processing correctness
Verify processing completeness
Detect and correct errors

Data storage and retrieval requires full protection of classified data and application programs (see Table 17.5). Classified data must be protected so that only those with a need to know have access to it. Application program source code and object code must always be treated as classified information. Program changes are similar to data changes and are handled in much the same manner. Special procedures for reviewing classified data and for altering it will be discussed later.

TABLE 17.5 Data Storage and Retrieval

Full protection for classified data
Source program update procedures

Finally, the data output and distribution process requires control mechanisms (see Table 17.6). In order to verify that the processing occurred as planned, totals should balance, input and output volumes should reconcile, and manual processes on output must be under document control. The output documents should contain control totals and transaction counts which can be reconciled with the documented input records. Mechanisms within the application validating that the processing was accurate should produce reports for the application owner. Routine output documents and these processing reports should not be delivered to the same individual in the organization. Some of these records must be retained for future audits or for other purposes.

Output records must describe how the error-handling activity was conducted. All error-recovery actions must be documented and recorded for future review by the problem management team. Control of the data output activity is vital because it offers managers the opportunity to reconcile and validate the entire process.

TABLE 17.6 Data Output Handling

Reconciliation of output to input
Maintenance of transaction records
Balancing of transaction volumes
Control of error handling
Records retention requirements
Output distribution control

Operations personnel must exert control over output distribution. Output must be distributed to authorized individuals only—it must not be delivered to unauthorized individuals. Proper distribution of computer output involves privacy, confidentiality, and security.

Some information is personal and must be divulged only to authorized individuals. For example, employee salary or performance records are private information and should only be seen by the employee's manager and certain other individuals. Some information is confidential, such as marketing information, new product designs, and proposed pricing actions. Disclosure of this information to the wrong people may harm the firm's competitive position. Other information must be secured because of its intrinsic value. Checks and stock certificates are examples of this kind of information. Because they describe the location and value of tangible property, inventory records should also be protected.

Application Program Audits

Auditability features in business application programs are essential to the control posture of the firm. What makes an application system auditable?

An application system is auditable if the application owner can establish easily and with high confidence that the system performs the specified function on a continual basis. In addition, the owner must be able to verify that exceptional conditions, discrepancy handling, and error conditions are processed according to prescribed specifications and procedures. Auditability, therefore, includes manual and automated procedures in the IT organization and in the user organization.

A system is auditable if it contains functions and features that make it easy for the owner to determine whether on not the system is processing data correctly. Features that provide editing capabilities on input as well as journaling or logging functions throughout the processing stages are important ingredients of an auditable system. The system must provide documented verification that processing was performed according to specifications and that the results can be compared to known or expected standards of operation. For example, the payroll program will produce a check register detailing each of the checks produced by identification number. The owner of the payroll program has physical possession of the blank checks and can account for each check processed, including any used to align the printer. Additionally, control totals are available to ensure that the payroll was calculated as expected.

Application owners must have reliable mechanisms to determine that authorized users used the system for legitimate business purposes. They must have assurance that their data, used in conjunction with the application system, is protected as they prescribe. These mechanisms, assurances, features, and functions must be available to the owner in the form of documentation or reports as a routine function of the system. A system that includes these features is auditable and can be considered to be under control.

Auditability begins with sound system development techniques and practices. Well-documented programs written in high-level languages are easier to audit than others. Sound application development techniques, such as structured design, modular programming, and complete documentation, are particularly important for programs slated for maintenance or enhancement. Program modification, a part of nearly every program's life, greatly complicates the task of maintaining systems integrity.

Programming disciplines and programming standards make the task of maintaining program integrity much easier. The program testing process is vital; preserving the test results and the test data makes subsequent maintenance and enhancement easier. The best time to insert auditability features into an application or to add reports to improve audit trails is during program development. Program maintenance and enhancement must preserve and, if possible, augment the original auditability features.

Controls in Production Operations

The application production process must also operate in a controlled environment. Internal application controls are only effective if the production process itself is disciplined. Application controls and audits must be complemented by a well-controlled and highly disciplined production environment.

The disciplined management processes for production operations are important to application integrity. In particular, problem, change, and recovery management are essential. A well-disciplined production operation will display sound control over performance objectives. It will do a good job of maintaining sufficient system capacity for operating the applications. Batch and on-line operations execute applications in a structured and planned manner. Detailed scheduling and rigorous on-line management provide a controlled environment in which the applications can be processed. The disciplines form an important base for a complete state of confidence regarding application control and auditability. They are required for other reasons as well; they are vital for business controls.

Controls in Client/Server Operations

Distributed processing environments can lack application processing controls or audit trails and can be vulnerable to unauthorized penetration, or lack adequate separation of duties. In some firms, distributed processing has been motivated in part by a desire of operating departments to escape the bureaucracy of central IT organizations, including that imposed for control and security reasons. However, client/server or other distributed processing implementations have the same control requirements as centralized operations, and for all the same reasons.

The issues discussed in this chapter and the supporting figures and tables equally apply to interactive distributed processing operations. Owners of distributed systems are responsible for ensuring satisfactory controls and asset protection of their operations.

Because there are more points of vulnerability in client/server operations, control and asset protection is necessarily more extensive. But because there are fewer manual operations, fewer paper documents, data tapes, and removable storage media in well-designed client/server systems, controls can concentrate on programmed tasks using program audits and controls. For example, each transaction originating at a client workstation can be identified by workstation number, operator code, and date and time stamped automatically. Recordkeeping at the server can be fully automated as well. Well-designed client/server applications can contain the same kinds of audits used in centralized systems and can inform management of successful operation, too.

As firms move applications from the relative security of centralized operations, they must be constantly aware of the increased exposures and vulnerabilities of distributed processing. Client/server operations need not be less secure than centralized operations. They can and must be secured and protected according to the value of their assets and their worth to the firm. The chief risk of distributed operations is that increased access is granted first and controls and security are added later. As with all business activities, control is a fundamental management responsibility and must not be an afterthought.

NETWORK CONTROLS AND SECURITY

Network managers face many challenges concerning business controls and security. The network operations they supervise are growing rapidly with the installation of departmental LANs and interdepartmental WANs, connections to common carriers and value-added networks, and the use of long-haul national and international networks. These interconnections provide faster and more direct coupling between organizations. Telecommunication facilities dispatch valuable information and assets within the firm and between it and its customers and suppliers. In the financial industry, for example, the daily electronic funds transfers dwarf the total dollar reserves of the central banks.[7] Because of threats to networks, the importance of controls and security in today's environment is rapidly increasing.

Networks are subject to two types of security threats: passive and active. Passive threats consist of attempts to monitor network data transmission in order to read messages or obtain information on network traffic. The intruder hopes to profit from the information or to identify information sources. Active threats consist of attempts to alter, destroy, or divert message data, or to pose as a network node. As we learned in the Business Vignette, Kevin Mitnick posed as a network node and actively diverted valuable information. In this extreme case, the intruder became an active network participant, exchanging information with legitimate sites and obtaining free network services.

In order to minimize these threats, network managers must control access to the system and its data and protect data in transit. The first step in controlling system access is to secure the physical system. This means that the facility housing user devices such as workstations, facsimile machines, and phones must be secured so that only authorized individuals are admitted.

Rooms containing controllers, routers, bridges, or servers must also be secured tightly. Cables connecting user devices to network devices should pass through restricted access passageways. Transmission media between bridges, routers, servers, network gateways, and external communication links such as the phone system must also be protected against damage or intrusion.

The second step in securing access to the system is to establish user identification and verification processes. For most systems, this means that users sign on to the system with their name and then supply a password. The user identification/password scheme can be quite effective when properly implemented and used. For example, users can be required to establish passwords that have six characters, two of which must be numbers, with alphabetic characters separating the numbers. In this case, a2gj3b would be a valid password while cwfa12 would not. Unfortunately, even with these restrictions, people tend to develop passwords that are easy to remember and therefore easy to duplicate. Well-protected operations have system-generated passwords that frequently change. Password systems alone, however, are insufficient protection.

Network systems should have automatic disconnect capability. If password validation fails after five attempts, the system disconnects the user and the event is logged. This restricts hackers from using automatic password-creation programs. Other security measures are available. All attempts to gain unauthorized system access or to use the system in an unauthorized manner should be recorded and reviewed by management.

All system users must be trained in the need for system protection and in company security procedures. Network and user managers should not hesitate to withdraw access from employees who abuse system controls or use the system for unauthorized purposes. In some firms, employee use of information systems for personal or unauthorized activities is cause for dismissal.

The third step toward system access security is data security. Properly secured data means that each dataset in the system has a designated owner and that the owner's identification and the dataset classification are part of the data. Each owner must specify who can access the dataset and what kind of access is available to the user. The types of dataset access are read, write, alter, delete, and execute. For data security to work properly, the owner authorizes each valid user to access the dataset. Then, the user's authorized actions are identified. When a user tries to open a dataset, the system validates the user and permits only the authorized level of access. Unauthorized attempts to access data are recorded and investigated by management. Since application programs are also datasets, the execute authorization permits the user to operate the application.

Data Encryption

For several reasons, access security can never be perfect, as the Business Vignette demonstrated. For example, common carrier links can be tapped and microwave or satellite transmissions can be intercepted. Therefore, it is necessary to protect date enroute between locations. The most important tool for protecting data in transit and for maintaining network security is message encryption. Prior to transmission, encryption uses an algorithm and a key to change message characters into a different stream of characters. The message must be decoded or deciphered after reception using the algorithm and key. The encryption process may operate at the character level but, usually, it is applied to the message bit stream.

In 1977, the National Bureau of Standards established a data encryption standard (DES) which has been widely adopted and used in commercial applications. Proper use of data encryption can make telecommunication very secure. There are means for authenticating transmissions and for validating signatures. For especially sensitive traffic, a third party can validate the traffic and guard against lost or stolen keys. In addition, there is research underway to improve the encryption algorithms by increasing their speed of operation and making them more secure. Network managers must secure critical network traffic against passive and active threats by using encryption and authentication techniques. This is especially important if the firm uses insecure facilities such as the Internet.

Data encryption is the ultimate protection for stored information too, and it protects against sophisticated hackers such as Kevin Mitnick. For critical data, some firms and agencies are encrypting locally stored information.

Control of today's business operations and the security of the firm's intangible assets depend heavily on control and security of telecommunication systems. They are the link between users and applications; they link systems and applications with customers and suppliers; and they integrate the firm's physically dispersed operations. They must operate under control, and they must be secure. Table 17.7 lists the network security considerations we have discussed.

Protection of information assets tied to networks, either public or private, requires special attention. The Business Vignette that opened this chapter describes an extreme threat to networked systems.

TABLE 17.7 Network Security Considerations

Physically secured workstation devices
Physically secured network components
User identification and verification
Processes to deal with unauthorized use
Dataset protection mechanisms
Data encryption and authentication processes

There are several clear lessons to be learned from this vignette and from other similar incidents. First, there are a small number of people who prefer to intrude on your network and invade your systems and applications for pleasure or for profit. They are intelligent and persistent; you must protect your network and your systems and applications from them. Second, networks and systems connected to them can be secured from most intrusions; the degree of protection depends on the amount of resources expended and the type of protection obtained. Third, in spite the obvious threats, many organizations still fail to take simple precautions.[8] IT managers have considerable responsibility for the firm's assets; they must actively protect the vital assets entrusted to them.

Network control and security directly relate to controls and audits in applications because networks are the sources and sinks of much of the application data. Network control and security directly relate to the other network management disciplines such as problem, change, and recovery management because effective network control reduces network problems and network security reduces damage to information assets. The disciplines of network management are an integral part of the IT management system.

ADDITIONAL CONTROL AND PROTECTION CONSIDERATIONS

Some of the IT manager's most important security and control responsibilities concerning applications and networks have been discussed above, but others must be addressed. In particular, physical protection of major processors, control and protection of critical applications, and security of unusually important datasets are issues that IT managers must handle.

The data center containing the firm's central CPU complex, vast stores of information, network gateways and bridges, servers, communication controllers, and telephone equipment requires extraordinary physical protection.

Some practices that help data center managers secure their operations are listed below:

1. Only people who work in the data center should be allowed routine access to the facility.
2. Data center workers must wear special badges that identify them on sight.
3. The identity of all visitors to the center must be validated and they must document their visit by signing in and out.
4. Duties within the center should be separated so that operators who initiate or control programs cannot access the data stores.

Under certain circumstances, additional actions may be necessary to fully secure the data center. For example, in some critical centers, visitors are prohibited entirely and the center itself is protected by specially constructed floors, walls, and ceilings. In all cases, however, the degree of protection must be consistent with the value of the assets housed in the data center.

Downsizing the mainframe requires IT managers to apply these considerations to the distributed operations. Users who operate their applications on their own processors must take many of the security and control precautions required of mainframe operators. Thus, managers of distributed operations assume considerable control and security responsibilities.

Systems programs such as operating systems, file handling utilities, password generation programs, and data management systems must also be specially protected. Managers must take careful precautions with system programmers and network specialists who have access to restricted utility programs and control elements. For example, the file of authorized users and their passwords is available to system programmers. System support personnel use utilities that can copy or rename datasets, or that can alter executable modules or tape labels. Most systems also have a superuser capability—one or more privileged passwords that system programmers can use to access any dataset on the system. System programmers need these capabilities in order to do their jobs.

IT managers must develop individualized security measures for the few individuals in the data center who have these special privileges. To meet this need, managers can take several routine precautions. Individuals in these responsible positions should rotate or change duties frequently. Their actions should be routinely recorded and managers should review the records frequently. When system programmers use privileged passwords or access restricted utility programs, they should do so only after obtaining clearance for their actions in advance from the center manager. The manager should validate their actions upon completion; this increases security and reduces errors to critical information.

Privileged network technicians have access to codes, keys, utilities, and passwords for many remote operations. Their work must be handled in the same manner as system programmer's activity.

Managing Sensitive Programs

In addition to the controls and procedures required for the operation of the applications, some programs in the portfolio require special handling. Application programs that permit or authorize the transfer of cash or valuable inventory items need careful and specialized controls. For example, payroll, accounts payable, and inventory control programs are of this nature; they need special protection. Managers must identify these programs and maintain an inventory of them. Their owners must prescribe protection and security conditions covering storage, operation, and maintenance for these programs, and must ensure that these special considerations are implemented satisfactorily.

First, program source code, load modules, and test data must be classified as sensitive information and protected accordingly. In most cases the protection should be the highest available on the system. For instance, the owner of the accounts payable program may classify the source code as confidential and restrict its access to a single maintenance programmer. Load modules used for program execution may be restricted to another individual. This ensures that maintenance programmers do not have the ability to operate the program with live data. The test data should be entrusted to yet another individual who will operate the modified program against it and deliver the results to the owner. When program testing is complete, the executable load modules are updated and protected. Change control for these special applications must be carefully managed.

Second, these critical programs usually operate differently from routine applications. For example, control over input and output documents is tighter for accounts payable than for most other applications. Checks are hand carried to the computer center after the sequence numbers have been recorded and verified. When processing is complete, all the output is returned to the accounting manager who verifies the check count, returns unused checks to the safe, takes the checks and stubs to the distribution center, and gives the check register and other control information to the appropriate accounting manager. These operations are verified in the accounting department and recorded for later reference.

Some datasets for these applications are also highly sensitive. The vendor name and address file for accounts payable is an example of one such dataset. Anyone scheming to create a fictitious vendor to whom payments will be sent fraudulently must access it. With appropriate controls in place, the firm should be able to prevent the fraudulent act from taking place.

Accounting managers must validate all transactions against this file to protect it from this type of breech in security.

Application owners must be especially vigilant during periods of maintenance or enhancement. The owner ensures that only authorized changes are made and then reviews all modifications. The owner must control maintenance through close supervision and through documentation and testing procedures. It is important that the routine audit and control features in these programs function flawlessly, without modification. Additions and changes to these applications must incorporate additional audit and control features. These sensitive applications must be guarded more carefully than the assets they control.

If a firm decides to downsize these applications to distributed departmental systems, it must first ensure that all the precautions noted above are in place prior to moving the systems. Some firms, facing the daunting task of securing distributed systems, decide that critical systems should remain centralized. One organization that owned and operated a midrange computer for financial applications requested to move the system to the centralized secure facility instead of facing the formidable task of securing the departmental system.

THE KEYS TO EFFECTIVE CONTROL

In order to operate under control, managers must have a complete understanding of their control responsibilities. They must know which assets are theirs, they must know the value of them, and they must classify and protect them accordingly. Managers must be actively involved in the control process. Their involvement must be timely and responsive to changing conditions; they must follow up to ensure that their actions have been effective.

The operation of information assets such as systems and applications must routinely produce measurements and reports that reveal the state of control. Managers must review these reports frequently. They must be able to determine that specific operating procedures have been followed and that the operation complies with defined control practices.

Managers should separate duties to disperse information access and handling, and frequently rotate employees in critical positions to different jobs. IT managers must audit their operations periodically, and use their findings to improve their business controls. Information assets are usually very valuable; their owners must carefully protect them according to their relative worth.

Business controls in application systems are of interest to user managers, IT managers, and to the firm's senior executives, too. The capstone discipline—management reporting—must function well. IT managers, their peers, and their superiors require knowledge of correct application performance periodically. Astute IT managers take extra steps to keep all interested individuals informed of the sound controls applied to application assets. For example, IT managers may summarize problem management actions for senior executives, or they may present trend information from internal audits and reviews of their operations.

These periodic reports should highlight the major routine actions that ensure correct and valid performance of the applications. A summary of the deficiencies in operation, if any, and the manager's corrective actions must also be reported. The results of routine tests and audits of the control mechanisms performed since the last reporting period should be included in this report. Additional steps taken to augment existing controls or to secure further the development or operations areas will be important to these executives. Comprehensive reporting is an offensive tactic because, in its absence, senior executives are likely to seek information through outside audits, independent reviews, or other less welcome means.

SUMMARY

Computer crime has reached epidemic proportions costing business and industry somewhere between $500 million and $5 billion per year, according to *Forbes* magazine.[9] These figures may be low because most computer crime goes unreported, and some goes undetected. Numerous incidents prove that almost no organization is safe from computer crime, yet some fail to take even rudimentary precautions. Computer hackers have bulletin boards, several magazines, and regular meetings to exchange new information. These activities should give corporate executives the shudders. If that isn't enough, laws in Eastern Europe and Russia legitimize the export of software virus programs. Virus factories are as common as software publishers in the U.S.[10]

With these astonishing facts in mind, remember that business controls and asset protection are fundamental to business operations. They are part of every manager's primary responsibilities. Controls are more important now because of changing business conditions, new technology, and increased attention demanded by law. Controls in IT are especially important because the firm depends on computerized systems for control throughout its operations.

Managers must know their assets and their estimated value. Assets must be classified and protected in keeping with their relative worth. For application program assets, individual responsibilities are assigned to owners, users, programming managers, and providers of the computing environment. IT managers must ensure that these individuals effectively discharge their responsibilities.

IT managers must exert control over data handling and computer processing; reports must provide evidence that the programs operate as specified. The production environment, whether centralized or distributed, must be controlled through operational disciplines and by separating duties within the center. Physical network elements and data in storage and in transit must be protected. User identification and passwords, dataset classification and access protection, and data encryption all help secure the network from intrusion, or protect data from unauthorized viewing or use.

System control programs, utility and data management programs, and cash dispensing programs must be tightly controlled during storage, operation, and maintenance. The manual operations surrounding these applications are critical, too, and must be carefully controlled. Corporate information of all kinds is valuable: IT managers must ensure that it is not vulnerable to loss or damage.

Review Questions

1. What lessons did you learn from the Business Vignette?
2. What is the first thing that managers must know in order to establish effective controls? What are some other things they must know?
3. Why are business controls important to IT and user managers?
4. Why does new technology usually require new and different business controls? What new controls might be required for a firm introducing office automation?
5. What staff responsibilities do IT managers have regarding business controls?
6. What are some physical information assets that must be controlled? What are the most important intangible assets that must be controlled and protected?
7. Intangible assets usually have great value to the firm. Can you give an example of a firm in which intangible assets are more valuable that all other assets?
8. Define the concept of separation of duties. How might this concept work in managing physical inventory?
9. What are the characteristics of efficient and effective controls?
10. Identify the participants who are required for complete control over the development and use of application programs.

11. What security and control responsibilities does the application owner have? How are these related to user responsibilities?
12. How do the duties of the IT manager relate to the duties of other participants in the business control process?
13. Identify the control points in computerized data processing.
14. What are the control features of transaction origination?
15. What are the similarities and differences in controls between batch and interactive data input?
16. Who specifies control features in applications? What management processes ensure the correct and complete implementation of these features?
17. Identify the main elements in computer processing designed to provide protection and control.
18. The final data processing control point is data output. What control activities take place at this point in the processing cycle?
19. Describe system auditability. Why do application owners require auditability features as part of system controls?
20. Why is a disciplined production process essential to a well-controlled application portfolio? What constitutes a disciplined production process in a client/server operation?
21. To what kind of threats are networks exposed? What actions can managers take to minimize these threats?
22. What actions can managers take to protect data center assets? Downsizing changes the physical arrangement of data center assets and rearranges the control and security responsibilities. How does this impact the user manager's responsibilities? What are some considerations which govern their actions in these circumstances?
23. System programmers have almost unlimited access to information system assets. How do managers control this situation to protect system program assets?
24. What special precautions must be taken with critical programs?
25. What are the keys to effective system controls and asset protection?

Discussion Questions

1. Why are application business controls not listed as a critical success factor in Chapter 1? In what instances do you think they should be?
2. Why are control issues in applications more important now than ten years ago?
3. The Business Vignette illustrated some of the many things that can go wrong in networked systems. Considering all that you learned in this chapter, discuss the actions you would take to protect your networked system.
4. Discuss the control issues that accompany the introduction of facsimile machines to the firm. Who is responsible for resolving these issues?
5. Intangible property is very valuable in most firms. Suppose you were the manager in Merrill Lynch who controlled the name and address file for the Cash Management Account program. Discuss how you might protect and control this asset from loss or damage.
6. Along with separation of duties, the text discussed the need to rotate employees regularly. Discuss why these two actions work together to improve security and control.
7. In some firms, the IT manager owns and controls the application programs and databases. Discuss the advantages and disadvantages of this arrangement.
8. Discuss the changes in control responsibilities accompanying end-user computing and downsizing. What changes occur in the provider-of-services role? With end-user computing, the owner, users, development manager, and provider of services are all in one department. What are the implications of this?
9. Describe the IT manager's responsibilities to departments that have adopted end-user computing.
10. Discuss the special control precautions needed during the application enhancement process.
11. If you were the payroll manager, what control actions might you take when the payroll program is being altered to handle changes in the tax law?

12. System control and auditability must exist across the automation boundary. Using the accounts payable program as an example, discuss the shifts in responsibility at this interface.
13. Discuss the reasons why the disciplines of problem, change, and recovery management are business controls issues?
14. Discuss any ethical issues that might arise in testing business controls in applications.
15. Discuss the conflicting issues surrounding ease of use and business controls. How can an analysis of risk help resolve these issues?
16. Develop a list of critical applications you believe exist in many firms. Why are these applications critical? How should they be protected?
17. What special control considerations are involved when considering purchased applications? How do you think purchased applications should be tested?
18. Discuss the important business controls issues that arise when a service bureau is employed to process payroll.
19. Describe the role that the firm's controller plays in controlling and auditing application systems. If you were the controller in a firm contemplating downsizing the firm's data processing operations, what information would you want to know before you approved the plan?

Assignments

1. Assume that you are the manager of the workstation store in a firm starting to implement end-user computing. Develop the list of business control actions for the store's operation. As store manager, what reports and audit trails do you think are necessary, when should they be produced, and how would you handle separation of duties among your three employees?
2. One of the applications serving your department is being replaced by a newly developed program. The prototyping methodology is used for the development process. Devise a process by which you, the manager, can feel confident that business controls issues receive sufficient attention during the development process.

ENDNOTES

[1] Mitch Betts and Gary H. Anthes, "FBI Nabs Notorious Hacker," *Computerworld*, February 20, 1995, 4 and John Markoff, "To Catch a Cyberthief," New York Times News Service to the *Gazette Telegraph*, February 18, 1995, A1.

[2] *The Wall Street Journal*, November 7, 1988, 1.

[3] The book, *Cyberpunk: Outlaws and Hackers on the Computer Frontier*, by Katie Hafner and John Markodd, Simon and Schuster, 1991, describes the activities of Mitnick and his accomplice against DEC, a West Berlin cracker called Pengo who sold loot to the Soviets, and a lot more about Robert T. Morris.

[4] Annual Report, 1994, SBC Communications Inc., 33.

[5] Greene, Richard, "Inequitable Equity," *Forbes*, July 11, 1988, 83.

[6] Violation of this principle at the CIA was at the heart of the Aldrich Ames case.

[7] Walter B. Wriston, *The Twilight of Sovereignty* (New York: Charles Scribner's Sons, 1991), 59.

[8] Gary H. Anthes, "Is Your Data Secure?" *Computerworld*, November 28, 1994, 65. In a survey of 1,271 corporate officials—61 percent of whom were IS managers—42 percent said senior management in their organizations believes that information security is "somewhat important" or "not important."

[9] William G. Flanagan and Bregid McMenamin, "The Playground Bullies Are Learning How To Type," *Forbes*, December 21, 1992, 184.

[10] Peter Stephenson, "Beefing Up Your Anti-Virus Strategy," *Lan Times*, June 28, 1993, 62. About six new viruses are created and released every day and, for amateurs, do-it-yourself virus-creation tools are available according to Stephenson.

Part Six

Preparing for IT Advances

Technology introduction often changes the work people do and their organizational structures. As transitions occur for individuals, departments, and organizations, astute managers deal skillfully with the people issues. Organizations and effective leaders facilitate people working together to achieve new technology benefits. The chapters in this part are People, Organizations, and Management Systems, and The Chief Information Officer's Role.

Managing human resources effectively is an IT manager's most important success factor. Effective information technology managers display an abundance of people management skills.

18 People, Organizations, and Management Systems

A Business Vignette

Denver Utility Outsources Data Center to IBM's ISSC[†] *Reciprocal Outsourcing—An Emerging Trend*

In a deal intended to help both companies move into the era of deregulated utilities, IBM's subsidiary Integrated Systems Solutions Corporation (ISSC) recently entered a 10-year outsourcing arrangement with Denver-based Public Service Company of Colorado.

At the core of the deal is a classic information technology operations contract that calls for ISSC to take over Public Service's Denver data center, its workstations and help desk operation, and its intrastate network. ISSC will also pick up the company's application development. Del Hock, the utility's chairman, said that the $500 million technology project will save Public Service of Colorado $190 million over 10 years.

In a separate but related contract, Public Service will provide IBM with energy consulting to help IBM trim its gas and electric costs at 16 U.S. plants. The utility will provide those services through a new subsidiary called E Prime, which will market its services to other companies. Completing the partnership circle, E Prime and IBM formed an alliance under which IBM will develop applications to help E Prime customers manage energy procurement and consumption. IBM is not purchasing equity in E Prime.

Analysts note that the exchange of services between IBM and E Prime illustrates an emerging trend in which outsourcers play the role of both vendor and customer with their clients. In another recent example, Electronic Data Systems Corporation signed a 10-year outsourcing deal with Lake Forest, Ill.–based Moore Corporation Ltd., in which Moore is supplying EDS with business forms and commercial printing.

And like IBM's alliance with E Prime, AT&T Global Information Solutions and Delta Air Lines are jointly marketing services to the airline industry as part of AT&T GIS's 10-year, $2.8 billion outsourcing deal with the airline. In this case the companies established a 50/50 jointly owned venture.

"These relationships are extending themselves with real interesting nuances," said Allie Young, an analyst at Dataquest, Inc., based in San Jose, California. "These types of contracts show a clear direction for the future of outsourcing. The key is the strategic involvement of the two companies."

[†] Mark Helper, *Computerworld*, February 6, 1995, 14. Copyright 1994 by *Computerworld*, Inc., Framingham, MA 01701. Reprinted from *Computerworld*.

The Public Service contract is one of many outsourcing contracts expected as utilities try to blast out of their complacent environments and into a competitive market. Recent energy industry outsourcing deals include the following:

Energy Company	**Outsourcer**	**Length of Contract/Amount**
Philadelphia Electric	ISSC	10 years/ $450 million
San Diego Gas & Electric	Computer Sciences Corp.	5 years/ $60 million
Halliburton Energy Services	Andersen Consulting, Power Computing, I-Net	10 years/ $500 million
Public Service of Colorado	ISSC	10 years/ $500 million

Late last year, San Diego Gas & Electric Company and PECO Energy Company also entered outsourcing pacts. For Public Service, the $190 million cost savings will be vital in the new competitive environment, where utilities are vying to offer services at the lowest price.

The Federal Energy Policy Act of 1992 requires utilities to allow other power companies into their power distribution grid so that wholesale customers can shop for the lowest rates. This policy set the stage for market competition, and utilities are striving to lower costs in order to retain their large customers. Outsourcing is expected to be very popular with utilities.

Philadelphia Electric outsourced its routine mainframe operations, help desk, and LAN and desktop administration to ISSC so it could concentrate on the future. PECO plans to connect its meters with wireless and fiber-optic networks, replacing manual meter reading, and permitting it to record electricity consumption from its offices. The utility believes that using these connections may permit PECO to cycle home appliances on and off, thus achieving cost savings from low-rate, off-peak electricity.

INTRODUCTION

This text emphasizes the management of information technology assets and focuses on the tools, techniques, processes, and procedures required for effectively controlling and using these assets. The processes of strategizing and planning set the stage for understanding and embracing software and hardware technology trends. Management and control of application assets, and the production facility in which applications are processed, follow technology trends. This book turns now to the most important IT assets, IT people and their organizations and structures.

People and their organizations are critical to the successful functioning of the modern business firm. But the effective use of firm personnel requires an environment or corporate culture in which they can thrive and be highly productive for themselves and their organizations. Employees and managers need to know "how we do things around here." To a large extent, how we do things around here is a function of the firm's management system. The management system provides an intellectual framework, not only for management actions, but for employee actions, too.

As Peter Drucker's quote in Chapter 1 and the Business Vignette in this chapter indicate, dramatic changes are taking place in conventional business organizations. These changes are not confined to the firm, but involve its customers and its suppliers: In most cases, the entire industry is undergoing rapid and dramatic change. Deregulation in the utility industry is one current example and, in many instances, changes within the firm result from collaboration with others. In all cases, information technology enables new structures of business activity and is itself a change agent.

TECHNOLOGY SHAPES ORGANIZATIONS

We are witnessing tremendous, fundamental, pioneering changes in the business environment. Rapidly, the environment is becoming increasingly complex; tremendous advances in communication and information processing capability drive this complexity. These driving forces are significantly affecting our business structures. "There is no question that the new information technologies (IT) are having a major impact on the range of strategic options open to an organization. IT is not only creating an environment that is making it imperative for organizations to evolve, it is also helping them adjust to the enormous changes outside their own boundaries," so states Michael S. Scott-Morton.[1]

Organizational Transitions

The 1980s ushered in major organizational transitions for many corporations worldwide—the 1990s promises more of the same. In the United States, mergers, acquisitions, and leveraged buyouts took place at a frenzied pace. According to *Business Week*, "before the decade had ended, a towering $1.3

trillion was spent on shuffling assets—an amount on a par with the annual economic output of West Germany."[2] With some variations, the 1990s continue the trend of the 1980s. Activity in the computer-telecommunications sector is especially high. The AT&T/McCaw merger for $12.6 billion, British Telecom/MCI investment of $4.3 billion, US West/Time Warner investment of $2.5 billion, and Tele-Communications/Liberty Media merger worth $8.3 billion are some that occurred during an 18-month period. The phenomenon is not isolated to the U.S. as the investments of US West and SBC Communications and many others illustrate.

Mergers, acquisitions, and leveraged buyouts have caused turbulent times for IT executives and for employees, too. Studies reveal that companies are rapidly reorganizing as they downsize, outsource, and reengineer business processes. Much of this activity is driven by mergers or acquisitions, or results from business partnerships or joint ventures. Corporate reorganization frequently leads to unemployment—in many cases workforce reductions are the motivation for restructuring. Information systems professionals are not immune to this plight.

Through purchases, joint ventures, and planned internal expansion, many firms are attaining a significant international presence. Some of the world's largest, most successful corporations are losing much of their national identity as they become truly international businesses. Some examples illustrate the point: Honda manufactures in several countries and sells in many; IBM manufactures in many countries and sells globally; Phillips, the huge Dutch electronics firm, has 10 percent of the world's color TV market and is the world's largest manufacturer of light bulbs. Royal Dutch-Shell, among the world's largest industrial corporations, is of Dutch and British origin but remains cognizant of the interests of many nations in its worldwide operations. The trend toward dispersed national, transnational, and international business enterprises is certain to accelerate driven by communication technology and political considerations.

Centralized Control—Decentralized Management

Not only are firms growing in size through mergers acquisitions, but alliances between firms are becoming increasingly common. There are many examples of alliances in business today. The semiconductor, computer, and telecommunications industries demonstrate the value of cooperative efforts. Alliances and partnerships, once considered a sign of corporate weakness, are becoming necessary to speed new products to market, to improve operational efficiency, and to obtain skills necessary to meet business goals.

The adoption and use of alliances and joint ventures to capture the advantage of time, to employ critical skills, and to access distribution channels or new markets add complexity to the firm's value chain. Value is added to the firm's product through inbound logistics (the firm imports raw materials or subassemblies from suppliers), internal operations (assembly, manufacturing, or process activities), outbound logistics (distribution activities), sales and marketing, and service. These processes are supported by activities such as procurement and technology development, and they depend on human resources and the firm's infrastructure.[3] For firms operating internationally or firms with alliances, one or more parts of the product value chain may lie outside the firm itself.

For example, consider the five firms in five countries which manufacture jet engines, as mentioned in Chapter 15. Value is added to the final product in the customer's plane because the five manufacturers produce engine subassemblies, develop engineering and technical documentation, and produce maintenance spare parts for delivery to final-assembly locations. There are many other examples in diverse industries, but the auto industry and the computer industry provide some additional well-known examples. In all these examples, managers rely upon the firm's infrastructure and information technology architecture to add value to the end product.

Information technology is essential and in some cases absolutely critical to the success of firms engaged in forming partnerships and alliances or in internationalizing their businesses. The Business Vignette noted that Public Service of Colorado and IBM shared skills, technology, and resources to help each other solve problems. Their partnership enabled these firms to share resources used in helping others. Information and guidance from the hub of the corporation are the intellectual resources that allow firms in five countries to develop and build jet engines. Similarly, these assets enable others to produce oil on several continents and to refine, distribute, and market it in many countries. Executives use information technology extensively to keep the networked business operating as a cohesive unit.

Centralization versus decentralization has been a long-standing corporate strategic issue. The issue looms larger now because of the recent trends toward corporate restructuring and global strategies. As firms grow in size, they tend to decentralize operational control and to put their decision-making capability closer to operational centers. For many firms, particularly the conglomerates, growth via acquisition accompanied by decentralized profit centers has become the norm. Decentralized operations with limited centralized control have become popular. However, information technology, especially telecommunication systems, offers executives the potential to have the best of both worlds. Decision making and operational control can be delegated to operational units; control information can be available to headquarters on a real-time basis.

The Impact of Telecommunications

Advances in telecommunication technology and rapid implementation of these advances to achieve important business goals are the hallmarks of information technology today. The union of information systems and networks gives business executives enormous capability to expand their operations, to form alliances and joint ventures, and to restructure their assets for greater effectiveness. Telecommunication systems shrink time and distance and help establish new relationships: They improve organizational efficiency and effectiveness, and they promote innovation.[4] The following examples illustrate how companies capture these advantages.

Volvo uses an international computer-aided-design network to design and manufacture heavy trucks. IBM uses its engineering design system and programming support systems through national and international networks to design and manufacture computer systems for a global market. Network systems allow GM to design cars in Germany, build them in Korea, and sell them in the U.S. More than 12,000 companies in North America now use Electronic Data Interchange (EDI) to transmit information from one computer system to another. Electronic communication can take place between the decentralized operations of the firm or from the firm to the offices of a supplier or customer. The concepts of just-in-time manufacturing and the paperless factory critically depend on telecommunication systems.[5] Adopting sophisticated telecommunication technology is a critical success factor for most firms today.

The Span of Communication

Telecommunication systems greatly increase the span of communication and remove the need for filtering layers of management. Span of communication facilitated by information systems has made the traditional span of control irrelevant.[6] Executives who communicate throughout the organization electronically can understand the details of their operations essential in maintaining control. The reduced number of middle managers must expand their spheres of influence in order to attain corporate goals and objectives. They must also use advanced communication techniques to increase their effectiveness over a broader range of organizational activity. They serve little other function if they cannot do this.

The span of communication can be very large in modern industrial firms. For example, technicians in California, working with technologists in New York, help manufacturing engineers in Germany solve difficult production problems on advanced disk drives for European customers. On a smaller scale, application software specialists diagnose customer problems and provide solutions electronically. The customers are located throughout North America, but the specialists are located in Colorado. On yet a smaller scale, employees at one computer development and manufacturing facility unite product designs with manufacturing technology and procurement operations to build customer solutions when the customer needs them, just in time. Telecommunication technology enables these operations. Indeed, telecommunication enables parallel processing at the firm level, whether the firm is local, national, or international in scope.

Organizational impacts of technology are not limited to user organizations; the IT organization itself is undergoing substantial change in most firms. The IT organization must adapt even further if it is to remain relevant. The days of a centralized IT organization are rapidly disappearing and, in its place, a new structure is emerging. The new structure recognizes that some information processing functions are best performed centrally, while others must be performed locally. The new IT organization must have strong leadership to develop, operate, and maintain systems vital to the firm's centralized operations. The IT leader must ensure that both central and dispersed IT activities conform to and support the firm's strategic direction.

The new IT structure must also form partnerships and develop alliances with users throughout the firm to ensure effective use of the technology at all levels in the firm. IT must facilitate user adoption of appropriate information systems and must support their growth and development. Important user tools such as computer-aided-design systems, computer-integrated-manufacturing systems, office systems, and end-user systems for marketing, sales, or service are all part of the firm's information infrastructure. IT managers are responsible in part for the success of these user systems. IT organizations must be tightly coupled to the using organizations; they must share common values, goals, and objectives.

Organization and structure facilitate people working together to achieve their personal goals and the goals and objectives of the firm. Cooperative, innovative endeavors between individuals or groups are the lifeblood of business enterprise. Improved technology, sophisticated management systems, and effective people management enable employees, managers, and the enterprise itself to adapt to the steady stream of changes required for success. But, as the following passage illustrates, the difficulties for the innovator are an age-old problem:

> It should be borne in mind that there is nothing more difficult to arrange, more doubtful of success, and more dangerous to carry through than initiating changes. The innovator makes enemies of all those who prospered under the old order, and only lukewarm support is forth-coming from those who prosper under the new. Their support is lukewarm, partly from fear of their adversaries, who have the existing laws on their side, and partly because men are generally incredulous, never really trusting new things unless they have tested them by experience.

In consequence, whenever those who oppose the changes can do so, they attack vigorously, and the defense made by the others is ineffective.

So both the innovator and his friends are endangered together.

A. Machiavelli, *The Prince*, 1513

PEOPLE ARE THE ENABLING RESOURCE

Managers universally recognize that human resources are the most important assets contributing to business success. Most CEOs pay tribute to this fact by noting the importance of people resources in letters to stockholders or in annual reports. It is fairly common to find a paragraph in the CEO's letter to stockholders thanking the employees for their contributions to the success of the business and acknowledging that "people are our most important asset."[7]

When R. J. Stegmeier assumed the position of CEO at Unocal, he published the following statement in his first letter to employees and stockholders:

> In my view, this company's greatest asset has always been its human resource. Without the talent and teamwork of Unocal's 18,000 employees, our oil would stay in the ground and our facilities would stand idle. We face some tough challenges in the months and years ahead, but together—by our creativity and hard work—I'm certain that we can meet those challenges and turn them into opportunities for productivity and growth.

This is a good example of the often-expressed thought by CEOs regarding the enormous value of human resources.

While most of the corporation's functional areas vitally depend on a base of skilled employees, the overall performance of some organizations may be limited by their ability to recruit, train, and retain individuals. Individual performance directly relates to the overall performance of the unit. The revenue of a marketing unit may be limited by the availability of people trained to sell the firm's products. New product development directly depends on the creativity and ingenuity of the engineering force engaged in development activities. The firm's ability to grow and prosper hinges on the skills, abilities, and energy of the management team in establishing and implementing strategies and plans for success.

Not only is it generally agreed that human resources may limit unit performance but, in many instances, very significant contributions by a small number of individuals, or by an individual working mostly alone, materially improve the bottom-line performance of the organization. An engineer whose brilliant idea spawns one or more new products, or a salesman who clinches a big sale, are very important to the firm. Thomas Musmanno, who invented the Cash Management Account, and the managers at Merrill Lynch who led the implementation of the CMA, are truly valuable assets.

A section of GTE's 1994 annual report titled *Our People*, states that "GTE will continue to differentiate itself in the marketplace through the quality of its employees. We're constantly striving to ensure that we attain the best-in-class workforce in corporate America." "We subscribe to the belief that out employees can provide a key competitive advantage," is the subtitle leading the passage.[8]

People and Information Technology

Since the beginning of electronic data processing in the 1950s, the effective implementation of information systems in business and industry has been limited to some extent by the availability of highly skilled people. Recently, the shortage of trained and experienced managers to lead the introduction of complex and rapidly evolving technology has limited progress. The job opportunities for programmers and systems analysts has remained firm for thirty years; these skills are predicted to be in short supply in the near future. Numerous articles have detailed the consequences of inexperienced or ill-prepared individuals leading organizations to increasing automation in the face of high corporate expectations. While it is clear that twenty years of experience has taught us much about coping with these difficulties, there remain plenty of examples to remind us that we still have a long way to go.

Not only have we not closed the skills gap in the field of information handling, but current trends in telecommunications, client/server computing, downsizing, and outsourcing are placing increasing demands on skills currently in short supply. Talented individuals who are able to capitalize on advances in hardware, software, and telecommunications are in great demand. Additionally, the skill and experience required to implement strategies in these growing and potentially profitable areas differ in many ways from the current skill base. For example, object-oriented programming in C++ is significantly different from programming in COBOL 74, and writing mainframe batch systems differs greatly from programming client/server applications. Managing IT people has always been difficult; new and emerging challenges promise to keep it that way.

The time of personnel shortages in the information arena is far from over. Prudent IT managers and their superiors are planning to cope with the situation lest they find themselves on the critical path of their parent organization's roadmap to success. Some alternatives are available. Effectively managing the present employees, who are familiar with the organization and its mission, must be the highest priority. An organization that fails to manage its present staff in a superior fashion stands little chance of obtaining and retaining outstanding individuals.

ESSENTIAL PEOPLE MANAGEMENT SKILLS

The key to using information processing technology effectively is management's ability to employ and retain skilled individuals. These individuals must be skilled in the technology itself; they must also be skilled in supporting the use of the technology within the firm. Talented people are the enabling resource that allows the firm to capture the benefits and capitalize on the opportunities offered by advanced technology. Employing skilled people, managing them with sensitivity, and providing effective leadership are necessary conditions for success. Managers who fail to lead their people skillfully and effectively set the scene for mediocre or marginal performance in most endeavors. Successful managers employ solid people management skills: They view these skills as the cornerstone of their success.

What do we mean by solid people management skills? How do employees perceive good people management traits, and what can employees reasonably expect of their managers? What distinguishes managers who possess these skills from those who do not?

The answers to these questions directly depend on the assumptions one makes about organized human effort and the values individuals derive from participating in organized activities. Abraham Maslow attempted to explain individual needs as a pyramidal hierarchy with basic physiological needs such food and drink at the bottom and self-actualization needs at the apex. The most essential needs such as physiological and safety needs are followed in order of importance by social needs and esteem needs which precede self-actualization needs. Individuals will try to satisfy the most basic needs first and, when these are satisfied, will be motivated to satisfy the next-most-important needs. For example, an individual whose physiological and safety needs have been met will be motivated to satisfy social needs such love or a sense of belonging. When these needs are met, the individual will be motivated to attain esteem and recognition. When all other needs are fulfilled, individuals strive for self-actualization according to Maslow.[9]

Most IT employees have satisfied their physiological and safety needs and are motivated toward fulfilling higher needs. Organized activity in the firm provides managers with the means to appeal to employees' higher needs and to accomplish objectives for the firm. Skillful managers believe that most employees prefer to engage in meaningful, self-satisfying work that offers them a chance for self-development. The ideal situation is one in which the individual's goals and the organization's goals are congruent. Managers who have a good understanding of their employees as individuals, and who can arrange for this goal congruence, have taken a giant step toward improving morale and increasing productivity.

Effective People Management

It follows that effective people management centers around the concept of respect for individual dignity. Good managers recognize the enormous, frequently unrealized potential in each individual. Effective people managers work hard at understanding each individual's preferences and motivations. They use their knowledge of each individual in a manner that displays respect for each employee's unique personal characteristics. Talented professionals expect to be treated as special individuals; good managers meet this expectation.

Good people managers make appropriate assumptions about individuals and act with these beliefs in mind until proven wrong. These assumptions form the basis for managers' behavior toward employees and, in turn, influence employees attitudes toward managers.[10] The assumptions are very important, not only because they govern attitudes, but because there is a certain self-fulfilling character about them. Therefore, the assumptions and the attitudes they imply have an important bearing on individual and group performance.

Attitudes and Beliefs of Good People Managers

Good people managers believe that employees are honest and industrious and act with the best interests of the firm in mind. Good managers also believe that employees are intelligent, willing to learn, and that they desire self-fulfillment.

In keeping with these beliefs, effective managers establish an environment of high but reasonable expectations for their employees. Based on the belief that professionals want to contribute meaningfully to the organization, managers work with individual employees to establish challenging goals. The manager and employee agree on objectives that are aligned with the needs of the organization while leading to the employee's self-fulfillment. Employees who want to achieve challenging, fulfilling goals that are congruent with organizational goals are ideally positioned to contribute significantly to the firm. And in the process, they achieve self-satisfaction for themselves. They are productive and self-fulfilled.

Management equals leadership within the most successful organizations. Employees expect their managers to lead with clear goals and objectives for the organization based on a vision of the future. Good managers must clearly articulate this vision to the organization and must establish organizational goals and objectives, provide pathways, and set directions that lead employees toward meeting these challenging goals, for the organization's sake and for their own. The best managers are true leaders who provide their employees with the vision, and motivate them with the desire, to achieve challenging corporate and personal objectives.

Managers must also set high and challenging performance standards for themselves. By setting good examples for their employees to follow, effective managers establish an environment of high performance and high productivity for the organization. High-performance organizations are led by managers who have high expectations for themselves and who have attained agreement on challenging goals for their employees. Managers achieve self-satisfaction and self-fulfillment from attaining difficult goals just as employees do, and their morale and productivity rises with that of their employees.

When managers and employees have reached mutually acceptable standards of accomplishment, employees should assume responsibility for meeting the standards. They should be given the authority to accomplish the associated tasks. Managers must also provide their employees with the tools to do the job and should train them to use the tools effectively. In addition, good managers delegate responsibility to individual employees and give them the requisite authority to carrying out the objectives. Through delegation, managers empower employees to meet their organizational objectives. Managers must also pinpoint accountability and responsibility so that employees are clear about what is expected of them. In this way, employees can earn rewards for their individual accomplishments.

All organizations face obstacles and obstructions that impede progress. Frequently, the bureaucracy appears to stand in the way of progress, while meaningless rules make it difficult to get the job done. Sometimes the bureaucracy is used as an excuse for lack of progress. One of the tasks of good people managers is to remove or reduce the barriers to accomplishment. If the rules are truly meaningless, managers should remove or revise them. If the rules are meaningful and required, managers should explain why they are needed and should assist employees in complying with them. Managers represent the firm to employees; they should enforce effective regulations and should eliminate ineffective ones.

Good managers expect employees to be thoughtful about their work, but they also expect action. In most situations, there exists a balance between exhaustive analysis and thoughtless action. Frequently, in their effort to obtain all the facts surrounding a decision, employees and managers spend far more time analyzing problems than solving them. It may be impossible to understand all the minute consequences of certain actions, but occasionally, action must be taken despite incomplete information. That's the essence of decision making, after all. Effective managerial and staff decision making means striving for the middle ground, which balances expended effort against potential risk.

Effective managers expect employees to solve problems for the firm, but they also expect employees to prevent them. Often, solving problems requires more time and energy than preventing them in the first place. Good people managers encourage and reward problem prevention. In many cases, preventing problems can be a difficult task. Often, visible external signs, such as distressed operations and frenetic activity, accompany problem solution actions. Conversely, problem prevention usually consists of thoughtful actions taken without fanfare while operations are proceeding smoothly. Good managers must take special care to acknowledge and reward the subtle but valuable efforts of those who work to prevent problems in the first place, or who have the ability to keep small problems from becoming large.

Through their actions and their attitudes, superior people managers establish an environment of high productivity and good morale. In this environment, individuals can accomplish challenging objectives and experience high degrees of self-fulfillment. The environment encourages creativity, innovation, and invention. By providing an atmosphere of productivity filled with opportunities for self-satisfaction, managers create an environment that fosters innovation (for many, innovation is the highest form of self-fulfillment). But managers' actions must also support those whose innovations were less than expected or whose attempts at invention failed. Managers know that not all innovative endeavors result in success. They must also remember that the surest way to stifle innovation is to belittle or criticize individuals whose innovative ideas were less than totally successful.

Achieving High Morale

Studies have shown that employee morale and employee opinions of their immediate manager's performance strongly correlate with the level of trust and confidence employees have in their manager.[11] Employee trust and confidence is granted to managers whose behavior is reliable, based on predictable norms for the organization, and consistent across similar situations. Managers who maintain steady, consistent patterns of behavior inspire trust and confidence in their employees. Their employees view them more favorably. Employee morale in departments headed by these managers is also predictably high.

High employee trust and confidence in management stems from many factors of managerial behavior; factors that are under managers' control and that can become part of the management practice. Some of the actions that managers can take to increase employee trust and confidence are:

1. Maintain two-way communication with employees to understand their needs and desires and to share company information.
2. Provide training and complete information so that employees can do their jobs effectively and efficiently.

3. Inform employees of opportunities for promotion and career advancement.
4. Listen to employee suggestions for improving the work environment and respond encouragingly to all suggestions.
5. Sponsor teamwork and cooperation among department members.
6. Be available to employees when they seek consultation.
7. Understand the amount and quality of each employee's work.
8. Use the knowledge of each employee's work to grant fair salary increases and promotions.
9. Provide enthusiastic leadership to the department in achieving its goals and objectives.

Managers who inspire trust and confidence in their employees are good people managers. They are effective communicators. They share information candidly and consistently, and they welcome dialogue intended to improve business operations. Effective managers treat individuals with respect and dignity. They understand individual wants, needs, and contributions. Good people managers understand good performance, and they reward it both publicly and privately. Good people management is the hallmark of effective executives and a critical success factor for IT managers.

Ethical and Legal Considerations

Organized human activities involve individual and group interactions implicitly governed by rules of conduct. These rules or disciplines of what is good, bad, or morally obligated are termed *ethics*. Ethics is an important topic for all organizations. It is important to managers because they represent the organization to employees and help establish the rules of conduct. Through their words and actions, managers govern how these rules and obligations are implemented.

Individual ethical standards are based on many factors such as culture, law, religion, and nationality among others. In actual practice, there are wide differences in individual standards, even among cohorts. These cultural differences stem from parentage, heritage, religious belief, and training and experience. Managers must assume that employees intend to behave ethically, but that their initial sense of ethics may not be consistent with that of the firm or fellow employees. Managers have special responsibilities toward the company and its employees regarding ethical conduct.

Well-managed organizations take steps to ensure that managers and employees understand the basic beliefs and policies governing behavior within the organization and in external business relationships. They conduct training sessions regularly and provide specially trained people who can answer specific questions regarding the organization's policies and practices.

These organizations take steps to ensure that a code of conduct exists and that all employees follow it.

Some actions that can be taken to foster ethical behavior in an organization are displayed in Table 18.1.

TABLE 18.1 An Ethics Guide for Managers

1. Develop a statement of ethical principles for the firm.
2. Establish employee and manager rules of conduct.
3. State and enforce penalties for rule violations.
4. Emphasize ethics as a critical factor in the firm.
5. Inform employees of applicable laws and regulations.
6. Recognize ethical behavior in performance evaluations.
7. Maintain business controls, thus removing temptation.
8. Support a confidential forum for answering ethical questions.
9. Lead by outstanding example.

Managers must understand the firm's code of conduct and must present it to employees as needed. Managers, themselves, must develop a personal code of behavior that embodies the firm's policies and leaves no doubt that ethical behavior is valued. Managers must answer employee questions thoughtfully; there may be subtleties. In case of doubt, seek guidance from more senior executives; discuss ethical dilemmas with them and seek their wisdom. The posture of ethical behavior in an organization cannot be determined on a manager-by-manager basis without developing serious internal inconsistencies that undermine the firm's rules of business conduct and behavior.

Ethical issues raised by the information age are numerous and increasing; information systems managers must be especially concerned about them. Consider the issue of privacy: Is it ethical, for example, for a firm to read the outgoing E-mail from its employees? Or the issue of property: Is it ethical for employees to use the firm's resources such as E-mail for personal business?[12] These and myriad other questions must be answered thoughtfully and clearly so that managers and employees are free of paranoia or guilt feelings. For some situations, the firm establishes rules of conduct. In others, laws and regulations are established by sources external to the firm.

The diffusion of information technology in organizations and society raises many issues that have both ethical and legal overtones. Issues such as intellectual property rights, licensing, copyrights, and royalty payments are governed by laws and regulations that individuals and firms are expected to know and follow. Although gray areas exist, especially regarding liability, employees and managers must remain informed. Desktop software and hardware are especially critical areas for most organizations.[13] Business controls established by the firm and implemented by the information center and the workstation store are important in controlling physical property and managing licenses and royalty payments.

Periodically, the firm's legal and audit departments should review the organization's position and recommend improvements. The firm's managers, IT managers included, should call upon the firm's legal and personnel staffs for advice and counsel; these specialized departments support management just like some IT departments do. Individual managers should verify the firm's legal and personnel policies with the firm's legal and personnel experts and not attempt to chart these difficult waters unaided.

THE COLLECTION OF MANAGEMENT PROCESSES

Effective operation of the IT management system is a critical success factor for IT managers. It provides a framework or background within which employees and managers operate and, in part, from which the norms for the IT organization are derived. The management system consists of tools, techniques, and processes that are exercised periodically. For example, the strategic planning process establishes long-term corporate directions for the use of information technology. And, the problem management system verifies the manager's intention to achieve service levels. The IT management system must be aligned with the firm's management system, must support and augment it, and must embrace the firm's values and basic beliefs.

It is essential that IT managers have systems to guide them in achieving critical goals for themselves, their organizations, and their firms. These critical factors can be organized into business issues, strategic and competitive issues, planning and implementation concerns, and operational items. The management systems presented in this text are designed to help managers achieve their critical goals and deal effectively with the issues facing them and their firms.

Strategizing and Planning

The management tasks of building IT strategies and developing long- and short-range plans to implement IT's strategic direction are critical first steps for IT managers. Strategizing and planning are the cornerstones of the IT management system, and link the IT organization to the firm's management system. Their results align the IT strategic direction with the firm's strategic business direction. The linkages and alignments are critical to IT and the firm. The relationship between these activities is shown in Figure 18.1.

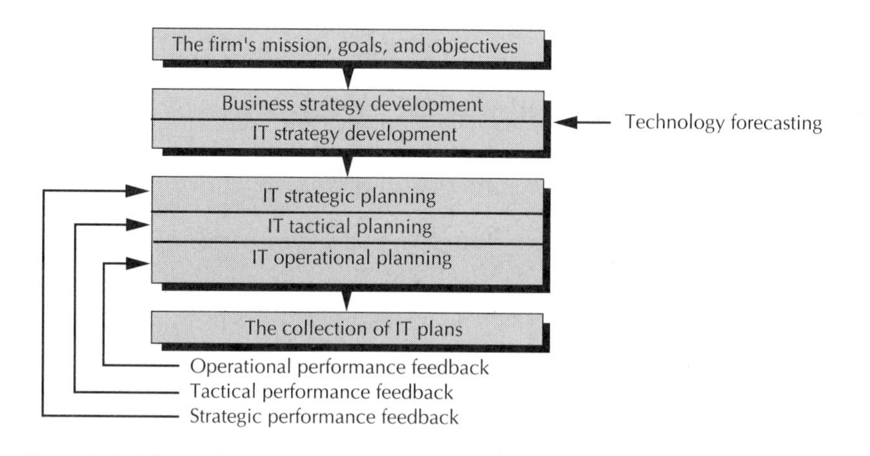

FIGURE 18.1 IT Planning and Control

Information technology planning and control begin with the firm's business strategy development process. The firm's mission statement and its goals and objectives are the foundation on which the firm's business strategy is built. The firm's business strategy and its IT strategy are linked intimately through shared goals, objectives, and processes. Business objectives for the firm that require information technology resources and actions are translated into IT strategy directions. Thus, the process of IT strategy development occurs in conjunction with the business strategy development process. Since the IT strategy is the basis for IT plan development, shared strategic directions and plans are developed within the firm, forging links between IT and the remainder of the firm. It is this interweaving and sharing of goals, objectives, and processes that ensures alignment between the IT long-range plan and the business strategy.

These cohesive strategizing and planning activities deal directly with many of the issues facing senior executives, IT managers, and their peers throughout the firm. Specifically, these processes force alignment of IT and corporate objectives, educate senior managers about the role of IT and its potential, and demonstrate IT's contribution to the business. If properly implemented, these processes eliminate strategic planning as an issue. They provide a mechanism through which the firm can exploit information technology for competitive advantage in a systematic manner.

IT strategizing and planning processes offer excellent opportunities to coach others in the firm, including executives, about how technology can be used to achieve objectives, including attaining competitive advantage. These processes must consider forecasts of technology capabilities. Technology capability is matched with the firm's requirements for new technology, and plans are developed to incorporate it into the business. Planning for the adoption of new technology and improved use of current technology can reduce costs, improve efficiency, and provide or sustain competitive advantage. If properly handled, these actions will also ensure realistic, long-term technological expectations.

There are several reasons why the process described above may not succeed, and the resulting IT strategic plan may not be aligned with the firm's business plan. For example, the process can only work when the firm itself has a well-defined mission statement and business strategy. In their absence, an IT strategy may not be well correlated with the intended business direction. In some firms, IT is not considered critical to the firm's success and the organization is not brought into the firm's planning process. In these firms, strong support for corporate goals and objectives from information technology is not required. In other cases, IT managers share in the process but lack the necessary business skills and knowledge to construct strategies and plans aligned with the firm's plans. To mitigate these difficulties, IT managers must take action to ensure that they understand what their role is and that they contribute to the firm in substantive ways.

Control in plan execution stems from operational, tactical, and strategic performance information. Performance information describes the variances between planned and actual results. It also includes variances between environmental and business assumptions and actual business and environmental conditions. For example, if competition threatens some of the firm's markets, the IT development team may need to improve the planned schedule of a marketing system under development. Depending on the nature of the variances, feedback may generate course corrections in operational, tactical, or strategic planning. Also, major environmental perturbations may cause the firm to adjust its strategic direction. In the extreme, the firm may adjust its mission to capture perceived opportunities or to avert potential threats.

The critical activities of strategizing and planning direct the firm's exploitation of its current and future assets and form the foundation on which its future depends. The management systems described in this text are an excellent framework for succeeding in these crucial activities.

Portfolio Asset Management

IT managers must deal successfully with many issues surrounding the important task of managing the application portfolio assets. In earlier chapters, this text described methods for prioritizing the application backlog, for managing the development process, and for finding alternatives to local application development. Figure 18.2 describes the relationships between these activities.

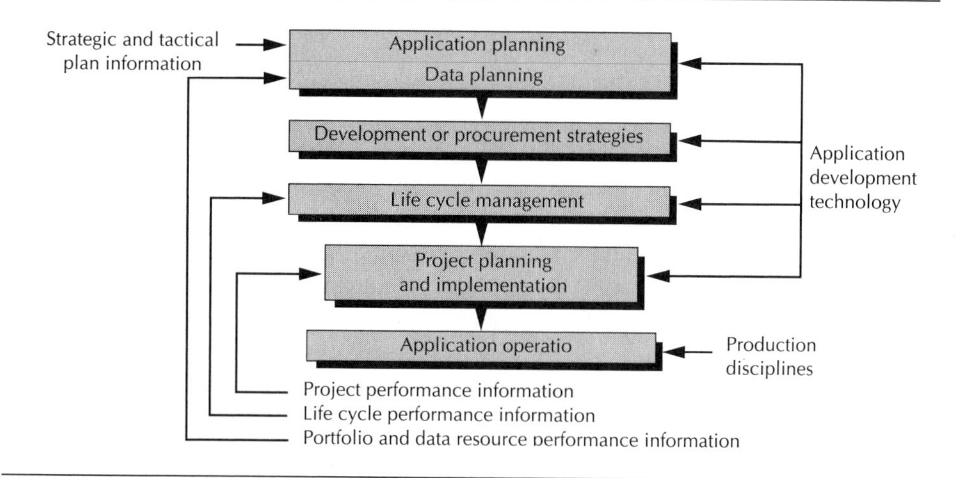

FIGURE 18.2 Application Management

Most firms spend a significant percentage of their IT budget maintaining and enhancing the applications and managing the associated data resources. The management systems for dealing with these challenges begin with IT strategizing and long-range planning. These processes establish the direction for enlarging and enhancing the application portfolio, and lay the foundation for incorporating technology advances. They help managers optimize and add value to the portfolio assets.

Application and data planning rely on information prepared during the strategy and plan development activities. Application performance and capability are compared with the firm's requirements, and the sufficiency of the data resources is analyzed. Using this information, the prioritization process identifies the applications that qualify for investment and specifies the preferred acquisition methodology for each application. This decision-making process combines the technical knowledge of computer experts with the business knowledge and vision of senior managers.[14] With information gleaned from this analysis, IT managers build an acquisition plan and establish an installation schedule.

Life-cycle management, project planning, and implementation govern the investment of resources in the selected applications. These management-oriented processes consist of business case development, phase reviews, resource allocation and control, risk analysis, and risk reduction. These processes ensure that the objectives embodied in the strategic and tactical plans for the portfolio are achieved in a controlled and disciplined manner. Skillful execution of these management processes results in an improved portfolio that satisfies the firm's functional and business objectives.

Application planning and management processes introduce new technology that may be valuable to the firm and enhance its business. Executives may become acquainted with hypertext or imaging systems, client/server implementations, CD-ROM data storage technology, or important data transfer technology such as ATM or wireless LANs. The management process may also incorporate new application development technology such as advanced languages, methodologies like object-oriented design, and new development tools and methodologies. Input regarding technological developments is a vital part of the application management system.

Application asset managers must consider alternative acquisition methods such as alliances, joint ventures, purchased application packages, and alternative strategies such as client/server computing, and outsourcing application development and operation. Businesses need modern applications and rapid solutions to business problems. Their availability is critical. IT managers must use a variety of techniques to improve productivity in bringing new solutions to the firm.

Management information is available during these processes to assess whether the activities are proceeding satisfactorily and to provide data for course corrections that might be required. Information derived from phase reviews, for example, serves to keep application development projects on schedule or to make necessary schedule corrections.

Application and data planning in conjunction with strategic and tactical planning for the firm lay the foundation for developing an information architecture. Information architecture is an important concept for IT and the firm. Strategic plan objectives coupled with assessments of present application and data performance measured against requirements allow an information architecture to develop from application and data planning. Developing and maintaining an information architecture is a critical issue in most firms today.

The systematic application of portfolio management processes focuses executives on application issues and gives them the tools and techniques needed to address their concerns. Through effective use of portfolio management systems, the firm acquires the applications and functions it needs, on schedule, and within established budgets, based on the business case. New and enhanced applications become available for productive use within the firm in a controlled and optimal fashion.

The Disciplines of Production Operations

Sound portfolio management delivers new and enhanced applications to improve the firm's business. These applications codify the firm's internal functions and link the firm to customers and suppliers through telecommunication systems. They provide the company with competitive advantage.

Successful operation of the applications is a critical success factor for the firm's managers.

IT customers relate to the production operations department through the disciplines of service-level agreements, batch operations, and on-line operations. These relationships are displayed in Figure 18.3.

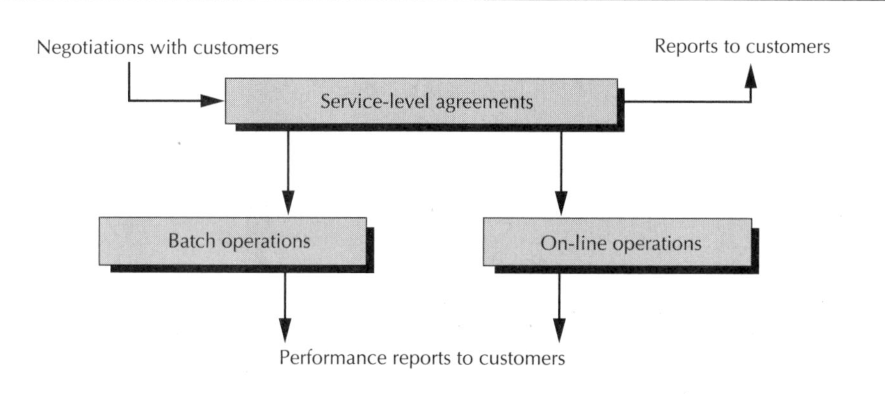

FIGURE 18.3 Production Operations—Customer Performance System

Customer expectations are developed by establishing service-level agreements between service users and service providers. Technically achievable and financially sound levels of service that meet the needs of the business are negotiated. Production service is delivered through batch processing, on-line processing, or a combination of both. The customer's service requirements are described in the service-level agreement and factored into the batch or on-line processes.

For example, routine inventory transactions are entered continuously throughout the day via client/server applications. Records accumulate for the nightly batch run. Nightly inventory processing is scheduled as a batch operation and the results are transmitted to the inventory department's server before 7:00 A.M. daily. In each activity, the inventory department achieved agreement on the service parameters it needs and can afford; IT's ability to meet committed service levels is measured and reported on a regular basis.

These processes are the basis for achieving customer satisfaction in production operations. They establish requirements rigorously and provide the means to address them systematically. These management systems meet many, but not all, of the necessary conditions for success in production operation.

During program execution, service problems sometimes occur and changes must be made to the applications or to their operating environment. Disciplined processes to manage these activities are mandatory. Achieving service levels also depends on the capacity of hardware and software system components and their performance. The management systems to deal with these complexities is displayed in Figure 18.4.

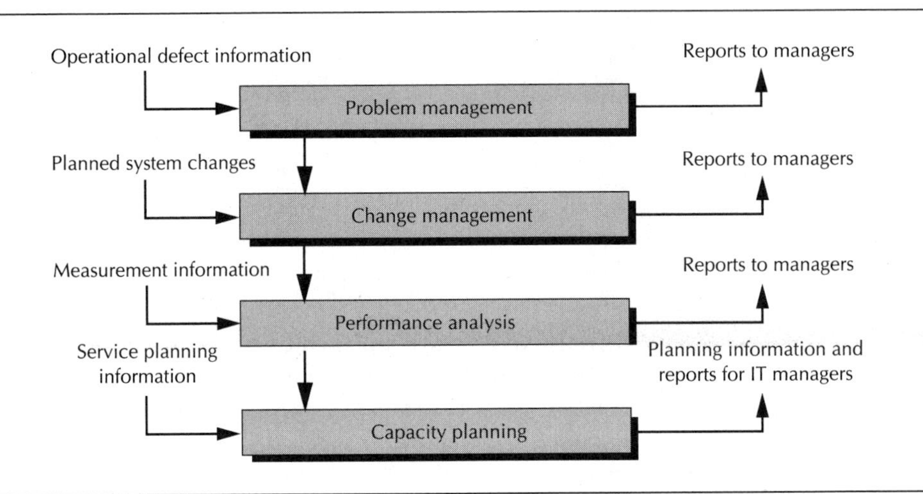

FIGURE 18.4 Production Operations—Internal Systems Management

Production operations occasionally experiences problems or defects that affect service or have the potential to do so. The problem management discipline handles these incidents, and the significant activities are reported to management for monitoring and control purposes. The resolution of customer problems is also confirmed with the management team.

Controlled system changes to complex data processing or telecommunications operations originate from two primary sources: planned changes and changes required to correct problems. The change management discipline obtains input from these sources, manages the changes through implementation, and reports the results to appropriate managers. Working together, problem management and change management provide a systematic way for service providers to correct inevitable operational faults and to implement changes required to correct problems or implement planned system alterations.

Performance management and capacity planning round out the management system for production operations. Since service levels are based on known or anticipated performance factors, ongoing performance measures must be available to validate system productivity. Analyses of system productivity is reported to service providers so they can take necessary action. Therefore, capacity planning receives input from performance analysis and from service-level planning. Plan input, customer requirements, and performance analysis generate future capacity requirements that become input for tactical and operational plans.

Production operations management is a critical success factor for service providers. The management systems for application operations is a platform for attacking issues and dealing successfully with them. This operational framework, combined with the tactical and strategic management systems, forms an important base for management success.

Network Management

Advances in various technologies for processing, transmitting, and storing binary data have led to systems that handle voice, data, and image information interchangeably. Consolidation of the hardware and software for processing all types of information has been followed by consolidation of the organizations supporting these operations. In most cases, these merged organizations have found it advantageous to consolidate previously separate management systems, too. This means that the management systems for centralized or distributed operations also support network management.

For example, network faults (defects that impair or have the potential to impair network service levels) are processed by the problem management system and network changes are handled by the change management system. Network performance and network capacity planning are also consolidated with application systems performance and capacity planning. This works well for the IT organization but also makes a great deal of sense for IT customers. Customers are much less able to distinguish network components from application components in their systems than IT people, and they are much less likely to care about these distinctions. They are interested in obtaining needed services and appreciate a management system that provides them without complications or bureaucracy.

In addition to the management systems for networks discussed so far, IT needs network configuration management to provide data for other disciplines. Figure 18.5 displays the sources and uses of the data developed and stored by configuration management.

Configuration management maintains several databases required to manage problems and changes and to plan recovery actions. Managers also are able to obtain information on the status of the network and its components from these databases. Automated network management systems use configuration data to optimize network operations and to update other databases. The management system for networks depends heavily on this information.

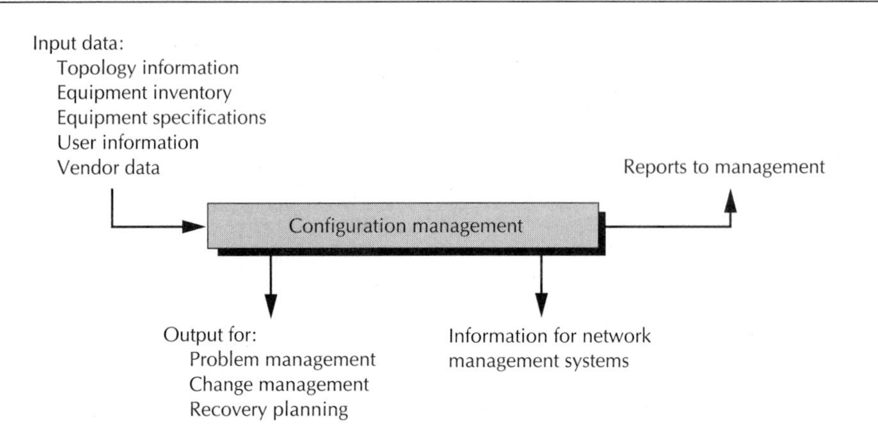

FIGURE 18.5 Configuration Management System

Financial and Business Controls

Effective IT functional operation requires that managers exercise control over all aspects of the IT business. This means that they understand the objectives of the organization and are managing IT to meet organizational objectives in a systematic manner. Disciplined strategizing and planning processes to meet established objectives are implemented and disciplines to manage the acquisition and operation of applications are installed. In addition to these activities, IT managers must maintain control over the financial and business aspects of their operations through appropriate systems and business controls.

The mechanism for monitoring IT finances and maintaining financial control begins with the tactical plan and its accompanying budget. Whether IT is organized as a cost center or a profit center, or operates as corporate overhead, the annual planning process yields an operating budget for IT. Usually, each organization in the firm receives monthly financial statements that describe the actual financial position versus the budgeted or planned position. IT managers review actual expenditures as compared with planned expenditures and act on the variances. Figure 18.6 displays this process.

If IT is organized as a cost center, the financial plan will contain planned cost center support and expenses, both of which are subject to variance. IT managers must analyze each and take corrective action if the variances are large or if the trend is unsatisfactory. In some cases, this will lead to negotiation with clients and may result in changes to the planned rates for some classes of services.

If IT is organized as a profit center, both revenue and expense must be analyzed for discrepancies. Variances can be corrected by changing revenue, altering expenses, or permitting changes in profit margin. Revenue can be altered by increasing or decreasing business volumes or by changing prices. An IT profit center operates as a business within a business, and managers have many opportunities to make adjustments that affect revenue, expenses, or margins.

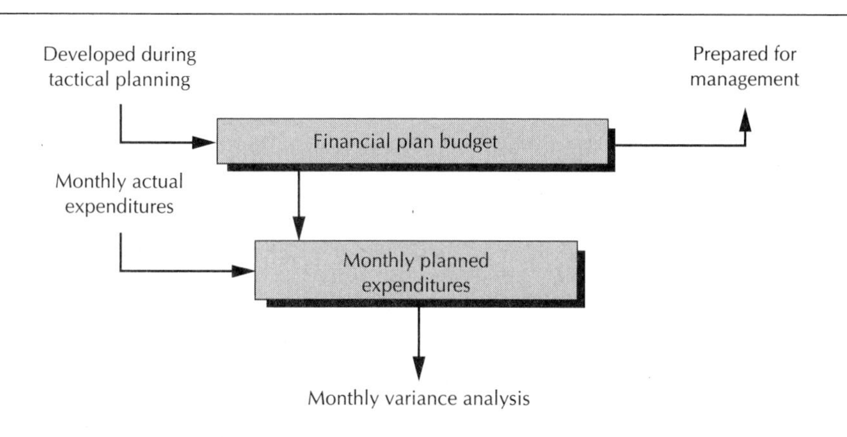

FIGURE 18.6 Budget Management

IT financial management systems that control to a budget developed in coordination with tactical and operational plans, account for user services, and recover costs from users, attack several of the issues of concern to top management. IT organizations using these techniques operate in a businesslike manner and spend resources on activities that the firm's managers agree are required.

Business Management

Sound management systems are essential because they provide a framework for accomplishing important firm objectives. Properly functioning systems guide managers and help them discharge their responsibilities in a logical and organized manner. But systems operate through people; good people relationships are required if the systems are to be effective. IT managers must execute their management systems harmoniously with others if they are to achieve maximum effectiveness.

There are several ways that IT managers can foster improved interaction with clients, increase coupling between customers and suppliers, and provide input for improving service. Many firms form an IT steering committee consisting of executives from client organizations who work with key IT executives. The steering committee acts as a high-level sounding board to help establish IT strategies and develop actions for the IT organization. In many ways, the steering committee acts as a sponsor for IT and helps build awareness of IT benefits to the firm. Through its actions and deliberations, the committee influences service levels by recommending and approving spending levels. In firms that have a cultural bias toward participative management, the steering committee can be particularly effective in communicating with top management and with functional managers.

Another effective mechanism for linking IT with clients is to assign individual IT managers liaison responsibility to client organizations. For example, an application programming manager may be assigned as the business service representative to manufacturing or an operations manager to product development. The manager represents IT to the individual function and represents the function to IT.

The service representative responsibilities include:

1. Obtaining information on new or changing client needs
2. Broadening communications between clients and IT
3. Answering user's questions about IT procedures
4. Facilitating problem resolution for clients
5. Advising on client-developed programs

In short, service representatives provide high-speed communication links between IT and other functions in the firm to everyone's mutual advantage. In a large, diversified firm, many service representatives interact with product development, manufacturing, marketing, sales, service, finance, administration, and possibly other departments.

Effective IT executives use every means at their command to maintain and improve communication between IT the rest of the organization. They understand the value of effective communication and work to maintain close contact with their peers throughout the firm. These formal and informal communication processes are an important ingredient of their management system.

THE IT MANAGEMENT SYSTEM

The management system described in this text is designed to focus IT managers on activities that improve their effectiveness and increase the firm's effectiveness. IT managers must do the things that the firm needs to accomplish to succeed. IT management processes focus managers on the activities with highest payoff for the firm. Through effective operations, IT improves the firm's competitive posture.

The management system also helps IT conduct activities in an efficient manner. It forces managers to focus on critical operational details, and on control and financial details. This focus enables efficiency improvements and productivity increases. When businesses are able to use information more effectively, decision making will improve, and the performance of IT and of the firm will increase dramatically.

SUMMARY

Today's business environment is undergoing rapid and fundamental changes and is becoming much more complex. Information technology, particularly telecommunications, is one of the driving factors behind these changes because it provides the means for altering competitive positions among firms. It gives firms opportunities to restructure in response to the changing environment. The restructuring of firms enables them to reduce costs, improve their value chain, or compete globally. It presents both problems and opportunities.

Organizations are comprised of people working together toward common goals directed by managers who provide leadership within the organization. People are the most important assets of the firm; they must be managed skillfully and with sensitivity. This means that effective people managers respect the dignity of their employees and make assumptions about people that reflect positive beliefs about them. Productive organizations operate in an environment of high expectations supported by managers and employees who encourage innovation and who recognize and reward accomplishment. Productive organizations are filled with people who are achieving self-satisfaction and self-actualization through accomplishing challenging tasks for the firm. Good people management is a trait of effective IT executives and is a critical success factor for them.

Management systems consisting of tools, techniques, and processes that support IT managers and guide them and their organizations in achieving corporate and organizational objectives within the firm's cultural norms. IT management systems assist in developing strategies and plans supporting the firm's strategic direction. IT planning processes ensure alignment between IT and firm objectives. Portfolio management leads to sound applications supporting the firm's operations.

The management systems for operations and networks establish an environment that fosters careful planning, service level achievement, and cost control. Financial management and control is accomplished through planning, budgeting, and review activities. Although these activities form an effective basis for IT management, sound communication channels between IT and users increases their effectiveness. Steering committees and IT service representatives play an important role in improving communication and increasing understanding and cooperation. Effective IT and client managers promote these communication techniques.

Review Questions

1. What role does information technology play in the dramatic changes taking place in the business environment today?
2. In what ways are alliances or joint ventures important to the firm's success?
3. How does telecommunication technology relate to the issue of centralization versus decentralization in business operations?
4. How do Volvo, IBM, and GM use telecommunications to improve their businesses?
5. Define the span of communications. What is the importance of span of communications to middle managers today?
6. What IT trends increase the importance of people skills today, more than ever before?
7. Describe Maslow's hierarchy of individual needs. Where on the hierarchy do most IT people exist, with respect to their working environment?
8. What are some of the assumptions that good people managers make concerning their employees?
9. What are some of the reasons that managers and employees should establish high expectations for themselves?
10. How can managers and employees achieve the required balance between thoughtfulness and action when making decisions?

11. The text states that managers should expect employees to solve problems, but that they should expect employees to prevent problems, too. What possible difficulties arise when managers expect their employees accomplish these simultaneous goals?
12. What should managers do when employees come to them with ethical dilemmas?
13. Why are strategizing and planning activities so important for IT managers?
14. Why might the management system for IT strategy and plan development fail in some firms? What are the consequences of plan failure?
15. Why is the management system for the application portfolio important? What critical issues does it address?
16. What possibilities exist for making additions to the applications portfolio?
17. What are some important steps in application life cycle management?
18. Review the process of developing service-level agreements and indicate its role in production operations.
19. What are the processes through which IT budgets are developed?
20. What are some activities that steering committees can perform? Why are these activities valuable for IT and for users?

Discussion Questions

1. Describe the arrangement between Public Service of Colorado and ISSC. Could part of this arrangement be called co-sourcing, and if so, explain why.
2. In the Business Vignette, Allie Young commented that these types of contracts (between Public Service and ISSC) show a clear direction for the future of outsourcing. The key is the strategic involvement of the two companies. Discuss the significance of these ideas.

3. Describe how firms use alliances or joint ventures to improve their value chain or expand their operations.
4. Discuss how the issue of centralization versus decentralization is related to the use and application of information technology.
5. Discuss how partnerships and alliances apply to IT organizations today. Describe the relevance of these concepts to the IT activities of downsizing and outsourcing.
6. Describe how effective people managers use their knowledge of human needs to improve their managerial performance.
7. In what ways do you think that managers' assumptions about people will impact employees' performances?
8. List the characteristics that good people managers display in dealing with their employees. Which of these characteristics is most important in your opinion? Which characteristic do you think is most difficult to attain in practice?
9. Describe the environment that good people managers attempt to create. What are the advantages to managers and to employees who work in this kind of environment?
10. What kinds of managerial behavior patterns improve employees' perceptions of managers' performance, and inspire trust and confidence in them?
11. What actions should a firm take to develop and foster an atmosphere of ethical behavior?
12. Describe the role that management systems play in managing the IT function. How do management systems relate to the corporate culture?
13. Describe how IT might implement control processes to ensure that plan implementation is proceeding correctly.
14. Describe the system for portfolio asset management. What processes provide input data for the system? What is the desired outcome of effective portfolio asset management?

15. How can the application management system discussed in the text manage the large data stores associated with the firm's portfolio?
16. Describe the management system for computer operations and relate it to other management systems.
17. Discuss the advantages and disadvantages of having separate operational processes for applications and for networks.
18. How is financial control maintained in an IT profit center?
19. Describe how steering committees and service representatives augment the IT management system.
20. How would the service representative concept differ between an organization that operated as a cost center and one that sold IT services to other firms for profit.

Assignments

1. Read "The Changing Value of Communications Technology" by Hammer and Mangurian (*Sloan Management Review,* Winter 1987, page 39) and prepare a two-page summary for the class.
2. Interview a human resources manager at your university and discuss his or her views on people management. Compare and contrast your interview observations with the approach toward people management advocated in this text.
3. Visit a computer center near your school and review its written operational procedures. Evaluate them in light of the management systems described in the text. Did you find any processes or procedures unique to the center you visited? If so, why are these important?

ENDNOTES

[1] Michael S. Scott-Morton, "Information Technology and Corporate Strategy," *Planning Review*, September-October 1988, 28.

[2] "The Best and Worst Deals of the '80s," *Business Week*, January 15, 1990, 52.

[3] The value chain is discussed in Michael Porter's *Competitive Advantage: Creating and Sustaining Superior Performance* (New York: Free Press, 1985), 12.

[4] Michael Hammer and Glenn E. Mangurian, "The Changing Value of Communications Technology," *Sloan Management Review*, Winter 1987, 39.

[5] Eric E. Sumner, "Telecommunication Technology in the 1990's," *Telecommunications*, January 1989, 38. Data traffic on U.S. telecommunications networks is growing about 20 percent per year and voice at about 5-6 percent per year. The result of these growth rate differentials is that data will account for about half of the business telecommunications traffic in the early 1990s.

[6] Peter Drucker, *The Frontiers of Management* (New York: Truman Talley Books, E. P. Dutton, 1986), 204.

[7] For example, see 1994 Annual Report, Southwestern Public Service Company, 5, in which the chairman states, "I am especially proud of the employees at your company. In a climate of uncertainty, they have never wavered in their commitment to our customers. ... They are responsible for our success, and I give them full credit."

[8] GTE 1994 Annual Report to stockholders, 23.

[9] Abraham H. Maslow, *Motivation and Personality* (New York: Harper & Row, 1954), 80.

[10] Douglas McGregor, *The Professional Manager* (Edited by Warren Bennis and Caroline McGregor, New York: McGraw-Hill Book Company, 1967), 16.

[11] M. L. Pesci, personal communication.

[12] For a discussion of some information age ethical issues, see Richard O. Mason, "Four Ethical Issues of the Information Age," *MIS Quarterly*, January 1986, 486. Mason discusses privacy, accuracy, property, and access.

[13] "Get Legal on the Desktop," *Datamation* May 1, 1995, 56.

[14] Thomas H. Davenport, Michael Hammer, and Tauno J. Metsisto, "How Executives Can Shape Their Company's Information Systems," *Harvard Business Review*, March-April 1989, 131.

19 *The Chief Information Officer's Role*

A Business Vignette

Caterpillar Uses IT to Remain Competitive

Tens of thousands of Caterpillar jobs in the U.S. have been saved due in part to an impressive injection of information technology by this leading construction equipment manufacturer. Facing strong foreign competition, Caterpillar invested in new computer and telecommunication technology to increase its global competitive position. Thanks largely to technology investments, Caterpillar's sales and profits are surging. "To survive in this new globally competitive world, we had to modernize," says vice chairman James Wogsland. "Information technology is the glue for everything we do."[1]

To improve its competitiveness, Caterpillar launched two massive initiatives. The first initiative is termed "Plant With A Future." This project is designed to continually increase plant floor automation, reduce manufacturing costs, improve marketplace responsiveness, and retain Caterpillar's market position lead. Led by Caterpillar chairman Donald Fites, the second initiative broadly restructures the company to improve internal and external customer responsiveness. Information technology plays an important role in both undertakings.

As part of the restructuring, Robert Hinds, the former director of computer-integrated manufacturing, was named CIO and corporate IT director. Hinds manages an annual budget of $180 million, including a corporate IT staff of 1,300 people and a centralized data center in Peoria, Illinois, home of many Caterpillar facilities. In addition to the corporate IT staff, the operating divisions employ another 1,300 IT people who manage divisional information systems. "Our corporate systems make us look like an integrated company—because we are—but giving autonomy to the business groups gives us a competitive edge," Hinds says.[2] Using workstations or PCs, 90 percent of Caterpillar's employees can access corporate data, validating Hinds' assertion that technology is truly a part of the job regardless of where you work.

As the first step in reengineering the company, plant and division managers were directed to find or create superior support systems. In many instances, Caterpillar's plant and divisional IT teams built their own, while critical corporate systems, including purchasing, logistics, and MRP II (manufacturing resource planning) systems, were being rebuilt by the Corporate IT staff. To ensure compatibility among these dispersed activities, Corporate IT established company-wide program development standards.

In addition, Caterpillar developed factory integration specifications that equipment suppliers are required to build into their controllers. Since nearly every vendor of manufacturing equipment supplies proprietary, built-in software, the specifications forced them to meet the Caterpillar manufacturing system interface. This strategy eliminates the "islands of information" problem at the outset.

Another critical IT activity is supporting Caterpillar's dealer network, the company's only direct link with customers. Cat's 70 dealers operate more than 300 outlets in the U.S. alone. "Our goal is to give the customer 100 percent of the parts he needs over the counter when he needs them," says Larry O'Neill, a $100 million Elmhurst, Illinois, Caterpillar dealer. "We can't carry every part, but by jointly sharing information, we can get any part in here by 6:30 A.M. the next day."[3]

Caterpillar built a tri-level telecommunications system to support its worldwide operations. Digital microwave links back up 70 miles of fiber optic cable that support company activities in the Peoria area. North American dealer operations and other company activities outside of Peoria are supported via leased fiber optics capacity and satellite links. Operations outside of North America are linked to Caterpillar systems via leased fiber and satellite services.

To advance its automation efforts, Caterpillar developed proprietary EDI software for its suppliers to use. More than 1,000 key suppliers, providing about 80 percent of its parts, exchange data electronically with Cat.

Several other important IT initiatives have been completed. Caterpillar's corporate IT staff developed and built a financial modeling system that enables executives to perform what-if analyses and estimate future project profitability. IT implemented an activity-based cost accounting system that helps managers obtain production cost contributions by plant for particular products. And, Cat engineers now use a computer-aided design system that reduces product development time by up to four years.

"Plant With A Future" is a great achievement. Some measures of success include:

1. Work-in-process inventory has been reduced by 60 percent.
2. Product build time has been reduced from 45 days to 10 days or less.
3. On-time customer delivery is up by 70 percent.

The results of the initiatives at Caterpillar reduce inventory carrying costs, save millions on Cat's annual $5 billion of purchases, improve customer satisfaction, and provide competitive leadership. In 1994, Caterpillar earned $4.50 per share on revenues of $65.20 per share and, according to *Value Line*, 1995 earnings are projected to be $5.50 on an 8.7 percent revenue rise.[4]

INTRODUCTION

The managers of information technology in most modern firms have considerable responsibility and authority. Power and responsibility flow from the line organization headed by the senior IT executives. Information technology widely deployed throughout the firm also presents IT executives with considerable staff responsibility and empowers them with additional authority. As the firm increases its dependence on IT for success, it also increases its dependence on the senior IT manager. Consequently, senior IT managers enjoy a position of opportunity—opportunity for great success or for substantial failure. To achieve success, IT managers must make sure that their superiors view their accomplishments positively.

The premise of this text is that superior IT managers are made, not born, and that superb management skills are learned and developed, not the result of genetics.[5] Therefore, many significant management issues and trends have been discussed throughout the text. Management systems and practices designed to cope with these issues and trends dominate the course of this book. CIOs and senior IT managers are in a good position to appreciate the management issues, to follow the emerging trends, and to implement well-designed management systems.

However, do senior IT executives enjoy success commensurate with their responsibilities? Can well-trained individuals succeed in the job consistently? Or is the evolving nature of the technology and its influence on the organization so dominant that success is mostly a matter of chance? Given the difficulties of the position and the changing character of the responsibilities, what are the critical success factors of the CIO's position. These and other topics will be explored in this chapter.

CHALLENGES FOR THE SENIOR IT EXECUTIVE

The role of the CIO or senior IT executive in modern firms encapsulates the opportunities and pitfalls inherent in the technology, embraces the ebb and flow of managers' and employees' opinions and motivations, and symbolizes the dominant effect of information technology on business and industry worldwide. The role is an extremely difficult one. Unfortunately, the realities of the job differ markedly from the philosophical foundation of the CIO title.[6] Consequently, the title and position are alternately praised and cursed.

Numerous articles have chronicled the dilemma of the CIO position. In 1986, for example, Gordon Bock writing in *Business Week* stated, "There's a new breed of manager surfacing in the executive suite. Some members of this new information elite sit behind such recognizable name-plates as senior vice president, vice president for information services, or information resources manager. Others are beginning to get a higher-sounding title to reflect their new status: chief information officer, or CIO."[7] As information technology shapes the future of many industries and countless firms, chief executives recognize the increasing value of IT to their firms and see the need for sound IT leadership. In 1990, Thomas Friel, managing partner for information technology at Heidrick & Struggles predicted that "in five years, virtually every major company will have a CIO who's a peer to the CEO."[8]

Current literature suggests that business firms have yet to sort out completely the CIO's role and position. Rather than a peer relationship between the CIO and CEO, for example, in nearly 80 percent of the cases among *Computerworld's* Premier 100 companies, there were one or more management levels between the CIO and the CEO. About 30 percent of these CIOs reported to an executive heading financial operations. Literature also suggests that part of the CIOs' identity problem may result directly from their behavior patterns in the firm. Analysis also reveals that, among the premier 100 companies, CIOs' average tenure exceeds six years whereas, in CIO ranks generally, the annual turnover rate approaches 20 percent, significantly greater than that of top executives generally.[9] In addition to the executive's actual performance, the perceived performance in relationship to expectations may be the critical aspect affecting the perceptions of the CIO position and title.

In discussing the CIO position, Professor Wetherbe at the University of Minnesota stated that "the title created such animosity that its bearers gave it up happily." But the title is not necessarily what makes the job risky for the incumbent—the task itself is hazardous. Some failures can be attributed to individual performance deficiencies, but others result from organizational consolidations, cost reduction exercises, or bad times for the industry or the firm.

Though many more firms now have an information chief than 10 years ago, few CIO's have the skills and the opportunity to advance further in the corporation. And, in many companies, information technology staffers are viewed with skepticism or apprehension. "With most of my peers, there tends to be a distrust with the information services organization and with the CIO for that matter," says Bob McLendon, CIO of Texas Instruments.[10] His remarks seem to characterize the feelings of many regarding the title and position of CIO.

Organizational Position of the CIO

Whether they are called CIOs, VPs of information systems, or information resource managers, senior IT executives have line and staff responsibility for information technology within the firm. With the coalescence of telecommunication and information processing technologies, telecommunications and information processing organizations are being consolidated under the senior IT executive. Because information technology is now being widely dispersed throughout the firm, the senior IT executive is gaining extensive staff responsibility too. Just as the firm's chief financial officer does not spend most of the money budgeted for the firm but is held responsible for the firm's expenditures, the firm's chief information officer is responsible for IT in the firm, though most IT resources are consumed outside the IT line organization.

CIOs at most firms have responsibility for the traditional data processing operation, telecommunications services, and in some cases, other responsibilities as well. For example, a survey of 137 health-care CIOs revealed that 99 percent were responsible for information systems, 65 percent for telecommunications, 31 percent for management systems, 15 percent for medical records, and 10 percent for admitting services.[11] They believed their most important role is to integrate information systems, telecommunications, and management systems. Knowledge of hospital systems ranked fourth among top attributes needed by CIOs, according to the respondents who placed more importance on leadership ability, vision and imagination, and business acumen.

The reporting relationship between the senior IT executive and the CEO varies considerably—from direct reporting to three levels removed. Many CIOs report to chief financial officers, executive or senior VPs, VPs, or division heads. Fewer than 20 percent report directly to the CEO. National Car Rental, however, feels that information technology is critically important. In this firm, the CEO and CIO share many responsibilities.[12] In this unusual situation, the CIO is National's second highest paid executive. "Any CEO who doesn't have the CIO as a direct-report is absolutely crazy," says Vincent Wasik, National's CEO. This unusual partnership pairs a business-wise CEO with a technology-smart CIO both committed to using automation fully to revolutionize the car rental business.

In addition to reporting relationships, most, but not all, CIOs exert significant influence through committee appointments and staff assignments. Regardless of their organizational level, however, CIOs must closely communicate with the top echelon of executives or must be an integral part of the executive management team in order to be effective.

But being in close contact with the executive management team is not enough; the executive management team must recognize the importance and value of information technology. Top managers may fail to appreciate IT because they lack awareness of the strategic importance of IT, or because they see only the operational importance of computers. Senior executives may not view information as a resource, or they may experience a credibility gap with the direction of IT. CIOs can overcome some of these difficulties by educating top management and by marketing IT accomplishments to management. It is especially effective to have users sell and promote IT's business image.[13]

Career Paths of the CIO

The most popular route to the CIO position is through the information systems ranks, although some CIOs have combined their IS experience with that gained in other disciplines. Consulting and telecommunications backgrounds combined with IS experience is rather common among CIOs. When telecommunication organizations are merged with IS, the former IS manager is most likely to head the consolidated organization. This occurs because the IS manager has experience developing or managing critical systems for the firm and is considered to have broad management skills. The telecommunications manager, on the other hand, is usually seen as someone who merely manages the phone system. Many other managers, however, move to the top IT position after having served in line or staff positions elsewhere in the company. Engineering, product development, or general management backgrounds are also common.

According to a Coopers & Lybrand survey, more that 75 percent of CEOs, COOs, and CFOs prefer that the chief information officer have both an IT and a general/line management background. Only 2 percent indicated that the CIO's background didn't matter.[14]

Firms are more likely to bring in CIOs from outside the IT organization when the IT function needs revamping. An IT organization that resists downsizing, outsourcing, or decentralization, or one that fails to respond rapidly to changing business conditions is likely to be reorganized under a former line manager. In many cases it is easier to change the IT culture by replacing the CIO than by taking other less drastic actions. Newly appointed CIOs from outside the IT fraternity most likely bring with them a strong mandate for change.

Abrupt changes in the top IT position are quite common. A 1994 survey of 400 CIOs revealed that nearly one-third of their predecessors had been dismissed or demoted, and that nearly one in five left voluntarily for other positions. The average tenure of CIOs is less than three years, according to some studies. Part of the reason for the turmoil at the top can be attributed to the relative newness of the IT profession and to frequent redefinitions of roles and corporate relationships. Reengineering business processes and the adoption of client/server modes of operation are straining financial, technical, and personnel resources beyond their limits. The difficulty of recruiting, training, and retaining competent IT professionals who can implement complex IT systems on new platforms within limited or no-growth IT budgets adds to the CIO's difficulties.

Successful senior IT executives, however, enjoy the option of considerable mobility. The careers of some of the top CIOs indicate that exceptional individuals are very much in demand, and that technology skills and general business management skills are largely transferable across industries. However, the position of the person to whom they report is usually beyond their reach, unless they work to become qualified to assume the higher position.

Performance Measures

Regardless of the title that the senior IT executive holds, the firm's top executives have expectations of the position and of the IT organization. Satisfying these expectations is the CIO's highest priority. What is expected of CIOs? What must they do to be successful?

CIOs' performance is measured by their ability to bring value to the firm through the application of information technology to the firm's goals and objectives. CIOs are expected to identify technological opportunities and provide leadership in attaining advantage for the firm through the use of technology. They must educate the management team on the opportunities and gain the team's agreement to incorporate these opportunities into the firm's strategies and plans. At the same time, CIOs are measured on their abilities to develop planned courses of action that balance opportunities, expectations, and risks.

CHALLENGES WITHIN THE ORGANIZATION

Businesses evolve as they change their business practices or adopt new practices to improve their competitiveness or to respond to changes in the external environment. Mergers, acquisitions, alliances, joint ventures, and new business formations are external indications of the evolution. Firms also evolve internally. Restructuring, downsizing, and internal reorganizations are taking place as firms seek to tighten up their internal operations, improve their business processes, and improve their results. Support functions within the firm must be flexible enough to respond to these evolving corporate structures and to changing business requirements.

Organizational changes and new ways of doing business bear heavily on the senior IT manager. New computing technology and advances in telecommunication systems and products are the tools that IT executives must use to facilitate organizational transitions and to deal with competitive threats. Corporate executives are deeply concerned about competition. They must consider computer and telecommunication systems as vital elements in their competitive struggles. Businesses need to understand technology and must develop a vision of how technology can improve competitiveness if they are to succeed in today's dynamic environment. The CIO is primarily responsible for making this happen.

As relentless advances in information technology continue, they create new opportunities for alert CIOs and new pitfalls for complacent ones. Today, advanced technology is leading to innovation in product development, marketing, sales, and service. The IT organization must support technological innovations if the firm is to remain competitive in the future. As firms explore the menu of opportunities, make selections, and introduce new technology, IT executives become change agents.

The need to change and improve the business is compelling for most firms in most industries. The CIO and the IT organization are also compelled by fundamental business forces to find new and better ways of doing business. Competitive pressures demand improved performance. CIOs have a mandate in most corporations to reexamine their methods of operation in search of improved productivity for the firm. The improvements may result from lowered costs, increased quality, or enhanced responsiveness. Many top executives have high expectations that the IT organization can add substantial value to the firm's output.

Some employees, however, may not view favorably the changes required to keep the firm healthy and viable in response to competitive threats. IT managers who encourage and facilitate these changes are perceived accordingly. In addition, the adoption of advanced technology may significantly alter the IT organization. Some IT professionals may resist these changes, and some may resent them.

THE CHIEF INFORMATION OFFICER'S ROLE

Formulating Information Policy

The most important task for CIOs is to formulate information policies and win the approval of the firm's senior officers. The CIO should ensure that policy instructions are well known and followed throughout the organization. In addition, CIOs must develop standards for information processing within the firm and make sure that all systems acquisition and operations activities adhere to them. Policies and standards are essential because they provide a framework for subsequent systems integration, data and program sharing, and for developing an information infrastructure.

As part of policy management, the process of linking IT and business strategy and planning activities should be developed and agreed to by the firm's senior management team. Chapters 3 and 4 provide guidance in how this should be accomplished. When a basis for strategy development and planning has been formulated, the CIO must ensure that it operates effectively and that IT is a partner in major plans for the firm. Attaining a sound relationship for doing this is a fundamental priority for CIOs, particularly recently appointed ones.

CIOs are responsible for ensuring that IT investments and returns on them are sound and reasonable, whether they are made in corporate systems or in local development and operations. The staff role of CIOs is to ensure the financial integrity of business investments in technology. To accomplish this, the CIO should evaluate, review, and give staff approval for major IT investments in hardware, telecommunications, and application software. The authority to do this must be embedded in company policy. The techniques for exposing costs, comparing them to outside providers (benchmarking), and achieving agreement on expenditures, as outlined in Chapter 16, should be employed in the firm under the direction of the CIO.

The CIO must continually seek to improve the firm's business systems in the most cost-effective manner. This includes making sound make-buy decisions, improving productivity in local development, and increasing operational efficiency. CEOs expect improvements in productivity, performance, schedules, and quality, and CIOs should help achieve them. But, at the same time, CIOs must keep expectations within achievable ranges.

The CIO must enforce business controls, security, and asset protection policies and standards. Problem, change, and recovery activities in all areas of telecommunications and computing should proceed in accordance with the guidelines that the CIO promotes. The preservation of the firm's hardware and software assets must be a high priority for the CIO.

Managing Technology

CIOs are responsible for providing technological leadership to the firm in the area of information handling. The firm's executives depend on them to develop forecasts of technical direction and to assess the technology's significance to the firm. The CIO must assemble the forecasts and develop assessments for the firm's senior officers to use in the strategic planning process. Because information technology is advancing so rapidly, the technology management task is difficult. Yet, because information technology is so important, technology management is a critical task.

The task is difficult because the pace of technological change is extremely rapid and the magnitudes of potential changes are great. For example, predictions are that computers will increase in processing power by two orders of magnitude with no increase in cost over the next two decades. Likewise, enormous bandwidth increases are occurring as utility companies, railroads, pipelines, and traditional communication companies install fiber optic cables at breakneck speed. These trends, along with dozens of others including new application development methodologies, advances in artificial intelligence, and sophisticated optical storage devices, make technology forecasting both difficult and risky.

Technology forecasting is risky and difficult, but it is critical for the firm. Decision making in the firm involves a vision of the firm's future. In particular, at the senior or executive level, strategic decision making attempts to capitalize on future opportunities while avoiding future pitfalls; strategic decision making also attempts to shape or define the future. The firm's CIO plays an important role in ensuring that technology and technology trends are understood by those who need to know. The increased pace of future technological innovation and the explosive deployment of information technology into all industries increase the critical nature of technology forecasting.

Technology forecasting is much more than extrapolating past trends into the future. For example, many forecasts indicate exponential increases in circuit density per chip or in bits stored per cubic centimeter. These forecasts lead to others that predict that substantial computing power will be available at reasonable prices in the form of individual workstations. Making these predictions requires considerable understanding of the technology itself, and of scientists' and technicians' ability to develop it. The ability to put the technological achievements into practice, and the capacity to manufacture the resulting products efficiently, are also major factors. The forecasts are based on reasonable estimates of the difficulties involved in attaining these capabilities and on past progress in doing so.

The user of the technology faces another kind of problem when trying to estimate the technology's potential for the firm. This is because new technology leads to new uses for it, not just extrapolations of present use. For example, experts predict that personal computers with voice and handwriting recognition capability and extensive optical storage for text and voice will have artificial intelligence interfaces so the computer can be taught how the operator wants to use it. In general, forecasting how capabilities like these can be valuable to the firm is a difficult task. But technologists and managers must search for emerging technologies and attempt to forecast innovative uses for existing and developing technological capability.

In one attempt to answer some of these questions, researchers at the University of Minnesota found that information management experts independently grouped technologies into primary areas, and rank-ordered them according to perceived impact on organizations. Topping the list of important technologies are those innovations designed to improve significantly the manner in which people relate to computing devices.[15] Very powerful individual workstations which employ natural language or voice input and output will permit workers to access the tremendous computing capacity and data stores predicted for the future.

Straub and Wetherbe found the key human interface technologies to be speech recognition, voice input-output, and natural language interfaces supported on high-end workstation. Key communications technologies for the 1990s are voice-mail, E-mail, FAX, VSAT, and EDI.[16] Powerful individual workstations will enable these technologies to be effective. ISDN, LANs, and desk top publishing were seen as important ingredients of the communications technology according to this research.

In addition to these, other rapidly growing technologies are multimedia applications, groupware, video conferencing, pen-based computing, mobile computing, and wireless LANs. Most expect these to increase in importance in the 1990s.

In reviewing the events of 1994 and forecasting trends for 1995 and beyond, *Computerworld* found that several major topics dominated their findings.[17] Most significant is the explosive growth in the development of the information infrastructure. Corporations are connecting servers to the Internet at the rate of 2000 per month, and this is just the beginning. During the next five years, telecommunications companies are planning to spend upwards of $100 billion on infrastructure in the U.S. and billions more linking the U.S. and the rest of the world. Numerous organizations are exploring how to give electronic access to their services through the emerging global information infrastructure. Business executives here and abroad are expected to exploit these trends for improved national and global business effectiveness.

Distributed computing is another dominant theme for the mid-90s. IT managers believe that the rapid growth in networked workstations and in client/server implementations that characterized the recent past will continue to grow significantly in the years ahead. Applications will continue to grow in the distributed environment, and migration from larger platforms will continue.[18] Additionally, CIOs and their firms must evaluate wireless networks and cellular systems as they become more widely available. If the technology is potentially valuable, firms must learn to capitalize on it. But first, firms must analyze the risks and plan risk containment actions.

Communication technologies, including data and personal interactions, will be of substantial importance according to most experts. Information from many sources and in large volumes will be easily available to the firm's knowledge workers. Communication technology and appropriate software (groupware) removes the barriers of time and distance and allows cooperative work groups to form, perform, and disband readily and rapidly. Improved human-computer interfaces and expanded communication capability are the technological foundations for the information-based organization postulated by Drucker.

The job of the CIO and the executive management group is to evaluate future technology directions, select appropriate technology and products and develop strategies to capitalize on them, and establish plans to adopt new technology and introduce it to improve business processes and operations. In the coming years, it will be critical for the firm to focus on technology and products that can enhance its business while avoiding the temptation to spend energy and resources chasing each new technology development.

Introducing New Technology

Technology adoption and introduction begins after the firm's strategic plan has been accepted. The adoption and introduction process is formulated in tactical and operational plans developed by the functions for which the technology or products are intended. Implementation occurs as the individual units carry out the plans.

An organization's adoption and acceptance of new ideas consists of many individual decisions to use the innovation or to adopt the new product. The process by which individuals become adopters is communication-based and consists of the following five steps:

1. Individuals become aware of the innovation.
2. They become interested and seek information about it.
3. They evaluate the innovation in light of their needs.
4. They give it a try.
5. If conditions are favorable, they adopt the innovation.

This process through which new ideas permeate an organization, first described by Everett Rogers, is called *innovation diffusion.*[19]

Individuals differ significantly in their propensities to accept new ideas or innovations. Those few individuals who are eager to accept a new idea become champions of the innovation and are called *pioneers*. A somewhat larger group who accept the innovation readily are *early adopters*. The adoption process continues until the majority accepts the innovation. The last individuals to adopt, those most resistant to the new idea, are called *laggards*. Adoption propensity and the percentage of individuals in each class is shown in Table 19.1.

CIOs must understand these individual adoption differences because they are fundamental to the success of technology introduction. Perceptive managers search for pioneers and early adopters who are opinion leaders when introducing new ideas. Opinion leaders are effective in establishing awareness and developing interest in others. They pave the way for the majority to adopt the innovation.

TABLE 19.1 Innovation Adoption Process

Category of Adopters	Percent in this Class	Cumulative Percent
Pioneers	2.5	2.5
Early adopters	13.5	16.0
Early majority	34.0	50.0
Late majority	34.0	84.0
Laggards	16.0	100.0

Individuals are not necessarily pioneers for all new ideas. Some managers, for example, readily adopt and implement new organization structures but resist changes in personnel policies. Some managers readily accept changes in product strategy but resist personal computers in the executive suite. Skillful CIOs understand that these considerations apply to top executives as well as to nonmanagerial employees. Wise CIOs acknowledge these individual differences and use their knowledge to its best advantage when implementing innovations and new technology.

CIOs themselves have varying propensities to accept changes that enhance or inhibit their effectiveness. In today's fast-paced world, CIOs who cannot remain flexible and adaptable probably cannot lead effectively through the turbulence of alliances, restructuring, and new business processes. Although CIOs must be cautious and must recognize risk, they must not be viewed as laggards if they are to survive.

FACILITATING ORGANIZATIONAL CHANGE

The need to streamline business processes and to improve efficiency and productivity frequently leads to technology introduction. Likewise, the adoption and use of new information technologies has brought steady and unrelenting change to the enterprise and to the IT organization. Thus, microcomputers, networked systems, and strategic applications have altered work patterns, streamlined organizational structures, and improved competitive positions. Indeed, IT managers have justly earned the reputation for being change agents.

Today, most firms have installed large numbers of microcomputers and workstations and have implemented some form of end-user computing throughout the enterprise. Most have installed employee workstations for database query, statistical analysis, decision support, and many other applications. User productivity has improved as new applications ranging from word processing to executive support systems have been installed. Employee computing tools have enabled both employees and managers to become more effective and more efficient.

As personal workstations become more powerful and are linked to larger systems, work once performed on large centralized computers can move even closer to using departments. This process of downsizing mainframe-based systems is an important trend in business and industry today.

But downsizing presents many problems for IT professionals and managers; it represents an organizational challenge for IT executives. As the firm migrates applications from centralized to decentralized processors, people, procedures, and data move with them. Skilled IT people migrate toward client organizations in support of the applications, and some IT professionals who can't make the transition to the new environment seek jobs elsewhere. Downsizing alters the function of the IT organization and changes its structure.

The new IT organization is more responsive to business needs and demands. The new IT structure is required to improve the firm's competitiveness. Structure is related to management style. Centralized companies with more autocratic management styles tend to use conservative competitive strategies while more aggressive companies tend to be less centralized. "IT structure is strongly related to competitive strategy, and specifically the degree of centralization of IT activities is significantly related to competitive strategies," according to Tavakolian.[20] Highly competitive companies require flexible structures, especially in critical areas such as IT. Thus, CIOs and IT organizations must remain flexible and responsive to changing business conditions.

IT leadership and expertise is critical when rebalancing workload between centralized operations and the distributed environment. IT professionals are needed after decentralization in order to preserve the good features of the centralized environment, such as data security and business controls. The assumption that downsizing will yield people savings in the IS shop may be valid, but if people reductions occur, they may come at the expense of increases in user departments.

CIOs must recognize that some IT professionals may not welcome the plan to downsize because they believe their mainframe experience will not be very useful when the plan is implemented. They may be reluctant to abandon their COBOL skills for new languages and may elect to leave the organization. Others enjoy the opportunity to acquire new skills and be part of a changing organization. They know their application knowledge will still be needed when the applications reside on LAN-based workstations. They also know that much of their knowledge of database operations, application controls, and business procedures will apply directly to the downsized environment. CIOs must remain alert to numerous people issues that arise during the process of downsizing.

Acquiring or building distributed systems or moving applications from the mainframe to LAN-based systems requires outstanding planning and implementation skills. As shown in the Business Vignette, Caterpillar exemplifies good planning and implementation.[21] Their corporate IS group established companywide rules, standards, and policies defining an information architecture. Caterpillar also defined specifications that vendors must meet in order to integrate their equipment into the established factory automation architecture. This strategy avoids isolated, nonintegrated systems and databases, and it forms an architectural base for future growth and development.

Decentralization can sometimes be carried too far. At times, the most appropriate strategy is to centralize. This was the case for the city of Cambridge, Massachusetts, which had 50 separate agencies each responsible for its own data processing. A centralized system has been installed that resulted in a 11 percent reduction in the data processing budget over a three-year period. Annual savings of more than $800,000 are anticipated, and the payback took less than two years.

Sometimes the firm's strategic direction is to centralize. When the company adopts a centralized corporate structure, the IT organization must follow the corporate strategy. This was the case at Trailer Train Company of Chicago. Trailer Train's corporate philosophy is to consolidate divisions that were once wholly owned subsidiaries, and to centralize the corporate structure. The MIS organization replaced small computers in the subsidiaries with large central mainframe support, and reduced costs in the process.

There are other reasons to centralize. Some firms have had bad experiences with security or control problems, or found that functional areas preferred not to manage computer operations. Some firms find that migrating some applications back to the mainframe simplifies their operations and returns application processing to data processing professionals. In some of these instances, the decision does not hinge directly on out-of-pocket costs.

The firm's top executives must consider many factors before striking the balance between centralized processing and downsized applications. The technology is generally available, but the decision to downsize is mostly driven by critical business factors such as improved response to customers, higher quality customer service, and lower costs for the firm and its customers. In order to strike the proper balance, the CIO must carefully blend business, organizational, and technical skills.

"Most companies run their data processing operations the wrong way," according to John Singleton, President, Security Pacific Automation Co. "DP is not about being centralized or decentralized. It's more that all managers have to be business executives first and technologists second.[22] John Singleton is an extraordinarily successful executive, and a glance at his approach to IT business management is presented in the Chapter 12 Business Vignette.

Finding Better Ways of Doing Business

Maintaining the status quo can be fatal to the CIO's position and career. The old adage "if it ain't broke, don't fix it" definitely does not apply to the IT organization or to the CIO. CIOs must constantly seek ways to improve the organization's productivity and to create or sustain business advantage. They must control costs and risks, and they must find innovative ways to improve organizational effectiveness.

There are several ways in which effective CIOs have reduced costs and improved performance. Many of these are already familiar to the reader. They include using high performance and cost-effective hardware, adopting alternative application acquisition methods such as joint development or purchased applications, downsizing and decentralizing IT operations, and using highly effective disciplines in managing production operations and networks. Some firms find that turning over some of their IT operations to third-party operators is cost effective. This practice is called *outsourcing*.

Outsourcing

Outsourcing is currently receiving considerable attention. The practice is not new. Computer service bureaus and subcontract application development have always been available as options for most firms. Outsourcing took the spotlight when major firms such as Kodak and Merrill Lynch made major outsourcing decisions. Kodak negotiated a deal with IBM to manage Kodak's data center operation, and Merrill Lynch gave management of Merrill's network to MCI and IBM. And EDS, partly owned by General Motors, performs the majority of GM's information processing. There are many other examples.

The reasons for outsourcing vary among firms, but most deals are motivated by strategy, technical, financial, or scale considerations. For example, a manufacturing firm concentrating on its core strengths may find that network management is a distraction requiring skills in short supply. Additionally, the firm may not be prepared to invest in the rapidly moving telecommunications field, so it finds outsourcing its network needs to a third party attractive. Although it relies on telecommunications, the firm's strategy is to buy network capability and concentrate on its manufacturing strengths.

Other firms already using service bureaus for payroll and related financial applications find that the outsourcer's economies of scale provide cost savings. Looking to capitalize on this finding, they discover additional applications to offload. By moving additional appropriate work outside, firms can sometimes reduce even further the costs of work already at the outsourcer through volume discounts. In most outsourcing decisions, cost reductions are the most important motivating factor, but the economics of outsourcing remain controversial; each case must be decided on its merits.

Outsourcing firms offer the advantages of economies of scale. By performing data processing for many firms, they are able to procure large processors and to keep these systems busy and fully loaded. Their economies extend to people and supporting functions, too. For instance, because outsourcing firms have large production operations, they can afford to hire talented system-support programmers and network specialists to support the operation. These highly skilled people are fully utilized in a large operation but may not be in a smaller one. These economies potentially give outsourcing firms cost advantages as compared to in-house operations.

Whether outsourcing grows in the future as predicted depends on more than the economics of the deal. Other, major considerations play a significant part in the decision to outsource.

This decision turns on several non-economic advantages and disadvantages. Outsourcing all or part of the centralized operation permits the IT organization to concentrate on other important issues such as strategic applications or client/server development instead of computer center or network management processes. IT can also concentrate on user support and business objectives with less distraction when technical considerations of hardware, systems software, and network management are the responsibility of the outsourcing firm. Advocates of outsourcing believe that computer centers and network operations are not the place to spend scarce technical resources. Even as the cost of hardware declines, the cost of acquiring, training, and retraining hardware, operating system, and network technicians is rising rapidly. More importantly, people with many of these technical skills are extremely difficult to acquire and retain.

Outsourcing has its critics as well. Though computer hardware and networks are considered mundane by some, others believe them to be vital to the firm's survival and prefer to retain firm control. Giving up data center and network operations to others, however capable they may be, may lead to long-term disadvantages, say the critics. The issues center around quality, strategic direction, and control of critical production activities. Can the vendor provide long-term, high-quality service? How will the firm and the vendor work together on problems, changes, recovery, and other operational issues? The decision to outsource is not easily reversed, and the firm may find itself locked into a computer architecture and management system ill-suited for its long-term goals and objectives.

Outsourcing is an alternative that must be considered by CIOs for all or part of their operations. In most cases, the primary decision is not a financial one, but financial considerations are always important. The decision to outsource is strategic in nature and probably depends on intangible factors such as the corporate culture as much as it does on strictly financial considerations.

Outsourcing is the current name given to an emerging long-term trend of turning to professionally managed information utilities for IT services. In many cases, local IT craftsmen are falling behind the IT utility professionals in both cost and risk containment. The computer utility or information utility, highly speculative during the past 20 years or so, is looming larger on the landscape and promises to become increasingly important in the next decade. CIOs and senior executive need to prepare themselves and their organization to take advantage of the trend.

SUCCESSFUL CIOS ARE GENERAL MANAGERS

Successful CIOs must demonstrate skills commensurate with the general management positions they hold. Additionally, skills they learn as IT general managers help them as they advance to more responsible positions beyond IT. The career of Katherine Hudson, now CEO of an international manufacturing company, is an example of these ideas. Her career is described in the following Business Vignette.

A Business Vignette

Exceptional Executive Advances

Katherine Hudson joined Kodak in 1969 and held various positions in legal, public affairs, finance, and investor relations. In 1987, she was promoted to vice president and tapped to head a newly formed corporate IS department overseeing a $500 million budget and more than 3000 employees.[23] In this position, corporate IS reports directly to the president's office.

During her tenure as director of corporate IS, Ms. Hudson worked to implement productive new technology in operating departments. Under her direction, Kodak installed local networks in its plants to reduce inventory and improve delivery performance. Inventory declined 90 percent and on-time deliveries improved to 98 percent with the new systems. Ms. Hudson also established

Centers of Excellence in the business units that worked with senior IS managers to identify and understand technology trends useful to Kodak. She considers IS alignment and technology forecasting to be critical aspects of her job.

In 1989, she upset the status quo by outsourcing significant portions of Kodak's data processing operations to IBM and Digital Equipment Corp. and to Businessland, Inc. This bold and successful endeavor was praised both inside and outside of Kodak. Later she was given more responsibility by being named to head Kodak's professional printing and publishing imaging division, a position she left in late 1993.

Ms. Hudson is now president and chief executive of W. H. Brady company, a Milwaukee firm that manufactures more than 20,000 industrial products. She believes her IS experience at Kodak gave her insight into multifunctional operations now valuable in her new role. She hopes to use her IS experience to grow the firm and to create value for its owners. Ms. Hudson is truly an exceptional executive.

WHAT CIOS MUST DO FOR SUCCESS

CIOs critically depend on many people throughout the firm. In particular, the people at the top are extremely important to CIOs.[24] They must gain the confidence of top executives by understanding the business from the executives' vantage point and must, themselves, identify with the business and be sensitive to business priorities. The CIO must be a fully contributing member of the executive suite and must provide leverage to the firm's senior executives through leadership and vision. This leadership includes executive education and salesmanship. CIOs must not assume that their design for the firm will be immediately adopted, or even that it will be completely understood.

The CIO must develop a vision for the firm's use of information technology that realistically adds substantial value to the firm. Reducing costs, saving headcount, and avoiding expenses are excellent activities, but the firm's executives expect much more. Executives expect IT to be a substantive contributor to the firm's value chain. They expect CIOs to contribute to bottom-line results and to have a shared vision for improving the firm's results that extends beyond IT activities; they expect the vision to attract advocates. The advocates and stakeholders must reside at all levels in the organization, not just at the top.

CIOs are action-oriented people; they want to make things happen. They are impatient and intolerant of mediocrity, and they set high expectations for themselves and their organizations. At the same time, they have the patience to work with the situation at hand. They recognize that progress usually occurs in many small increments. They realize that effective executives continually strive to improve their operations across a broad front. They know that business success rarely comes in one grand stroke.

The corporate culture is a powerful force in nearly all organizations: The CIO must understand the culture and must learn to work within it. Realistically, the CIO cannot singlehandedly change the firm's accounting procedures or corporate personnel policies. If the firm is conservative in its behavior, radical infusions of high technology will probably not be acceptable. The firm expects the CIO to operate the IT function within the norms of the larger organization. Thus, the CIO should pay careful attention to the corporate culture.

CIOs must manage expectations within the firm regarding information technology. Unfulfilled expectations are one of the leading causes of failure for managers at all levels. Unfortunately, IT managers suffer more than most from this difficulty. The problem arises naturally because many communication channels to the firm's executives, managers, and employees are biased toward inflating the benefits of information technology while understating the costs or implementation difficulties. Additionally, lack of discipline within the IT organization itself frequently creates unrealistic expectations.

CIOs depend on many people for their success. Effective CIOs cultivate good human relations and encourage innovation. They withhold criticism for well-intentioned but flawed inventions, and they are free with praise for accomplishments. Exceptional CIOs are unselfish. They believe that there is almost no limit to the amount of good they can do if they don't mind who gets the credit.

There four kinds of people in most firms: those who make things happen, those who prevent things from happening, those who watch things happen, and those who don't know or don't care what's happening. Astute CIOs can distinguish between them and have an approach for coping with all of them. CIOs must make things happen and must develop supporters from among those of similar persuasion. CIOs should ignore those who always vote for the status quo and should concentrate on motivating the onlookers to get involved. For those who don't know or don't care what's happening, CIOs should prescribe education. With some effort, most people will buy into an idea that is good for the firm and, therefore, probably good for them.

CIOs must understand where the firm is positioned in its use of information technology and what ability it has to assimilate new technology. CIOs must balance the availability of new technology with the firm's need for it, and with the firm's propensity to adopt it. Effective CIOs believe that the best place to work is where they are with what they have. They build on the capabilities currently in place. They understand the systems that run the firm, and they know the capabilities and limitations of the people who run the systems. Effective CIOs always try to improve the environment, but they use the present environment as the base for improvement within the culture.

SUMMARY

The position of the chief information officer is precarious in both theory and practice. The CIO position, derived from the accepted CFO title, stands on shaky ground because information, unlike money, cannot be quantified or measured. Information can be created, used, and discarded without the CIO's knowledge or approval. Therefore, to identify an officer of the firm as the chief information officer is somewhat misleading.

In practice, the job of senior IT executive, regardless of the title, is large and important. But it is hazardous because of the nature of the work. The CIO is expected to be the technological leader of the firm and to provide the business direction for the selection and introduction of new technology. Information technology is critical to the firm, but its adoption and implementation frequently cause fundamental changes within the firm. Changes are difficult to manage and their success may be doubtful. When things go astray, the CIO is vulnerable.

Chief information officers face many challenges as their firms alter strategic direction or adopt new ways of doing business. CIOs must assess technology changes and provide guidance to the firm on technology adoption and introduction. They must facilitate organizational changes within the firm and must find more effective ways of doing business for the firm and for the IT organization. They must play a strong general management role.

CIOs depend on many people for their success; they must be astute managers of people. CIOs are action oriented and must surround themselves with individuals with similar inclinations. They must study the convictions of the firm's executives, but they must also comply with the corporate culture if they are to be effective.

Successful CIOs have learned the practice of management. They have learned to develop and manage IT management systems, and they have learned to manage themselves. People and technology are incredibly complex. After extensive study, considerable introspection, and prolonged experience, wise managers maintain a respectable level of humility.

Review Questions

1. Why is the title of CIO surrounded by myth and confusion? Does your university have a CIO?
2. Regardless of the title, why is the senior IT executive in a position of power and responsibility? Distinguish between the CIO's line and staff responsibilities. Give examples of each type of responsibility.

3. Discuss the reasons why the CIO job is so difficult. List six challenges associated with the CIO position.
4. To whom do most CIOs report? What other avenues besides the direct reporting relationship do CIOs have to the executive management team?
5. Why do some CEOs fail to recognize the importance of information technology? What can the CIO do about this?
6. What are the career paths for most CIOs? Why is the CIO in a dead-end job in most firms? What, if anything, can the CIO do to improve this situation?
7. What kind of backgrounds do most CIOs have? What kind of backgrounds and experiences do most executives prefer in CIOs?
8. Under what circumstances is the firm most likely to import a CIO from outside its IT ranks? What do firms hope to accomplish by bringing in a CIO from outside the firm?
9. Describe some CIO performance measures. What things can the CIO do to improve these measurements?
10. What are some of the changes going on, both external to the firm and within it, that pose challenges for the CIO?
11. What is the principle driving force behind changes taking place in firms today? Describe the ways in which CIOs must adopt to these changes.
12. What is the CIO's role in policy matters?
13. What responsibilities do CIOs have regarding technology? Why is this a difficult task?
14. Why is technology forecasting risky, difficult, and important?
15. Why is forecasting the use of technology more difficult than forecasting the availability of technology?
16. What technologies promise to be important in the near future?
17. Describe the process through which the results of technology forecasting are incorporated into the firm's business.
18. Describe the process of innovation diffusion. How would you use these ideas if you wanted to install a new user-driven application?

19. Describe the process of downsizing the IT organization. What are the advantages of downsizing to the firm? What are the pros and cons of downsizing to IT professionals?
20. What is meant by outsourcing? What motivates most firms to consider outsourcing?
21. What are the disadvantages of outsourcing? Which disadvantage is most significant, in your opinion?
22. What must CIOs do to be successful? What role does the corporate culture play in the CIO's plan for success?
23. What philosophical difficulties surround the title of chief information officer?

Discussion Questions

1. Discuss the conceptual difficulties with the CIO title and describe how the title effects the practical realities of the job.
2. Discuss the significance of the comment by Keen in endnote number 6.
3. Why do you think Bob McLendon of TI finds a distrust of information-services organizations? Do you think this attitude is widespread and, if so, why?
4. Discuss the general line and staff responsibilities of the CIO position. How have these responsibilities changed over time? In large corporations, the CIO may not have any line responsibility but may have considerable staff responsibility. Discuss what this means for the firm organizationally and what it means for the CIO.
5. Where do individuals gain the experience necessary to become CIOs? Sketch the path that an MBA might take to become a CIO. What business experience would you recommend for this individual?
6. Discuss the balance a CIO needs between technical skills and general or line management skills. How would you recommend a potential CIO obtain these skills?
7. CIOs must respond to changes initiated by the organization and they must take action, which itself creates change. Discuss the interplay between these two activities.

8. Discuss why it is critically important for CIOs to engage in policy formulation and enforcement. What is the connection between policy formulation and the line and staff roles of the CIO?
9. If you were the CIO of a major firm, how would you obtain information on technology trends? Describe the process you would use to make sure the trends were evaluated by the appropriate people in the firm.
10. How do you believe the technology trends forecast by Straub and Wetherbe will impact businesses in the future? Give one example of how these trends might be used in practice.
11. How might astute CIOs use the theory of innovation diffusion in dealing with the firm's top executives? How would innovation diffusion apply during the development and installation of an executive information system? How might it apply if the firm wanted to install EDI to link the firm with its suppliers?
12. The term *downsizing* has two different meanings. What does firm downsizing mean? How does the meaning of the term change when applied to the IT organization? Under what conditions might both activities occur at the same time? What are the fundamental motivations for undertaking either type of downsizing?
13. Discuss the firmwide advantages and disadvantages of downsizing the IT operation.
14. Discuss the advantages and disadvantages of outsourcing the firm's telecommunication system. Under what conditions would a firm outsource its telecommunication system and not its production operation?
15. What considerations are important in deciding to outsource application development? What industry factors might influence this decision? What strategic considerations are important? What role does the corporate culture play in these decisions?
16. Discuss the balance the CIO must achieve among technology availability, the firm's need for it, and the firm's propensity to adopt it. What actions can the CIO take to affect the balance among these variables?
17. Describe Robert Hinds' general management responsibilities.

Assignments

1. "Conclusion: Effectiveness Must Be Learned" is the title of the final chapter in Peter Drucker's book *The Effective Executive,* 166–174. Read this chapter and summarize it in writing for the class.
2. Analyze the IT factors that contributed to Caterpillar's success as discussed in the first Business Vignette. Categorize them as IT line responsibilities, IT staff responsibilities, or client department responsibilities. Compare Caterpillar's actions with those you learned to be valuable in this text. How would you describe the relationships and roles of CIO and CEO as compared with what you consider to be desirable.
3. Conduct an interview with the senior information executive in a firm in your community. Determine from the interview what the job consists of and what the individual must do to be successful. How does this person balance technical skills with business skills in performing the job? Compare and contrast this person's view of the CIO title and job with that developed in the second Business Vignette.

ENDNOTES

[1] Doug Bartholomew, "How IT Is Helping Caterpillar Fend Off Japanese Competition and Keep Jobs in the U.S." *Information Week*, June 7, 1993, 36.

[2] See Note 1.

[3] See Note 1.

[4] *Value Line*, November 11, 1994, 1347.

[5] In speaking about executive effectiveness, Drucker states that "effectiveness, in other words, is a habit; that is, a complex of practices. And practices can always be learned." Peter Drucker, *The Effective Executive* (New York: Harper & Row, Publishers Inc., 1986), 23.

[6] Peter Keen states that "The CIO position is a relationship, not a job. If the CIO/top management team relationship is effective, the title doesn't matter. If it is ineffective, the title doesn't matter." *Every Manager's Guide to Information Technology* (Boston, MA: Harvard Business School Press, 1991), 55.

[7] Gordon Bock, "Management's Newest Star," *Business Week*, Special Report, October 13, 1986, 160.

[8] Jeffrey Rothfeder, "CIO is Starting to Stand for 'Career is Over'," *Business Week*, February 26, 1990, 78.

[9] "Impossible Dream," *Computerworld*, November 7, 1994, 34. Some studies show turnover to be less in small firms. See Deloitte & Touch survey in *Computerworld*, August 9, 1993, 86.

[10] "Information Chiefs Get Plaudits but Rarely Promotions," *Wall Street Journal*, November 10, 1994, B1.

[11] "Trends," *Computerworld*, March 5, 1990, 118.

[12] Dan Richman, "National's Steering Committee," *CIO*, June 1991, 74.

[13] Albert L. Lederer and Aubrey L. Mendelow, "Convincing Top Management of the Strategic Potential of Informations Systems," *MIS Quarterly*, December 1988, 525.

[14] Allan E. Alter, "Good News, Bad News...," *CIO*, January 1990, 18.

[15] Detmar W. Straub and James C. Wetherbe, "Information Technologies for the 1990s: An Organizational Impact Perspective," *Communications of the ACM*, November 1989, 1328.

[16] See Note 15.

[17]"Forecast 95," *Computerworld*, December 26, 1994/January 2, 1995, 15.

[18] Robert I. Benjamin and Jon Blunt, "Critical IT Issues: The Next Ten Years," *Sloan Management Review*, Summer 1992, 7.

[19] Everett M. Rogers, *Diffusion of Innovations* (New York: Free Press, 1962), 79.

[20] Hamid Tavakolian, "Linking the Information Technology Structure With Organizational Competitive Strategy: A Survey," *MIS Quarterly*, September 1989, 309.

[21] "How IT Is Helping Caterpillar Fend off Japanese Competition and Keep Jobs in the U.S." *Information Week*, June 7, 1993, 36.

[22] Deborah Cooper, "High Five," *CIO*, December 1988, 55.

[23] "Former Kodak IS director to Head Global Plastics Company," *Computerworld*, December 13, 1993, 10.

[24] David F. Feeney, Brian R. Edwards, and Keppel M. Simpson, *MIS Quarterly*, December 1992, 435.

Bibliography

PART ONE FOUNDATIONS OF IT MANAGEMENT

Anthony, R. N. *Planning and Control Systems: A Framework for Analysis.* Cambridge, MA: Harvard University Press, 1965.

Boar, Bernard H. *The Art of Strategic Planning for Information Technology.* New York: John Wiley & Sons, Inc. 1993.

Boynton, Andrew C., and Robert W. Zmud. "An Assessment of Critical Success Factors." *Sloan Management Review*, Summer 1984.

Browning, John. "A Survey of Information Technology." *The Economist*, June 16, 1990.

Drucker, Peter F. "The Coming of the New Organization." *Harvard Business Review*, January-February, 1988, 45.

Earl, Michael J. "Experiences in Strategic Information Systems Planning." *MIS Quarterly*, No. 1, 1993, 1.

Hefferon, George J. "Taking the Mystery out of IS Strategic Planning." *Information Executive*, Vol. 1, Fall, 1988, 18.

Hopper, Max. "Rattling SABRE: New Ways to Compete on Information." *Harvard Business Review*, May-June, 1990, 118.

Information Week, Special Issue: Strategic Systems, May 26, 1986.

Izzo, Joseph E. *The Embattled Fortress.* San Francisco: Jossey-Bass Publishers, 1989.

Jarvenpaa, S. L., and Blake Ives. "Executive Involvement and Participation in the Management of IT." *MIS Quarterly*, No. 2, 1991, 205.

Keen, Peter G. W. *Competing in Time: Using Telecommunications for Competitive Advantage.* Cambridge, MA: Ballinger Publishing Company, 1988.

Kemerer, Chris F., and Glenn L. Sosa. "Barriers to Successful Strategic Information Systems." *Planning Review*, September-October, 1988, 20.

King, William R. "Strategic Planning for Information Systems." *MIS Quarterly*, March, 1978, 27.

Marrus, Stephanie K. *Building the Strategic Plan.* New York: John Wiley & Sons, Inc., 1984.

McFarlan, F. Warren. "Information Technology Changes the Way You Compete." *Harvard Business Review*, May-June, 1984, 98.

Millar, Howard W. "Developing Information Technology Strategies." *Journal of Systems Management*, September, 1988, 28.

Nolan, Richard L. "Plight of the EDP Manager." *Harvard Business Review*, May-June, 1973, 143.

Ohmar, Kenechi. "Getting Back to Strategy." *Harvard Business Review*, November-December, 1988, 149.

Parker, Marilyn M., H. Edgar Trainor, and Robert J. Benson. *Information Strategy and Economics*. Englewood Cliffs, NJ: Prentice Hall, 1989.

Pearson, Andrall E. "Tough-Minded Ways to Get Innovation." *Harvard Business Review*, May-June, 1988, 99.

Porter, Michael E. *Competitive Advantage*. New York: Free Press, 1985.

Porter, Michael E., and Victor E. Miller. "How Information Gives You Competitive Advantage." *Harvard Business Review*, July-August, 1985, 149.

Rackoff, Nick, Charles Wiseman, and Walter Ullrich. "Information Systems for Competitive Advantage: Implementation of a Planning Process." *MIS Quarterly*, December, 1985, 285.

Stalk, George Jr. "Time—The Next Source of Competitive Advantage." *Harvard Business Review*, July-August, 1988, 41.

Sullivan, Cornelius H. "Systems Planning in the Information Age." *Sloan Management Review*, Winter, 1985, 3.

Sullivan, Cornelius H. "The Changing Approach to Systems Planning." *Journal of Information Systems Management*, Summer, 1988, 8.

Wiseman, Charles. *Strategic Information Systems*. Homewood, IL: Irwin, 1988.

PART TWO TECHNOLOGY AND INDUSTRY TRENDS

Blumberg, Donald F. "Looking Ahead: Network Planning for the 1990s." *Data Communications*, February, 1988, 185.

"The Computer in the 21st Century." *Scientific American,* Special Issue, Volume 6, No. 1, 1995.

Elmer-Dewitt, Philip. "The Electronic Superhighway." *Time*, April 12, 1993, 50-58.

Gelernter, David. "The Metamorphosis of Information Management." *Scientific American*, August, 1989, 66.

"The Global 100, Outstanding Users of Information Technology from Around the World." *Computerworld*, Special Issue, May 1, 1995, 6-64.

Hyman, Leonard S., Richard C. Toole, and Rosemary M. Avellis. *The New Telecommunications Industry: Evolution and Organization*. Volume 1, Arlington, VA: Public Utilities Reports, Inc., 1987.

Meindl, James D. "Chips for Advanced Computing." *Scientific American*, October, 1987, 78.

Peterson, James L., and Abraham Silberschatz. *Operating Systems Concepts*. 2nd ed. Reading, MA: Addison-Wesley Publishing Company, 1985.

"Premier 100, the Productivity Payoff." *Computerworld*, Special Issue, September 19, 1994, 4-55.

Stallings, William. *ISDN: An Introduction*. New York: Macmillan Publishing Company, 1989.

Stamper, David A. *Business Data Communications*, 2nd ed. Redwood City, CA: The Benjamin/Cummings Publishing Company, Inc., 1989.

"Technology: Eyes on the Future." *Newsweek*, May 31, 1993, 38-50.

Terplan, Kornel. *Communications Network Management*. Englewood Cliffs, NJ: Prentice Hall, Inc., 1987.

Tobias, Randall L. "Telecommunications in the 1990s." *Business Horizons*, January-February, 1990.

PART THREE MANAGING APPLICATION PORTFOLIO RESOURCES

Alavi, Maryam, and Ira R. Weiss. "Managing the Risks Associated with End-User Computing." *Journal of Management Information Systems*, Winter, 1985, 5.

Bachman, Charles W. "A Personal Chronicle: Creating Better Information Systems, with Some Guiding Principles." *IEEE Transactions on Knowledge and Data Engineering*, March, 1989, 17.

Bacon, C. James. "The Use of Decision Criteria in Selecting Information Systems/Technology Investments." *MIS Quarterly*, No. 3, 1992, 335.

Benjamin, Robert I., and Jon Blunt. "Critical IT Issues: The Next Ten Years." *Sloan Management Review*, Summer, 1992, 7.

Brooks, Frederick P. *The Mythical Man Month—Essays on Software Engineering*. Reading, MA: Addison-Wesley Publishing Company, 1974.

Burch, John G. *Systems Analysis, Design, and Implementation*. Danvers, MA: boyd & fraser publishing company, 1991.

Clemons, Eric K. "Evaluation of Strategic Investments in Information Technology." *Communications of the ACM*, January 1991.

Davenport, Thomas H., and James E. Short. "The New Industrial Engineering: Information Technology and Business Process Redesign." *Sloan Management Review*, Summer 1990.

Davenport, Thomas H. *Process Innovation*. Boston: Harvard Business School Press, 1993.

Gerrity, Thomas P., and John F. Rockart. "End-User Computing: Are You a Leader or a Laggard?" *Sloan Management Review*, Summer, 1986, 25.

Gralla, Preston. "Something Ventured, Something Gained." *CIO Magazine*, May, 1988, 12.

Humphrey, Watts. *Managing the Software Process*. Reading, MA: Addison-Wesley Publishing Company, 1993.

"IS Shops Form Alliances as Development Costs Rise." *Datamation*, September 1, 1988, 19.

Keen, Peter G. W. *Shaping the Future: Business Design through Information Technology*. Boston: Harvard Business School Press, 1991.

Keider, Stephen P. "Managing Systems Development Projects." *Journal of Information Systems Management*, Summer, 1984, 33.

Martin, James. *Application Development without Programmers*. Englewood Cliffs, NJ: Prentice Hall, Inc., 1982.

Martin, James, and Carma McClure. *Structured Techniques: The Basis For CASE*. Englewood Cliffs, NJ: Prentice Hall, Inc., 1988.

McFarlan, F. Warren. "Portfolio Approach to Information Systems." *Harvard Business Review*, September-October, 1981, 142.

Pyburn, Phillip J. "Managing Personal Computer Use: The Role of Corporate Management Information Systems." *Journal of Management Information Systems*, Winter, 1986-87, 49.

Schendler, Brenton. "How to Break the Software Logjam." *Fortune*, September 25, 1989, 100.

Strassmann, Paul A. "The Real Cost of Office Automation." *Datamation*, February 1, 1985.

Swider, Gaile A. "Ten Pitfalls of Information Center Implementation." *Journal of Information Systems Management*, Winter, 1988, 22.

Verity, John, and Evan I. Schwartz. "Software Made Simple." *Business Week*, September 30, 1991.

Weinberg, Randy S. "Prototyping and the Systems Development Life Cycle." *Journal of Information Systems Management*, 1991.

Wojtkowski, W. Gregory, and Wita Wojtkowski. *Applications Software: Programming with Fourth-Generation Languages*. Danvers, MA: boyd & fraser publishing company, 1990.

Yourdon, Edward. *Managing the Structured Techniques*. New York: Yourdon Press, 1986.

Yourdon, Edward. *Modern Structured Analysis*. Englewood Cliffs NJ: Yourdon Press, 1989.

PART FOUR TACTICAL AND OPERATIONAL CONSIDERATIONS

Borovits, Israel. *Management of Computer Operations*. Englewood Cliffs, NJ: Prentice Hall, 1984.

FitzGerald, Jerry. *Business Data Communications*, 3rd ed. New York: John Wiley & Sons, 1990.

Orazine, Ron. "Why MIS Managers Are Becoming Network Experts." *Telecommunications*, January, 1988, 57.

Rothstein, Phillip J. "Up and Running: How to Ensure Disaster Recovery." *Datamation*, October 15, 1988, 86.

Shafe, Laurence. *Client/Server: A Manager's Guide*. Reading, MA: Addison-Wesley Publishing Company, 1995.

Singleton, John P., Ephraim R. McLean, and Edward N. Altman. "Measuring Information Systems Performance: Experience with the Management by Results System at Security Pacific Bank." *MIS Quarterly*, June, 1988, 325.

Stallings, William. *Business Data Communications*. New York: Macmillan Publishing Company, 1990.

Toigo, Jon William. *Disaster Recovery Planning: Managing Risk & Catastrophe in Information Systems*. Englewood Cliffs, NJ: Prentice Hall, 1989.

Van Schaik, Edward A. *A Management System for the Information Business: Organizational Analysis*. Englewood Cliffs, NJ: Prentice Hall, 1985.

PART FIVE CONTROLLING INFORMATION RESOURCES

Allen, Brandt. "Make Information Services Pay Its Way." *Harvard Business Review*, January-February, 1987, 57.

Bruns, William J. Jr., and F. Warren McFarlan. "Information Technology Puts Power in Control Systems." *Harvard Business Review*, September-October, 1987, 89.

Clemons, Eric K. "Evaluation of Strategic Investments in Information Technology." *Communications of the ACM*, January 1991.

Connell, John Jr. "A Measured Response." *CIO Magazine*, January-February, 1988, 52.

Coughlin, John W. "The Fairfax Embezzlement." *Management Accounting*, May 1983.

Deardon, John. "Measuring Profit Center Managers." *Harvard Business Review*, September-October, 1987, 84.

FitzGerald, Jerry, and Ardra FitzGerald. *Fundamentals of Systems Analysis*, 3rd ed. New York: John Wiley & Sons, 1987. See Chapter 9, Designing New System Controls, and Appendix 2, Controls for Inputs, Data Communications, Programming, and Outputs, A15-55.

Mason, Richard O. "Four Ethical Issues of the Information Age." *MIS Quarterly*, January 1986.

Perry, William E. "Evaluating Data Center Controls." *Handbook of MIS Management*, 2nd ed. Robert E. Umbaugh, Ed, Boston: Auerbach Publishers, 1988.

Perry, William E. "User Chargeback Procedures for Distributed Systems." *A Practical Guide to Distributed Processing Management*. James Hannon, Ed. New York: Van Nostrand Reinhold Company, 1982, 73.

Roach, Stephen S. "Services Under Siege." *Harvard Business* Review, September-October, 1991, 82.

PART SIX PREPARING FOR IT ADVANCES

Benjamin, R. I., C. Dickinson, and J. F. Rockart. "The Changing Role of the Corporate Information Systems Officer." *MIS Quarterly*, September, 1985, 177.

Brancheau, James C., and James C. Wetherbe. "Understanding Innovation Diffusion Helps Boost Acceptance Rates of New Technology." *Chief Information Officer Journal*, Fall 1989.

Cash, James I. Jr., and Poppy L. McLeod. "Managing the Introduction of Information Systems Technology in Strategically Dependent Companies." *Journal of Management Information Systems*, Spring, 1985, 5.

Clermont, Paul. "Outsourcing without Guilt." *Computerworld*, September 9, 1991.

Cougar, D. "Key Human Resource Issues in the 1990s: Views of IS Executives versus Human Resource Executives." *Information & Management*, Volume 14, 1988, 161.

Donovan, John J. "Beyond Chief Information Officer to Network Manager." *Harvard Business Review*, September-October, 1988, 134.

Emory, James. "What Role for the CIO?" *MIS Quarterly*, No. 2, 1991, vii.

Feeny, David F., Brian R. Edwards, and Keppel M. Simpson. "Understanding the CEO/CIO Relationship." *MIS Quarterly*, No. 4, 1992, 435.

Feigenbaum, E., et al. *The Rise of the Expert Company: How Visionary Companies Are Using Computers to Make Huge Profits*. New York: Times Books, 1988.

Ives, Blake. "Transformed IS Management." *MIS Quarterly*, No. 4, 1992, iix.

Orazine, Ron. "Why MIS Managers Are Becoming Network Experts." *Telecommunications*, January, 1988, 57.

Owen, Darrell E. "Information Systems Organizations Keeping Pace with the Pressures." *Sloan Management Review*, Spring, 1986, 59.

Pensouneault, Alain, and Kenneth Kraemer. "The Impact of IT on Middle Managers." *MIS Quarterly*, No. 3, 1993, 271.

Walton, Richard E. *Up and Running: Integrating Information Technology and the Organization*. Boston: Harvard Business School Press, 1989.

OTHER REFERENCES AND READINGS

Boone, Mary E. *Leadership and the Computer*. Rocklin, CA: Prime Publishing, 1991.

Brynjolfsson, Erik, and Loren Hitt. "The Big Payoff from Computers." *Fortune*, March 7, 1994, 28.

Davis, Stan, and Bill Davidson. *2020 Vision*. New York: Simon & Schuster, 1992.

Fried, Louis. *Managing Information Technology in Turbulent Times*. New York: John Wiley & Sons, Inc., 1995.

Gilder, George. *Microcosm, the Quantum Revolution in Economics and Technology*. New York: Simon & Schuster, 1989.

Hammer, Michael, and James Champy. *Reengineering the Corporation*. New York: HarperCollins, 1993.

Kanter, John, Stephen Schiffman, and J. Faye Horn. "Let the Customer Do It." *Computerworld*, August 27, 1990, 75.

McKinnon, Sharon M., and William J. Bruns, Jr. *The Information Mosaic*. Boston: Harvard Business School Press, 1992.

Negroponte, Nicholas. *Ones and Zeros*. New York: Knopf, 1995.

Peters, Tom. *Liberation Management*. New York: Knopf, 1993.

Strassmann, Paul A. *The Politics of Information Management*. New Canann, CT: The Information Economics Press, 1995.

Strassmann, Paul A. *Information Payoff*. New York: The Free Press, 1985.

Wang, Charles B. *Techno Vision*. New York: McGraw-Hill, Inc., 1994.

Womack, James P., Daniel T. Jones, and Daniel Roos. *The Machine that Changed the World*. New York: Rawson Associates, 1990.

Wriston, Walter B. *The Twilight of Sovereignty*. New York: Charles Scribner's Sons, 1991.

Zuboff, Shoshana. *In the Age of the Smart Machine*. New York: Basic Books, Inc., 1988.

Index

A

B

C

D

E

F

N

O

P

Q

R

S

U

V

W

X

Y